T0205848

Analysis of Pesticides in Food and Environmental Samples

Second Edition

Analysis of Pesticides in Food and Environmental Samples

Second Edition

Edited by

José L. Tadeo

CRC Press is an imprint of the
Taylor & Francis Group, an **informa** business

CRC Press
Taylor & Francis Group
6000 Broken Sound Parkway NW, Suite 300
Boca Raton, FL 33487-2742

First issued in paperback 2021

ISBN 13: 978-1-138-48603-4 (hbk)
ISBN 13: 978-1-03-224129-6 (pbk)

DOI: 10.1201/9781351047081

Library of Congress Cataloging-in-Publication Data

Names: Tadeo, José L., editor.
Title: Analysis of pesticides in food and environmental samples / editor, José L. Tadeo.
Description: Second edition. | Boca Raton : Taylor & Francis, a CRC title, part of the Taylor & Francis imprint, a member of the Taylor & Francis Group, the academic division of T&F Informa, plc, 2019. | Includes bibliographical references and index.
Identifiers: LCCN 2018051427 | ISBN 9781138486034 (hardback : alk. paper)
Subjects: LCSH: Pesticide residues in food. | Food--Analysis. | Pesticides.
Classification: LCC TX571.P4 A52 2019 | DDC 664/.07--dc23
LC record available at https://lccn.loc.gov/2018051427

Contents

Preface..vii
Editor ix
Contributors xi

Chapter 1 Pesticides: Classification and Properties...............................1
José L. Tadeo, Beatriz Albero, and Rosa Ana Pérez

Chapter 2 Sample Handling of Pesticides in Food and Environmental Samples.....41
Francisco Barahona, Esther Turiel, and Antonio Martín-Esteban

Chapter 3 Analysis of Pesticide Residues by Chromatographic Techniques Coupled with Mass Spectrometry 73
Wang Jing, Jin Maojun, Jae-Han Shim, and A.M. Abd El-Aty

Chapter 4 Immunoassays and Biosensors...105
Jeanette M. Van Emon, Tanner Drum, and Kilian Dill

Chapter 5 Quality Control and Quality Assurance.............................135
Árpád Ambrus and Gabriella Suszter

Chapter 6 Determination of Pesticides in Food of Vegetal Origin...................175
Jon W. Wong, Kai Zhang, Douglas G. Hayward, Alexander J. Krynitsky, Frank J. Schenck, James B. Wittenberg, Jian Wang, and Paul Yang

Chapter 7 Determination of Pesticide Residues in Food of Animal Origin...207
Giulia Poma, Marina López-García, Roberto Romero González, Antonia Garrido Frenich, and Adrian Covaci

Chapter 8 Determination of Pesticides in Soil.................................245
Beatriz Albero, Rosa Ana Pérez, and José L. Tadeo

Chapter 9   Determination of Pesticides in Water..271

Rosa Ana Pérez, Beatriz Albero, and José L. Tadeo

Chapter 10   Sampling and Analysis of Pesticides in the Atmosphere.................301

Maurice Millet

Chapter 11   Levels of Pesticides in Food and Food Safety Aspects....................329

Kit Granby, Annette Petersen, Susan Strange Herrmann, and
Mette Erecius Poulsen

Chapter 12   Monitoring and Assessment of Pesticides and
Transformation Products in the Environment..................................365

Ioannis Konstantinou, Dimitra Hela, Dimitra Lambropoulou,
and Triantafyllos Albanis

Index...**415**

Preface

You should go on learning for as long as your ignorance lasts; and, if the proverb is to be believed, *for the whole of your life.*

Lucius Annaeus Seneca

Consumer concerns on food safety and society awareness of chemical contaminants in the environment have increased in recent years. As a result, more restrictions have been imposed on the use of chemical products at national and international levels.

Pesticides are widely applied for the control of weeds, diseases, and pests of cultivated plants all over the world. Their use has increased 50-fold since 1950, when the discovery of some organic compounds with good insecticide or herbicide activity took place after the Second World War. At present, around 3 million tons of pesticides are used annually and the number of registered active substances is higher than 500.

However, as pesticides are toxic substances that may have undesirable effects, their use has to be regulated. The risk assessment of pesticides requires information on the toxicological and ecotoxicological properties of these compounds as well as on their levels in food and environmental compartments. Therefore, reliable analytical methods are needed to carry out the monitoring of pesticide residues in those matrices.

Analysis of Pesticides in Food and Environmental Samples, Second Edition updates the information published in recent years on the different pesticide compounds used, sample preparation methods, quality assurance, chromatographic techniques, immunoassays, pesticide determination in food, soil, water, and air, and the results of their monitoring in food and environmental compartments. I think that this timely and up-to-date work can significantly improve the information in this research area and contribute to a better understanding of the behavior of pesticides, which will lead to an improvement of their use.

My sincere thanks to everyone who has contributed and particularly to all the contributors of the different chapters of *Analysis of Pesticides in Food and Environmental Samples, Second Edition.*

This work is dedicated to Teresa, my wife.

José L. Tadeo

Editor

José L. Tadeo, PhD in chemistry, is a Full Research Professor at the National Institute for Agricultural and Food Research and Technology in Madrid, Spain. He graduated in chemistry in June 1972 from the University of Valencia and began his research career at the Institute of Agrochemistry and Food Technology, Spanish Council for Scientific Research, in Valencia, investigating natural components of plants with insecticide activity. In 1976, he was engaged in the research of analytical methodologies for the determination of pesticide residues in food, water, and soil at the Jealott's Hill Research Station in the UK.

In 1977, Dr. Tadeo became a research scientist at the Institute for Agricultural Research, in Valencia, where his work focused on the study of chemical composition of citrus fruits and the behavior of fungicides used during postharvest of fruits.

In 1988, he became a senior researcher at the National Institute for Agricultural and Food Research and Technology. During his stay at the Plant Protection Department, the main research lines were the analysis of herbicide residues and the study of their persistence and mobility in soil.

Since 1998 he has led the Environmental Chemistry Laboratory at the Environment Department of the National Institute for Agricultural and Food Research and Technology. His current research is the analysis of pesticides and other contaminants in food and environmental matrices and the evaluation of the exposure to biocides and existing chemicals. His expertise is documented by the publication of more than 200 research papers, monographs, and book chapters on these topics, an H index of 36, and the supervision of ten doctoral theses. Dr. Tadeo has been a member of national and international working groups for the evaluation of chemicals, and he is currently involved in the assessment of biocides at European and international levels.

Contributors

A.M. Abd El-Aty*
Department of Pharmacology
Faculty of Veterinary Medicine
Cairo University
Giza, Egypt

and

Department of Medical Pharmacology
Medical Faculty
Ataturk University
Erzurum, Turkey

Triantafyllos Albanis
Department of Chemistry
University of Ioannina
Ioannina, Greece

Beatriz Albero*
Department of Environment
Instituto Nacional de Investigacion y
    Tecnologia Agraria y Alimentaria
Madrid, Spain

Árpád Ambrus*
National Food Chain Safety Office
Budapest, Hungary (retired)

Francisco Barahona
Department of Environment
Instituto Nacional de Investigacion y
    Tecnologia Agraria y Alimentaria
Madrid, Spain

Adrian Covaci*
Toxicological Centre
University of Antwerp
Wilrijk, Belgium

Kilian Dill
Multerra Bio
University of California Medical
    School
San Francisco, California, USA

Tanner Drum
National Exposure Research Laboratory
U.S. Environmental Protection Agency
Las Vegas, Nevada, USA

Antonia Garrido
Frenich Department of Chemistry and
    Physics, Analytical Chemistry Area
University of Almería
Almería, Spain

Kit Granby*
National Food Institute
Technical University of Denmark
Lyngby, Denmark

Douglas G. Hayward
U.S. Food and Drug Administration
Center for Food Safety and Applied
    Nutrition
College Park, Maryland, USA

Dimitra Hela
Department of Chemistry
University of Ioannina
Ioannina, Greece

Susan S. Herrmann
National Food Institute
Technical University of Denmark
Lyngby, Denmark

Wan Jing
Institute of Quality Standards & Testing
    Technology for Agro-Products
Chinese Academy of Agricultural
    Sciences
Beijing, China

Ioannis Konstantinou
Department of Chemistry
University of Ioannina
Ioannina, Greece

Dimitra Lambropoulou
Department of Chemistry
Aristotle University of Thessaloniki
Thessaloniki, Greece.

Alexander J. Krynitsky
U.S. Food and Drug Administration
Office of Regulatory Affairs Southeast
    Regional Laboratory
Atlanta, Georgia, USA

Marina López-García
Department of Chemistry and Physics,
    Analytical Chemistry Area
University of Almería
Almería, Spain

Jin Maojun
Institute of Quality Standards & Testing
    Technology for Agro-Products
Chinese Academy of Agricultural
    Sciences
Beijing, China

Antonio Martín-Esteban*
Department of Environment
Instituto Nacional de Investigacion y
    Tecnologia Agraria y Alimentaria
Madrid, Spain

Maurice Millet*
Laboratoire de Physico-Chimie de
    l'Atmosphère
Institut de Chimie et Procédés pour
    l'Energie
l'Environnement et la Santé Université
    de Strasbourg
Strasbourg, France

Rosa Ana Pérez*
Department of Environment
Instituto Nacional de Investigacion y
    Tecnologia Agraria y Alimentaria
Madrid, Spain

Annette Petersen
National Food Institute
Technical University of Denmark
Lyngby, Denmark

Giulia Poma
Toxicological Centre
University of Antwerp
Wilrijk, Belgium

Mette Erecius Poulsen
National Food Institute
Technical University of Denmark
Lyngby, Denmark

Roberto Romero González
Department of Chemistry and Physics,
    Analytical Chemistry Area
University of Almería
Almería, Spain

Frank J. Schenck
U.S. Food and Drug Administration
    Office of Regulatory Affairs
Southeast Regional Laboratory
Atlanta, Georgia, USA

Jae-Han Shim
Natural Products Chemistry Laboratory
College of Agriculture and Life
    Sciences
Chonnam National University
Gwangju, Republic of Korea

Gabriella Suszter
Wessling Hungary Kft.
Budapest, Hungary

José L. Tadeo*
Department of Environment
Instituto Nacional de Investigacion y
    Tecnologia Agraria y Alimentaria
Madrid, Spain

Esther Turiel
Department of Environment
Instituto Nacional de Investigacion y
    Tecnologia Agraria y Alimentaria
Madrid, Spain

Jeanette M. Van Emon*
National Exposure Research Laboratory
U.S. Environmental Protection Agency
Las Vegas, Nevada, USA

Jian Wang
Canadian Food Inspection Agency
Calgary Laboratory
Calgary, Alberta, Canada

Jon W. Wong*
U.S. Food and Drug Administration
Center for Food Safety and Applied
    Nutrition
College Park, Maryland, USA

James B. Wittenberg
Alcohol and Tobacco Tax and Trade
    Bureau
Beltsville, Maryland, USA

Paul Yang
Ontario Ministry of the Environment
Toronto, Ontario, Canada

Kai Zhang
U.S. Food and Drug Administration
Center for Food Safety and Applied
    Nutrition
College Park, Maryland, USA

* denotes lead contributor

1 Pesticides
Classification and Properties

José L. Tadeo, Beatriz Albero, and Rosa Ana Pérez

CONTENTS

1.1 Introduction 2
1.2 Herbicides 5
    1.2.1 Amides 5
    1.2.2 Benzoic Acids 6
    1.2.3 Carbamates 7
    1.2.4 Imidazolinones 7
    1.2.5 Nitriles 10
    1.2.6 Nitroanilines 10
    1.2.7 Organophosphorus 11
    1.2.8 Phenoxy Acids 11
    1.2.9 Pyridines and Quaternary Ammonium Compounds 13
    1.2.10 Pyridazines and Pyridazinones 15
    1.2.11 Triazines 16
    1.2.12 Ureas 18
        1.2.12.1 Phenylureas 18
        1.2.12.2 Sulfonylureas 18
1.3 Insecticides 19
    1.3.1 Benzoylureas 19
    1.3.2 Carbamates 19
    1.3.3 Neonicotinoids 21
    1.3.4 Organochlorines 23
    1.3.5 Organophosphorus 25
    1.3.6 Pyrethroids 28
    1.3.7 Miscellaneous 28
1.4 Fungicides 30
    1.4.1 Azoles 30
    1.4.2 Benzimidazoles 32
    1.4.3 Dithiocarbamates 32
    1.4.4 Morpholines 33
    1.4.5 Miscellaneous 34
1.5 Mode of Action 36
    1.5.1 Herbicides 36
        1.5.1.1 Amino Acid Synthesis Inhibitors 36
        1.5.1.2 Cell Division Inhibitors 36

          1.5.1.3   Photosynthesis Inhibitors...36
     1.5.2   Insecticides 37
          1.5.2.1   Signal Interference in the Nervous System.......................37
          1.5.2.2   Inhibitors of Cholinesterase...37
          1.5.2.3   Inhibitors of Chitin Synthesis...37
     1.5.3   Fungicides..37
          1.5.3.1   Sulfhydryl Reagents...37
          1.5.3.2   Cell Division Inhibitors..38
          1.5.3.3   Inhibitors of Ergosterol Synthesis...................................38
1.6   Toxicity and Risk Assessment 38
References...40

1.1   INTRODUCTION

A pesticide is any substance or mixture of substances, natural or synthetic, formulated to prevent, destroy, repel, or mitigate any pest that competes with humans for food, destroys property, or spreads disease [1]. The term pest includes insects, weeds, mammals, and microbes, among others.

Pesticides are usually chemical substances, although they can sometimes be biological agents such as virus or bacteria. The active portion of a pesticide, known as the active ingredient, is generally formulated by the manufacturer as emulsifiable concentrates or in solid particles (dust, granules, soluble powder, or wettable powder). Many commercial formulations have to be diluted with water before use and contain adjuvants to improve pesticide retention and absorption by leaves or shoots. Recently, nanotechnology has been used to design and prepare nano-formulations that are more soluble, less degradable, and have new delivery mechanisms of controlled release.

There are different classes of pesticides organized according to their type of use. The main pesticide groups are herbicides, used to kill weeds and other plants growing in places where they are unwanted; insecticides, employed to kill insects and other arthropods; and fungicides, used to kill fungi. Other types of pesticides are acaricides, molluscicides, nematicides, pheromones, plant growth regulators, repellents, and rodenticides. Chemical substances have been used by humans to control pests from the beginning of agriculture. Initially, inorganic compounds such as sulfur, arsenic, mercury, and lead were used. The discovery of dichlorodiphenyltrichloroethane (DDT) as an insecticide by Paul Müller in 1939 caused a great impact in the control of pests and soon became widely used worldwide. At that time, pesticides had a good reputation mainly due to the control of diseases like malaria, transmitted by mosquitoes, and the bubonic plague, transmitted by fleas, both killing millions of people over time. Nevertheless, this opinion changed after the discovery of the toxic effects of DDT on birds, particularly after the publication of the book *Silent Spring* by Rachel Carson in 1962 [2]. At present, due to the possible toxic effects of pesticides on human health and on the environment, there are strict regulations for their registration and use all over the world, especially in developed countries. However, although some progress is achieved in the biological control and in the development of resistance of plants to pests, pesticides are still indispensable for feeding and protecting the world population from diseases. It has been estimated that around one-third of crop production would be lost if pesticides were not applied.

Pesticide use has increased 50-fold since 1950 and around three million tons of industrial pesticides per year are used nowadays. Figure 1.1 shows the time course of pesticide sales during the last years.

The main agricultural areas of pesticide usage are Europe, Asia, Latin America, and North America, with 27.6%, 26.1%, 21.9%, and 20.7%, respectively, in 2010 (Figure 1.2). These percentages of pesticide sales are expressed in millions of US dollars, and although the mentioned regions are the most important agricultural

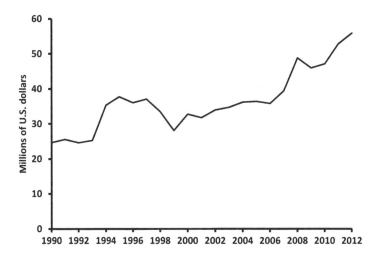

FIGURE 1.1 World market of pesticides since 1990. Values are expressed in millions of US dollars. (From EPA. Available at http://www.epa.gov/pesticides.)

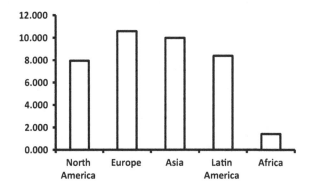

FIGURE 1.2 Regional pesticide sales expressed in millions of US dollars. (From European Parliament. Pesticide legislation in the EU. *Library of the European Parliament.* Available at http://www.europarl.europa.eu/RegData/bibliotheque/briefing/2012/120291/ LDM_BRI%282012%29120291_REV1_EN.pdf.)

areas in the global pesticide market, their relative position may vary due to changes in the currency exchange rates, climatic conditions, and national policies on agricultural support and regulations.

The amount of pesticides applied in a determined geographical area depends on the climatic conditions and on the outbreak of pests and diseases in a particular year. Nevertheless, herbicides are the main group of pesticides used worldwide, followed by insecticides and fungicides (Figure 1.3).

The development of a new chemical as a pesticide takes at present about ten years and around $286 million and, on average, only one compound out of 150,000 compounds initially tested might reach final commercial production (http://www.ecpa. eu/regulatory-policy-topics/registration-and-placement-pesticides-eu-market). The registration of a pesticide for its application on a particular crop requires a complete set of data to prove its efficacy and safe use. This normally includes data on physicochemical properties, analytical methods, efficacy, toxicology, ecotoxicology, and fate and behavior in the environment. Residues left on crops after pesticide application have been restricted in developed countries to guarantee safe food consumption. The maximum residue levels (MRLs) in different foods have been established according to good agricultural practices, the observed toxic effects of the pesticide, and the amount of food consumed. MRLs are normally fixed in relation to the admissible daily intake (ADI) of pesticides, which is the amount of pesticide that can be ingested daily during the whole life without showing an appreciable adverse effect. MRLs are proposed by the Joint FAO/WHO Meeting on Pesticide Residues (JMPR) and recommended for adoption by the Codex Committee on Pesticide Residues [3,4].

In the following sections of this chapter, the main classes of pesticides (herbicides, insecticides, and fungicides) will be described together with their main physicochemical properties and principal uses. These data have been gathered from *The Pesticide Manual* [5] as well as from the primary manufacture sources [6,7] and other available publications [8].

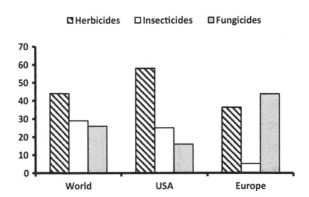

FIGURE 1.3 Distribution of the market (%) per pesticide type. (From Environmental Protection Agency (EPA). Pesticides industry sales and usage, 2016. Available at http://www. epa.gov. Pesticide sales by major groups, EU-28, 2014 Eurostat.)

1.2 HERBICIDES

The implementation of mechanization in agriculture has increased the ability of humans to control weeds and cultivate crops; herbicides have played a major part in this development, and a higher proportion of farmers would be needed if herbicides were not used.

Herbicides can be classified as soil- or foliage-applied compounds, which are normally absorbed by roots or leaf tissues, respectively. These compounds can be total or selective herbicides. Total herbicides can kill all vegetation, whereas selective herbicides can control weeds without affecting the crop. These chemical substances may be applied at different crop stages, such as pre-sowing and pre- or post-emergence, and these different treatments will be used depending on the weed needed to be controlled in a particular crop. The selectivity of a herbicide may depend on a differential plant uptake, translocation, or metabolism, as well as on differences at the site of action. A knowledge of physicochemical properties, that is, vapor pressure, octanol–water partition coefficient (K_{ow}, expressed in the logarithmic form log P), and solubility in water, allows the fate and behavior of such chemicals in the environment to be predicted.

In addition, herbicides can be classified according to their chemical composition. The principal physicochemical properties, together with the field persistence and major uses of representative herbicides, grouped in their main chemical classes, are described below.

1.2.1 AMIDES

A large variety of compounds form this group of herbicides, which have the following general formula: R_1–CO–N–(R_2,R_3).

The key components of this group are the N-substituted chloroacetamides and the substituted anilides.

Propanil   Alachlor

Chloroacetamides are effective pre-emergence herbicides for annual grasses and annual broad-leaved weeds, but they also have foliar contact activity. In general, these compounds are soil applied and used in various horticultural crops, such as maize, soybean, and sugarcane. These herbicides are normally absorbed by shoots and roots and they are, in general, nonpersistent compounds in soil (Table 1.1).

TABLE 1.1
Chemical Names and Properties of Amide Herbicides

Common Name	IUPAC Name	Vapor Pressure mPa (25°C)	K_{ow} log P	Water Solubility mg/L (25°C)	Half-Life in Soil (days)
Acetochlor $C_{14}H_{20}ClNO_2$	2-chloro-*N*-ethoxymethyl-6'-ethylacet-*o*-toluidide	0.05	4.14	282*	8–18
Alachlor $C_{14}H_{20}ClNO_2$	2-chloro-2',6'-diethyl-*N*-methoxymethylacetanilide	5.5	3.09	170*	8–20
Butachlor $C_{17}H_{26}ClNO_2$	N-butoxymethyl-2-chloro-2',6'-diethylacetanilide	0.25	–	16*	10–20
Metolachlor $C_{15}H_{22}ClNO_2$	2-chloro-6-ethyl-*N*-(2-methoxy-1-methylethyl)acet-*o*-toluidide	4.2	2.9	488	11–31
Propachlor $C_{11}H_{14}ClNO$	2-chloro-*N*-isopropyl acetanilide	10	1.4–2.3	580	4–6
Propanil $C_9H_9Cl_2NO$	3',4'-dichloro propionanilide	0.05	3.3	130*	0.2–2

Sources: Data from MacBean, C. (Ed.) in *The Pesticide Manual*, British Crop Protection Council, 2012.

University of Hertfordshire, The Pesticide Properties DataBase (PPDB) developed by the Agriculture & Environment Research Unit (AERU), University of Hertfordshire, 2006–2013, 2015. Available at http://sitem.herts.ac.uk/aeru/ppdb/en/atoz.htm.

EFSA Scientific Publications. Available at https://www.efsa.europa.eu/en/publications. European Commission. *EU Pesticides Database*. Available at http://ec.europa.eu/food/plant/pesticides/eu-pesticides-database.

*20°C.

1.2.2 BENZOIC ACIDS

This group is mainly formed by chlorinated derivatives of substituted benzoic acids.

Dicamba

Benzoic acid herbicides are known to have growth regulating and auxin activity properties. These compounds are especially used to control deep-rooted perennial weeds and applied as salts or esters (Table 1.2).

TABLE 1.2

Chemical Names and Properties of Benzoic Acid Herbicides

Common Name	IUPAC Name	Vapor Pressure mPa (25°C)	K_{ow} log P	Water Solubility g/L (25°C)	Half-Life in Soil (days)
Chlorthal-dimethyl $C_{10}H_6Cl_4O_4$	dimethyl tetrachloroterephthalate	0.21	4.28	0.5×10^{-3}	6–100
Dicamba $C_8H_6Cl_2O_3$	3,6-dichloro-*o*-methoxybenzoic acid	1.67	−1.88	>250	3–14

Sources: Data from MacBean, C. (Ed.) in *The Pesticide Manual*, British Crop Protection Council, 2012. University of Hertfordshire, The Pesticide Properties DataBase (PPDB) developed by the Agriculture & Environment Research Unit (AERU), University of Hertfordshire. Available at http://sitem.herts.ac.uk/ aeru/ppdb/en/atoz.htm. EFSA Scientific Publications. Available at https://www.efsa.europa.eu/en/publications. European Commission. *EU Pesticides Database.* Available at. http://ec.europa.eu/food/plant/pesticides/eu-pesticides-database.

1.2.3 CARBAMATES

Carbamates are esters of the carbamic acid (R_1-O-CO-NR_2R_3) and together with thiocarbamates (R_1-S-CO-NR_2R_3) represent a broad group of herbicides frequently applied to soil in pre-emergence.

Propham
—NHCO$_2$CH(CH$_3$)$_2$

EPTC
[CH$_3$(CH$_2$)$_2$]$_2$NC(O)SCH$_2$CH$_3$

These compounds are root or shoot absorbed and are frequently used to control annual grasses and broad-leaved weeds in peas, beets, and other horticultural crops. These herbicides are normally decomposed by soil microorganisms in three to five weeks. Their main physicochemical properties are summarized in Table 1.3.

1.2.4 IMIDAZOLINONES

Imidazolinones are herbicides that are applied either in pre- or post-emergence to control broadleaf weeds in soybean, barley, wheat, and alfalfa crops. These herbicides are very potent weed killers and are used in doses that are substantially lower than those of conventional herbicides. Their mode of action is the inhibition of the acetohydroxy acid synthase (AHAS), an enzyme that is key in the biosynthesis of branched-chain amino acids in plants. These chemicals block the normal function of AHAS, formerly referred to as acetolactate synthase (ALS). This enzyme is essential in amino acid (protein) synthesis. Their main physicochemical properties are summarized in Table 1.4.

TABLE 1.3
Chemical Names and Properties of Carbamate Herbicides

Common Name	IUPAC Name	Vapor Pressure mPa (25°C)	K_{ow} log P	Water Solubility mg/L (25°C)	Half-Life in Soil (days)
Chlorpropham $C_{10}H_{12}ClNO_2$	isopropyl-3-chlorocarbanilate	24*	3.79	89	3–30
Desmedipham $C_{16}H_{16}N_2O_4$	ethyl-3-phenylcarbamoyloxy phenylcarbamate	4×10^{-5}	3.39	7*	5–30
EPTC $C_9H_{19}NOS$	S-ethyl dipropylthiocarbamate	1×10^4	3.2	375	6–30
Molinate $C_9H_{17}NOS$	S-ethyl azepane-1-carbothioate	500	2.86	1.100	3–35
Phenmedipham $C_{16}H_{16}N_2O_4$	methyl-3-(3-methylcarbaniloyloxy) carbanilate	7×10^{-7}	3.59	4.7	6–40
Propham $C_{10}H_{13}NO_2$	isopropyl phenylcarbamate	Sublimes slowly	–	250*	5–15
Thiobencarb $C_{12}H_{16}ClNOS$	S-4-chlorobenzyl diethylthiocarbamate	2.39	4.23	16.7*	14–21
Tri-allate $C_{10}H_{16}Cl_3NOS$	S-2,3,3-trichloroallyl diisopropyl(thiocarbamate)	16	4.6	4	8–82

Sources: Data from MacBean, C. (Ed.) in *The Pesticide Manual*, British Crop Protection Council, 2012. University of Hertfordshire, The Pesticide Properties DataBase (PPDB) developed by the Agriculture & Environment Research Unit (AERU), University of Hertfordshire. Available at http://sitem.herts.ac.uk/aeru/ppdb/en/atoz.htm.
EFSA Scientific Publications. Available at https://www.efsa.europa.eu/en/publications. European Commission. *EU Pesticides Database.* Available at http://ec.europa.eu/food/plant/pesticides/eu-pesticides-database.
*20°C.

TABLE 1.4

Chemical Names and Properties of Imidazolinones Herbicides

Common Name	IUPAC Name	Vapor Pressure mPa (25°C)	K_{ow} log P	Water Solubility mg/L (25°C)	Half-Life in Soil (days)
Imazamox $C_{15}H_{19}N_3O_4$	(RS)-2-(4-isopropyl-4-methyl-5-oxo-2-imidazolin-2-yl)-5-methoxymethylnicotinic acid	< 1.3×10^{-2}	−2.4	626000*	5–41
Imazapic $C_{14}H_{17}N_3O_3$	(RS)-2-(4-isopropyl-4-methyl-5-oxo-2-imidazolin-2-yl)-5-methylnicotinic acid	< 1×10^{-2}	2.47	2150	31–410
Imazapyr $C_{13}H_{15}N_3O_3$	2-(4-isopropyl-4-methyl-5-oxo-2-imidazolin-2-yl)nicotinic acid	< 0.013	0.11	11300	24–143
Imazaquin $C_{17}H_{17}N_3O_3$	(RS)-2-(4-isopropyl-4-methyl-5-oxo-2-imidazolin-2-yl) quinoline-3-carboxylic acid	7×10^{-10}	0.34	102012	2–60
Imazethapyr $C_{15}H_{19}N_3O_3$	(RS)-5-ethyl-2-(4-isopropyl-4-methyl-5-oxo-2-imidazolin-2-yl) nicotinic acid	< 0.013	1.49	1400	14–290

Sources: Data from University of Hertfordshire, The Pesticide Properties DataBase (PPDB) developed by the Agriculture & Environment Research Unit (AERU), University of Hertfordshire. Available at http://sitem.herts.ac.uk/aeru/ppdb/en/atoz.htm.

EFSA Scientific Publications. Available at https://www.efsa.europa.eu/en/publications. European Commission. *EU Pesticides Database.* Available at http://ec.europa.eu/food/plant/pesticides/eu-pesticides-database.

*20°C.

Imazaquin   Imazapyr

1.2.5 NITRILES

Bromoxynil and ioxynil are hydroxybenzonitriles used as herbicides.

Bromoxynil   Ioxynil

They are formulated as salts or octanoate esters and foliage applied to control broad-leaved weeds in cereals and horticultural crops. These compounds are used in post-emergence and frequently applied in combination with other herbicides to extend the spectrum of weed species to be controlled. They have a low persistence in soil (Table 1.5).

1.2.6 NITROANILINES

These compounds are derivatives of 2,6-dinitroaniline.

Pendimethalin

Nitroanilines are a group of herbicides with similar physicochemical properties, such as a low water solubility and a high octanol–water partition coefficient.

TABLE 1.5
Chemical Names and Properties of Nitrile Herbicides

Common Name	IUPAC Name	Vapor Pressure mPa (20°C)	K_{ow} log P	Water Solubility mg/L (20°C)	Half-Life in Soil (days)
Bromoxynil $C_7H_3Br_2NO$	3,5-dibromo-4-hydroxybenzonitrile	1.7×10^{-1}	1.04	90	0.2–1
Ioxynil $C_7H_3I_2NO$	4-hydroxy-3,5-diiodobenzonitrile	2.04×10^{-3}	2.5	64.3	1–10

Sources: Data from MacBean, C. (Ed.) *The Pesticide Manual*, British Crop Protection Council, 2012. University of Hertfordshire, The Pesticide Properties DataBase (PPDB) developed by the Agriculture & Environment Research Unit (AERU), University of Hertfordshire. Available at http://sitem.herts.ac.uk/aeru/ppdb/en/atoz.htm.
EFSA Scientific Publications. Available at https://www.efsa.europa.eu/en/publications. European Commission. *EU Pesticides Database*. Available at http://ec.europa.eu/food/plant/pesticides/eu-pesticides-database.

These compounds are soil-applied herbicides used to control annual grasses and many broad-leaved weeds in a wide variety of crops. The 2,6-dinitroanilines possess a marked general herbicide activity. Substitution at the third and/or fourth position of the ring or on the amino group modifies the degree of herbicidal activity. In general, they have a certain persistence in soil and are normally soil incorporated due to their significant vapor pressure (Table 1.6).

1.2.7 ORGANOPHOSPHORUS

$$HO_2CCH_2NHCH_2P(OH)_2$$

Glyphosate

$$CH_3\overset{O}{\underset{OH}{P}}CH_2CH_2\underset{NH_2}{CH}CO_2H$$

Glufosinate

Glyphosate and glufosinate are broad spectrum, nonselective, post-emergence contact herbicides, active only for foliar application. They are extensively used in various applications for weed control in aquatic systems and vegetation control in non-crop areas. Aminomethylphosphonic acid (AMPA) is the major degradation product of glyphosate found in plants, water, and soil. The main properties of these compounds are shown in Table 1.7.

1.2.8 PHENOXY ACIDS

Phenoxy acids is a common name given to a group of compounds formed by a phenoxy radical linked to a low carbon number alkanoic acid, such as 2,4-D (acetic acid)

TABLE 1.6
Chemical Names and Properties of Nitroaniline Herbicides

Common Name	IUPAC Name	Vapor Pressure mPa (25°C)	K_{ow} log P	Water Solubility mg/L (25°C)	Half-Life in Soil (days)
Butralin $C_{14}H_{21}N_3O_4$	N-sec-butyl-4-tert-butyl-2, 6-dinitroaniline	0.77	4.93	0.3	19–190
Ethalfluralin $C_{13}H_{14}F_3N_3O_4$	N-ethyl- α,α,α -trifluoro-N- (2-methylallyl)-2,6-dinitro- p-toluidine	11.7	5.11	0.3	25–46
Pendimethalin $C_{13}H_{19}N_3O_4$	N-(1-ethylpropyl)-2, 6-dinitro-3,4-xylidine	1.94	5.18	0.3*	40–187
Trifluralin $C_{13}H_{16}F_3N_3O_4$	α,α,α-trifluoro-2,6-dinitro- N,N-dipropyl-p-toluidine	6.1	4.83*	0.22	35–375

Sources: Data from MacBean, C. (Ed.) in *The Pesticide Manual*, British Crop Protection Council, 2012.
University of Hertfordshire, The Pesticide Properties DataBase (PPDB) developed by the Agriculture & Environment Research Unit (AERU), University of Hertfordshire. Available at http://sitem.herts.ac.uk/aeru/ppdb/en/atoz.htm.
EFSA Scientific Publications. Available at https://www.efsa.europa.eu/en/publications. European Commission. *EU Pesticides Database*. Available at http://ec.europa.eu/food/plant/pesticides/eu-pesticides-database.
*20°C.

TABLE 1.7
Chemical Names and Properties of Organophosphorus Herbicides

Common Name	IUPAC Name	Vapor Pressure mPa (25°C)	K_{ow} log P	Water Solubility g/L (25°C)	Half-Life in Soil (days)
Glyphosate $C_3H_8NO_5P$	N-(phosphonomethyl) glycine	1.3×10^{-2}	<–3.2	10.5*	1–130
Glufosinate-ammonium $C_5H_{15}N_2O_4P$	ammonium 4-[hydroxy(methyl) phosphinoyl]-DL- homoalaninate	3.1×10^{-2}	–4.1	500.000*	7–20

Sources: Data from MacBean, C. (Ed.) in *The Pesticide Manual*, British Crop Protection Council, 2012. University of Hertfordshire, The Pesticide Properties DataBase (PPDB) developed by the Agriculture & Environment Research Unit (AERU), University of Hertfordshire. Available at http://sitem.herts.ac.uk/aeru/ppdb/en/atoz.htm.
EFSA Scientific Publications https://www.efsa.europa.eu/en/publications. European Commission. *EU Pesticides Database*. Available at http://ec.europa.eu/food/plant/pesticides/eu-pesticides-database.
*20°C.

or mecoprop (propionic acid). Some herbicides of this group are formed by stereo-isomers, which are commercialized as single enantiomers or as racemic mixtures.

2,4–D   Diclofop

These hormone type herbicides were discovered during the Second World War and, some years later, the phenoxy-phenoxy acids like diclofop were introduced to overcome the problem of selective control of grass weeds in cereal crops. These compounds are active by contact and by translocation from leaves to roots of perennial weeds and they are also used in pre-emergence applications to soil for control of young seedlings. The chlorophenoxy compounds are selective against broad-leaved annual weeds in cereal and grass crops. In general, they have a short persistence in soil (Table 1.8).

1.2.9 PYRIDINES AND QUATERNARY AMMONIUM COMPOUNDS

The herbicide group of pyridines, also named bipyridylium, is formed by paraquat and diquat. These compounds were developed as the result of observations that quaternary ammonium germicides, such as cetyltrimethylammonium bromide, desiccated young plants. Other quaternary ammonium compounds, like chlormequat and mepiquat, were developed and used as plant growth regulators to increase yields in cereals, promote flowering in ornamental plants, and improve fruit setting in horticultural plants and trees.

Paraquat   Diquat

Paraquat and diquat are broad-spectrum herbicides absorbed by leaves, but they are not translocated in sufficient quantities to kill the roots of perennial weeds. These compounds are very strong bases because of their quaternary ammonium structures and are rapidly adsorbed and inactivated in soil. Therefore, these compounds are not effective as pre-emergence herbicides. They have a high water solubility and a low octanol–water partition coefficient (Table 1.9), and are available commercially as dibromide or dichloride salts. These herbicides are strongly adsorbed in soil, requiring acid digestion during several hours for their desorption.

TABLE 1.8
Chemical Names and Properties of Phenoxy Acid Herbicides

Common Name	IUPAC Name	Vapor Pressure mPa (25°C)	K_{ow} log P	Water Solubility mg/L (20°C)	Half-Life in Soil (days)
2,4-D $C_8H_6Cl_2O_3$	2,4-dichlorophenoxy acetic acid	1.86×10^{-2}	-0.75	23180	22–38
Diclofop $C_{15}H_{12}Cl_2O_4$	(RS)-2-[4-(2,4-dichloro phenoxy)phenoxy]propionic acid	9.7×10^{-6}	1.61	122700	1–57
Fenoxaprop-P $C_{16}H_{12}ClNO_5$	(R)-2-[4-(6-chloro-1, 3-benzoxazol-2-yloxy) phenoxy]propionic acid	3.5×10^{-2}	1.83	61000*	1–10
Fluazifop-P $C_{15}H_{12}F_3NO_4$	(R)-2-[4-(5-trifluoromethyl-2-piridyloxy) phenoxy] propionic acid	7.9×10^{-4}	-0.8	780	<2–38
MCPA $C_9H_9ClO_3$	4-chloro-2-methylphenoxy) acetic acid	2.3×10^{-2}	-0.71	293.9**	7–41
Mecoprop-P $C_{10}H_{11}ClO_3$	(R)-2-(4-chloro-o-tolyloxy)propionic acid	0.23*	0.02	860	5–17
Quizalofop-P-ethyl $C_{19}H_{17}ClN_2O_4$	ethyl(R)-2-[4-(6-chloroquinoxalin-2-yloxy) phenoxy] propionate	8.65×10^{-4}*	4.28	0.3	0.5–8
Triclopyr $C_7H_4Cl_3NO_3$	3,5,6-trichloro-2-pyridyloxyacetic acid	0.2	-0.45	8.10	7–54

Sources: Data from MacBean, C. (Ed.) in *The Pesticide Manual*, British Crop Protection Council, 2012. University of Hertfordshire, The Pesticide Properties DataBase (PPDB) developed by the Agriculture & Environment Research Unit (AERU), University of Hertfordshire. Available at http://sitem.herts.ac.uk/aeru/ppdb/en/atoz.htm.

EFSA Scientific Publications. Available at https://www.efsa.europa.eu/en/publications. European Commission. *EU Pesticides Database.* Available at http://ec.europa.eu/food/plant/pesticides/eu-pesticides-database.

*20°C; **25°C.

TABLE 1.9

Chemical Names and Properties of Pyridine Herbicides and Quaternary Ammonium Compounds

Common Name	IUPAC Name	Vapor Pressure mPa (20°C)	K_{ow} log P	Water Solubility g/L (20°C)	Half-Life in Soil (days)
Diquat dibromide $C_{12}H_{12}Br_2N_2$	1,1'-ethylene-2,2'-bipyridyldiylium dibromide	0.01*	–4.6	718	365– >1000**
Paraquat dichloride $C_{12}H_{14}Cl_2N_2$	1,1'-dimethyl-4,4'-bipyridinium dichloride	<0.01*	–4.5	620	>365**
Chlormequat chloride $C_5H_{13}Cl_2N$	2-chloroethyl trimethyl ammonium	<0.001*	–1.59	>1000	1–32
Mepiquat chloride $C_7H_{16}ClN$	1,1'-dimethyl-piperidinium chloride	<1x10^{-5}*	–3.55	>500	11–40

Sources: Data from MacBean, C. (Ed.) in *The Pesticide Manual*, British Crop Protection Council, 2012. University of Hertfordshire, The Pesticide Properties DataBase (PPDB) developed by the Agriculture & Environment Research Unit (AERU), University of Hertfordshire. Available at http://sitem.herts.ac.uk/aeru/ppdb/en/atoz.htm.
EFSA Scientific Publications. Available at https://www.efsa.europa.eu/en/publications. European Commission. *EU Pesticides Database*. Available at http://ec.europa.eu/food/plant/pesticides/eu-pesticides-database.
*25°C; **Strongly adsorbed to soil.

1.2.10 PYRIDAZINES AND PYRIDAZINONES

Pyridate and pyridazinones, like norflurazon and chloridazon, are included in this group.

Norflurazon   Pyridate

They are contact-selective herbicides with foliar activity and are used in pre- or post-emergence to control annual grasses, broad-leaved weeds, and grassy weeds on cereals, maize, rice, and some other crops. In general, the pyridazinone herbicides are long lasting in soil (Table 1.10).

TABLE 1.10

Chemical Names and Properties of Pyridazine and Pyridazinone Herbicides

Common Name	IUPAC Name	Vapor Pressure mPa (25°C)	K_{ow} log P (25°C)	Water Solubility mg/L (20°C)	Half-Life in Soil (days)
Chloridazon $C_{10}H_8ClN_3O$	5-amino-4-chloro-2-phenylpyridazin-3(2H)-one	<0.01*	1.19	340	3–97
Norflurazon $C_{12}H_9ClF_3N_3O$	4-chloro-5-methylamino-2- (α,α,α-trifluoro-m-tolyl) pyridazin-3(2H)-one	3.8×10^{-3}	2.45	34**	90–270
Pyridate $C_{19}H_{23}ClN_2O_2S$	6-chloro-3-phenylpyridazin-4-yl-S-octyl thiocarbonate	4.8×10^{-4}*	4.01	0.32	0.3–7.7

Sources: Data from MacBean, C. (Ed.) in *The Pesticide Manual*, British Crop Protection Council, 2012. University of Hertfordshire, The Pesticide Properties DataBase (PPDB) developed by the Agriculture & Environment Research Unit (AERU), University of Hertfordshire. Available at http://sitem.herts.ac.uk/aeru/ppdb/en/atoz.htm.
EFSA Scientific Publications. Available at https://www.efsa.europa.eu/en/publications. European Commission. *EU Pesticides Database*. Available at http://ec.europa.eu/food/plant/pesticides/eu-pesticides-database.
*20°C; **25°C.

1.2.11 TRIAZINES

Simazine   Metribuzin

A wide range of triazines have been synthesized over time to control annual and broad-leaved weeds in a variety of crops as well as in non-cropped land. They are effective, at low dosages, in killing broad-leaved weeds in corn and other crops and they can be used in high dosages as soil sterilants. In general, these herbicides are applied in pre- or post-emergence and they are absorbed by the roots or by the foliage, respectively. In some cases, they are used in combination with other herbicides to broaden the spectrum of activity. These compounds have an appreciable persistence in soil (Table 1.11).

TABLE 1.11

Chemical Names and Properties of Triazine Herbicides

Common Name	IUPAC Name	Vapor Pressure mPa (25°C)	K_{ow} log P (25°C)	Water Solubility mg/L (25°C)	Half-Life in Soil (days)
Atrazine $C_8H_{14}ClN_5$	6-chloro-N^2-ethyl-N^4-isopropyl-1,3,5-triazine-2,4-diamine	3.8×10^{-2}	2.5	33*	6–150
Cyanazine $C_9H_{13}ClN_6$	2-(4-chloro-6-ethylamino-1,3,5-triazin-2-ylamino)-2-methylpropionitrile	2.0×10^{-4}*	2.1	171	12–25
Metribuzin $C_8H_{14}N_4OS$	4-amino-6-tert-butyl-4,5-dihydro-3-methylthio-1,2,4-triazin-5-one	0.058*	1.6*	1050*	5–30
Prometryn $C_{10}H_{19}N_5S$	N^2,N^4-di-isopropyl-6-methylthio-1,3,5-triazine-2,4-diamine	0.165	3.1	33	14–158
Simazine $C_7H_{12}ClN_5$	6-chloro-N^2,N^4-diethyl-1,3,5-triazine-2,4-diamine	2.9×10^{-3}	2.1	6.2*	27–102
Terbutryn $C_{10}H_{19}N_5S$	N^2-tert-buthyl-N^4-etil-6-methylthio-1,3,5-triazine-2,4-diamine	0.225	3.65	22	9–47

Sources: Data from MacBean, C. (Ed.) in *The Pesticide Manual*, British Crop Protection Council, 2012. University of Hertfordshire, The Pesticide Properties DataBase (PPDB) developed by the Agriculture & Environment Research Unit (AERU), University of Hertfordshire. Available at http://sitem.herts.ac.uk/aeru/ppdb/en/atoz.htm.

EFSA Scientific Publications. Available at https://www.efsa.europa.eu/en/publications. European Commission. *EU Pesticides Database*. Available at http://ec.europa.eu/food/plant/pesticides/eu-pesticides-database.

*20°C.

1.2.12 UREAS

1.2.12.1 Phenylureas

The urea herbicides may be considered as derivatives of urea, $H_2NC(=O)NH_2$.

Fenuron   Linuron

Phenylureas belong to a large group of substituted ureas directly applied to soil in pre-emergence to control annual grasses in various crops. These compounds have a range of specific selectivity as well as variable persistence in soil according to their chemical composition (Table 1.12).

1.2.12.2 Sulfonylureas

Chlorsulfuron   Triasulfuron

TABLE 1.12

Chemical Names and Properties of Phenyl Urea Herbicides

Common Name	IUPAC Name	Vapor Pressure mPa (25°C)	K_{ow} log P (25°C)	Water Solubility mg/L (25°C)	Half-Life in Soil (days)
Chlorotoluron $C_{10}H_{13}ClN_2O$	3-(3-chloro-p-tolyl)-1, 1-dimethylurea	0.005	2.5	74	30–40
Diuron $C_9H_{10}Cl_2N_2O$	3-(3,4-dichlorophenyl)-1,1-dimethylurea	1.1×10^{-3}	2.85	37.4	20–231
Isoproturon $C_{12}H_{18}N_2O$	3-(4-isopropylphenyl)-1,1-dimethylurea	8.1×10^{-3}	2.5*	65	6–28
Linuron $C_9H_{10}Cl_2N_2O_2$	3-(3,4-dichlorophenyl)-1-methoxy-1-methylurea	0.051*	3.0	63.8*	13–82

Sources: Data from MacBean, C. (Ed.) in *The Pesticide Manual*, British Crop Protection Council, 2012. University of Hertfordshire, The Pesticide Properties DataBase (PPDB) developed by the Agriculture & Environment Research Unit (AERU), University of Hertfordshire. Available at http://sitem.herts.ac.uk/aeru/ppdb/en/atoz.htm.

EFSA Scientific Publications. Available at https://www.efsa.europa.eu/en/publications. European Commission. *EU Pesticides Database*. Available at http://ec.europa.eu/food/plant/pesticides/eu-pesticides-database.

*20°C.

This group of substituted ureas was developed more recently and they have, in general, a herbicidal activity higher than the phenylurea herbicides, with application rates in the range of g/ha instead of kg/ha. They can be absorbed by foliage and roots. They are normally applied in postemergence and may have in some cases a noticeable field persistence (Table 1.13).

1.3 INSECTICIDES

Horticultural crops may be affected by various pests, causing serious damages to plants and consequently important yield reductions. Therefore, insecticides are widely used to control pests in crops. These compounds may be applied to the soil to kill soil-borne pests or to the aerial part of the plant.

A major part of the applied insecticide reaches the soil, whether from the direct applications to the soil or indirectly by run-off from leaves and stems.

1.3.1 BENZOYLUREAS

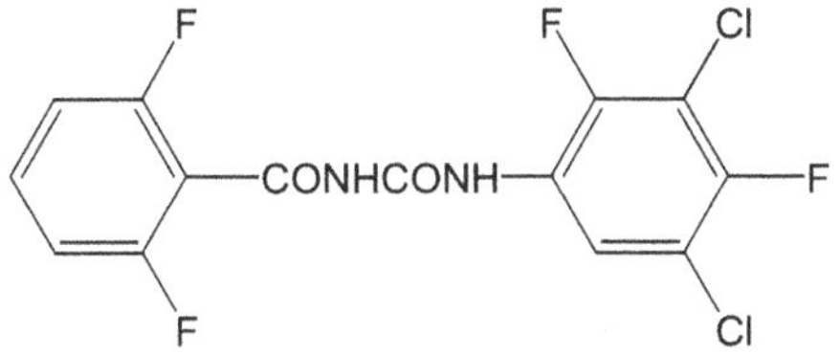

Teflubenzuron

A new insecticide activity acting on the molting process of insects was discovered in the study of biological activity of some benzoylurea derivatives. Benzoylureas act as insect growth regulators, interfering with the chitin formation in the vital insect exoskeleton. Most benzoylureas used as insecticides contain fluorine atoms and have high molecular weights. Table 1.14 summarizes the physicochemical properties of these compounds.

1.3.2 CARBAMATES

The N-methyl and N,N-dimethyl carbamic esters of a variety of phenols possess useful insecticidal properties. Aromatic N-methylcarbamates are derivatives of phenyl N-methylcarbamate with a great variety of chloride, alkyl, alkylthio, alkoxy, and dialkylamino side chains. Some carbamate insecticides contain a sulfur atom in their molecule.

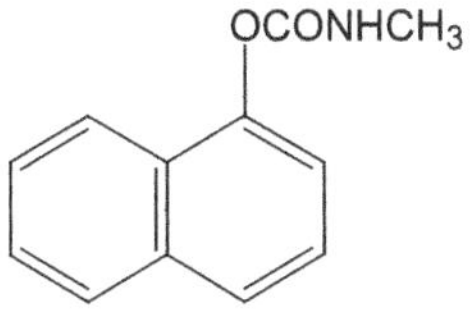

Carbaryl 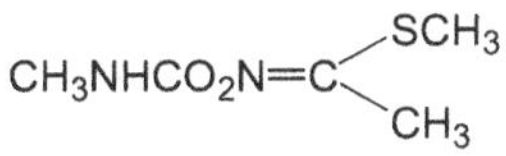   Methomyl

TABLE 1.13

Chemical Names and Properties of Sulfonylurea Herbicides

Common Name	IUPAC Name	Vapor Pressure mPa (25°C)	K_{ow} log P (25°C)	Water Solubility mg/L (25°C)	Half-Life in Soil (days)
Azimsulfuron $C_{13}H_{16}N_{10}O_5S$	1-(4,6-dimethoxypyrimidin-2-yl)-3-[1-methyl-4-(2-methyl-2H-tetrazol-5-yl)-pyrazol-5-ylsulfonyl]urea	4.0×10^{-6}	-1.4	1050*	12–51–
Chlorsulfuron $C_{12}H_{12}ClN_5O_4S$	1-(2-chlorophenylsulfonyl)-3-(4-methoxy-6-methyl-1,3,5-triazin-2-yl)urea	3×10^{-6}	-0.99	31.800	11–70
Flazasulfuron $C_{13}H_{12}F_3N_5O_5S$	1-(4,6-dimethoxypyrimidin-2-yl)-3-(3-trifluoromethyl-2-pyridilsulfonyl)urea	<0.013	-0.06	2100	2–18
Imazosulfuron $C_{14}H_{13}ClN_6O_5S$	1-(2-chloroimidazo[1,2-a]pyridin-3-ylsulfonyl)-3-(4,6-dimethoxy pyrimidin-2-yl)urea	3.5×10^{-3}	-0.07	429	21–91–
Metsulfuron-methyl $C_{14}H_{15}N_5O_6S$	methyl 2-(4-methoxy-6-methyl-1,3,5-triazin-2-ylcarbamoylsulfamoyl)benzoate	3.3×10^{-7}	$-1.87*$	2790	7–37
Rimsulfuron $C_{14}H_{17}N_5O_7S_2$	1-(4,6-dimethoxypyrimidin-2-yl)-3-(-3-ethylsulfonyl-2-pyridylsulfonyl)urea	1.5×10^{-3}	-1.47	7300	5–18
Thifensulfuron-methyl $C_{12}H_{13}N_5O_6S_2$	methyl 3-(4-methoxy-6-methyl-1,3,5-triazin-2-ylcarbamoylsulfamoyl)thiophen-2-carboxylate	1.7×10^{-5}	0.02	2240	3–20
Triasulfuron $C_{14}H_{16}ClN_5O_5S$	1-[2-(2-chloroethoxy)phenylsulfonyl]-3-(4-methoxy-6-methyl-1,3,5-triazin-2-yl)urea	$<2 \times 10^{-3}$	-0.59	815	16–92
Tribenuron-methyl $C_{15}H_{17}N_5O_6S$	methyl 2-[4-methoxy-6-methyl-1,3,5-triazin-2-yl(methyl)carbamoylsulfamoyl]benzoate	5.2×10^{-5}	0.78	2040*	2–10

Sources: Data from MacBean, C. (Ed.) in *The Pesticide Manual*, British Crop Protection Council, 2012. University of Hertfordshire, The Pesticide Properties DataBase (PPDB) developed by the Agriculture & Environment Research Unit (AERU), University of Hertfordshire. Available at http://sitem.herts.ac.uk/aeru/ppdb/en/atoz.htm. EFSA Scientific Publications. Available at https://www.efsa.europa.eu/en/publications. European Commission. *EU Pesticides Database*. Available at http://ec.europa.eu/food/plant/pesticides/eu-pesticides-database.

*20°C.

TABLE 1.14
Chemical Names and Properties of Benzoylurea Insecticides

Common Name	IUPAC Name	Vapor Pressure mPa (20°C)	K_{ow} log P	Water Solubility mg/L (25°C)	Half-Life in Soil (days)
Diflubenzuron $C_{14}H_9ClF_2N_2O_2$	1-(4-chlorophenyl)-3-(2,6-difluorobenzoyl)urea	1.2×10^{-4}*	4	0.08	2–6.7
Hexaflumuron $C_{16}H_8Cl_2F_6N_2O_3$	1-[3,5-dichloro-4-(1,1,2,2-tetrafluoroethoxy)phenyl]-3-(2,6-difluorbenzoyl)urea	5.9×10^{-2}*	5.68	0.027**	100–280
Teflubenzuron $C_{14}H_6Cl_2F_4N_2O_2$	1-(3,5-dichloro-2,4-difluorophenyl)-3-(2,6-difluorbenzoyl)urea	$1.3 10^{-5}$*	4.3	0.01**	8–16
Triflumuron $C_{15}H_{10}ClF_3N_2O_3$	(1-(2-chlorobenzoyl)-3-(4-trifluoromethoxyphenyl)urea)	4×10^{-5}	4.91	0.025**	22

Sources: Data from MacBean, C. (Ed.) in *The Pesticide Manual*, British Crop Protection Council, 2012. University of Hertfordshire, The Pesticide Properties DataBase (PPDB) developed by the Agriculture & Environment Research Unit (AERU), University of Hertfordshire. Available at http://sitem.herts.ac.uk/aeru/ppdb/en/atoz.htm.
EFSA Scientific Publications. Available at https://www.efsa.europa.eu/en/publications. European Commission. *EU Pesticides Database*. Available at http://ec.europa.eu/food/plant/pesticides/eu-pesticides-database.
*25°C; **20°C.

These compounds have a very broad spectrum of action, and they are particularly effective against lepidopterous larvae and on ornamental pests including snails, slugs, and household pests. Some of them exhibit systemic characteristics (Table 1.15).

1.3.3 NEONICOTINOIDS

Imidacloprid

Thiamethoxam

TABLE 1.15
Chemical Names and Properties of Carbamate Insecticides

Common Name	IUPAC Name	Vapor Pressure mPa (20°C)	K_{ow} log P	Water Solubility mg/L (20°C)	Half-Life in Soil (days)
Aldicarb $C_7H_{14}N_2O_2S$	2-methyl-2-(methylthio) propionaldehyde O-methylcarbamoyloxyme	3.87*	1.15	4930	2–12
Carbaryl $C_{12}H_{11}NO_2$	1-naphthyl methylcarbamate	4.1×10^{-2}*	1.85	120	7–28
Carbofuran $C_{12}H_{15}NO_3$	2,3-dihydro-2,2-dimethyl benzofuran-7-yl methylcarbamate	0.031	1.52	320	1–27
Carbosulfan $C_{20}H_{32}N_2O_3S$	2,3-dihydro-2,2-dimethyl benzofuran-7-yl(dibutylaminothio) methylcarbamate	0.041*	5.4	0.11	0.3–72
Fenoxycarb $C_{17}H_{19}NO_4$	ethyl-2-(4-phenoxyphenoxy) ethylcarbamate	8.67×10^{-4}*	4.07	7.9*	4–9
Methomyl $C_5H_{10}N_2O_2S$	S-methyl N-(methylcarbamoyloxy) thioacetamidate	0.72*	0.093	55000	4–9
Oxamyl $C_7H_{13}N_3O_3S$	N,N'-dimethyl-2-methyl carbamoyloxyimino-2-(methylthio) acetamide	0.051*	−0.44	148100	3–11.5
Pirimicarb $C_{11}H_{18}N_4O_2$	2-dimethylamino-5,6-dimethyl pyrimidin-4-yl dimethylcarbamate	0.4	1.7	3100	5–13

Sources: Data from MacBean, C. (Ed.) in *The Pesticide Manual*, British Crop Protection Council, 2012. University of Hertfordshire, The Pesticide Properties DataBase (PPDB) developed by the Agriculture & Environment Research Unit (AERU), University of Hertfordshire. Available at http://sitem.herts.ac.uk/aeru/ppdb/en/atoz.htm.

EFSA Scientific Publications. Available at https://www.efsa.europa.eu/en/publications. European Commission. *EU Pesticides Database.* Available at http://ec.europa.eu/food/plant/pesticides/eu-pesticides-database.

*25°C.

Neonicotinoids are a relatively new class of insecticides with a common mode of action affecting the central nervous system of insects. Neonicotinoids mimic the action of neurotransmitters. Upon their uptake by plants, these compounds and their metabolites circulate throughout plant tissues and provide a period of protection against a number of sap-feeding insects/arthropods. Nevertheless, some uncertainties have been identified since their initial registration, as data suggest that their residues can accumulate in the pollen and nectar of treated plants and may represent a potential exposure to pollinators. Recently, the use in of several neonicotinoids was prohibited in bee-attractive crops (including maize, oilseed rape, and sunflower) in the EU.

The low potential for volatilization of these substances indicates that these pesticides will most likely only be present in gaseous form for a short period during spray applications. Seed coating is the leading delivery method for neonicotinoids in agriculture. This method of pesticide application was considered to be a safer option for minimizing impacts on non-target organisms by reducing drift.

The solubility of neonicotinoids in water depends on multiple factors such as water temperature and pH as well as the physical state of the pesticide applied. The solubility ranges between 185 (moderate) and 590,000 mg/L (high) for thiacloprid and nitenpyram, respectively, at 20°C and at pH 7 (Table 1.16).

1.3.4 ORGANOCHLORINES

p,p'-DDT   Endosulfan

These insecticides are characterized by three kinds of chemicals: DDT analogs, benzene hexachloride (BHC) isomers, and cyclodiene compounds. DDT is one of the most persistent and durable of all contact insecticides because of its insolubility in water and its very low vapor pressure. DDT has a wide spectrum of activity among different families of insects and related organisms. BHC isomers are active against a great variety of pests. Cyclodiene compounds are effective where contact action and long persistence is required. These compounds have a broad-spectrum insecticide and have been used for the control of insect pests of fruits, vegetables, and cotton, as soil insecticides, and for seed treatment. Due to their persistence and toxicity, most of them have been banned or their use as pesticide restricted (Table 1.17).

TABLE 1.16

Chemical Names and Properties of Neonicotinoid Insecticides

Common Name	IUPAC Name	Vapor Pressure mPa (25°C)	K_{ow} log P	Water Solubility mg/L (25°C)	Half-Life in Soil (days)
Acetamiprid $C_{10}H_{11}ClN_4$	(E)-N^1-[(6-chloro-3-pyridyl)methyl]-N^2-cyano-N^1-methylacetamidine	1.73×10^{-4}	0.80	4250	0.8–5.6
Clothianidin $C_6H_8ClN_5O_2S$	(E)-1-(2-chloro-1,3-thiazol-5-ylmethyl)-3-methyl-2-nitroguanidine	1.3×10^{-7}	0.70	340 *	13–1386
Dinotefuran $C_7H_{14}N_4O_3$	(RS)-1-methyl-2-nitro-3-((tetrahydro-3-furylmethyl)guanidine	$< 1.7 \times 10^{-3}$	−0.549	39800*	50–100
Imidacloprid $C_9H_{10}ClN_5O_2$	(E)-1-(6-chloro-3-pyridylmethyl)-N-nitroimidazolidin-2-ylideneamine	4×10^{-7}*	0.57	610*	104–228
Nitenpyram $C_{11}H_{15}ClN_4O_2$	(E)-N-(6-chloro-3-pyridylmethyl)-N-ethyl-N'-methyl-2-nitrovinylidenediamine	1.1×10^{-3}	−0.66	590000*	1–15
Sulfoxaflor $C_{10}H_{10}F_3N_3OS$	[methyl(oxo){1-[6-(trifluoromethyl)-3-pyridyl]ethyl]-λ^6-sulfanylidene]cyanamide	1.4×10^{-3}	0.8	809	1–7
Thiacloprid $C_{10}H_9ClN_4S$	(Z)-3-(6-chloro-3-pyridylmethyl)-1,3-thiazolidin-2-ylidenecyanamide	3×10^{-7}*	0.74	185*	7–27
Thiamethoxam $C_8H_{10}ClN_5O_3S$	3-(2-chloro-1,3-thiazol-5-ylmethyl)-5-methyl-1,3,5-oxadiazinan-4-ylidene(nitro)amine	6.6×10^{-6}	−0.13	4100	7–109

Sources: Data from University of Hertfordshire, The Pesticide Properties DataBase (PPDB) developed by the Agriculture & Environment Research Unit (AERU), University of Hertfordshire. Available at http://sitem.herts.ac.uk/aeru/ppdb/en/atoz.htm.

EFSA Scientific Publications. Available at https://www.efsa.europa.eu/en/publications. European Commission. *EU Pesticides Database.* Available at http://ec.europa.eu/food/plant/pesticides/eu-pesticides-database.

*20°C.

TABLE 1.17
Chemical Names and Properties of Organochlorine Insecticides

Common Name	IUPAC Name	Vapor Pressure mPa (20°C)	K_{ow} log P	Water Solubility mg/L (25°C)	Half-Life in Soil (days)
p,p'-DDT $C_{14}H_9Cl_5$	1,1,1-trichloro-2,2-bis (4-chlorophenyl)ethane	0.025	6.91	0.006*	2012
Dicofol $C_{14}H_9Cl_5O$	2,2,2-trichloro-1,1-bis (4-chlorophenyl)ethanol	0.25**	4.3	0.8	40–80
Endosulfan $C_9H_6Cl_6O_3S$	(1,4,5,6,7,7-hexachloro-8,9, 10-trinorborn-5-en-2,3-ylenebismethylene)sulfite	0.83	4.74	0.32*	62–126
γ-HCH $C_6H_6Cl_6$	1,2,3,4,5 ,6-hexachlorocyclohexane	4.4**	3.5	8.5	400
Tetradifon $C_{12}H_6Cl_4O_2S$	4-chlorophenyl-2,4,5-trichlorophenylsulfone	3.2×10^{-5}**	4.61	0.078*	112

Sources: Data from MacBean, C. (Ed.) in *The Pesticide Manual*, British Crop Protection Council, 2012. University of Hertfordshire, The Pesticide Properties DataBase (PPDB) developed by the Agriculture & Environment Research Unit (AERU), University of Hertfordshire. Available at http://sitem.herts.ac.uk/aeru/ppdb/en/atoz.htm.
EFSA Scientific Publications. Available at https://www.efsa.europa.eu/en/publications. European Commission. *EU Pesticides Database.* Available at http://ec.europa.eu/food/plant/pesticides/eu-pesticides-database.
*20°C; **25°C.

1.3.5 ORGANOPHOSPHORUS

Organophosphorus insecticides are hydrocarbon compounds which contain one or more phosphorus atoms in their molecule. They are relatively short lived in biological systems.

Fenitrothion   Chlorpyrifos

The diversity of organophosphorus insecticides types means that they form the most versatile group. There are compounds with nonresidual action and others with prolonged residual action, compounds with a broad spectrum and others with very specific action that can have activity as systemic insecticides for plants, seed, and soil treatments, as well as for animals. In general, they are soluble in water and readily hydrolyzed and they dissipate from soil within a few weeks after application. Because of their low persistence and high effectiveness, these compounds are widely used as systemic insecticides for plants, animals, and soil treatments (Table 1.18).

TABLE 1.18

Chemical Names and Properties of Organophosphorus Insecticides

Common Name	IUPAC Name	Vapor Pressure mPa (25°C)	K_{ow} log P	Water Solubility mg/L (20°C)	Half-Life in Soil (days)
Azinphos-methyl $C_{10}H_{12}N_3O_3PS_2$	S[3,4-dihydro-4-oxobenzo [d]-[1,2,3]triazin-3-ylmethyl] O,O-dimethyl phosphoro dithioate	1×10^{-3}	2.96	28	10–30
Chlorfenvinphos $C_{12}H_{14}Cl_3O_4P$	2-chloro-1-(2,4-dichlorophenyl)vinyl diethyl phosphate	1	3.85	145	10–45
Chlorpyrifos $C_9H_{11}Cl_3NO_3PS$	O,O-diethyl O-3,5,6-trichloro-2-pyridyl phosphorothioate	2.7	4.7	1.05	7–56
Chlorpyrifos-methyl $C_7H_7Cl_3NO_3PS$	O,O-dimethyl O-3,5,6-trichloro-2-pyridyl phosphorothioate	3	4.24	2.6	1–7
Coumaphos $C_{14}H_{16}ClO_5PS$	O,O-3-chloro-4-methyl-2-oxo-2H-chromen-7-yl O,O-didiethyl phosphorothioate	0.013**	4.13	1.5	24
Diazinon $C_{12}H_{21}N_2O_3PS$	O,O-diethyl O-2-isopropyl -6-methylpyrimidin-4-yl phosphorothioate	12	3.30	60	7–30
Dichlorvos $C_4H_7Cl_2O_4P$	2,2-dichlorovinyl dimethyl phosphate	2.1×10^3	1.9	$18 \times 10^{3*}$	0.5–2
Dimethoate $C_5H_{12}NO_3PS_2$	O,O-dimethyl S-methyl carbamoylmethyl phosphoro dithioate	0.25	0.7	25900	5–10
Fenitrothion $C_9H_{12}NO_5PS$	O,O-dimethyl O-4-nitro-m-tolyl phosphorothioate	1.57	3.43	19	12–28
Fenthion $C_{10}H_{15}O_3PS_2$	O,O-dimethyl O-4-methylthio-m-tolyl phosphorothioate	1.4	4.84	4.2	34
Malathion $C_{10}H_{19}O_6PS_2$	S-1,2-bis(ethoxycarbonyl) ethyl O,O-dimethyl phosphoro dithioate	3.1	2.75	145	1

(Continued)

TABLE 1.18 (CONTINUED)
Chemical Names and Properties of Organophosphorus Insecticides

Common Name	IUPAC Name	Vapor Pressure mPa (25°C)	K_{ow} log P	Water Solubility mg/L (20°C)	Half-Life in Soil (days)
Methamidophos $C_2H_8NO_2PS$	O,S-dimethyl phosphoramido thioate	4.7	-0.8	>2×10⁵	6
Methidathion $C_6H_{11}N_2O_4PS_3$	S-2,3-dihydro-5-methoxy-2-oxo-1,3,4-thiadiazol-3-ylmethyl O,O-dimethyl phosphoro dithioate	0.25**	2.2	240	3–18
Oxydemeton-methyl $C_6H_{15}O_4PS_2$	S-2-ethylsulfinylethyl O,O-dimethyl phosphorothioate	2	-0.74	1.2×10⁶	1–4
Phosmet $C_{11}H_{12}NO_4PS_2$	O,O-dimethyl S-phtalimidomethyl phosphorodithioate	0.065	2.95	15.2	1–11
Pirimiphos-methyl $C_{11}H_{20}N_3O_3PS$	O-2-diethylamino-6- methylpyrimidin-4-yl O,O-dimethyl phosphorothioate	2×10⁻⁰³	4.2	10	18–67
Profenofos $C_{11}H_{15}BrClO_3PS$	O-4-bromo-2-chlorophenyl O-ethyl S-propyl phosphorothioate	0.12	4.44	28*	7
Trichlorfon $C_4H_8Cl_3O_4P$	dimethyl 2,2,2-trichloro-1-hydroxyethylphosphonate	0.5	0.43	1.2×10⁵	7–30

Sources: Data from MacBean, C. (Ed.) in *The Pesticide Manual*, British Crop Protection Council, 2012. University of Hertfordshire, The Pesticide Properties DataBase (PPDB) developed by the Agriculture & Environment Research Unit (AERU), University of Hertfordshire. Available at http://sitem.herts.ac.uk/aeru/ppdb/en/atoz.htm.

EFSA Scientific Publications. Available at https://www.efsa.europa.eu/en/publications. European Commission. *EU Pesticides Database.* Available at http://ec.europa.eu/food/plant/pesticides/eu-pesticides-database.

*25°C; **20°C.

1.3.6 PYRETHROIDS

Permethrin

Pyrethrins are natural insecticides obtained from pyrethrum, extracted from the flowers of certain species of chrysanthemum. The insecticide properties are due to five esters that are present mostly in the flowers. These esters have asymmetric carbon atoms and double bonds in both alcohol and acid moieties. The naturally occurring forms are esters from (+)-*trans* acids and (+)-*cis* alcohols. Synthetic pyrethrins, called pyrethroids, present better activity for a larger spectrum of pests than natural ones. They show selective activity against insects and present low toxicity to mammals and birds. Pyrethroids are considered as contact poisons, affecting the insect nervous system and depolarizing the neuronal membranes. These compounds are degraded in soil and have no detectable effects on soil microflora. They have also been used in the household to control flies and mosquitoes. Piperonyl butoxide $(C_{19}H_{30}O_5)$ is used as a synergist for pyrethrins and related insecticides (Table 1.19).

1.3.7 MISCELLANEOUS

Fipronil   Spirotetramat

Fipronil is a blocker of the GABA-regulated chloride channel, toxic by contact or ingestion and able to control insects tolerant to pyrethroid, organophosphate, and carbamate insecticides. Spirotetramat is a lipid biosynthesis inhibitor, acting after ingestion or on contact with phloem and xylem mobility, and able to control a variety of sucking insects. Spirodiclofen is also a lipid biosynthesis inhibitor non-systemic insecticide used for control of mite pests. Table 1.20 summarizes the properties of various frequently used insecticides belonging to different chemical classes.

TABLE 1.19

Chemical Names and Properties of Pyrethroid Insecticides

Common Name	IUPAC Name	Vapor Pressure mPa (20°C)	K_{ow} log P	Water Solubility mg/L (25°C)	Half-Life in Soil (days)
Acrinathrin $C_{26}H_{21}F_6NO_5$	(S)-α-cyano-3-phenoxybenzyl (Z)-(1R,3S)-2,2-dimethyl-3-[2-(2,2,2-trifluoro-1-trifluoromethyl ethoxycarbonyl) vinyl] cyclopropanecarboxylate	4.4×10^{-5}	5.6	≤ 0.02	8–111
Cyfluthrin $C_{22}H_{18}Cl_2FNO_3$	(RS)-α-cyano-4-fluoro-3-phenoxybenzyl(1RS,3RS; 1RS,3SR)-3-(2,2-dichlorovinyl)-2,2-dimethyl cyclopropanecarboxylate	Diastereoisomer I: 9.6×10^{-4} II: 1.4×10^{-5} III: 2.1×10^{-5} IV: 8.5×10^{-5}	I: 6 II: 5.9 III: 6 IV:5.9	I: $2.2 \times 10^{-3**}$ II: 1.9×10^{-3} III: 2.2×10^{-3} IV: 2.9×10^{-3}	26–40
Cypermethrin $C_{22}H_{19}Cl_2NO_3$	(RS)-α-phenoxybenzyl(1RS,3RS; 1RS,3SR)-3-(2,2-dichlorovinyl)-2,2-dimethyl cyclopropanecarboxylate	2×10^{-4}	6.6	0.004	9–31
Deltamethrin $C_{22}H_{19}Br_2NO_3$	(S)-α-cyano-3-phenoxybenzyl(1R,3R)-3-(2,2-dibromovinyl)-2,2-dimethyl cyclopropanecarboxylate	$1.24 \times 10^{-5*}$	4.6	$<0.2 \times 10^{-3}$	8–28
Esfenvalerate $C_{25}H_{22}ClNO_3$	(S)-α-cyano-3-phenoxybenzyl-(S)-2-(4-chlorophenyl)-3-methylbutirate	1.17×10^{-6}	6.24	0.002	36–199
Tau-fluvalinate $C_{26}H_{22}ClF_3N_2O_3$	(RS)-α-cyano-3-phenoxybenzyl-N-(2-chloro-α,α,α-trifluoro-p-tolyl)-D-valinate	9×10^{-8}	4.26	$0.002**$	12–92
Permethrin $C_{21}H_{20}Cl_2O_3$	3-phenoxybenzyl-(1RS,3RS; 1RS,3SR)-3-(2,2-dichlorovinyl)-2,2-dimethylcyclopropane carboxylate	cis: $0.0029*$ trans: $9.2 \times 10^{-4*}$	$6.1**$	$6 \times 10^{-3**}$	<38

Sources: Data from MacBean, C. (Ed.) in *The Pesticide Manual*, British Crop Protection Council, 2012. University of Hertfordshire, The Pesticide Properties DataBase (PPDB) developed by the Agriculture & Environment Research Unit (AERU), University of Hertfordshire. Available at http://sitem.herts.ac.uk/aeru/ppdb/en/atoz.htm.

EFSA Scientific Publications. Available at https://www.efsa.europa.eu/en/publications. European Commission. *EU Pesticides Database*. Available at http://ec.europa.eu/food/plant/pesticides/eu-pesticides-database.

*25°C; **20°C.

TABLE 1.20

Chemical Names and Properties of Miscellaneous Insecticides

Common Name	IUPAC Name	Vapor Pressure mPa (25°C)	K_{ow} log P	Water solubility mg/L (25°C)	Half-Life in Soil (days)
Fipronil $C_{12}H_4Cl_2F_6N_4OS$	(±)-5-amino-1-(2,6-dichloro-α,α,α-trifluoro-p-tolyl)-4-trifluoromethylsulfinylpyrazole-3-carbonitrile	2×10^{-3}	4.0	3.78*	6–135
Spirotetramat $C_{21}H_{27}NO_5$	cis-4-(ethoxycarbonyloxy)-8-methoxy-3-(2,5-xylyl)-1-azaspiro[4.5]dec-3-en-2one	1.5×10^{-5}	2.51	29.9*	0.3–1
Spirodiclofen $C_{21}H_{24}Cl_2O_4$	3-(2,4-dichlorophenyl)-2-oxo-1-oxaspiro[4.5]dec-3-en-4-yl 2,2-dimethylbutyrate	3×10^{-4}	5.1	0.19	1–13

Sources: Data from University of Hertfordshire, The Pesticide Properties DataBase (PPDB) developed by the Agriculture & Environment Research Unit (AERU), University of Hertfordshire. Available at http://sitem.herts.ac.uk/aeru/ppdb/en/atoz.htm.
EFSA Scientific Publications. Available at https://www.efsa.europa.eu/en/publications. European Commission. *EU Pesticides Database.* Available at http://ec.europa.eu/food/plant/pesticides/eu-pesticides -database.
*20°C.

1.4 FUNGICIDES

Fungicides used in agriculture to control plant diseases belong to various chemical classes. A wide variation of physicochemical properties of these substances can be observed, according to the different chemical structures of fungicides. Some fungicides are stereoisomers and they are normally commercialized as mixtures of these isomers. Fungicides can be applied pre or post-harvest for the protection of cereals, fruit, and vegetables from fungal diseases.

1.4.1 AZOLES

Cyproconazole

The imidazole ring is present in several biologically active compounds, while others have a triazole ring. These compounds are fungicides with systemic action, effective against several phytopathogenous fungi and recommended for seed dressing, as well as foliage fungicide and postharvest application in fruits. They are scarcely soluble in water, although their salts are soluble in water (Table 1.21).

TABLE 1.21
Chemical Names and Properties of Azole Fungicides

Common Name	IUPAC Name	Vapor Pressure mPa (20°C)	K_{ow} log P	Water Solubility mg/L (20°C)	Half-Life in Soil (days)
Cyproconazole $C_{15}H_{18}ClN_3O$	(2RS,3RS;2RS,3SR)-2-(4-chlorophenyl)-3-cyclopropyl-1-(1H-1,2,4-triazol-1-yl)butan-2-ol	0.026*	3.1	93	45–191
Flusilazole $C_{16}H_{15}F_2N_3Si$	Bis(4-fluorophenyl)(methyl)(1H-1,2,4-triazol-1-ylmethyl)silane	0.04*	3.74	54	63–240
Hexaconazole $C_{14}H_{17}Cl_2N_3O$	(RS)-2-(2,4-diclorophenyl)-1-(1H-1,2,4-triazol-1-yl)hexan-2-ol	0.018	3.9	17	49–200
Imazalil $C_{14}H_{14}Cl_2N_2O$	(±)-1-(b-allyloxy-2,4-dichlorophenylethyl)imidazole	0.158	2.56	184	44–128
Prochloraz $C_{15}H_{16}Cl_3N_3O_2$	N-propyl-N-[2-(2,4,6-trichlorophenoxy)etyl]imidazole-1-carboxamide	0.09	3.53	26.5	2–73
Propiconazole $C_{15}H_{17}Cl_2N_3O_2$	(±)-1-[2-(2,4-dichlorophenyl)-4-propyl-1,3-dioxolan-2-ylmethyl]-1H-1,2,4-triazole	0.056	3.72	100	15–96
Tebuconazole $C_{16}H_{22}ClN_3O$	(RS)-1-p-chlorophenyl-4,4-dimethyl-3-(1H-1,2,4-triazol-1-ylmethyl) pentan-3-ol	0.002	3.7	36	26–92
Triadimefon $C_{14}H_{16}ClN_3O_2$	1-(4-chlorophenoxy)-3,3-dimethyl-1-(1H-1,2,4-triazol-1-yl)butan-2-one	0.02	3.11	64	6–26

Sources: Data from MacBean, C. (Ed.) in *The Pesticide Manual*, British Crop Protection Council, 2012. University of Hertfordshire, The Pesticide Properties DataBase (PPDB) developed by the Agriculture & Environment Research Unit (AERU), University of Hertfordshire. Available at http://sitem.herts.ac.uk/aeru/ppdb/en/atoz.htm.
EFSA Scientific Publications. Available at https://www.efsa.europa.eu/en/publications. European Commission. *EU Pesticides Database*. Available at http://ec.europa.eu/food/plant/pesticides/eu-pesticides-database.
*25°C.

1.4.2 Benzimidazoles

Thiabendazole

Fungicides of the benzimidazole type have a systemic action. Generally, they are taken up by the roots of the plants, and the active substances are then acropetally translocated through the xylem to the leaves. These compounds have been used in plant protection in the form of their insoluble salts. They are foliage and soil fungicides with a specific and broad spectrum of action, also used for seed treatment and in postharvest (Table 1.22).

1.4.3 Dithiocarbamates

Ethylenebisdithiocarbamates are prepared from ethylenediamine $H_2N\text{-}CH_2\text{-}CH_2\text{-}NH_2$.

$$\left[SC(S)NHCH_2CH_2NHCSSMn \right]_x Zn_y$$

Mancozeb

TABLE 1.22

Chemical Names and Properties of Benzimidazole Fungicides

Common Name	IUPAC Name	Vapor Pressure mPa (25°C)	K_{ow} log P	Water Solubility mg/L (25°C)	Half-Life in Soil (days)
Benomyl $C_{14}H_{18}N_4O_3$	Methyl 1-(butyl carbamoyl) benzimidazol-2-ylcarbamate	<0.005	1.37	2	1–100
Carbendazim $C_9H_9N_3O_2$	Methyl benzimidazol-2-yl carbamate	0.15	1.51	8	11–78
Thiabendazole $C_{10}H_7N_3S$	2-(thiazol-4-yl) benzimidazole	5.3×10^{-4}	2.39	30*	33–500

Sources: Data from MacBean, C. (Ed.) in *The Pesticide Manual*, British Crop Protection Council, 2012. University of Hertfordshire, The Pesticide Properties DataBase (PPDB) developed by the Agriculture & Environment Research Unit (AERU), University of Hertfordshire. Available at http://sitem.herts.ac.uk/aeru/ppdb/en/atoz.htm.

EFSA Scientific Publications. Available at https://www.efsa.europa.eu/en/publications. European Commission. *EU Pesticides Database.* Available at http://ec.europa.eu/food/plant/pesticides/eu-pesticides-database

*20°C.

These compounds are heavy metals salts of ethylenebisdithiocarbamate and these salts are unusually stable and suitable as fungicides. The dithiocarbamate fungicides are the most widely used of the organic fungicides and have a wide spectrum of activity as foliar sprays for fruits, vegetables, and ornamentals, and as seed protectants (Table 1.23).

1.4.4 MORPHOLINES

Fenpropimorph

TABLE 1.23
Chemical Names and Properties of Dithiocarbamate Fungicides

Common Name	IUPAC Name	Vapor Pressure mPa (20°C)	K_{ow} log P	Water Solubility mg/L (25°C)	Half-Life in Soil (days)
Mancozeb $(C_4H_6\ MnN_2S_4)_x$ $(Zn)_y$	Manganese ethylenebis(dithiocarbamate) (polymeric) complex with zinc salt	0.01	0.26	6.2×10^{-3}	<1
Maneb $C_4H_6\ MnN_2S_4$	Manganese ethylenebis(dithiocarbamate)	0.01	−0.45	178*	<7
Metiram $(C_{16}H_{33}\ N_{11}S_{16}Zn_3)_x$	Zinc ammoniate ethylenebis(dithiocarbamate)-poly(ethylenethiuram disulfide)	0.01**	0.3	2*	7
Nabam $C_4H_6N_2Na_2S_4$	Disodium ethylenebis(dithiocarbamate)	1×10^{-7}**	−4.24	2×10^5*	1–4
Zineb $(C_4H_6N_2S_4Zn)_x$	Zinc ethylenebis(dithiocarbamate) (polymeric)	0.008**	1.3	10	16–23
Ziram $C_6H_{12}N_2S_4Zn$	Zinc bis(dimethyldithiocarbamate)	1.8×10^{-2}**	1.65	0.97*	2–30

Sources: Data from MacBean, C. (Ed.) in *The Pesticide Manual*, British Crop Protection Council, 2012. University of Hertfordshire, The Pesticide Properties DataBase (PPDB) developed by the Agriculture & Environment Research Unit (AERU), University of Hertfordshire. Available at http://sitem.herts.ac.uk/aeru/ppdb/en/atoz.htm.
EFSA Scientific Publications. Available at https://www.efsa.europa.eu/en/publications. European Commission. *EU Pesticides Database*. Available at http://ec.europa.eu/food/plant/pesticides/eu-pesticides-database.
*20°C; **25°C.

Morpholines are specific systemic fungicides against powdery mildew fungi and are used to control these diseases in cereals, cucumbers, apples, etc. These compounds are distributed in the plants by translocation from the root and foliage and protect the plants against infection by phytopathogenic fungi. They have a certain persistence in soil (Table 1.24).

1.4.5 MISCELLANEOUS

Captan   Chlorothalonil

Captan and Folpet are fungicides used in foliar treatment of fruits, vegetables, and ornamentals, in soil and seed treatments, and in post-harvest applications.

Procymidone is a dicarboximide-derived fungicide with moderate systemic action. It is rapidly absorbed through the roots but also through the stem or the leaves. It is used for the control of storage rots of fruits and vegetables and it is effective for seed dressing of cereals. Table 1.25 summarizes the properties of various frequently used fungicides belonging to different chemical classes.

TABLE 1.24
Chemical Names and Properties of Morpholine Fungicides

Common Name	IUPAC Name	Vapor Pressure mPa (20°C)	K_{ow} log P	Water solubility mg/L (20°C)	Half-Life in Soil (days)
Dodemorph $C_{18}H_{35}NO$	4-cyclododecyl-2,6-dimethyl morpholine	0.48	4.14	<100	24–125
Fenpropimorph $C_{20}H_{33}NO$	(±)-cis-4-[3-(4-tert-buthylphenyl) -2-methylpropyl]-2,6- dimethylmorpholine	3.5	4.1	4.3	9–124
Tridemorph $C_{19}H_{39}NO$	2,6-dimethyl-4-tridecyl morpholine	12	4.2	1.1	13–130

Sources: Data from MacBean, C. (Ed.) in *The Pesticide Manual*, British Crop Protection Council, 2012. University of Hertfordshire, The Pesticide Properties DataBase (PPDB) developed by the Agriculture & Environment Research Unit (AERU), University of Hertfordshire. Available at http://sitem.herts.ac.uk/aeru/ppdb/en/atoz.htm.
EFSA Scientific Publications. Available at https://www.efsa.europa.eu/en/publications. European Commission. *EU Pesticides Database*. Available at http://ec.europa.eu/food/plant/pesticides/eu-pesticides-database.

TABLE 1.25
Chemical Names and Properties of Miscellaneous Fungicides

Common Name	IUPAC Name	Vapor Pressure mPa (25°C)	K_{ow} log P	Water Solubility mg/L (25°C)	Half-Life in Soil (days)
Azoxystrobin $C_{22}H_{17}N_3O_5$	(E)-2-{2-[6-(2-cyanophenoxy) pyrimidin-4-yloxy] phenyl}-3-methoxyacrylate	1.1×10^{-7}*	2.5	6.7*	35–248
Captan $C_9H_8Cl_3NO_2S$	N-(trichloromethylthio) ciclohex-4 -ene-1,2-dicarboximide	0.004	2.8	3.3	1–7
Chlorothalonil $C_8Cl_4N_2$	Tetrachloroisophthalonitrile	0.076	2.92	0.81	1–32
Cyprodinil $C_{14}H_{15}N_3$	4-cyclopropyl-6-methyl-N- phenylpyrimidin-2-amine	0.51	4	13	11–98
Fenhexamid $C_{14}H_{17}Cl_2NO_2$	N-(2,3-dichloro-4-hydroxy phenyl)-1- methylcyclohexanecarboxamide	4.0×10^{-4}*	3.51	20*	≤1–8
Folpet $C_9H_4Cl_3NO_2S$	N-(trichloromethylthio) phtalimide	2.1×10^{-2}	3.11	0.8	0.2–4
Iprodione $C_{13}H_{13}Cl_2N_3O_3$	3-(3,5-dichlorophenyl)-N- isopropyl-2,4dioxoimidazo lidine-1-carboxamide	5×10^{-4}	3	13*	3–35
Metalaxyl-M $C_{15}H_{21}NO_4$	methyl-N-(methoxy acetyl)-N- (2,6-xylyl)-D-alaninate	3.3	1.71	26000	9–31
Ofurace $C_{14}H_{16}ClNO_3$	(±)-α-(2-chloro-N-2,6-xylyl) acetamido)-γ-butyrolactone	2×10^{-2}*	1.39	146*	26
Orthophenylphenol $C_{12}H_{10}O$	biphenyl-2-ol	474	3.18	700	1–7
Procymidone $C_{13}H_{11}Cl_2NO_2$	N-(3,5-dichlorophenyl)-1,2- dimethylcyclopropane-1,2- dicarboximide	0.023	3.14	4.5	17–158
Pyrimethanil $C_{12}H_{13}N_3$	N-(4,6-dimethylpyrimidin-2-yl) aniline	2.2	2.84	121	23–54
Tolylfluanid $C_{10}H_{13}Cl_2FN_2O_2S_2$	N-dichlorofluoromethylthio-N', N'dimethyl-N-p-tolylsulfamide	0.2*	3.90	0.9*	0.5–11
Triforine $C_{10}H_{14}Cl_6N_4O_2$	N-N' [piperazine-1,4-diylbis [(trichloromethyl)methylene]] diformamide	80	2.2	9*	21

Sources: Data from MacBean, C. (Ed.) in *The Pesticide Manual*, British Crop Protection Council, 2012. University of Hertfordshire, The Pesticide Properties DataBase (PPDB) developed by the Agriculture & Environment Research Unit (AERU), University of Hertfordshire. Available at http://sitem.herts.ac.uk/ aeru/ppdb/en/atoz.htm.
EFSA Scientific Publications. Available at https://www.efsa.europa.eu/en/publications. European Commission. *EU Pesticides Database*. Available at http://ec.europa.eu/food/plant/pesticides/eu-pesticides -database.
*20°C.

1.5 MODE OF ACTION

The control of pests by pesticides depends on several factors like the mode of action of these compounds, the crop stage and the environmental conditions, moisture, soil type and temperature, among others, and numerous works have been published on these subjects [9–12]. The main pesticide modes of action are summarized below.

1.5.1 HERBICIDES

1.5.1.1 Amino Acid Synthesis Inhibitors

Amino acid synthesis inhibitors act on a specific enzyme to prevent the production of certain amino acids, which are key building blocks for normal plant growth and development.

One type of herbicides causes the inhibition of acetolactate synthase (ALS), the first common enzyme in the branched-chain amino acid biosynthetic pathway. ALS inhibitors include, among others, herbicides of the sulfonylurea family. These compounds vary greatly in selectivity, some of them being extremely active.

The aromatic ring amino acids, tryptophan, phenylalanine, and tyrosine, are synthesized by plants through the shikimic acid pathway. Only one herbicide, glyphosate, inhibiting that pathway has been commercialized. The mode of action of glyphosate is the inhibition of the enzyme EPSPS, 5-enolpyruvoyl-shikimate-3-phosphate synthase. This enzyme is present in plants, fungi, and bacteria but absent in animals, which need to ingest those amino acids in their diet.

Another enzyme involved in amino acid synthesis and used as a target for herbicides is the glutamine synthetase (GS), which makes glutamine from glutamate and ammonia. This enzyme is present in plants, where it plays an important role in nitrogen assimilation, as well as in animals, as glutamate is a neurotransmitter that can be inactivated by GS. The mode of action of the herbicide glufosinate is the inhibition of the enzyme GS.

1.5.1.2 Cell Division Inhibitors

This type of herbicides reacts with tubulin, a protein essential for building the intracellular skeleton in eukaryotic cells forming the wall of microtubules. These compounds disturb normal cell division by binding with tubulin.

Inhibitors of cell division are herbicides belonging to various chemical classes, such as dinitroanilines, benzoic acids, and pyridines.

1.5.1.3 Photosynthesis Inhibitors

Photosynthesis is a key process for plants and consequently is a main target for many herbicides. There are different mechanisms involved in the inhibition of photosynthesis, such as free radical generation, blockage of the electron transport system, and inhibition–destruction of protective pigments but, in general, most herbicides interfere with the transfer of electrons to the plastoquinone pool by binding to a specific protein that regulates electron transfer.

The herbicides acting as photosynthesis inhibitors are all nitrogen-containing compounds with a diversity of chemical composition. These compounds, including phenylureas, triazines, pyridazines, phenyl carbamates, nitriles, and amides, are represented by various herbicide families, although some of these chemical classes also have specific herbicides that do not act as photosynthesis inhibitors.

1.5.2 INSECTICIDES

1.5.2.1 Signal Interference in the Nervous System

Chemicals that disturb signal systems are frequently potent poisons. Pyrethroids and organochlorines are the most important insecticides in this category. Their mode of action is to inhibit the proper closing of the channels by acting at the voltage-gated sodium channels. Pyrethroids modify axonal conduction within the central nervous system of insects by altering the permeability of the nerve membrane to sodium and potassium ions. Organochlorines may interact with the pores of the lipoprotein structure of the insect nerve causing distortion and consequent excitation of nerve impulse transmission. The toxic properties of chlorinated cyclodiene insecticides, such as lindane, reside in the blockade of the gamma-aminobutyric acid-gated chlorine channels, inducing convulsions in insects. Neonicotinoids and fipronil operate by disrupting neural transmission in the central nervous system of invertebrates.

1.5.2.2 Inhibitors of Cholinesterase

The target for many insecticides is an enzyme called acetylcholinesterase (AChE). This enzyme is an essential constituent of the nervous system and plays an important role in animals, but not in plants as they lack a nervous system. AChE hydrolyzes acetylcholine, an ester released when nerve impulses are transmitted. Synapses, myoneural functions, and ganglia of the nervous system transmit neural impulses by the mediation of acetylcholine (ACh).

Organophosphorus insecticides have the capacity to phosphorylate the esteratic active site of the AChE. The phosphorylated enzyme is irreversibly inhibited and is not able to carry out its normal function of the rapid removal of ACh. As a result, ACh accumulates and disrupts the normal functioning of the nervous system. Carbamates are also strong inhibitors of AChE and may also have a direct effect on ACh receptors.

1.5.2.3 Inhibitors of Chitin Synthesis

Chitin is a very abundant polysaccharide in nature; although it is present in arthropods and fungi, it is absent in plants and mammals.

Benzoylureas affect chitin synthesis in the insect cuticle by disrupting the process of connecting the *N*-acetylglucosamine units to the chitin chain, preventing in this way the normal molting process of insects.

1.5.3 FUNGICIDES

1.5.3.1 Sulfhydryl Reagents

Sulfhydryl (SH) groups are important reactive groups often found in the active sites of many enzymes. Dithiocarbamate fungicides react with the SH-containing enzymes and coenzymes of fungal cells. Enzyme inhibition may also occur by complex formation of the active substance with the metal atoms of metal-containing enzymes.

The perhalogenmercaptans, captan and folpet, are good examples of pesticides that react with SH groups in many enzymes. These fungicides affect the structure and functions of the cell membranes and inhibit the enzyme system causing tumors in the mitochondria.

1.5.3.2   Cell Division Inhibitors

Benzimidazole fungicides react with tubulin, a protein that is the building block of the intracellular skeleton in cells. The impairment of cell division is produced in most cases by inhibiting the formation of the microtubules. Benzimidazoles, such as benomyl, carbendazim, and thiabendazole, as well as other fungicide groups like carbamates, have this mode of action.

1.5.3.3   Inhibitors of Ergosterol Synthesis

Ergosterol-inhibitor fungicides are active against many different fungi. Although they disturb sterol synthesis in higher plants, as well as the synthesis of gibberellins, their phytotoxicity is low. The synthesis of sterols is very complex and various groups of fungicides act on different targets of that synthesis. One large group of fungicides, called demethylase inhibitors, includes various compounds having a heterocyclic N-containing ring, such as azoles, morpholines, pyridines, and piperazines.

1.6   TOXICITY AND RISK ASSESSMENT

Pesticides are toxic compounds that may cause adverse effects on the human and the environment. Toxicity has been defined as the capacity of a substance to produce harmful effects, and other terms used in the risk assessment of chemicals are hazard, defined as the potential to cause harm, and risk, defined as the likelihood of harm. Risk characterization is then the estimation of the incidence and severity of the adverse effects likely to occur in a human population, animals, or environmental compartments due to actual or predicted exposure to any active substance.

Humans can be exposed to pesticides by direct or indirect means. Direct or primary exposure normally occurs during the application of these compounds and indirect or secondary exposure can take place through the environment or the ingestion of food. Figure 1.4 summarizes the main routes of indirect exposure to pesticides.

A complete set of data is needed for the toxicological and ecotoxicological evaluation of pesticides.

Regarding the human health assessment, toxicological studies need to include the following tests:

A: Acute toxicity, which involves harmful effects in an organism through a
   single or short-term exposure, should be studied by:
   • Dermal toxicity test: rabbits are employed more often than any other species for studies of skin toxicity, although guinea pigs, rats, or mice are also used. The results are expressed in terms of the LD_{50}, the dose which under the conditions stated will cause death in 50% of a group of test animals.

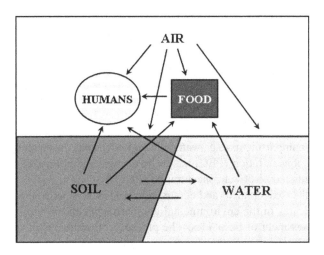

FIGURE 1.4   Routes of indirect exposure to pesticides.

- Mucous membrane and eye toxicity test: the conjunctiva of the eye and the vaginal vault of experimental animals (rabbits and monkeys) have been employed in tests of the toxic or irritant effects of chemical substances on mucous membrane.
- Inhalation toxicity test: the procedures for the evaluation of potential hazards of gases, dusts, mists, or vapors via the inhalation route vary depending on the physical nature (solubility, particle size) of the pesticide.
- Oral toxicity test: the procedure normally employs the administration of compounds in the diet or intragastrically by gavage. The advantage is that it allows precise measurement of daily dosage to body weight. The oral toxicity generally takes place in three stages, acute (short-term), subacute (subchronic) and chronic (long-term).

B: Repeated dose test, that comprise the adverse effects occurring in experimental animals as a result of repeated daily dosing with, or exposure to, a pesticide for a short part of their expected life-span.

C: Reproduction and teratology, the endocrinological changes associated with the reproductive cycle in the female and the anabolic systems involved in embryologic and fetal growth constitute a challenging background against which to test the toxic potentiality of a pesticide where the enzymatic system plays an important role.

D: Carcinogenesis, to identify the carcinogenicity potential of pesticides in laboratory animals. The studies must be sufficient to establish the species and organ specificity of tumors induced, and to determine the dose–response relationship. For non-genotoxic carcinogens, they identify doses that cause no adverse effects.

Estimation of a no observed adverse effect level (NOAEL), when possible, or of a lowest observable adverse effect level (LOAEL) is a critical step in the toxicological risk assessment of pesticides [13].

Concerning ecotoxicity, the estimation of hazard to wildlife involves the determination of the effects on different species. Toxicity data should be gathered for soil organisms, for beneficial arthropods (such as honeybees), for aquatic species (fish, invertebrates, algae, and microorganisms), for terrestrial vertebrates (mammals and birds), and for plants. Several endpoints are of interest, depending on the species, namely acute toxicity, growth and activity inhibition, bioconcentration, and effects on reproduction. Recently, endocrine disruption effects have also been included. The poisoning of the species depends on the concentration of the pesticide in the different environmental compartments (e.g., water, air, and soil), and consequently a predicted environmental concentration (PEC) has to be derived and compared with the corresponding predicted no effect concentration (PNEC).

Therefore, the toxicological and ecotoxicological effects, as well as the concentration of pesticides in the environmental compartments and in food, are required for the risk assessment of pesticides. The presence of pesticides in these matrices are normally referred to as pesticide residues, which are defined as any original or derived residue, including relevant metabolites, from a chemical. Analytical methods to determine pesticide residues in the mentioned matrices with adequate sensitivity and selectivity are then needed.

The analysis of pesticide residues in food and environmental samples, together with their monitoring in those matrices and the implication of their levels in risk assessment, will be described in the following chapters of the present book.

REFERENCES

1. EPA. *Pesticides*. Available at http://www.epa.gov/pesticides.
2. Carson, R.L. *Silent Spring*, The Riverside Press, Boston, MA, 1962.
3. FAO. *The Joint FAO/WHO Meeting on Pesticide Residues (JMPR)*. Available at http://www.fao.org/agriculture/crops/thematic-sitemap/theme/pests/jmpr/en/.
4. FAO. *Codex Committee on Pesticide Residues (CCPR)*. Available at http://www.fao.org/fao-who-codexalimentarius/committees/committee/en/?committee=CCPR.
5. MacBean, C. (Ed.) *The Pesticide Manual*, British Crop Protection Council, Hampshire, UK, 2012.
6. European Commission. *EU Pesticides Database*. Available at http://ec.europa.eu/food/plant/pesticides/eu-pesticides-database.
7. European Food Safety Authority. Available at http://www.efsa.europa.eu/.
8. PPBD. *A to Z List of Pesticide Active Ingredients*. Available at http://sitem.herts.ac.uk/aeru/ppdb/en/atoz.htm.
9. White-Stevens, R. *Pesticides in the Environment*, Marcel Dekker, New York, NY, 1971.
10. Hutson, D.H. and Roberts, T.R. *Herbicides*, John Wiley & Sons, Chichester, UK, 1987.
11. Matolcsy, G. Nádasy, M. and Andriska, V. *Pesticide Chemistry*, Elsevier, Amsterdam, 1988.
12. Stenersen, J. *Chemical Pesticides. Mode of Action and Toxicology*, CRC Press, Boca Raton, FL, 2004.
13. European Commission. *Technical Guidance Document on Risk Assessment, Part 1 and Part 2*. Available at https://ec.europa.eu/jrc/en/publication/eur-scientific-and-technical-research-reports/technical-guidance-document-risk-assessment-part-1-part-2.

2 Sample Handling of Pesticides in Food and Environmental Samples

Francisco Barahona, Esther Turiel,
and Antonio Martín-Esteban

CONTENTS

2.1 Introduction 42
2.2 Sample Pretreatment..42
    2.2.1 Drying...43
    2.2.2 Homogenization...43
2.3 Extraction and Purification..44
    2.3.1 Solid–Liquid Extraction 44
        2.3.1.1 Shaking..45
        2.3.1.2 Soxhlet Extraction..47
        2.3.1.3 Microwave-Assisted Extraction 48
        2.3.1.4 Pressurized Solvent Extraction 48
    2.3.2 Supercritical Fluid Extraction..49
    2.3.3 Liquid–Liquid Extraction 50
    2.3.4 Liquid-Phase Microextraction 51
    2.3.5 Solid-Phase Extraction...54
        2.3.5.1 Polar Sorbents 55
        2.3.5.2 Non-Polar Sorbents 56
        2.3.5.3 Ion-Exchange Sorbents 58
        2.3.5.4 Affinity Sorbents..58
    2.3.6 Solid-Phase Microextraction 61
        2.3.6.1 Extraction 62
        2.3.6.2 Desorption...63
    2.3.7 Solid–Solid Extraction: Matrix Solid-Phase Dispersion 63
    2.3.8 Stir-Bar Sorptive Extraction 64
    2.3.9 QuEChERS..64
    2.3.10 Nanoparticles-Based Extractions..65
        2.3.10.1 Metal–Organic Frameworks (MOFs) 66
        2.3.10.2 Carbon Nanotubes..66
        2.3.10.3 Graphene...67
        2.3.10.4 Magnetic Nanoparticles 67
2.4 Future Trends..68
References..68

2.1 INTRODUCTION

The determination of pesticides in food and environmental samples at low concentrations is always a challenge. Ideally, the analyte to be determined would be already in solution and at a concentration level high enough to be detected and quantified by the selected final determination technique (i.e., high-performance liquid chromatography [HPLC] or gas chromatography [GC]). Unfortunately, the reality is far from this ideal situation. Firstly, the restrictive legislation from the European Union and the World Health Organization devoted to preventing contamination of food and environmental compartments by pesticides make necessary the development of analytical methods suitable for detecting target analytes at very low concentration levels. Besides, from a practical point of view, even when the analyte is already in solution (i.e., water or juice), there are several difficulties related to the required sensitivity and selectivity of the selected determination technique that must be overcome, since the concentration of matrix interfering compounds is much higher than that of the analyte of interest. Consequently, the development of an appropriate sample preparation procedure involving extraction, enrichment, and clean-up steps becomes mandatory in order to obtain a final extract concentrated on target analytes and as free as possible of matrix compounds.

In this chapter, the different sample treatment techniques currently available and most commonly used in analytical laboratories for the analysis of pesticides in food and environmental samples are described. Depending upon the kind of sample (solid or liquid) and the specific application (type of pesticide, concentration level, multi-residue analysis), the final procedure might involve the use of only one or the combination of several of the different techniques described below.

2.2 SAMPLE PRETREATMENT

Generally, sampling techniques provide amounts of sample, much higher (2–10 L of liquid samples and 1–2 kg of solid samples) than those needed for the final analysis (just a few mg). Thus, it is always necessary to carry out some pretreatments in order to get a homogeneous and representative sub-sample. Even if the sample is apparently homogeneous, i.e., an aqueous sample, it will be at least necessary to perform a filtration step to remove suspended particles, which could affect the final determination of target analytes. However, some hydrophobic analytes (i.e., organochlorine pesticides) could be adsorbed onto the particles' surface and thus, depending upon the objective of the analysis, it might be necessary to analyze such particles. This simple example demonstrates the necessity of establishing clearly the objective of the analysis, since it will determine the sample pretreatments to be carried out, and highlights the importance of this typically underrated analytical step.

Usually, environmental water samples just require filtration, whereas liquid food samples might be subjected to other kinds of pretreatments depending upon the objective of the analysis. However, solid samples (both environmental and food samples) need to be more extensively pretreated in order to get a homogeneous sub-sample. The wide variety of solid samples prevents an exhaustive description of the different procedures in this chapter; however, some general common procedures will be described below.

2.2.1 DRYING

The presence of water or moisture in solid samples has to be taken into account since it might produce alterations (i.e., hydrolysis) of the matrix and/or analytes, which will obviously affect the final analytical results. Besides, water content varies depending upon atmospheric conditions and thus it is recommended to refer the content of target analytes to the mass of dry sample.

Sample drying can be carried out before the crushing and sieving steps, although it is recommended to dry again before final determination since re-hydration might occur. Typically, the sample is dried inside an oven at temperatures of about 100°C. It is important to stress that higher temperatures can be used in order to decrease the time devoted to this step but losses of volatile analytes might occur. In this sense, it is important to know *a priori* the physicochemical properties of target analytes in order to preserve the integrity of the sample. A more conservative approach, using low temperatures, can be followed but it will unnecessarily increase the drying time. Alternatively, lyophilization is recommended if a high risk of analyte losses exists and it is an appropriate procedure for food, biological material, and plant samples drying. However, even following this procedure, losses of analytes might occur depending upon their physical properties (i.e. solubility, volatility).

It is therefore evident that it is not possible to establish a general rule on how to perform sample drying. Thus, studies on the stability of target analytes in spiked samples should be carried out in order to guarantee the integrity of the sample before final determination of the analytes.

2.2.2 HOMOGENIZATION

As mentioned above, samples are heterogeneous in nature and thus, they must be treated in order to get a homogeneous distribution of target analytes.

Generally, soil samples are crushed, ground, and sieved through 2 mm mesh. Grinding can be done manually, or automatically using specially designed equipment (e.g., ball mills). It is important to stress that this procedure might provoke the local heating of the sample and thus thermolabile or volatile compounds might be affected. In this sense, it is recommended to grind the sample at short time intervals in order to minimize sample heating. Also, due to heating, water content may vary, making it necessary to re-calculate sample moisture.

Food samples can be cut down to small pieces with a laboratory knife before further homogenization with automatic instruments (e.g., a blender). Sample freezing is a general practice to ease blending, especially recommended for samples with high fat content (e.g., cheese) and soft samples with high risk of phase separation during blending (e.g., liver, citrus fruits).

Apart from these general guidelines, especially in food analysis, the determination of pesticides might be restricted to the edible part of the sample or to samples previously cooked and thus sample pretreatments will vary depending upon the objective of the analysis.

Finally, it is important to point out that, in most of the cases, samples need to be stored for certain periods of time before performing the analysis. In this sense, although sample storage cannot be considered a sample pre-treatment, the addition of preservatives, as well as the establishment of the right conditions of storage (e.g., at room temperature or in the fridge) to minimize analyte/sample degradation are typical procedures carried out at this stage of the analytical process. Hence, they must be also taken into account to guarantee the accuracy of the final result.

2.3 EXTRACTION AND PURIFICATION

The main aim of any extraction process is the isolation of analytes of interest from the selected sample by using an appropriate extracting phase. Pesticides from liquid samples (i.e., environmental waters) are preferably extracted using solid phases by solid-phase extraction (SPE) or solid-phase microextraction (SPME) procedures, although for low volume samples, liquid–liquid extraction (LLE) can be also carried out. Extraction of pesticides from environmental or food solid samples is usually performed by mixing the sample with an appropriate extracting solution, and subjecting that mixture to some process (e.g., agitation, microwaves) to assist migration of analytes from the sample matrix to the extracting solution. For certain applications, matrix solid-phase dispersion (MSPD) can be also a good alternative. In all cases, once a liquid extract has been obtained, it is subsequently subjected to a purification step (namely clean-up), which is usually performed by solid-phase extraction or liquid–liquid extraction. In some cases, extraction and clean-up procedures can be performed in a single step (e.g., SPE with selective sorbents), which enormously simplifies the sample preparation procedure.

2.3.1 SOLID–LIQUID EXTRACTION

As mentioned above, solid–liquid extraction is probably the most widely used procedure in the analysis of pesticides in solid samples. Solid–liquid extraction includes various extraction techniques based on the contact of a certain amount of sample with an appropriate solvent. Figure 2.1 shows a scheme of the different steps that

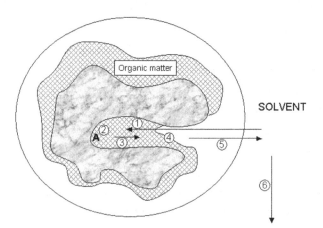

FIGURE 2.1 Scheme of the different steps involved in the extraction of a target analyte A from a solid particle.

take place in a solid–liquid extraction procedure and influence the final extraction efficiency. In the first stage (step 1), the solvent must penetrate inside the pores of the sample particulates to achieve desorption of the analytes bound to matrix active sites (step 2). Subsequently, analytes have to diffuse through the matrix (step 3) to be dissolved in the extracting solvent (step 4). Again, the analytes must diffuse through the solvent to leave the sample pores (step 5) and finally be swept away by the external solvent (step 6). Obviously, the proper selection of the solvent to be used is a key factor in a solid–liquid extraction procedure. However, other parameters such as pressure and temperature have an important influence on the extraction efficiency. Working at high pressure facilitates the solvent to penetrate sample pores (step 1) and, in general, increasing temperature increases solubility of the analytes in the solvent. Moreover, high temperatures increase diffusion coefficients (steps 3 and 5) and the capacity of the solvent to disrupt matrix–analyte interactions (step 2). Depending upon the strength of the interaction between the analyte and the sample matrix, the extraction will be performed in soft, mild, or aggressive conditions. Table 2.1 shows a summary and a comparison of drawbacks and advantages of the different solid–liquid extraction techniques (which will be described below) most commonly employed in the analysis of pesticides in food and environmental samples.

2.3.1.1 Shaking

It is a very simple procedure to extract pesticides weakly bound to the sample and is very convenient for the extraction of pesticides from fruits and vegetables. It just consists of shaking (manually or automatically) the sample in the presence of an appropriate solvent for a certain period of time. The most commonly used solvents are acetone and acetonitrile, due to their miscibility with water making easy the diffusion of analytes from the solid sample to the solution, although immiscible solvents such as dichloromethane or hexane can also be used for the extraction depending on the properties of target analytes. In a similar manner, the use of mixtures of solvents is a typical practice when analytes of different polarity are extracted in multiresidue analysis. Once analytes have been extracted, the mixture needs to be filtered before further treatments. Besides, since the volume of organic solvents used following this procedure is relatively large, it is usually necessary to evaporate the solvent before final determination.

However, shaking might not be effective enough to extract analytes strongly bound to the sample. In order to achieve a more effective shaking, the use of ultrasound-assisted extraction is recommended. Ultrasound radiation provokes molecule vibration and eases the diffusion of the solvent to the sample favoring the contact between both phases. Thanks to this improvement, both the time and the amount of solvents in the shaking process are considerably reduced.

An interesting and useful modification for reducing both the amount of sample and organic solvents is the so-called ultrasound-assisted extraction in small columns proposed by Sánchez-Brunete et al. [1, 2] for the extraction of pesticides from soils. Briefly, this procedure just consists in placing the sample (~ 5 g) in a glass column equipped with a polyethylene frit. Subsequently, samples are extracted with around 5–10 mL of an appropriate organic solvent in an ultrasonic water bath. After extraction, columns are placed on a multiport vacuum manifold where the solvent is filtered and collected for further analysis.

TABLE 2.1
Solid–Liquid Extraction Techniques

Technique	Description	Advantages	Drawbacks
Shaking	Samples and solvent are placed in a glass vessel. Shaking can be done manually or mechanically	• Simple • Fast (15–30 min) • Low cost	• Filtration of the extract is necessary • Dependent on kind of matrix • Moderate solvent consumption (25–100 mL)
Soxhlet	Sample is placed in a porous cartridge and solvent recirculates continuously by distillation–condensation cycles	• Standard method • No further filtration of the extract necessary • Independent of kind of matrix • Low cost	• Time consuming (12–48 h) • High solvent volumes (300–500 mL) • Solvent evaporation needed
USE	Samples and solvent are placed in a glass vessel and introduced in an ultrasonic bath	• Fast (15–30 min.) • Low solvent consumption (5–30 mL) • Bath temperature can be adjusted • Low cost	• Filtration of the extract is necessary. • Dependent on kind of matrix
MAE	Sample and solvent are placed in a reaction vessel. Microwave energy is used to heat the mixture	• Fast (15 min approx.) • Low solvent consumption (15–40 mL) • Easily programmable	• Filtration of the extract is necessary • Addition of a polar solvent is required • Moderate cost
PSE	Sample is placed in a cartridge and pressurized with a high temperature solvent	• Fast (20–30 min) • Low solvent consumption (30 mL) • Easy control of extraction parameters (temperature, pressure) • High temperatures achieved • High sample processing	• Initial high cost • Dependent on the kind of matrix

Abbreviations: USE, ultrasound-assisted extraction; MAE, microwave-assisted extraction; PSE, pressurized solvent extraction.

2.3.1.2 Soxhlet Extraction

As indicated above, in some cases shaking is not enough to disrupt interactions between analytes and matrix components. In this regard, an increase in the temperature of the extraction is recommended. The simpler approach to isolate analytes bound to solid matrices at high temperatures is the Soxhlet extraction, introduced by F. Soxhlet in 1879, which is still nowadays the more used technique and of reference in the new techniques introduced during the last number of years.

The sample is placed in an apparatus (Soxhlet extractor) and extraction of analytes is achieved by means of a hot condensate of a solvent distilling in a closed circuit. Distillation in a closed circuit allows the sample to be extracted many times with fresh portions of solvent, and exhaustive extraction can be performed. Its weak points are the long time required for the extraction and the large amount of organic solvents used.

In order to minimize the mentioned drawbacks, several attempts toward automation of the process have been proposed. Among them, Soxtec systems (Foss, Hillerød, Denmark) are the most extensively accepted and used in analytical laboratories and allow a reduction in extraction times of about five times compared to the classical Soxhlet extraction.

Table 2.2 shows a comparison of the recoveries obtained for several pesticides in soils after extraction using different techniques. In this case, it is clear that ultrasound-assisted extraction allows the isolation of target analytes, whereas simple shaking is not effective enough to quantitatively extract the selected pesticides [3]. It is important to stress that recoveries after Soxhlet extraction were too high which means that a large amount of matrix components were co-extracted with target analytes. In this regard, it is clear that an exhaustive extraction is not always required and a balance between the recoveries obtained of target analytes and the amount of matrix components co-extracted needs to be established.

TABLE 2.2
Recoveries (%) of Pesticides in Soils Obtained by Different Extraction Techniques

Pesticide	Concentration (mg.mL^{-1})	Ultrasound-assisted extraction	Soxhlet extraction	Shaking
Atrazine	0.04	103.5 ± 2.8	201.9 ± 14.6	108.3 ± 6.2
Pyropham	0.05	79.7 ± 6.3	143.0 ± 18.6	65.1 ± 9.3
Chlorpropham	0.05	93.6 ± 7.9	155.6 ± 20.4	88.1 ± 10.0
α-Cypermethrin	0.12	97.2 ± 4.4	128.4 ± 16.4	90.1 ± 9.1
Tetrametrin	0.26	83.4 ± 4.2	64.3 ± 16.0	52.0 ± 8.3
Diflubenzuron	0.02	92.8 ± 4.0	182.5 ± 17.4	98.1 ± 8.9

Experimental conditions: 10 g of soil sample spiked at indicated concentration level. Ultrasound-assisted extraction: 20 mL of acetone, 15 min; Soxhlet extraction: 250 mL of acetone, 4 h; Shaking: 20 mL of acetone, 2 h.

Source: Reproduced from Babić, S., Petrović, M., Kaštelan-Macan, M., *J. Chromatogr. A*, 823, 3, 1998, with permission from Elsevier.

2.3.1.3 Microwave-Assisted Extraction

Microwave-assisted extraction (MAE) has appeared during the last number of years as a clear alternative to Soxhlet extraction due to the ability of microwave radiation to heat the sample–solvent mixture in a fast and efficient manner. Besides, the existence of several instruments commercially available able to perform the sequential extraction of several samples (up to 14 samples in some instruments), allowing extraction parameters (pressure, temperature, and power) to be perfectly controlled, has made MAE a very popular technique.

Microwave energy is absorbed by molecules with a high dielectric constant. In this regard, hexane, a solvent with a very low dielectric constant, is transparent to microwave radiation and thus does not heat whereas acetone will be heated in few seconds due to its high dielectric constant. However, solvents with low dielectric constant can be used if the compounds contained in the sample (e.g., water) absorb microwave energy.

A typical practice is the use of solvent mixtures (especially for the extraction of pesticides of different polarities) combining the heating ability of one of the components (e.g., acetone) with the solubility of the more hydrophobic compounds in the other solvent of the mixture (e.g., hexane). As an example, a mixture of acetone–hexane (1:1) was used for the microwave-assisted extraction of atrazine, parathion-methyl, chlorpyriphos, fenamiphos, and methidathion in orange peel with quantitative recoveries in less than 10 min [4].

As a summary, in general, the recoveries obtained are quite similar to those obtained by Soxhlet extraction but the important decrease of the extraction time (~ 15 min) and of the volume of organic solvents (25–50 mL) have made microwave-assisted extraction extensively used in analytical laboratories.

2.3.1.4 Pressurized Solvent Extraction

Pressurized solvent extraction (PSE), also known as accelerated solvent extraction (ASE), pressurized liquid extraction (PLE), and pressurized fluid extraction (PFE), uses solvents at high temperatures and pressures to accelerate the extraction process. The higher temperature increases the extraction kinetics, while the elevated pressure keeps the solvent in liquid phase above its boiling point leading to rapid and safe extractions [5].

Figure 2.2 shows a scheme of the instrumentation and the procedure used in PSE. Experimentally, the sample (~ 10 g) is placed in an extraction cell and filled with an appropriate solvent (15–40 mL). Subsequently, the cell is heated in a furnace up to temperatures below 200°C, increasing the pressure of the system (up to 20 MPa) to perform the extraction. After a certain period of time (10–15 min), the extract is directly transferred to a vial without the necessity of subsequent filtration of the obtained extract. Then, the sample is rinsed with a portion of pure solvent and finally, the remaining solvent is transferred to the vial with a stream of nitrogen. The whole process is automated and each step can be programmed, allowing the sequential unattended extraction of up to 24 samples.

This technique is easily applicable for the extraction of pesticides from any kind of sample and the high temperature used allows very efficient extraction in a short time. In addition, the considerable reduction of the amount of organic solvents used

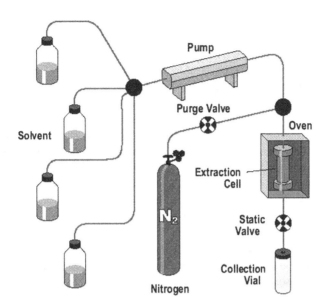

FIGURE 2.2   Pressurized solvent extraction equipment. (Courtesy of Dionex Corporation.)

makes PSE a very attractive technique for the extraction of pesticides. The main limitations of this technique are the high cost of the apparatus and the unavoidable necessity of purifying obtained extracts, which is common to other efficient extraction techniques based on the use of organic solvents as mentioned above.

2.3.2 SUPERCRITICAL FLUID EXTRACTION

Supercritical fluid extraction has been widely used for the isolation of a great variety of organic compounds from almost any kind of solid sample. Supercritical fluids can be considered a hybrid between liquids and gases, and possess ideal properties for the extraction of pesticides from solid samples. Supercritical fluids have in common with gases the ability to diffuse through the sample, which facilitates the extraction of analytes located in not easily accessible pores. In addition, the solvation power of supercritical fluids is similar to that of liquids, allowing the release of target analytes from the sample to the fluid.

Carbon dioxide has been widely used in SFE as it can be obtained with high purity, it is chemically inert, and its critical point (31.1°C and 71.8 atm) is easily accessible. Its main drawback is its apolar character, limiting its applicability to the extraction of hydrophobic compounds. In order to overcome, at least to a certain extent, this drawback, the addition of a small amount of an organic solvent modifier (i.e., methanol) has been proposed and permits varying the polarity of the fluid, thus increasing the range of extractable compounds. However, the role of the modifier during the extraction is not well understood. Figure 2.3 shows schematically the possible mechanisms taking place during the SFE of the herbicide diuron form soil samples using CO_2 as supercritical fluid modified with methanol [6]. Some authors

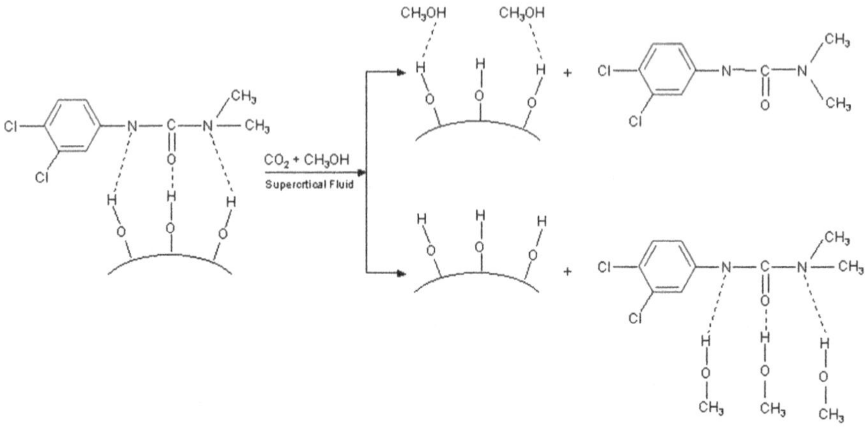

FIGURE 2.3 Mechanisms of the extraction of the herbicide diuron from sediments by SFE (CO_2 + methanol). (Reproduced from Martín-Esteban, A., Fernández-Hernando P., *Toma y tratamiento de muestra*, C. Cámara (Ed.), Editorial Síntesis S.A., Madrid, 2002, Chap. 6, with permission from Editorial Síntesis.)

propose that methanol molecules are able to establish hydrogen bounds with the phenolic moieties of the humic and fulvic acids present in soil samples and thus diuron is displaced from active sites. However, other authors consider that the modifier is able to interact with the target analyte, releasing it from the sample.

Once target analytes are in the supercritical fluid phase, they have to be isolated for further analysis, which is accomplished by decompression of the fluid through a restrictor, trapping analytes in a liquid trap or on a solid surface. With a liquid trap, the restrictor is immersed in a suitable liquid and thus, the analyte is gradually dissolved in the solvent while CO_2 is discharged to the atmosphere. In the solid surface method, analytes are trapped on a solid surface (e.g., glass vial, glass beads, solid-phase sorbents) cryogenically cooled directly by the expanding of the supercritical fluid or with the aid of liquid N_2. Alternatively, SFE can be directly coupled to gas chromatography or to supercritical fluid chromatography with the success of such on-line coupling dependent on the interface used, which determines the quantitative transfer of target analytes to the analytical column [7].

As mentioned above, SFE has been widely used for the extraction of pesticides from solid samples thanks to the effectiveness and selectivity of the extraction and to the possibility of on-line coupling to chromatographic techniques. However, the costs of the instrumentation and the appearance in the market of new, less sophisticated extraction instruments means that SFE is being displaced by other extraction techniques, especially by PSE.

2.3.3 LIQUID–LIQUID EXTRACTION

Liquid–liquid extraction (LLE) has been widely used for the extraction of pesticides from aqueous liquid samples and, although to a lesser extent, for the purification of organic extracts. LLE is based on the partitioning of target analyte between two immiscible liquids. The efficiency of the process depends on the affinity of the analyte for the solvents, on the ratio of volumes of each phase and on the number of successive extractions.

Most of the LLE applications deal with the extraction of pesticides from environmental waters. Hexane or cyclohexane are typical organic solvents used for extracting non-polar compounds such as organochlorine and organophosphorus pesticides, and dichoromethane or chloroform for medium polarity organic compounds such as triazines or phenylurea herbicides. However, quantitative recoveries for relatively polar compounds by LLE are difficult to achieve. As an example, a recovery for atrazine of 90% was obtained by LLE of 1 L of water with dichloromethane, whereas the recoveries for its degradation products des-isopropyl-, des-ethyl- and hydroxy-atrazine were 16, 46, and 46%, respectively [8].

In order to increase the efficiency and thus the range of application, the partition coefficients may be increased by using mixtures of solvents, changing the pH (preventing ionization of acids or bases) or by adding salts ("salting-out effect"). In this regard, the recoveries for the atrazine degradation products of the aforementioned example were 62, 87, and 63%, respectively by carrying out the extraction with a mixture of dichloromethane and ethylacetate with 0.2 M ammonium formate.

The high number of possible combinations of solvents and pHs makes ideally possible the isolation of any pesticide from water samples by LLE, which has been traditionally considered a great advantage of LLE. However, LLE is not exempt from important drawbacks. One of the most important is that the toxicity of the organic toxic solvents used leads to a large number of toxic residues. In this sense, the costs of the disposal of toxic solvents are rather high. However, it is important to mention that this problem is minimized when LLE is used for clean-up steps where low volumes are usually employed. Besides, the risk of exposure of the chemist to toxic solvents and vapors always exists. From a practical point of view, the formation of emulsions, which are sometimes difficult to break up, the handling of large water samples, and the difficulties of automation of the whole process mean that LLE is considered a tedious, time-consuming, and costly technique.

2.3.4 LIQUID-PHASE MICROEXTRACTION

The interest of the modern laboratories in the need to develop new methods that are less harmful to the environment has driven the miniaturization of traditional analytical processes. The reduction of scale during the extraction typically involves a series of advantages over conventional analytical techniques:

- Reduction or absence of organic solvents.
- Increase of the speed of the technique and possibility of simultaneous processing of several samples.
- Reduction of the systems' size that facilitates the implementation in portable devices and field work.
- Higher sensitivity due to the concentration of the analytes in a small extractant phase. Although not exhaustive, in many cases the transfer of the extracted analytes to the detection system is complete.

In 1996, independently, Dasgupta's [9] and Cantwell's [10] groups introduced the first formats of liquid-phase microextraction (LPME), followed by the work of Lee and co-workers [11]. These formats are based on the extraction of analytes from aqueous samples into a small drop of organic solvent by passive diffusion (single drop microextraction, SDME). A common system consists of a small drop of solvent suspended from a syringe for gas chromatography (Figure 2.4 A). The recoveries of the extraction are essentially determined by the partition coefficients of the analytes between the aqueous sample and the organic solvent. After the extraction, the drop is withdrawn into the syringe and injected in a gas chromatograph. This LPME format is simple, effective, and reduces the amount of organic solvents to a few µL per sample. However, the poor stability of the drop suspended in the syringe often hinders the analysis, thus limiting the use of the methodology.

To improve the stability and functionality of LPME, in 1999 Pedersen-Bjergaard and Rasmussen presented a new LPME format based on the use of porous polypropylene (PP) hollow fibers (hollow fiber-LPME, HF-LPME) [12]. In this conception of LPME, the extractant phase is located inside the lumen of a hollow and porous PP fiber previously soaked into an organic solvent immiscible with water. This solvent penetrates the pores of the walls of the polypropylene capillary, creating a liquid membrane that is supported by the polymeric network by capillary forces (supported liquid membrane, SLM). The acceptor phase inside the fiber is mechanically protected by the walls of the fiber and separated from the sample by the liquid membrane immiscible with the water, thus preventing the dissolution of the acceptor phase in the sample.

During the LPME process, the analytes are extracted from the aqueous sample through the organic SLM to the acceptor solution located inside the capillary. Then,

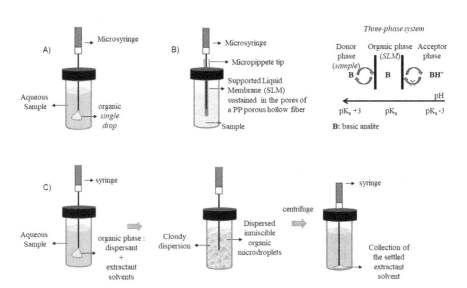

FIGURE 2.4   Liquid-phase microextraction: A) scheme for SDME; B) scheme for HF-LPME and three-phase extraction principle; C) scheme for DLLME.

the acceptor solution is collected with the aid of a microsyringe to be subjected to analysis. If the acceptor phase is an organic solvent, the extracts of the HF-LPME biphasic system are directly compatible with liquid and gas chromatography. On the contrary, when the acceptor phase is an aqueous solution, it is then a three-phase system and the extracts are directly compatible with liquid chromatography and electrophoresis.

The LPME is an equilibrium extraction in which the concentration of extracted analytes increases until a certain level, after which it remains constant with time. The chemical principle that governs the HF-LPME is basically the same as the one described in the pioneering works of the Audunsson and Jönsson groups on the extraction with liquid membranes [13, 14]. However, the instrumentation required for this type of extraction was not simple and included different pumps operating under controlled flow rates on each side of the membranes.

The efficiency of the extraction depends on the volumes and nature of the solvents and analytes involved, since these will determine the values of the partition coefficients that govern the system. As in any other type of extraction, the value of these coefficients can be modified by altering some of the variables of the system:

- Volume of the sample and acceptor phase. Generally, the use of small volumes of sample allows obtaining greater recoveries in extractions in the equilibrium, since the diffusion path of analytes is smaller. Obviously, the smaller the acceptor phase used, the higher the concentration of the sample.
- Organic phase. Whether in two- or three-phase mode, it is essential to select the appropriate organic phase, which must be totally immiscible with water, stable in the walls of the fiber, and possess an adequate partition coefficient. The most used organic phase for extraction is 1-octanol, and in particular, it has been used in food analysis in the extraction of carbaryl, thiabendazole, as well as other pesticides and herbicides [15–17]. Other organic phases such as trioctylphosphine oxide (TOPO) [18] or dihexyl ether have also been used to extract pesticides from vegetable samples [19]. Other commonly used organic phases are dodecyl acetate, chloroform, cyclohexane, or toluene, although the latter generates some problems as SLM due to its instability and volatility.
- pH conditions and saline effect. When working with two phases, it is essential to regulate the pH of the sample according to the acidic or basic nature of the analytes. In the case of three-phase systems, the pH control is even more important, since it can help to improve the extraction performance significantly. As Figure 2.4 B illustrates, for basic analytes, the pH of the sample must be high (preferably three units above the pKa of the target compound), while in the accepting aqueous phase the pH must be acidic (three units below pKa). The opposite can be applied for acidic compounds. The addition of NaCl to the sample can also improve the recovery of the process due to the *salting-out* effect.
- Shaking/stirring. As a technique in which the analytes are extracted by passive diffusion, it is necessary to shake the sample to facilitate the process. Generally, the recoveries increase with higher stirring, until a point

in which the stability of the system is compromised. The agitation can be performed by magnetic or orbital stirring. Operating with orbital stirring avoids possible contamination effects caused by the magnet.

Dispersive liquid–liquid microextraction (DLLME) is another simple LPME format proposed by Rezaee et al. [20]. In DLLME, a binary organic solvent mixture (containing the extraction and disperser solvents) is injected into an aqueous sample forming a cloudy suspension. The large surface area between the fine droplets and the aqueous phase facilitates the quick transfer of analytes from the sample solution into the extraction phase. After centrifugation of the mixture, the microdroplets sediment at the bottom of the conical tube so they can be collected using a syringe (Figure 2.4 C). Due to its simplicity, rapidity, high enrichment factors, and low organic solvent consumption, DLLME has gained popularity in recent years.

The key experimental parameters affecting the extraction efficiency of target analytes are the types and volumes of organic solvents (extractant and disperser). First, selection of an appropriate extraction solvent is essential in the DLLME procedure. The extraction solvent is chosen according to its density, miscibility with water, and extraction capability of the analyte. The lower the volume of the extractant, the higher the concentration factor. However, as an equilibrium extraction technique, the partition coefficient can be affected by the ratio between acceptor volume and sample volume. On the other hand, whereas the extractant solvent must be immiscible with water, the miscibility of the disperser solvent in both the extractant and the aqueous sample is necessary. Organic solvents with moderate/high polarity such as acetone, acetonitrile, ethanol, and methanol are appropriate disperser solvents. The amount of dispersant must be enough to form a cloudy state stable during the extraction time, without enhancing the solubility of the analytes in the aqueous sample. Examples of the application of this type of methodology include the analysis of chlorpyrifos in environmental waters and the multiresidue analysis of several pesticides in tea beverages [21, 22]. The centrifugation is the crucial step that can shorten the duration of the DLLME procedure. In order to avoid the centrifugation of the sample, Chen and co-workers used a water-acetonitrile-toluene emulsion in the extraction and analysis of carbamate pesticides in water samples, so that the addition of acetonitrile broke up the emulsion allowing the easy separation of the immiscible phases [23].

2.3.5 Solid-Phase Extraction

Solid-phase extraction (SPE), as LLE, is based on the different affinity of target analytes for two different phases. In SPE, a liquid phase (liquid sample or liquid sample extracts obtained following the techniques mentioned above) is loaded onto a solid sorbent (polar, ion exchange, non-polar, affinity), which is packed in disposable cartridges or enmeshed in an inert matrix of an extraction disk. Those compounds with a higher affinity for the sorbent will be retained on it whereas others will pass through it unaltered. Subsequently, if target analytes are retained, they can be eluted using a suitable solvent with a certain degree of selectivity.

The typical SPE sequence involving several steps is depicted in Figure 2.5. Firstly, the sorbent needs to be prepared by activation with a suitable solvent and

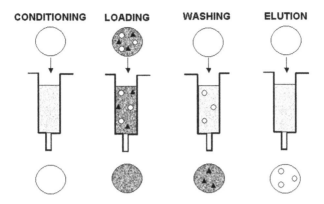

FIGURE 2.5 Solid-phase extraction steps.

by conditioning with the same solvent in which analytes are dissolved. Then, the liquid sample or a liquid sample extract are loaded onto the cartridge. Usually, target analytes are retained together with other components of the sample matrix. Some of these compounds can be removed by application of a washing solvent. Finally, analytes are eluted with a small volume of an appropriate solvent. In this sense, by SPE, it is possible to obtain final sample extracts ideally free of co-extractives, thanks to the clean-up performed, with high enrichment factors due to the low volume of solvent used for eluting target analytes. These aspects together with the simplicity of operation and the easy automation have made SPE a very popular technique widely used in the analysis of pesticides in a great variety of samples.

The success of an SPE procedure depends on the knowledge of the properties of target analytes and the kind of sample, which will help the proper selection of the sorbent to be used. Understanding the mechanism of interaction between the sorbent and the analyte is a key factor in the development of an SPE method since it will ease choosing the right sorbent from the wide variety of them available in the market.

2.3.5.1 Polar Sorbents

The purification of organic sample extracts is usually performed by SPE onto polar sorbents. Within this group, the sorbent most used is silica, which possesses active silanol groups in its surface able to interact with target analytes. This interaction is stronger for pesticides with base properties due to the slight acidic character of silanol groups. Other common polar sorbents are alumina (commercially available in its acid, neutral, and base form) and Florisil.

In the loading step, analytes compete with the solvent for the adsorption active sites of the sorbent, and elution is performed by displacing of analytes from the active sites by an appropriate solvent. In this sense, the more polar the solvent is, the higher elution power it has. The elution power is established by the eluotropic strength ($\varepsilon°$) which is a measure of the adsorption energy of a solvent in a given sorbent. The eluotropic series of different common solvents in alumina and silica are shown in Table 2.3. In this way, by a careful selection of solvents (or mixture of them), analytes (or interferences) will be retained on the sorbent by loading in

TABLE 2.3
Eluotropic Series

Solvent	ε° Al_2O_3	ε° SiOH
Pentane	0.00	0.00
Hexane	0.00–0.01	0.00–0.01
Iso-octane	0.01	0.01
Cyclohexane	0.04	0.03
Carbon tetrachloride	0.17–0.18	0.11
Xylene	0.26	–
Toluene	0.20–0.30	0.22
Chlorobenzene	0.30–0.31	0.23
Benzene	0.32	0.25
Ethyl ether	0.38	0.38–0.43
Dichloromethane	0.36–0.42	0.32–0.32
Chloroform	0.36–0.40	0.26
1,2-Dichloroethane	0.44–0.49	–
Methyl ethyl ketone	0.51	–
Acetone	0.56–0.58	0.47–0.53
Dioxane	0.56–0.61	0.49–0.51
Tetrahydrofuran	0.45–0.62	0.53
Methyl t-butyl ether	0.3–0.62	0.48
Ethyl acetate	0.58–0.62	0.38–0.48
Dimethyl sulfoxide	0.62–0.75	–
Acetonitrile	0.52–0.65	0.50–0.52
1-Butanol	0.7	–
n-Propyl alcohol	0.78–0.82	–
Isopropyl alcohol	0.78–0.82	0.6
Ethanol	0.88	–
Methanol	0.95	0.70–0.73

a non-polar solvent and subsequently eluted using a second solvent with a higher eluotropic strength. Obviously, the selection of these solvents will be determined by the polarity of the analytes. Thus, after loading, hydrophobic pesticides such as pyrethroids can be eluted with a mixture of hexane–diethylether, whereas for eluting carbamates a more polar mixture such as hexane–acetone is necessary.

The number of developed methods based on SPE using polar sorbents for the determination of pesticides in food and environmental solid samples is huge, and thus, for specific examples, the interested reader should consult Chapters 6, 7, and 8 of the present book.

2.3.5.2   Non-Polar Sorbents

This kind of sorbent is appropriate for the trace-enrichment and clean-up of pesticides in polar liquid samples (i.e., environmental waters). Traditionally, n-alkyl-bonded silicas, mainly octyl- and octadecyl-silica, both in cartridges and disks, have been used due to their ability to retain non-polar and moderate polar pesticides from liquid samples. The retention mechanism is based on van der Waals forces and hydrophobic interactions, which allows the handling of large sample volumes and the subsequent elution of target analytes in a small volume of a suitable organic solvent (e.g., methanol, acetonitrile, ethyl acetate), getting high enrichment factors. However, for more polar pesticides, the strength of the interaction is not high enough and low recoveries are obtained as the corresponding breakthrough volume is easily reached.

An easy manner of increasing breakthrough volumes is to increase the amount of sorbent used, which will increase the number of interactions that take place. A second option is the addition of salts to the sample, diminishing the solubility of target analytes (salting-out effect) and thus favoring their interactions with the sorbent. Table 2.4 shows the obtained recoveries of several triazines by the SPE of 1 L of water spiked at 1 $\mu g \cdot L^{-1}$ concentration level of each analyte in different experimental conditions [24]. It is clear that the combination of using two C_{18} disks and the addition of a 10% NaCl to the water sample allows the obtainment of quantitative recoveries for all the tested analytes including the polar degradation products of atrazine. However, these approaches do not always provide satisfactory results. In that case, the most direct way of increasing breakthrough volumes of most polar pesticides is the use of sorbents with higher affinities for target analytes. These sorbents include styrene-divinylbenzene based polymers with a high specific surface (~ 1000 $m^2 \cdot g^{-1}$), which are commercialized by several companies under different trademarks (e.g., Lichrolut, Oasis, Envichrom). The interaction of analytes with these sorbents is also based on hydrophobic interactions, but the presence of aromatic rings within the polymeric network leads to strong $\pi-\pi^*$ interactions with the aromatic rings present in the chemical structure of many pesticides. Another alternative is the use of graphitized carbon cartridges or disks, which have a great capacity for the preconcentration of highly polar pesticides (acidic, basic, and neutral) and transformation products

TABLE 2.4
Recoveries (R%) and Relative Standard Deviations (RSD) of Several Triazines Obtained by SPE of 1 L of LC Grade Water Spiked with 1 $\mu g \cdot L^{-1}$ of Each Triazine

| | 1 C_{18} disk | | | | 2 C_{18} disk | | | |
| | Without NaCl | | 10% NaCl | | Without NaCl | | 10% NaCl | |
Triazine	R%	RSD	R%	RSD	R%	RSD	R%	RSD
Desisopropylatrazine	21.5	18.6	42.3	13.6	35.8	18.2	89.2	8.7
Desethylatrazine	50.4	9.3	98.4	6.2	60.5	13.1	95.4	6.3
Simazine	100.2	6.1	93.5	7.8	96.4	8.7	91.7	6.2
Atrazine	94.6	8.7	98.3	4.9	104.3	4.6	97.0	4.1

Source: Adapted from Turiel, E., Fernández, P., Pérez-Conde, Cámara, C., *J. Chromatogr. A*, 872, 299, 2000, with permission from Elsevier.

such as oxamyl, aldicarb sulfoxide, and methomyl thanks to the presence of various functional groups, including positively charged active centers on its surface.

2.3.5.3   Ion-Exchange Sorbents

Ionic or easily ionizable pesticides can be extracted by these sorbents. Sorption occurs at a pH in which the analyte is in its ionic form and then it is eluted by a change of the pH value with a suitable buffer. The mechanism involved provides a certain degree of selectivity. Phenoxyacid herbicides can be extracted by anion-exchangers and amines or n-heterocycles using cation-exchangers. However, its use is rather limited due to the presence of high amounts of inorganic ions in the samples, which overload the capacity of the sorbent leading to low recoveries of target analytes.

2.3.5.4   Affinity Sorbents

The sorbents described above are able to successfully extract pesticides from a great variety of samples. However, the retention mechanisms (hydrophobic or ionic interactions) are not selective, leading to the simultaneous extraction of matrix compounds, which can negatively affect the subsequent chromatographic analysis. For instance, the determination of pesticides (especially polar pesticides) in soil and water samples by liquid chromatography using common detectors is affected by the presence of humic and fulvic acids. These compounds elute as a broad peak or as a hump in the chromatogram, hindering the presence of target analytes and thus making it difficult in some cases to reach the required detection limits. Even using selective detectors (e.g., mass spectrometry), the presence of matrix compounds can suppress or enhance analyte ionization, hampering accurate quantification.

The use of antibodies immobilized on a suitable support, so-called immunosorbent (IS), for the selective extraction of pesticides from different samples appeared some years ago as a clear alternative to traditional sorbents [25, 26]. In this approach, only the antigen which produced the immune response, or very closely related molecules, will be able to bind the antibody. Thus, theoretically, when the sample is run through the IS, the analytes are selectively retained and subsequently eluted free of co-extractives. The great selectivity provided by immunosorbents has allowed the determination of several pesticides in different matrices such as carbofuran in potatoes, or triazines and phenylureas in environmental waters, sediments, and vegetables. However, this methodology is not free of important drawbacks. The obtainment of antibodies is time-consuming and expensive, and few antibodies for pesticides are commercially available. Also, it is important to point out that after the antibodies have been obtained, they have to be immobilized on an adequate support, which may result in poor antibody orientation or even complete denaturation. Because of these limitations, the preparation and use of molecularly imprinted polymers has been proposed as a promising alternative.

Molecularly imprinted polymers (MIPs) are tailor-made macroporous materials with selective binding sites able to recognize a particular molecule [27]. Their synthesis, depicted in Figure 2.6, is based on the formation of defined (covalent or non-covalent) interactions between a template molecule and functional monomers during a polymerization process in the presence of a cross-linking agent. After

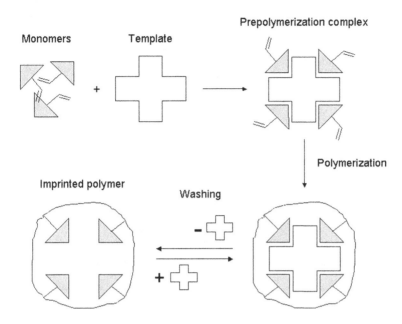

FIGURE 2.6 Scheme for the preparation of molecularly imprinted polymers.

polymerization the template molecule is removed, leaving cavities complementary in size and shape to the analyte. Thus, theoretically, if a sample is loaded on it, in a solid-phase extraction procedure, the analyte (the template) or closely related compounds will be able to selectively rebind the polymer, being subsequently eluted free of co-extractives. This methodology, namely molecularly imprinted solid-phase extraction (MISPE), has been successfully employed in the determination of pesticides such as triazines, phenylureas, and phenoxyacids herbicides, among others, in environmental waters, soils, and vegetable samples. As an example of the selectivity provided by MIPs, Figure 2.7 shows the chromatograms obtained in the analysis of fenuron in potato sample extracts without and with MISPE onto a fenuron-imprinted polymer. It is clear that the selectivity provided by the MIP allows the determination of fenuron at very low concentration levels [28].

Because of their easy preparation, excellent physical stability, and chemical characteristics (high affinity and selectivity for the target analyte), MIPs have received special attention from the scientific community in several fields, not only in pesticide residue analysis. Besides, there are already MISPE cartridges commercially available for the extraction of certain analytes (i.e., triazines) and some companies offer custom synthesis of MIPs for SPE, which will ease the implementation of MISPE in analytical laboratories.

The wide variety of available sorbents as well as the reduced processing times and solvent savings have made SPE a clear alternative to LLE. Besides, automation is possible using special sample preparation units that sequentially extract the samples and clean them up for automatic injections. However, the typical drawbacks associated with offline procedures, such as the injection in the chromatographic system

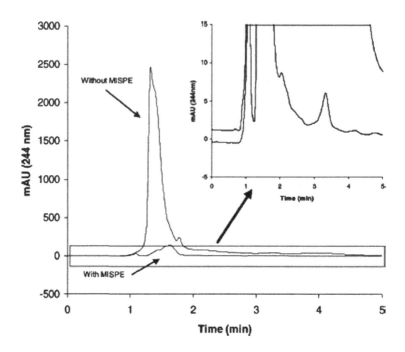

FIGURE 2.7   Chromatograms obtained at 244 nm without and with MISPE of potato sample extracts spiked with fenuron (100 ng.g⁻¹). Graph insert shows the same chromatograms with different absorbance scale. (Reproduced from Tamayo, F.G., Casillas, J.L., Martin-Esteban, A., *Anal. Chim. Acta*, 482, 165, 2003, with permission from Elsevier.)

of an aliquot of the final extract or the necessity of including an evaporation step, remain, which affects the sensitivity of the whole analysis.

The use of SPE coupled online with liquid and gas chromatography can solve the aforementioned drawbacks. The coupling of SPE to liquid chromatography is especially simple to perform in any laboratory and has been extensively described for the on-line preconcentration of organic compounds in environmental water samples [29]. The simplest way of SPE–LC coupling is shown in Figure 2.8, where a precolumn (1–2 cm×1–4.6 mm i.d.) filled with an appropriate sorbent, is inserted in the loop of a six-port injection valve. After sorbent conditioning, the sample is loaded by a low-cost pump and the analytes are retained in the precolumn. Then, the precolumn is connected online to the analytical column by switching the valve, so that the mobile phase can desorb the analytes prior to their separation in the chromatographic column. Apart from a considerable reduction in sample manipulation, the main advantage is the fact that the complete sample is introduced in the analytical column. Besides, there is equipment commercially available for the whole automation of the process.

Alkyl-bonded silicas (mainly C₁₈-silica) have been widely used as precolumn sorbent although they are being replaced by styrene-divinylbenzene copolymers which offer higher affinity for polar analytes and so permit the utilization of larger sample volumes without exceeding the breakthrough volumes of analytes. Other materials

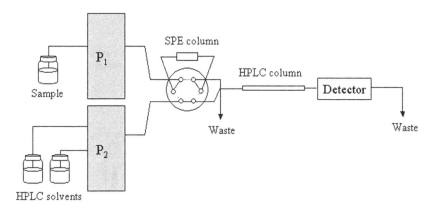

FIGURE 2.8 SPE–LC coupling set-up.

successfully employed have been small extraction disks and graphitized carbons; and in order to provide selectivity to the extraction, precolumns packed with yeast cells immobilized on silica gel [30] or with immunosorbents have been proposed for the extraction of polar pesticides from environmental waters [31, 32].

The coupling of SPE with GC is also possible thanks to the ability to inject large volumes into the gas chromatograph using a column of deactivated silica (retention gap) located between the injector and the analytical column. SPE–GC uses the same sorbents employed in SPE–LC but, in this case, after the preconcentration step, the analytes are desorbed with a small volume (50–100 µL) of an appropriate organic solvent, which is directly introduced into the chromatograph. In general, using only 10 mL of water sample, it is possible to reach detection limits at the µg·L^{-1} level employing common detectors.

2.3.6 SOLID-PHASE MICROEXTRACTION

As it has been stated previously, solid-phase extraction has shown itself to be a very useful procedure for the extraction of a great variety of pesticides in food and environmental analysis. However, although to a smaller extent than liquid–liquid extraction, this technique still requires the use of toxic organic solvents and its applicability is restricted to liquid samples. With the aim of eliminating these drawbacks, Arthur and Pawliszyn introduced solid-phase microextraction (SPME) in 1989 [33]. Its simplicity of operation, solvent-less nature, and the availability of commercial fibers have meant that SPME has been rapidly implemented in analytical laboratories.

As depicted in Figure 2.9, the SPME device is quite simple, and just consists of a silica fiber coated with a polymeric stationary phase similar to those used in gas chromatography columns. The fiber is located inside the needle (protecting needle) of a syringe specially designed to allow exposure of the fiber during sample analysis. As in any solid-phase extraction procedure, SPME is based on the partitioning of target analytes between the sample and the stationary phase and consists of two consecutive steps, extraction and desorption. An intermediate washing step can also be performed.

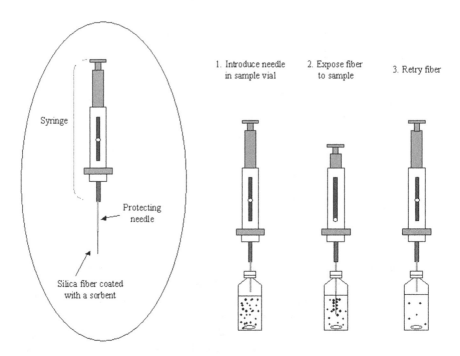

FIGURE 2.9   SPME device and typical mode of operation.

2.3.6.1 Extraction

The extraction step can be performed both by exposure of the fiber to the head-space (restricted to volatile compounds in liquid or solid samples) or by direct immersion of the fiber into the sample (aqueous-based liquid samples). As described in Figure 2.9, the experimental procedure is very simple. Firstly, the fiber is inside the protecting needle which is introduced into the sample vial. Then, the fiber is exposed to the sample to perform extraction by sorption of the analytes to the stationary phase. Finally, the fiber is retried inside the needle for further desorption and the whole device removed.

Obviously, a proper selection of the SPME sorbent is a key factor in the success of the analysis. In general, the polarity of the fiber should be as similar as possible to that of the analyte of interest. In this sense, there are nowadays a great variety of fibers commercially available that cover a wide range of polarities (e.g., carbowax/DVB for polar compounds or polydimethylsiloxane for hydrophobic compounds). Also, both the fiber thickness and the porosity of the sorbent will influence the final extraction efficiency. Besides, other physical and chemical parameters such as temperature, exposition time, agitation, pH, or ionic strength ("salting out" effect) of the sample can be optimized. An example of that is the extraction of dinoseb, an alquil-substituted dinitrophenol, in water. The SPME of this compound can be favored by using a poliacrylate fiber and by adding 10% of NaCl at pH=2 due to the produced "salting out" effect and the lower ionization of dinoseb at low pH values.

Finally, concerning extraction, it is interesting to mention that from the mathematical model governing SPME, it can be concluded that when the sample volume is much higher than the fiber volume, the extraction efficiency becomes independent of the sample volume. Although that is not applicable for laboratory samples (low volumes), this fact makes SPME a very interesting tool for in-field sampling procedures, since the fiber can be exposed to the air or directly immersed into a lake or a river regardless of sample volume.

2.3.6.2 Desorption

Desorption can be performed thermally in the injection port of a gas chromatograph, or by elution of the analytes by means of a suitable solvent. In the latter case, desorption can be carried out in a vial containing a small volume of the solvent to be further analyzed by chromatographic techniques or eluted with the mobile phase on a specially designed SPME-HPLC interface.

Thermal desorption of the analytes in the injector port of the GC instrument is based on the increase of the partition coefficient gas–fiber with the increasing temperature. In addition, a constant flow of carrier gas inside the injector facilitates removal of the analytes from the fiber. The main advantage of thermal desorption is the fact that the total amount of extracted analytes is introduced in the chromatographic system and analyzed, thus compensating the low recoveries usually obtained in the extraction step. However, unfortunately, thermal desorption cannot be used for non-volatile or thermolabile compounds, it being necessary, in those cases, to use desorption with solvents. The procedure is similar to SPE elution but, in this case, the fiber is immersed in a small volume of elution solvent and agitated or heated to favor the transfer of the analytes to the solvent solution. A fraction of this extract or, for some applications, an evaporated and re-dissolved extract, is subsequently injected into the chromatographic system.

Recently, interfaces have been made commercially available allowing the direct coupling of SPME with liquid chromatography. The coupling is similar to that described in Figure 2.8 for SPE–HPLC but a specially designed little chamber, instead of a precolumn, is placed in the loop of a six-port injection valve. This interface allows desorption of the analytes by the chromatographic mobile phase with the total amount of compounds extracted introduced in the chromatographic system.

2.3.7 SOLID–SOLID EXTRACTION: MATRIX SOLID-PHASE DISPERSION

Matrix solid-phase dispersion (MSPD), introduced by Barker et al. in 1989 [34], is based on the complete disruption of the sample (liquid, viscous, semi-solid, or solid) while the sample components are dispersed into a solid sorbent. Most methods use C_8- and C_{18}-bonded silica as solid support. Other sorbents such as Florisil and silica have been also used although to a lesser extent.

Experimentally, the sample is placed in a glass mortar and blended with the sorbent until a complete disruption and dispersion of the sample on the sorbent is obtained. Then, the mixture is directly packed into an empty cartridge, the same as those used in SPE. Finally, analytes are eluted after a washing step for the removal of interfering compounds. The main difference between MSPD and SPE is that with

MSPD the sample is dispersed through the column instead of only onto the first layers of sorbent, which typically allows the obtainment of rather clean final extracts, avoiding the necessity of performing a further clean-up.

MSPD has been successfully applied for the extraction of several pesticide families in fruit juices, honey, oranges, cereals, and soil, among others, and the achieved performance, compared to other classical extraction methods, has been found superior in most cases [35]. The main advantages of MSPD are the short extraction times needed, the small amount of sample, sorbent, and solvents required, and the possibility of performing extraction and clean-up in one single step.

2.3.8 STIR-BAR SORPTIVE EXTRACTION

Stir bar sorptive extraction (SBSE) is based on the partitioning of target analytes between the sample (mostly aqueous-based liquid samples) and a stationary phase-coated stir bar [36]. Until now, only polydimethylsiloxane (PDMS)-coated stir bars are commercially available, restricting the range of applications to the extraction of hydrophobic compounds (organochlorine and organophosphorus pesticides) due to the apolar character of PDMS.

The experimental procedure followed in SBSE is quite simple. The liquid sample and the PDMS-coated magnetic stir bar are placed in a container. Then, the sample is stirred for a certain period of time (30–240 min) until no additional recovery for target analytes is observed even when the extraction time is increased further. Finally, the stir bar is removed and placed in an especially designed unit, in which thermal desorption and transfer of target analytes to the head of the GC column take place.

SBSE is usually compared to and proposed as an alternative to SPME. The use of a PDMS-coated stir bar (10 mm length, 0.5 mm coating thickness) results in a significant increase in the volume of the extraction phase from approximately 0.5 μL for an SPME fiber (100 μm PDMS) to about 24 μL for a stir bar. Consequently, the yield of the extraction process is much greater when using a stir bar rather than an SPME fiber, both coated with PDMS. However, the greater coating area of magnetic stir bars is simultaneously its main drawback since the extraction kinetics are slower than for SPME fibers, and a high amount of interfering matrix compounds are co-extracted with target analytes. Nevertheless, the simplicity of operation and its solvent-less nature make SBSE a very attractive technique. Recently, the development of new stir bars coated with more polar and selective sorbents have extended the applicability of the SBSE, such as the cases of stirrers coated by MIPs or monoliths composites formed by MIPs and magnetic nanoparticles, which allow the extraction of thiabendazole and triazines in samples of citrus and environmental soils, respectively [37, 38].

2.3.9 QuEChERS

The QuEChERS method, proposed in 2003 by Anastassiades, Lehotay, and collaborators, is based on a solid–liquid extraction with acetonitrile from solid or semi-solid samples, followed by a cleaning step of the extract by extraction with solid-phase matrix dispersion using primary and secondary amines adsorbents (PSA adsorbents) [39]. The QuEChERS name refers to some of the advantageous properties attributed to the methodology: Quick, Easy, Cheap, Effective, Rugged, and Safe. QuEChERS methodologies are particularly suitable for multiresidue studies applied to solid samples of soil, sludge, tissue, or food, where the dispersion of a semi-solid material is necessary. Different studies have compared the efficacy and convenience of the QuEChERS method versus matrix solid-phase dispersion extraction and the suitability of each would depend on the specific case. In this regard, the European Committee for Standardisation (CEN) and the Association of Analytical Communities (AOAC International) have established official methods for the analysis of a broad spectrum of pesticide residues in foods applying the QuEChERS methodology [40, 41]. Regardless, as with any other analytical procedure, depending on the performance in a particular matrix such methodologies might be customized to tune the range of target analytes, for example, by acidifying the extractant solution or adding/removing a clean-up step with PSA.

Mezcua et al. applied this type of methodology in an illustrative work with the aim of performing the multiresidue screening of a series of pesticides in vegetable and fruit samples [42]. For this purpose, 15 mL of previously homogenized sample was weighed in a 200 mL centrifuge tube. Then, 15 mL of acetonitrile was added and the tube vigorously stirred for 1 minute. After this, 1.5 g of NaCl and 6 g of $MgSO_4$ were added, and the stirring process was repeated. The extract was then centrifuged at 3700 rpm for 1 minute and an aliquot of 5 mL of the supernatant was poured over 250 mg of PSA and 750 mg of $MgSO_4$ in a 15 mL centrifuge tube. After vigorous stirring for 20 seconds, the extract was centrifuged again (3700 rpm) for 1 minute. Finally, the researchers obtained an extract in acetonitrile equivalent to 1 g of sample for each mL of acetonitrile, which they analyzed by liquid chromatography time-of-flight mass spectroscopy (LC–TOF–MS). The results of the analysis performed by this method were compared with an accurate mass database, allowing the screening of about 300 different pesticides in several plant samples.

2.3.10 NANOPARTICLES-BASED EXTRACTIONS

Nanomaterials (NMs) are manufactured chemical substances containing particles, in an unbound state or as an aggregate or as an agglomerate where, for 50% or more of the particles in the number size distribution, one or more external dimensions is in the size range 1 nm–100 nm [43]. In addition to this, fullerenes, graphene flakes, and single wall carbon nanotubes (SWCNT) with at least one external dimension below 1 nm are also considered NMs. Nanoparticles (NPs) exhibit novel characteristics such as enhanced diffusion, greater specific surface, increased chemical reactivity, strength, conductivity, etc. compared to the same material without nanoscale features. In recent years, NMs have attracted great attention from industry and academia, who have revised classical processes and products to develop new applications that take advantage of NMs' extraordinary features. This is the case with chemical analysis in general, and sample preparation in particular. As a crucial stage in the problem of chemical analysis, the preparation of the sample has also benefited from the application of NPs to develop new procedures. In general, NPs can be used during sample treatment either as advanced sorbents, taking advantage of their high surface and increased chemical reactivity, or as a support for other adsorbents. Among the last type of use, superparamagnetic NPs represent one of the most interesting materials since they can be easily isolated from the matrix by using an external magnetic field. The different types of NMs have been frequently implemented in combination with well-established extraction techniques, such as solid-phase extraction (SPE), dispersive solid-phase extraction (d-SPE), matrix solid-phase dispersion (MSPD), stir bar sorptive extraction (SBSE), and hollow-fiber liquid-phase microextraction techniques (HF-LPME), among many others.

The application of NMs as sorbents for the extraction of analytes in complex food and environmental samples has increased, together with the development of novel tunable advanced materials, because of their versatility and enhanced performance. Moreover, due to the reduced scale of the sorbents, these can be implemented in miniaturized structures and devices, hence limiting the use of solvents (and consequently limiting waste) and facilitating the preconcentration of analytes in a small acceptor phase.

Different types of NMs with diverse nature have been developed and tested as sorbents in chemical analysis, such as magnetic NPs (MNPs), single-wall/multi-wall carbon nanotubes (SWCNTs and MWCNT, respectively), metal–organic frameworks (MOFs), and recently graphene, among others. It is not rare to see applications using combinations of them, resulting in novel customized materials.

2.3.10.1 Metal–Organic Frameworks (MOFs)

Metal–organic frameworks are coordination compounds based on the interaction between metal cations and organic electron donors. The great variety of possible combinations of *metallic ion–organic donor* permits the obtaining of many different materials with different properties such as particle size, pore diameter, and mechanical resistance. Moreover, the surface of MOFs can be easily functionalized using a suitable organic ligand. As an example, MOFs have been directly used as sorbent in MSPD to extract pesticide residues in fruits prior to their analysis by HPLC–UV/DAD. The tested material was $_\infty[(Nd_{0.9}Eu_{0.1})_2(DPA)_3(H_2O)_3]$, and it showed an excellent extraction capability for the pesticides thiamethoxam, thiacloprid, thiophanate-methyl, teflubenzuron, and bifenthrin. The method exhibited low limits of detection (0.03 mg kg^{-1}) and good linearity (correlation coefficients from 0.9995 to 0.9999) in the experimental range (0.08–100 mg kg^{-1}), for the six different pesticides studied, comparable to conventional Florisil [44]. The same group tested $_\infty[(La_{0.9}Eu_{0.1})_2(DPA)_3(H_2O)_3]$ MOF as sorbent in the MSPD extraction of pyrimicarb, procymidone, malathion, methyl parathion, and α- and β-endosulfan from lettuce, with analysis using GC–MS. Similarly, the analytical figures of merit were satisfactory and the nanosorbent showed improved performance compared to conventional silica-gel [45].

2.3.10.2 Carbon Nanotubes

Carbon nanotubes (CNTs) were first reported in 1991 [46]. CNTs show particular physical-chemical characteristics, such as high tensile strength, excellent thermal conductivity and stability, and switchable electronic properties depending on their conformation, etc. Because of their very small dimensions and large specific surface area, CNTs are able to interact with both inorganic and organic compounds. CNTs have been applied in the preparation of polymeric membranes for filtration (carbon nanotube film microextraction, CNTF) [47], as stationary phases in chromatographic applications [48], as well as sorbents in pre-treatment procedures. In this regard, CNTs have been applied to MSPD, SBSE, and SPME extraction techniques in complex matrices such as vegetables, fruit, honey, fish, tea or coffee, among others [49]. It is also common to disperse the CNT-based sorbent into a liquid sample to reduce the extraction time, in the so-called dispersive solid phase extraction (d-SPE). In this regard, Xu et al. [50] prepared hydroxyl functionalized MWCNTs to extract the fungicide thiabendazole from juices and to perform HPLC–DAD analysis, obtaining an LOD of 2.6 μg L^{-1} and recoveries ranging between 93 and 104%. Similarly, MWCNTs have also been implemented in a QuEChERS methodology to remove interferences in the determination of 171 pesticides in cowpea samples followed by GC–tandem mass (MS/MS) analysis [51].

2.3.10.3 Graphene

Graphene is a planar allotrope of carbon consisting of a single layer of carbon atoms arranged in a hexagonal lattice. Graphene has high thermal and electron conductivity, elasticity, flexibility, and ultra-high surface area. The use of graphene in sorbent-based extraction methodologies has recently increased, primarily due to its great hydrophobicity and the possibility of establishing π–π interactions thanks to delocalized electrons. Moreover, graphene can be chemically functionalized, making it an even more versatile material. In this regard, as an example, Qi and co-workers used a graphene-modified polymer to extract by μ-SPE and to analyze by HPLC–UV four pesticides in fruits and vegetables [52].

2.3.10.4 Magnetic Nanoparticles

Magnetic materials, when scaled down to nano-size, show different magnetization behavior than bulk magnetic materials. The combination of their magnetic behavior in nanometric scale (nm) and their utility in miniaturization make MNPs a very attractive tool for analytical chemists. MNPs, as well as NPs in general, can be used directly as sorbents or as a core for later surface coating. Because of the greater chemical reactivity, NPs tend to be sensitive to small changes of pH, ionic strength, and redox species. Therefore, the process of surface functionalization stabilizes the NPs, preventing them from potential dissolution or aggregation, while simultaneously providing the NPs with new adsorption capabilities. The nature of the coatings is diverse and depends on the types of sample and analyte. This is the case in the work of Tian and co-workers [53] to extract 1,1-bis(4-chlorophenyl)-2,2,2- trichloroethane (DDT) from aqueous media. The authors loaded Fe_3O_4 NPs onto mesoporous silica, which attached via Fe–O–Si bonds. The thickness of the silica shell, the volume of the pores, and the surface area can be modified by adjusting the silica–magnetite ratio. A magnetic core-shell structure of this type can be further functionalized by a ligand, as it is the case of ocatdecyl-silica functionalization, which has been used to determine several organophosphorus and pyrethroid pesticides in environmental samples at low concentration levels in combination with GC–MS analysis [54]. The use of carbon, CNT (both single and multi-wall), and graphene as functional coatings is also common thanks, among other properties, to their stability and excellent affinity for organic molecules [55]. In recent years, the use of MIP-functionalized MNPs has increased as they improve notably the selectivity of the extraction. Core-shell magnetic MIPs (MMIPs) have been applied to the extraction of the pesticide dicofol present in tea samples [56], imidacloprid residues in eggplant and honey [57], sulfonylurea herbicide residues in rice and environmental waters [58], and triazines in vegetables and soils [59, 60].

2.4 FUTURE TRENDS

In this chapter, a description of the different techniques developed during the last number of years for the extraction and clean-up of pesticides from environmental and food samples has been made. It is evident that a great effort has been made to improve the techniques and procedures used for sample preparation. However, still today, sample preparation is the limiting step of the analysis. Even using very powerful detection techniques such as LC–MS (MS), some sample preparation (including clean-up) is still necessary since otherwise interferences and signal suppression can occur. It is also convenient to keep the instruments in good condition, avoiding the introduction of unwanted material, and thus facilitating their maintenance and reducing costs and service interruptions.

Thus, since sample preparation cannot be avoided, further studies toward simplification and on-site applications are expected in the near future. In this regard, environmentally friendly, cost-effective, and selective procedures are required. In parallel, advances in miniaturization and automation will ease the integration of sample preparation and instrumental analysis leading to faster procedures with improved performance in terms of accuracy, precision, and traceability.

REFERENCES

1. C. Sánchez-Brunete, R.A. Pérez, E. Miguel, J.L. Tadeo, Multiresidue herbicide analysis in soil samples by means of extraction in small columns and gas chromatography with nitrogen-phosphorus and mass spectrometric detection, *Journal of Chromatography A*, 823 (1998) 17–24.
2. J. Castro, C. Sánchez-Brunete, J.L. Tadeo, Multiresidue analysis of insecticides in soil by gas chromatography with electron-capture detection and confirmation by gas chromatography-mass spectrometry, *Journal of Chromatography A*, 918 (2001) 371–80.
3. S. Babić, M. Petrović, M. Kaštelan-Macan, Ultrasonic solvent extraction of pesticides from soil, *Journal of Chromatography A*, 823 (1998) 3–9.
4. A. Bouaid, A. Martín-Esteban, P. Fernández, C. Cámara, Microwave-assisted extraction method for the determination of atrazine and four organophosphorus pesticides in oranges by gas chromatography (GC), *Fresenius' Journal of Analytical Chemistry*, 367 (2000) 291–4.
5. E. Björklund, T. Nilsson, S. Bøwadt, Pressurised liquid extraction of persistent organic pollutants in environmental analysis, *TrAC - Trends in Analytical Chemistry*, 19 (2000) 434–45.
6. A. Martín-Esteban, P. Fernández-Hernando, Preparación de la muestra para la determinación de analitos orgánicos, in: C. Cámara (Ed.), *Toma y tratamiento de muestra*, Editorial Síntesis, Madrid, 2002, 271–325.

7. M. Zougagh, M. Valcárcel, A. Ríos, Supercritical fluid extraction: A critical review of its analytical usefulness, *TrAC - Trends in Analytical Chemistry*, 23 (2004) 399–405.

8. G. Durand, D. Barcelo, Liquid chromatographic analysis of chlorotriazine herbicides and its degradation products in water samples with photodiode array detection. I.- evaluation of two liquid–liquid extraction methods, *Toxicological & Environmental Chemistry*, 25 (1989) 1–11.

9. H. Liu, P.K. Dasgupta, Analytical chemistry in a drop. Solvent extraction in a microdrop, *Analytical Chemistry*, 68 (1996) 1817–21.

10. M.A. Jeannot, F.F. Cantwell, Solvent microextraction into a single drop, *Analytical Chemistry*, 68 (1996) 2236–40.

11. Y. He, H.K. Lee, Liquid-phase microextraction in a single drop of organic solvent by using a conventional microsyringe, *Analytical Chemistry*, 69 (1997) 4634–40.

12. S. Pedersen-Bjergaard, K.E. Rasmussen, Liquid–liquid–liquid microextraction for sample preparation of biological fluids prior to capillary electrophoresis, *Analytical Chemistry*, 71 (1999) 2650–56.

13. G. Audunsson, Aqueous/aqueous extraction by means of a liquid membrane for sample cleanup and preconcentration of amines in a flow system, *Analytical Chemistry*, 58 (1986) 2714–23.

14. J.A. Jonsson, L. Mathiasson, Supported liquid membrane techniques for sample preparation and enrichment in environmental and biological analysis, *TrAC Trends in Analytical Chemistry*, 11 (1992) 106–14.

15. F. Barahona, A. Gjelstad, S. Pedersen-Bjergaard, K.E. Rasmussen, Hollow fiber-liquid-phase microextraction of fungicides from orange juices, *Journal of Chromatography A*, 1217 (2010) 1989–94.

16. S.P. Huang, S.D. Huang, Dynamic hollow fiber protected liquid phase microextraction and quantification using gas chromatography combined with electron capture detection of organochlorine pesticides in green tea leaves and ready-to-drink tea, *Journal of Chromatography A*, 1135 (2006) 6–11.

17. D.A. Lambropoulou, T.A. Albanis, Application of hollow fiber liquid phase microextraction for the determination of insecticides in water, *Journal of Chromatography A*, 1072 (2005) 55–61.

18. M.A. Farajzadeh, M.R. Vardast, J.Å. Jönsson, Liquid-gas-liquid microextraction as a simple technique for the extraction of 2,4-di-tert-butyl phenol from aqueous samples, *Chromatographia*, 66 (2007) 415–19.

19. R. Romero-González, E. Pastor-Montoro, J.L. Martinez-Vidal, A. Garrido-Frenich, Application of hollow fiber supported liquid membrane extraction to the simultaneous determination of pesticide residues in vegetables by liquid chromatography/mass spectrometry, *Rapid Communications in Mass Spectrometry*, 20 (2006) 2701–8.

20. M. Rezaee, Y. Assadi, M.R. Milani Hosseini, E. Aghaee, F. Ahmadi, S. Berijani, Determination of organic compounds in water using dispersive liquid–liquid microextraction, *Journal of Chromatography A*, 1116 (2006) 1–9.

21. X. Wang, J. Cheng, H. Zhou, M. Cheng, Development of a simple combining apparatus to perform a magnetic stirring-assisted dispersive liquid–liquid microextraction and its application for the analysis of carbamate and organophosphorus pesticides in tea drinks, *Analytica Chimica Acta*, 787 (2013) 71–7.

22. M.R. Khalili Zanjani, Y. Yamini, S. Shariati, J.A. Jönsson, A new liquid-phase microextraction method based on solidification of floating organic drop, *Analytica Chimica Acta*, 585 (2007) 286–93.

23. H. Chen, R. Chen, S. Li, Low-density extraction solvent-based solvent terminated dispersive liquid–liquid microextraction combined with gas chromatography-tandem mass spectrometry for the determination of carbamate pesticides in water samples, *Journal of Chromatography A*, 1217 (2010) 1244–8.

24. E. Turiel, P. Fernández, C. Pérez-Conde, C. Cámara, Trace-level determination of tri-azines and several degradation products in environmental waters by disk solid-phase extraction and micellar electrokinetic chromatography, *Journal of Chromatography A*, 872 (2000) 299–307.
25. V. Pichon, L. Chen, M.C. Hennion, R. Daniel, A. Martel, F. Le Goffic, J. Abian, D. Barcelo, Preparation and evaluation of immunosorbents for selective trace enrichment of phenylurea and triazine herbicides in environmental waters, *Analytical Chemistry*, 67 (1995) 2451–60.
26. A. Martín-Esteban, P. Fernández, C. Cámara, Immunosorbents: A new tool for pes-ticide sample handling in environmental analysis, *Fresenius' Journal of Analytical Chemistry*, 357 (1997) 927–33.
27. B. Sellergren, *Molecularly Imprinted Polymers: Man-Made Mimics of Antibodies and Their Applications in Analytical Chemistry*, 1st ed., Elsevier Science BV, Amsterdam, 2001.
28. F.G. Tamayo, J.L. Casillas, A. Martin-Esteban, Highly selective fenuron-imprinted polymer with a homogeneous binding site distribution prepared by precipitation poly-merisation and its application to the clean-up of fenuron in plant samples, *Analytica Chimica Acta*, 482 (2003) 165–73.
29. M.C. Hennion, P. Scribe, Sample handling strategies for the analysis of organic compounds from environmental water samples, in: D. Barceló (Ed.), *Environmental Analysis: Techniques, Applications and Quality Assurance*, Elsevier Science BV, Amsterdam, 1993, 23–77.
30. A. Martín-Esteban, P. Fernández, C. Cámera, Baker's yeast biomass (Saccharomyces cerevisae) for selective on-line trace enrichment and liquid chromatography of polar pesticides in water, *Analytical Chemistry*, 69 (1997) 3267–71.
31. V. Pichon, L. Chen, M.C. Hennion, On-line preconcentration and liquid chromato-graphic analysis of phenylurea pesticides in environmental water using a silica-based immunosorbent, *Analytica Chimica Acta*, 311 (1995) 429–36.
32. A. Martin-Esteban, P. Fernández, D. Stevenson, C. Cámara, Mixed immunosorbent for selective on-line trace enrichment and liquid chromatography of phenylurea herbicides in environmental waters, *Analyst*, 122 (1997) 1113–17.
33. C.L. Arthur, J. Pawliszyn, Solid phase microextraction with thermal desorption using fused silica optical fibers, *Analytical Chemistry*, 62 (1990) 2145–8.
34. S.A. Barker, A.R. Long, C.R. Short, Isolation of drug residues from tissues by solid phase dispersion, *Journal of Chromatography A*, 475 (1989) 353–61.
35. E.M. Kristenson, L. Ramos, U.A.T. Brinkman, Recent advances in matrix solid-phase dispersion, *TrAC - Trends in Analytical Chemistry*, 25 (2006) 96–111.
36. E. Baltussen, P. Sandra, F. David, C. Cramers, Stir bar sorptive extraction (SBSE), a novel extraction technique for aqueous samples: Theory and principles, *Journal of Microcolumn Separations*, 11 (1999) 737–47.
37. E. Turiel, A. Martín-Esteban, Molecularly imprinted stir bars for selective extraction of thiabendazole in citrus samples, *Journal of Separation Science*, 35 (2012) 2962–9.
38. M. Díaz-Álvarez, E. Turiel, A. Martín-Esteban, Molecularly imprinted polymer mono-lith containing magnetic nanoparticles for the stir-bar sorptive extraction of triazines from environmental soil samples, *Journal of Chromatography A*, 1469 (2016) 1–7.
39. M. Anastassiades, S.J. Lehotay, D. Štajnbaher, F.J. Schenck, Fast and easy multiresi-due method employing acetonitrile extraction/partitioning and "dispersive solid-phase extraction" for the determination of pesticide residues, *Journal of AOAC International*, 86 (2003) 412–31.
40. CEN, Foods of Plant Origin—Determination of Pesticide Residues Using GC–MS and/or LC–MS/MS Following Acetonitrile Extraction/Partitioning and Clean-up by Dispersive SPE—QuEChERS Method, in: EN 15662:2008 Method.

41. AOAC, Pesticide Residues in Foods by Acetonitrile Extraction and Partitioning with Magnesium Sulfate, in: AOAC Official 2007.01 Method.
42. M. Mezcua, O. Malato, J.F. García-Reyes, A. Molina-Díaz, A.R. Fernández-Alba, Accurate-mass databases for comprehensive screening of pesticide residues in food by fast liquid chromatography time-of-flight mass spectrometry, *Analytical Chemistry*, 81 (2009) 913–29.
43. European Commission, Commission Recommendation of 18 October 2011 on the definition of nanomaterial (2011/696/EU), Official Journal of the European Union, L275/38.
44. R. dos Anjos de Jesus, L.F.S. Santos, S. Navickiene, M.E. de Mesquita, Evaluation of metal-organic framework as low-cost adsorbent material in the determination of pesticide residues in soursop exotic fruit (*Annona muricata*) by liquid chromatography, *Food Analytical Methods*, 8 (2014) 446–51.
45. A.S. Barreto, R.L. Da Silva, S.C.G. Dos Santos Silva, M.O. Rodrigues, C.A. De Simone, G.F. De Sá, S.A. Júnior, S. Navickiene, M.E. De Mesquita, Potential of a metal–organic framework as a new material for solid-phase extraction of pesticides from lettuce (*Lactuca sativa*), with analysis by gas chromatography–mass spectrometry, *Journal of Separation Science*, 33 (2010) 3811–16.
46. S. Iijima, Helical microtubules of graphitic carbon, *Nature*, 354 (1991) 56–8.
47. D. Chen, Y.-Q. Huang, X.-M. He, Z.-G. Shi, Y.-Q. Feng, Coupling carbon nanotube film microextraction with desorption corona beam ionization for rapid analysis of Sudan dyes (I–IV) and rhodamine B in chili oil, *Analyst*, 140 (2015) 1731–8.
48. L.M. Ravelo-Pérez, A.V. Herrera-Herrera, J. Hernández-Borges, M.T. Rodríguez-Delgado, Carbon nanotubes: Solid-phase extraction, *Journal of Chromatography A*, 1217 (2010) 2618–41.
49. J. González-Sálamo, B. Socas-Rodríguez, J. Hernández-Borges, M.Á. Rodríguez-Delgado, Nanomaterials as sorbents for food sample analysis, *TrAC - Trends in Analytical Chemistry*, 85 (2016) 203–20.
50. N.L.X. Xu, J. Lv, L. Wang, M. Zhang, X. Qi, L. Zhang, Functionalized multiwalled carbon nanotube as dispersive solid-phase extraction materials combined with high-performance liquid chromatography for thiabendazole analysis in environmental and food samples, *Food Analytical Methods*, 9 (2016) 30–7.
51. L.S.Y. Han, N. Zou, R. Chen, Y. Qin, C. Pan, Multi-residue determination of 171 pesticides in cowpea using modified QuEChERS method with multi-walled carbon nanotubes as reversed-dispersive solid-phase extraction materials, *Journal of Chromatography B*, 1031 (2016) 99–108.
52. R. Qi, H. Jiang, S. Liua, Q. Jia, Preconcentration and determination of pesticides with graphene-modified polymer monolith combined with high performance liquid chromatography, *Analytical Methods*, 6 (2014) 1427–34.
53. H. Tian, J. Li, Q. Shen, H. Wang, Z. Hao, L. Zou, Q. Hu, Using shell-tunable mesoporous Fe_3O_4@HMS and magnetic separation to remove DDT from aqueous media, *Journal of Hazardous Materials*, 171 (2009) 459–64.
54. Z. Xiong, L. Zhang, R. Zhang, Y. Zhang, J. Chen, W. Zhang, Solid-phase extraction based on magnetic core-shell silica nanoparticles coupled with gas chromatography–mass spectrometry for the determination of low concentration pesticides in aqueous samples, *Journal of Separation Science*, 35 (2012) 2430–7.
55. L. Xie, R. Jiang, F. Zhu, H. Liu, G. Ouyang, Application of functionalized magnetic nanoparticles in sample preparation, *Analytical and Bioanalytical Chemistry*, 406 (2014) 377–99.
56. H. Yan, X. Cheng, N. Sun, Synthesis of multi-core–shell magnetic molecularly imprinted microspheres for rapid recognition of dicofol in tea, *Journal of Agricultural and Food Chemistry*, 61 (2013) 2896–901.

57. N. Kumar, N. Narayanan, S. Gupta, Application of magnetic molecularly imprinted polymers for extraction of imidacloprid from eggplant and honey, *Food Chemistry*, 255 (2018) 81–8.
58. S.S. Miao, M.S. Wu, H.G. Zuo, C. Jiang, S.F. Jin, Y.C. Lu, H. Yang, Core-shell magnetic molecularly imprinted polymers as sorbent for sulfonylurea herbicide residues, *Journal of Agricultural and Food Chemistry*, 63 (2015) 3634–45.
59. M.J. Patiño-Ropero, M. Díaz-Álvarez, A. Martín-Esteban, Molecularly imprinted core-shell magnetic nanoparticles for selective extraction of triazines in soils, *Journal of Molecular Recognition*, 30 (2017) article number e2593.
60. Y. Hu, R. Liu, Y. Zhang, G. Li, Improvement of extraction capability of magnetic molecularly imprinted polymer beads in aqueous media via dual-phase solvent system, *Talanta*, 79 (2009) 576–82.

3 Analysis of Pesticide Residues by Chromatographic Techniques Coupled with Mass Spectrometry

Wang Jing, Jin Maojun, Jae-Han Shim, and A.M. Abd El-Aty

CONTENTS

3.1 Overview of Chromatography–Mass Spectrometry 74
3.2 The Basic Theory of Mass Spectrometry 74
    3.2.1 The Working Principle of Mass Spectrometry 74
    3.2.2 Main Performance Parameters Used to Characterize a Mass
        Spectrometer 75
        3.2.2.1 Mass Measurement Range 75
        3.2.2.2 Resolution 76
        3.2.2.3 Sensitivity..76
    3.2.3 The Basic Structure of the Mass Spectrometer 77
        3.2.3.1 Vacuum System..77
        3.2.3.2 Ion Source 77
        3.2.3.3 Mass Analyzer 81
        3.2.3.4 Detector..84
3.3 Chromatography–Mass Spectrometry Techniques...84
    3.3.1 Gas Chromatography–Mass Spectrometry (GC–MS).......................84
        3.3.1.1 Qualitative Analysis by GC–MS 86
        3.3.1.2 Quantitative Analysis by GC–MS 87
    3.3.2 Liquid Chromatography–Mass Spectrometry (LC–MS)...................89
3.4 Other Chromatography–Mass Spectrometry Techniques 91
    3.4.1 Supercritical Fluid Chromatography–Mass Spectrometry
        (SFC–MS) 91
    3.4.2 Comprehensive Two-Dimensional Gas Chromatography–Time
        of Flight Mass Spectrometry (GC×GC–TOFMS)...............................92
3.5 Application of Chromatography–Mass Spectrometry in Pesticide Residue.......93

3.5.1   Determination of Tebuconazole Enantiomers in Water and
        Zebrafish by SFC–MS/MS 96
3.5.2   Multi-Residue Determination of 334 Pesticides in Vegetables by
        GC–MS and LC–MS 97
3.5.3   Comprehensive Suspect Screening of 185 Pesticides and their
        Transformation Products (TPS) in Surface Water Samples Using
        Liquid Chromatography–High Resolution Mass Spectrometry
        (LC–HRMS) 98
3.6   Conclusion and Future Trends 101
References 101

3.1   OVERVIEW OF CHROMATOGRAPHY–MASS SPECTROMETRY

Samples are required to be pure or relatively pure for direct analysis prior to MS detection. The complex nature of food and environmental samples, including overlapping and interferences derived from impurity and debris peaks, makes identification of multiple compounds difficult. So far, chromatography is one of the most common and effective methods in the separation of compounds from complex mixtures. However, chromatography itself possesses a poor qualitative capability. Thence, the combination of chromatographic separation and the qualitative and structural identification capability of mass spectrometry can ensure the accurate analysis of pesticide residues in complex matrices. Therefore, chromatography–mass spectrometry is an ideal combination [1, 2].

There are different kinds of chromatography–mass spectrometry techniques and, for pesticide residue analysis, the main combinations are as follows:

Gas chromatography–mass spectrometry (GC–MS). Owing to different working principles, GC–MS includes gas chromatography–triple quadrupole mass spectrometry, gas chromatography–time of flight mass spectrometry, gas chromatography–ion trap mass spectrometry, etc.

Liquid chromatography–mass spectrometry (LC–MS). Similarly, liquid chromatography–triple quadrupole mass spectrometry, liquid chromatography–ion trap mass spectrometry, liquid chromatography–time of flight mass spectrometry, in addition to multiple different data acquisition modes, have been developed [3]. Other mass spectrometry couplings include matrix assisted laser desorption time of flight–mass spectrometry (MALDITOF–MS), electrostatic field orbit trap–mass spectrometry (Orbitrap–MS), and Fourier transform ion cyclotron resonance–mass spectrometry (FT-ICR MS) [4, 5].

3.2   THE BASIC THEORY OF MASS SPECTROMETRY

3.2.1   THE WORKING PRINCIPLE OF MASS SPECTROMETRY

Mass spectrometers use electromagnetism to separate charged particle ions by their mass-to-charge ratio. The typical mode is to ionize the sample molecules and accelerate them into the magnetic field. Their kinetic energy is correlated to the acceleration voltage and the ion charge z, as stated in the formula below:

$$zeU = \frac{1}{2}mv^2$$

Where z is the charge number, e is the elementary charge ($e=1.60\times10^{-19}$), U is the accelerating voltage, m is the mass of the ion, and v is the motion rate of the ion. When charged ions having a velocity "v" enter through the electromagnetic field of the mass analyzer, they are finally separated by m/z.

According to the working principle of the mass analyzer, mass spectrometers are classified into two categories: dynamic and static. Static electric and/or magnetic fields are used in static mass spectrometers to separate and isolate ions with different m/z in spatial positions, such as single- and dual-focus mass spectrometers. In dynamic mass analyzers, changing the electromagnetic field was used to distinguish ions with different m/z according to time, such as time of flight and quadrupole mass spectrometers.

3.2.2 Main Performance Parameters Used to Characterize a Mass Spectrometer

3.2.2.1 Mass Measurement Range

The mass measurement range of a mass spectrometer represents the range of relative atomic masses (or relative molecular masses) of samples that can be analyzed using mass spectrometry, measured in atomic mass units (u). The atomic mass unit is defined by ^{12}C, i.e., 1/12 of the mass of a ^{12}C neutral atom in the ground state, that is:

$$1u = \frac{1}{12}\left(\frac{12.0000\text{g}/\text{mol}^{12}C}{6.02214\times10^{23}/\text{mol}^{12}C}\right)$$

$$= 1.66054\times10^{-24}\text{g}$$

$$= 1.66054\times10^{-27}\text{kg}$$

In case of inaccurate measurement, the total number of protons and neutrons contained in the nucleus is often used to represent the magnitude of mass, which is equal to the integer of its relative mass.

For most ionization sources, the ions are singly-charged and the mass range is the actual range of relative molecular masses that can be measured. By means of multiple charge techniques, modern mass spectrometers can evaluate samples with relative molecular masses of several hundreds of thousands.

3.2.2.2 Resolution

Resolution refers to the ability of the mass spectrometer to separate two adjacent mass spectral peaks. The general definition is: for two adjacent peaks of equal intensity, when the valley between the two peaks is not more than 10% of their peak height, the two peaks can be separated and the resolution is calculated with the formula below:

$$R = \frac{m_1}{m_2 - m_1} = \frac{m_1}{\Delta m}$$

Where m_1, m_2 are mass numbers, and $m_1 < m_2$. So, the smaller difference between the two peaks, the greater the resolution of the instrument.

In reality, it is sometimes difficult to find two adjacent peaks with equal heights, and their valleys no more than 10% of the peak height. In this case, a single peak may be optionally selected, and the peak width, $W_{0.05}$, is measured at 5% of peak height, which can be regarded as Δm in the formula above. The resolution, in this case, is defined as:

$$R = \frac{m}{W_{0.05}}$$

If the peak is Gaussian, both equations are the same.

The resolution of the mass spectrometer is mainly affected by: (a) the radius or length of the magnetic ion channel; (b) the slit width of the accelerator and collector or ion pulse; and (c) the nature of the ion source.

A mass spectrometer with a resolution of approximately 500 can meet the requirements of general organic analysis. The mass analyzers of such instruments are generally quadrupole mass analyzers and ion traps. For accurate measurement of isotopic mass and organic molecules, high-resolution mass spectrometers with a resolution greater than 10,000 are required. Dual-focusing magnetic mass analyzers are generally used in these types of mass spectrometers. The resolution of such instruments is currently up to 100,000.

3.2.2.3 Sensitivity

Sensitivity refers to the response of a certain mass peak generated by a selected unit sample under specified conditions. Alternatively, it is defined as the amount of sample needed to generate a molecular ion peak with a given signal-to-noise ratio at a certain resolution.

The sensitivity of the mass spectrometer is expressed in terms of absolute, relative, and analytical sensitivity. While absolute sensitivity is defined as the minimum amount of sample that can be detected, relative sensitivity refers to the ratio of large and small components simultaneously detected by the instrument, whereas analytical sensitivity refers to the ratio of true to measured value.

The factors affecting the sensitivity include resolution of the mass spectrometer, scanning mode, chemical noise, and tuning mode.

3.2.3 THE BASIC STRUCTURE OF THE MASS SPECTROMETER

3.2.3.1 Vacuum System

The ion source, mass analyzer, and detector of the mass spectrometer must be operated in a high vacuum environment (the vacuum degree of the ion source should be 10^{-3} to 10^{-5} Pa, and that of the mass analyzer should be 10^{-6} Pa). If the vacuum is low, the following risks can be encountered:

1. A large amount of oxygen will burn out the filament of the ion source.
2. It will increase the background noise and interfere with the mass spectra.
3. It will initiate additional ion–molecule reactions, change the fragmentation mode, and make the mass spectra complicated.
4. It will interfere with the normal adjustment of the electron beam in the ion source.
5. High voltages used to accelerate ions will cause electrical discharges.

Usually, the system is pre-evacuated with a mechanical pump and then pumped efficiently and continuously with a diffusion pump.

3.2.3.2 Ion Source

The role of the ion source is to provide energy to ionize the analyte forming an ion beam with different mass-to-charge ratio (m/z) ions. There are various kinds of ionization sources for mass spectrometers. The main ion sources are as follows:

(1) Electron ionization

Electron ionization (EI) is the most widely used ion source. It is mainly used for ionization of volatile organic samples. The sample is injected as a gas into the ion source. The sample molecules are ionized by colliding with the electrons emitted from the filament. The organic molecules may then form molecular ions. They may also form fragment ions, as chemical bonds may break down. The molecular weight of any compound can be validated from the molecular ion and its structure can be obtained from the fragment ions. The ionization potential of general organic compounds is approximately 10 eV, while the ionization energy commonly used for EI is 70 eV. Thus, sample molecules are ionized to molecular ions and further fragmented to produce abundant fragment ions under such high energy. EI is therefore called a "hard ionization" technology.

Ionization in the ion source is a very complicated process and there are different theories to explain and describe this process. Under the electron bombardment, the sample molecule may form ions in four different ways: (1) one electron is knocked out of the sample molecule to form the molecular ion; (2) the molecular ion undergoes further chemical bond cleavage to form a fragment ion; (3) the molecular ion undergoes structural rearrangement to form rearranged ions; and (4) adduct ions may be generated by reaction of the molecular ion.

The advantages of the EI source are: (1) EI is non-selective ionization, only requiring gasified samples for ionization; (2) it has high ionization efficiency and high sensitivity; (3) EI spectrum provides a wealth of structural information, as it is the "fingerprint spectrum"; and (4) there is a huge standard library for searching because the spectra in the library are all obtained at 70 eV. The spectra are reproducible, and called the classic EI spectra.

The disadvantages of the EI source: (1) the sample must be capable of gasification and EI is not suitable for samples that are difficult to evaporate and thermally unstable; (2) in the EI mode, the molecular ions of some compounds are unstable and easily broken. The information of molecular weight cannot be obtained and the spectrum is complicated; and (3) EI mode only detects positive ions, not negative ions.

(2) Chemical ionization

Molecular ions of some compounds with poor stability are not easily obtained in the EI mode, and thus, the information of molecular weight cannot be obtained. In order to obtain the molecular weight, a chemical ionization (CI) source can be used. CI is easier to control and is considered a "soft ionization" method. It is also mainly used in GC–MS. There is not much difference in structure with EI, or at least the main parts are shared. The main difference is that a reactive gas, which can be methane, isobutane, ammonia, etc., is used during the operation of the CI source. The concentration of the reaction gas is higher than that of the sample molecule. The electrons emitted by the filament first ionize the reaction gas, and then the reaction gas ions perform an ion–molecule reaction with the sample molecules to ionize them. Taking methane as the reaction gas, for example, the process of chemical ionization is described. Under the electron bombardment, methane is first ionized:

$$CH_4 \rightarrow CH_4^+ + CH_3^+ + CH_2^+ + CH^+ + C^+ + H^+$$

Methane ions react with molecules to generate adduct ions:

$$CH_4^+ + CH_4 \rightarrow CH_5^+ + CH_3$$

$$CH_3^+ + CH_4 C_2H_5^+ + H_2$$

Adduct ions react with sample molecules:

$$CH_5^+ + XH \rightarrow XH_2^+ + CH_4$$

$$C_2H_5^+ + XH \rightarrow X^+ + C_2H_6$$

The generated XH_2^+ and X^+ are one H more or less than the sample molecule XH and can be expressed as $(M\pm1)^+$, which is called a quasi-molecular ion. In fact, with methane as the reaction gas, in addition to $(M\pm1)^+$, there may also be $(M+17)^+$, $(M+29)^+$, and a large number of fragment ions.

The advantages of CI: (1) CI is not only an important means of obtaining the information of the molecular weight, but it also can be used for selective detection of different compounds by controlling the reaction and selecting different reagents according to the affinity of ions and electronegativity; and (2) for certain compounds with strong electronegativity (halogen, nitrogen, and oxygen compounds), CI provides negative ions that are not only of good selectivity but also very effective in improving detection sensitivity.

The disadvantages of CI: (1) Like EI, the sample must be capable of gasification. CI is not suitable for samples with poor evaporation and poor thermal stability; (2) the repeatability of CI spectra is not as good as that of EI spectra. For CI spectra, there are only a few dedicated libraries or you can build your own library; and (3) reaction reagents tend to produce higher backgrounds and affect the detection limit.

(3) Field ionization

Field ionization (FI) is also a soft ionization source. An FI source consists of an electrode and a set of focusing lenses, and the electrode with voltages up to several kV forms a strong electric field. When the gaseous sample is introduced into the ionization zone, the gaseous sample molecules are ionized under the action of the strong electric field. The formed ions do not have excessive energy, so the molecular ions usually do not undergo further fragmentation. The ion source vacuum is at 10^{-5} Torr, and the ion–molecule reaction does not occur. If the sample is entrained with salt, ions with Na^+ and K^+ are sometimes produced.

The advantages of FI: compared to EI, FI is a softer ionization method. It only produces molecular ions, almost no fragment ions. The FI spectrum is very clean because there is no reagent background. FI source is suitable for the determination of the molecular weight of polymers and homologues. Combined with high-resolution mass spectrometry, the elemental composition and the molecular formula of the compound can be obtained. FI source is therefore very advantageous for the identification of the compound.

The disadvantages of FI: Like EI and CI, FI is not suitable for analyzing samples that are difficult to gasify and thermally unstable. Compared with EI and CI, the sensitivity of FI is lower. Besides, the high voltage of FI source is prone to discharge reactions.

(4) Fast atom bombardment

The method of ionizing organic compounds homogenized with glycerol (substrate) and coated on a metal surface (target surfaces) by bombarding accelerated neutral atoms (fast atoms) is called fast atomic bombardment (FAB). FAB is an ionization source developed in the mid-1980s and is a soft ionization technology.

Argon gas produces argon ions through the discharge in the ionization chamber. The high-energy argon ions get a high-energy argon atom stream by charge exchange. The argon atom hits the sample placed on a target surface coated with a substrate (such as glycerin), and the sample molecules are ionized. Sample ions enter a vacuum chamber and then enter the analyzer under the action of the electric field. It is not necessary to vaporize them during ionization, so FAB is suitable for the analysis of samples with large molecular weight, difficult gasification, and poor thermal stability, such as peptides, oligosaccharides, natural antibiotics, and organometallic complexes. The mass spectrum obtained by the FAB source not only has strong excimer ion peaks, but also has abundant structural information. However, it is very different from the mass spectrum obtained by the EI source. First, its information of molecular weight is not obtained from the molecular ion peak M, but often from the quasi-molecular ion peak, such as $(M+H)^+$ or $(M+Na)^+$. Second, there are fewer fragmentation peaks than in the EI spectrum. The FAB source is mainly used for magnetic double-focus mass spectrometers. Due to the appearance of electrospray source and laser desorption ionization source, the importance of the FAB source has been greatly reduced.

(5) Electrospray ionization

Electrospray ionization (ESI) source is an ionization mode mainly used in liquid chromatography–mass spectrometry. It serves as both an interface between the liquid chromatograph and the mass spectrometer and as an ionization device. Its main component is an electrospray nozzle consisting of a two-layer casing. The inner layer of the nozzle is the liquid chromatography effluent, while the outer layer is the atomization gas, which is usually nitrogen gas with a large flow. The liquid sample is dispersed into droplets with the aid of the atomization gas. In addition, there is an auxiliary gas nozzle at the oblique front of the nozzle. The role of the auxiliary gas is to quickly evaporate the solvent of the droplet. The charge density on the surface of droplets gradually increases during evaporation, and when it reaches a critical value, ions can evaporate from the surface. Ions pass through the sampling hole and enter the analyzer by means of the voltage between the nozzle and the cone hole. The voltage applied to the nozzle can be positive or negative. By adjusting the polarity, a positive or negative ion mass spectrum can be obtained.

ESI is a soft ionization technique that produces quasi-molecular ions. Even compounds with large molecular weight and poor stability will not decompose during ionization. ESI has a high sensitivity for most compounds. It is suitable for the study of polar compounds and compounds with high molecular weight. ESI source is the most successful interface technology for liquid chromatography–mass spectrometry.

(6) Atmospheric pressure chemical ionization

The atmospheric pressure chemical ionization (APCI) source has a similar structure to the ESI source, except that an acicular discharge electrode is disposed downstream from the APCI nozzle. The neutral gas molecules in the air are ionized by the high-voltage discharge of the discharge electrode. Ionization produces H_3O^+, N_2^+, O_2^+, and O^+. In addition, the solvent molecules are also ionized. The sample molecules are ionized by the ion–molecule reactions with these ions from the solvent. Proton transfer and charge exchange produces positive ions whereas proton detachment and electron capture produce negative ions.

APCI is a type of soft ionization technology that produces quasi-molecular ions. APCI mainly produces single-charged ions, with few fragment ions, so the molecular weight of the analyte is generally less than 1000 Da. Compared to ESI, APCI is less affected by the matrix and is suitable for analysis of less polar compounds. Some analytes cannot generate enough strong ions in the ESI source due to their structure and polarity, so the APCI source can be used to increase the ion yield. It can be considered that the APCI source is a supplement of ESI and is also mainly used for liquid chromatography–mass spectrometry.

(7) Laser desorption

Laser desorption (LD) is an ionization method that uses a pulsed laser with a certain wavelength to irradiate and ionize the sample. The sample, placed on a substrate-coated target surface, is irradiated by the laser light. The substrate molecules absorb and transmit the laser energy, and ionize the sample molecules. The laser ionization source needs a suitable substrate to get a good ion yield. Therefore, this ionization source is often referred to as a matrix assisted laser desorption ionization (MALDI) source. MALDI is particularly suitable for time-of-flight mass spectrometers (TOFs). MALDI belongs to the category of soft ionization technology, mainly producing molecular ions, quasi-molecular ions, and fewer fragment ions and multi-charged ions. It is more suitable for the analysis of biological macromolecules. The common substrates used in MALDI are 2,5-dihydroxybenzoic acid, sinapic acid, nicotinic acid, and α-cyano-4-hydroxycinnamic acid.

3.2.3.3 Mass Analyzer

The function of the mass analyzer is to separate and detect the ions generated by the ion source according to their mass-to-charge ratio, m/z, to obtain the characteristic mass information of the compound. The mass analyzer is located between the ion source and the detector. The main types are single quadrupole, ion trap, triple quadrupole, magnetic sector, time-of-flight, and Fourier transform mass analyzers.

(1) Single quadrupole (Q)

The single quadruple mass analyzer is composed of four positive and negative sets of cylindrical or hyperbolic columnar electrodes, which are strictly parallel and are equally spaced from the central axis. DC and RF voltages are applied to the electrodes to generate a dynamic electric field, that is, a quadrupole field. After the ion enters this RF field, it will be affected by the electric field force, and only the proper m/z ion with a stable oscillation will enter the detector. For each moment of the voltage change, only one mass-to-charge ratio of ion can pass, so it is known as a "mass filter".

The resolution and m/z range of Q is roughly the same as that of a magnetic analyzer, with an ultimate resolution of 2000, typically about 700. Q is the most common mass analyzer used in GC–MS. It has two different scan modes: full scan and selected ion monitoring (SIM). The scanning speed is fast and the sensitivity is high. In particular, the SIM mode, with its maximum collection efficiency, selectively detects single or several mass ions, thereby reducing the signal-to-noise ratio and increasing the sensitivity. It is particularly suitable for various quantitative analyses to meet the needs of high-throughput analysis.

(2) Ion trap

The ion trap mass analyzer is a commodity instrument introduced in the 1980s. It consists of a ring electrode and two end cap electrodes that form a three-dimensional quadrupole field.

Unlike the Q, the movement of ions in the ion trap is controlled in three directions (x, y, z), while the Q controls only two directions (x, y). The Mathieu equation is also available to describe the movement of ions. Ions can exist steadily in the stable region, and maintain certain amplitude. In the unstable region, the ions rapidly increase in speed and hit the electrode, and disappear. By adding a negative pulse to the extraction electrode, stable ions in the "well" can be extracted and detected by the detector. That is, the ions are first stored in the "well", and then the electric field is changed to push the ions out of the "well" at different mass-to-charge ratios for detection.

Compared with other mass spectrometers, ion traps have a small size, a simple structure, and especially a low price. Therefore, ion traps are common instruments for qualitative analysis that requires the use of a multi-stage mass spectrometric method, and they are widely used in proteomics and the drug metabolism analysis field. For the qualitative analysis of GC–MS, EI ionization can provide abundant structural information and spectral library search. The advantage of ion trap multistage mass spectrometric analysis is not prominent. In the quantitative analysis, no matter what the detection limit, linear range, or stability is, the recognized quadrupole mass spectrometer is slightly better than ion trap.

(3) Triple quadrupole (Q-q-Q)

The triple quadrupole is a mass analyzer composed of three sets of quadrupoles. The first and third sets of quadrupoles are mass analyzers, and the middle set of quadrupoles is a collision activation chamber. It has several scanning modes, including product ion mode, precursor ion mode (also known as parent ion mode), neutral loss mode, and multiple reaction monitoring (MRM). The first three scanning modes are mainly used for the structural analysis of compounds by studying the fragmentation pathway of ions and the attribution of each ion. The MRM model is mainly used for quantitative analysis. It has better selectivity, stronger interference rejection ability, and a lower detection limit than the SIM model of the single quadrupole mass analyzer. The triple quadrupole mass spectrometer is commonly used in GC–MS/MS and LC–MS/MS. It is useful for the study of organic structure, and can also be used for the direct identification of mixed organics. It has a wide range of applications in pesticide multi-residue analysis.

(4) Magnetic sector

In the past few decades, the magnetic sector mass analyzer has always been an advantage in the application of high-resolution mass spectrometry, especially in organic analysis. It has many functions including mass dispersion, energy dispersion, and direction focusing. The ions generated in the ion source are accelerated by a high voltage of several kilovolts and pass through an electric field and magnetic field with a certain radius of curvature. The radius of curvature of the motion orbit depends on the momentum and the mass-to-charge ratio of the ions, the accelerating voltage, and the electric and magnetic field strengths. The ions are separated at varying electrical, magnetic, or accelerating voltages and arrive at the detector to be detected. Different combinations of electric and magnetic fields have different scanning modes, making it a mass analyzer with multiple MS/MS functions. One of the biggest advantages is that more structural information can be obtained when the magnetic sector mass analyzer with high collision energy is used. Although other mass analyzers have replaced the magnetic sector mass analyzer in many applications, they still have advantages in high-resolution applications and high-energy collision MS/MS applications. Its dynamic range and reliability make it competitive in quantitative analysis of dioxins and excited tracers.

(5) Time of flight

The time-of-flight (TOF) mass analyzer is developing rapidly due to great technological breakthroughs achieved since the 1990s. It has a promising application thanks to its fast scanning speed, extremely high ion collection efficiency, wide mass range, and high resolution (>10,000).

The linear coaxial TOF mass analyzer consists of a field-free flight tube. The ion beam is accelerated by high pressure to push the ion into the flight tube in a pulsed manner, and reaches the detector in "free drift" style. Different accelerations are obtained due to different ion masses. The ion with smaller mass has higher velocity than the ion with higher mass. The time it takes for the ion to arrive in the detector is related to the ion mass. Different ions achieve mass separation through the flight tube. Since TOF does not theoretically have an upper mass limit, the importance of analytical applications in compounds with high molecular weight (such as biomacromolecules and polymers) is undoubted. TOF is also an ideal analyzer for high-speed and high-efficiency separation. Due to the use of delayed extraction of ions, reflectors, and fast electronics, TOF offers high resolution and high-quality accuracy. At present, TOF is mainly used in the field of biomass spectrometry, and it is not widely used compared with quadrupole and ion trap mass spectrometry.

(6) Fourier transform ion cyclotron resonance

The Fourier transform ion cyclotron resonance analyzer (FTICR) uses chirp pulses to excite ions and performs fast frequency sweeps in a short period of time, resulting in exciting ions with a wide range of mass-to-charge ratios almost simultaneously. So, the scanning speed and sensitivity are much higher than those of ordinary cyclotron resonance analyzers. The FTICRMS signal depends on the number of ions. An image current is formed on the detection plate of the analysis cell. The image current is a sinusoidal time domain signal. The frequency of the sine wave is the same as the inherent cyclotron frequency of the ion, and the amplitude is proportional to the number of ions in the analysis chamber. If the ions of various masses in the analysis chamber satisfy the resonance conditions, the actually measured signal is a superposition of sine wave signals corresponding to various ions that perform coherent orbital motion at the same time. The measured time domain signals are repeatedly accumulated, amplified, and converted to an analog–digital signal that is then inputted to a computer for fast Fourier transform, so various frequency components can be detected and the common spectrum can be obtained.

The mass spectrometer made by the principle of FTICR is called Fourier transform ion cyclotron resonance mass spectrometer (FTICRMS), abbreviated as Fourier transform mass spectrometer (FTMS). FTMS has high resolution (>10^6) and measurement accuracy (reaching a few parts per million). Moreover, the FTMS analysis has a high sensitivity, and its sensitivity is four orders of magnitude higher than that of the ordinary cyclotron resonance mass spectrometer. High resolution can be obtained even at high sensitivity. In addition, FTMS also has the advantages of multi-stage mass spectrometry, fast scanning speed, stable and reliable performance, and a wide mass range, and it can be connected to any ion source to broaden the instrument function. However, FTMS requires liquid helium due to the high super-conducting magnetic field, which makes the instrument price and operating costs more expensive.

3.2.3.4 Detector

Mass spectrometric detection mainly uses electron multipliers, and some use pho-tomultiplier tubes. The ions from the quadrupole strike the high-energy dynodes to generate electrons. The electrons generate electrical signals through the electron multiplier. The signals of different ions are recorded to obtain the mass spectrum. The gain of the signal is related to the voltage of the multiplier. Increasing the volt-age of the multiplier can increase the sensitivity, but at the same time, it will reduce the life of the multiplier. Therefore, the multiplier voltage should be as low as pos-sible while ensuring the sensitivity of the instrument. The electrical signals from the multiplier are sent to the computer for storage. These signals can be then processed to obtain chromatograms, mass spectra, and other information.

3.3 CHROMATOGRAPHY–MASS SPECTROMETRY TECHNIQUES

Mass spectrometry (MS) can be used for effective qualitative analysis. However, it is not appropriate for the analysis of complex organic mixtures. Chromatography is an effective method for separation and analysis of organic compounds, and is particu-larly suitable for the quantitative analysis of organic compounds, however, qualita-tive analysis is difficult. Therefore, the effective combination of mass spectrometry and chromatography will provide an efficient qualitative and quantitative analysis tool for complex mixtures. The coupling of chromatography and mass spectrometry includes gas chromatography–mass spectrometry (GC–MS), gas chromatography–tandem mass spectrometry (GC–MS/MS), liquid chromatography–mass spectrom-etry (LC–MS), and liquid chromatography–tandem mass spectrometry (LC–MS/MS), among others.

3.3.1 Gas Chromatography–Mass Spectrometry (GC–MS)

Gas chromatography (GC) is the separation unit of the GC–MS. The sample is injected into the GC, and gasified in the inlet, then transferred to a chromatographic column after injection. The partition coefficient of the compounds in the two chro-matographic phases (stationary and mobile phases) are different, leading to the sepa-ration of the target compounds in the column that are eluted by the carrier gas. Then, the gaseous compounds are ionized by an ion source in the GC–MS interface, becoming an ion beam composed of different mass ions in a high vacuum state that enters a mass analyzer and an ion detector. The ions are separated in the mass analyzer, and the signal for each chromatographic peak is detected and amplified to obtain the total ion-current chromatogram.

The advantages of the GC and MS combination are as follows:

1. GC is an ideal "sampler" for MS. After being separated by GC, a pure substance enters into the MS and the expertise of MS can be fully utilized.
2. MS is an ideal "detector" for GC. The detectors used in GC, such as hydrogen flame ionization detectors, thermal conductivity detectors, electron capture detectors, etc., have their own limitations, but MS can detect almost all compounds with high sensitivity.

Therefore, the combination of both technologies not only exerts the high separation ability of the GC, but also exerts the accurate discrimination ability of MS. GC–MS is suitable for the qualitative identification of unknown components in multi-component mixtures. The molecular structure of compounds can be judged, the relative molecular mass of unknown components can be accurately determined, the misinterpretation of chromatographic analysis can be corrected, and the substance can be identified even for chromatographic peaks that are not completely separated.

The key of GC–MS is the interface device because the outlet of the GC column is usually at atmospheric pressure, whereas MS works under high vacuum. Therefore, a device is necessary for connecting GC and MS to transfer the sample and match the working pressure and working flow of the two parts. For packed column gas chromatography, due to the large carrier gas flow rate, a molecular separator must be used as an interface to separate the carrier gas from the sample to match the pressure. A jet-type molecular separator is a typical example. The gas flow at the outlet of the column is sprayed to the vacuum chamber with a certain pressure, by means of an ultrasonic expansion spray through a narrow nozzle hole, which produces a diffusion effect at the outlet end of the nozzle. The square root of the relative molecular mass is inversely proportional to the diffusion rate. The small mass of the carrier gas (helium, in the GC–MS instrument) is diffused in large quantities and is pumped out by the vacuum pump; the constituent molecules diffuse more slowly because of their much greater mass and most of them enter the mass spectrometer according to the original direction of motion, so the effect of separating the carrier gas and concentrating the components can be achieved. A double nozzle splitter can be used to improve efficiency.

Helium (He) is commonly used as a carrier gas in GC–MS, the reasons are as follows:

1. The ionization potential of He is 24.6 eV, which is the highest for carrier gases (H2 13.6 eV, N2 15.8 eV). It is difficult to ionize and does not affect the baseline of the chromatogram due to unstable gas flow.
2. The relative molecular mass of He is only four and is easily separated from other components. On the other hand, its mass spectral peak is very simple, mainly at $m/z=4$, which does not interfere with the subsequent mass peaks.

In GC–MS, information data for hundreds to thousands of mass ion currents per second are available, so a computer system is an important and necessary component to collect and process large amounts of data.

Derivatization methods are often used in GC–MS. Derivatization of functional groups such as hydroxyl, amine, and carboxyl groups often plays an important role in analyzing actual samples by GC–MS. There are some main benefits:

1. It improves the GC properties of the analyte. The GC properties of some polar groups in the analytes (like hydroxyl, carboxyl) are not good, leading to no peaks in common chromatograms. After derivatization, the situation is improved.
2. It improves the thermal stability of analyte. Some analytes have insufficient thermal stability and decompose easily or change during the vaporization and the chromatographic separation. After derivatization, the analytes are quantitatively converted to compounds that are stable under GC–MS conditions.
3. It changes the molecular mass of the analyte. Most of the derivatives of analytes have an increased molecular weight, which helps to separate the analytes from the substrate and reduce the influence of background chemical noise.
4. It improves the MS behavior of the analyte. In most cases, derivatized analytes produce regular and easily interpretable mass fragments.
5. The introduction of a halogen atom or an electron-withdrawing group allows the analyte to be detected by chemical ionization. In many cases, the detection sensitivity can be improved and the molecular weight of the analyte can be detected.
6. Some chiral compounds, which are difficult to separate, can be resolved through some special derivatization methods. The most used derivatization methods are silylation, acylation, and alkylation.

Due to the unique advantages of GC–MS, it has been widely used. In general, most of the samples that can be analyzed by GC can be qualitatively and quantitatively determined by GC–MS. Food and environmental analysis are the most important areas of GC–MS applications. GC–MS analytical methods for pesticide residues in fruits and vegetables, grain, water (surface water, wastewater, drinking water, etc.), soils, etc., have been adopted by many countries and have formed or will form a series of statutory or recognized standard methods.

3.3.1.1 Qualitative Analysis by GC–MS

Qualitative analysis using GC–MS is an important prerequisite for accurate quantitative analysis. GC is mainly used for the separation of targets, whereas MS has a strong qualitative ability. MS can provide a great deal of information regarding molecular structure, so MS is a powerful tool for the identification of compounds. Qualitative analysis by GC–MS is carried out for the determination of relative molecular mass, the determination of molecular formulas, and the identification of structures.

1. *Determination of relative molecular mass*

   The relative molecular mass of a compound can be accurately determined by the mass-to-charge ratio data from the molecular ion peak. Therefore, it is important to accurately identify the molecular ion peak.

Although the peak at the highest mass theoretically can be considered the molecular ion peak, with the exception of the isotope peak, sometimes this peak is not observed due to poor molecular stability.

2. *Determination of molecular formulas*

The mass-to-charge ratio of molecular or fragment ions can be accurately determined by high-resolution MS, so the elemental composition can be calculated using the exact mass and abundance ratio of the element. For compounds with relatively small molecular masses but strong molecular ion peaks, their chemical formulae can be deduced by relative isotope abundance on a low-resolution MS. Due to the different numbers of atoms in diverse samples, different numbers of isotopic peaks and a group of ion clusters with characteristic patterns are formed. Not only do the molecular ion peaks but also the fragment ion peaks have their corresponding isotopic peaks, which is a powerful basis for identifying spectra.

3. *Identification of structures*

Based on the mass of fragment ion peaks, various possible fragmentation patterns are applied and the molecular structure before fragmentation is deduced. In general, even-numbered molecular ions take off an odd number of neutral-free-radical-generated-fragment peaks, whereas odd-numbered molecular ions get even-numbered fragments.

Qualitative analysis using GC–MS must be confirmed by the following conditions: the mass spectrum of the unknown matches with the mass spectrum of a standard sample in the same instrumental conditions; the relative retention time of the unknown and the standard sample is the same under the same GC conditions; and the obtained spectrum should match the spectrum in the standard library. At the same time, it can also be matched with the spectra reported by the authoritative literature.

GC–MS full scan mode is generally used for qualitative analysis of unknown substances. Full scan mass spectra can be regarded as the fingerprint of the measured compound. It can be qualitatively determined by computer and by spectral analysis.

3.3.1.2 Quantitative Analysis by GC–MS

The quantitative analysis characteristics of GC–MS are qualitative firstly and then quantitative. A compound is first identified based on its retention time and mass spectrometric characteristic ions confirming to the target compound and then quantified, thus avoiding false positive detection. GC–MS quantification generally does not use total ion chromatograms. Instead, ion maps of characteristic ions are used because they are relatively stable and undisturbed, resulting in more reliable quantitative results.

Quantitative analysis is carried out to determine the exact content of each component in the tested sample. No matter what instrument is used, the content of the target compound is calculated based on the functional relationship between the response factor of the detector and the content of the test compound under certain conditions. The common quantitative methods for GC–MS are the normalized method, the external standard method, and the internal standard method.

(1) Normalized method

The sum of all the components in the sample is taken as 100, and the relative percentage of each component is calculated, which is called the normalization method. The formula to calculate the content of each component is as follows:

$$\omega i = \frac{f_i A_i}{\sum f_i A_i}$$

where:
    ωi is the component content
    A_i is the component peak area and
    f_i is the component correction factor.

The advantages of the normalization method: (1) it is not necessary to know the injection volume, especially when the injection volume is small and it cannot be measured accurately; (2) this method is robust, and slight changes in the instrument and operating conditions have little effect on the results; (3) it is more convenient than internal standard method, especially to analyze multiple components; and (4) if the correction factors of the components are similar or the same, it is not necessary to use a correction factor, and the area or peak height are directly normalized.

The disadvantages of the normalization method: (1) all components must flow out of the column and produce response signals and measured peak area; (2) the area of peaks that are not needed to be quantified also have to be measured; and (3) the correction factor for all components needs to be measured, otherwise this method cannot be applied.

(2) External standard method

Different concentrations of a standard sample are prepared. The peak area or peak height of each compound of the standard sample is measured under the same operating conditions as the sample to be tested. The response factor (*fi*) is obtained. A standard curve is obtained by representing the chromatographic response at different concentrations. The slope of the standard curve is the absolute correction factor. The content of the tested compound can be directly calculated from the linear equation of the standard curve.

(3) Internal standard method

The external standard method has some shortcomings, and each analytical condition will result in errors from sample handling to testing. In order to overcome unavoidable measurement errors, an appropriate reference substance (internal standard compound) is added to the standard sample and the test sample, and then the ratio of the response values of the test compound and the internal standard compound (referred to as relative response factor) is calculated. The quantitative method by relative response factor and the amount of compound added to the internal standard is called the internal standard method.

$$\frac{f_i}{f_s} = \frac{\omega i}{\omega s} \cdot \frac{A_s}{A_i}$$

where:

ωi   is the amount of the standard compound

Ai   is the peak area (or peak height) of the standard compound

ωs   is the amount of the internal standard compound and

As   is the peak area of the internal standard compound.

The requirement for the internal standard is that the purity is high and its structure is similar to the component to be tested. The internal standard peak is close to the component peak but can be well separated. The concentration of the internal standard and the measured component are close to each other.

The advantage of the internal standard method is that it is quantitatively accurate. When the injection volume and the instrument response value change, the response values ratio of the test compound to the internal standard in the same sample does not change. Therefore, the internal standard method can eliminate systematic errors due to fluctuations in the injection volume and instrument response values. The disadvantage of the internal standard method is that it is difficult to select an appropriate internal standard. It is necessary to weigh the internal standard and sample accurately each time, and the difficulty of chromatographic separation is increased.

3.3.2 Liquid Chromatography–Mass Spectrometry (LC–MS)

GC–MS is not appropriate for analyzing samples with poor thermal stability or uneasy vaporization. However, LC–MS is suitable for these types of samples. LC–MS combines the separation capacity of liquid chromatography with the molecular weight and structural information provided by mass spectrometry. LC–MS is an effective technique to analyze and separate complex organic mixtures, especially those of high molecular weight.

The key device of the LC–MS system is the interface between the LC and the MS. The main function of the interface device is to remove the solvent and ionize the target analyte. For the ionization, the electron ionization, chemical ionization, and other classic methods are not suitable for not very volatile and thermally unstable compounds. Since the 1980s, LC–MS technology has made great achievements with the development of the atmospheric pressure chemical ionization (APCI) and the electrospray ionization (ESI) interfaces. The use of 3 μm particulate stationary phase and narrow-diameter columns in LC improves column efficiency and greatly reduces mobile phase flow. All these advances have promoted the progress of LC–MS.

The analytes are ionized under atmospheric pressure and the ions are then transferred to a mass analyzer for mass spectrometric analysis. Due to ionization at room temperature, there is no pyrolysis of the sample. There are a variety of ionization methods, and the most commonly used are discussed below.

(1) Electrospray ionization interface (ESI)

LC–ESI-MS is the fastest growing and most widely used interface in LC–MS technology. The sample is injected into the column by a six-way valve. After the chromatographic separation, the LC effluent flows through the metal capillary nozzle, and a voltage of 3 to 8 kV is applied between the capillary and the counter electrode to form a highly dispersed effluent (sample solution) fan spray.

The ions formed under atmospheric pressure conditions are driven through a dry N_2 gas curtain into the vacuum zone of a mass spectrometer (quadrupole or TOF) by a potential difference (there is also a differential pressure effect). The role of the curtain gas is to further disperse the droplets to facilitate the evaporation of solvents and to block neutral solvent molecules, allowing ions to pass through a voltage gradient into the mass spectrometer. The gas curtain can increase the probability of collision of the molecules with the gas promoting their decomposition. The collision may also induce ion fragmentation that provides structural information of the compound. ESI technology is so far the most moderate ionization method. Spectrograms mainly provide information related to quasi-molecular ions. An important feature is that a large number of multiply-charged ions can be generated. Therefore, the molecular mass of macromolecular compounds can be determined; a maximum relative molecular mass of 200,000 can be reached.

(2) Atmospheric pressure chemical ionization interface (APCI)

The APCI interface widely used in LC–MS is called a heated pneumatic nebulizer interface. The LC effluent is injected by the atomization gas and the auxiliary gas into a heated normal pressure environment (100~120°C) through a central capillary. The droplets formed by the thermal spraying can evaporate into the gas phase, but the amount of analyte ions that evaporate directly into a gaseous state is insufficient to give mass spectrometric signals. Therefore, in APCI, the ionization of the analyte is mainly accomplished through a chemical ionization pathway. A needle-shaped corona discharge electrode is placed near the nozzle, and some neutral molecules in the air are ionized to generate abundant N^{2+}, O^{2+}, and O^- through its high-voltage discharge. When the sprayed aerosol mixture approaches the discharge electrode, a large number of solvent molecules will also be ionized. The aforementioned large number of ions and the analyte molecules undergo gaseous ion–molecule reactions to form proton transfer, adducts, and other molecular ions.

APCI produces mainly singly-charged ions and the compounds typically analyzed have a relative molecular weight of less than 1000. APCI is mainly used for the analysis of moderately polar compounds. Some analytes cannot produce strong enough ions with ESI because of their structure and polarity. APCI can be used to increase the ion yield so it can be considered a supplement of ESI.

Recently, the ultra-high performance liquid chromatography (UPLC) technology has been developed based on high performance liquid chromatography (HPLC). In comparison to conventional HPLC, UPLC not only has higher separation efficiency but also increases greatly the speed of analysis and the sensitivity of detection. UPLC technology has brought separation technology to a new stage. Ultra-high performance liquid chromatography–tandem mass spectrometry (UPLC–MS/MS) not only has the advantage of high column efficiency, rapid analysis, and low matrix interference, but also has high detection sensitivity and specificity to satisfy the trace analysis of targets in the sample.

In addition, it is also possible to determine the molecular weight of the compound as well as information on the abundance fragments in order to confirm the structure of the analyzed compound, which greatly improves the qualitative and quantitative analysis capabilities and application areas. UPLC–MS/MS has wide applications and can be used in food analysis, quality control, pesticide residue analysis, and environmental monitoring analysis.

3.4 OTHER CHROMATOGRAPHY–MASS SPECTROMETRY TECHNIQUES

3.4.1 SUPERCRITICAL FLUID CHROMATOGRAPHY–MASS SPECTROMETRY (SFC–MS)

SFC is a chromatographic technique that uses a supercritical fluid (SF) as the mobile phase. Supercritical carbon dioxide is the most frequently used mobile phase for SFC, because carbon dioxide can be easily converted to its supercritical state (critical temperature, $31.1°C$; critical pressure, 7.38 MPa). With the addition of modifiers and additives, rapid and efficient separation can be achieved. SF not only has the density and dissolution properties of a liquid, but also similar low-viscosity and high-diffusion properties to a gas. If an SF is used as the mobile phase in the chromatographic separation process, the high temperature of GC to analyze thermal unstable compounds can be avoided and faster analysis rates than HPLC for non-volatile and high-molecular compounds analysis can be achieved. In addition, SFC shows a superiority in the separation of enantiomers using packed columns. Through a certain interface technology, SFC–MS could supplement the GC–MS and HPLC–MS analyses.

The combination of SFC and MS is straightforward, since CO_2 is highly volatile so the SFC effluent can easily be converted into a gas phase during the ionization process. Similar to LC–MS, the most popular ionization sources for SFC–MS are ESI, APCI, and atmospheric pressure photoionization (APPI). But some specific conditions for interfacing SFC and MS may be required to enhance ionization and achieve a stable baseline.

The designed interfaces for hyphenating SFC and MS should be able to manage the compressibility of the SFC mobile phase and to preserve as much of the chromatographic separation integrity as possible. So far, four main types of interfaces successfully used for combining SFC and MS are as follows: direct coupling interface, pre-UV-BPR-split interface, pressure regulating fluid interface, and pre-BPR splitter with make-up pump interface. The last interface is the most widely used nowadays on modern SFC–MS instruments. The specific principle is shown in Figure 3.1. In this interface, the UV detector is located just after the column outlet. This interface is composed of two zero-dead volume T-unions located between the UV and MS detectors. The upstream T-union allows the addition of CO_2 miscible make-up liquid delivered by an isocratic pump. This make-up liquid is then mixed with the chromatographic effluent to enhance the ionization yield at low percentages

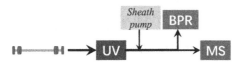

FIGURE 3.1 Schematic representations of pre-BPR splitter with sheath pump interface. (Reproduced from Desfontaine et al., 2017, with permission from Elsevier [6].)

of MeOH and to avoid analyte precipitation during CO_2 decompression. The downstream T-union acts as a flow splitter. Indeed, a fraction of the total flow is directed toward the MS, while the rest of the flow is directed to the back pressure regulator (BPR). Thanks to the BPR placed downstream of the SFC column, the chromatographic integrity is maintained. This interface is known as pre-BPR splitter with make-up pump interface. A good solution to neutralize the cooling effect of the eluant expansion at the column outlet is to heat the SFC–MS interface up to a temperature of 60°. Heating of the interface decreases peak broadening observed in SFC–MS, because of the cooling that accompanies decompression of the mobile phase [7]. This "pre-BPR-split + make-up" interface was reported as the most reliable in terms of retention time and peak area reproducibility. Due to its high sensitivity, linearity, and robustness, this configuration is therefore recommended for qualitative and quantitative analyses [8].

With the continuous development of technology, an ultra-high performance supercritical fluid chromatography (UHPSFC) system that possesses a faster separation rate integrating SFC and UPLC has been commercialized. UHPSFC coupled with MS has been widely used in various fields, such as food safety, environmental monitoring, medicine, and separation of chiral compounds. In particular, it is a powerful tool for the analysis and detection of complex matrices such as food.

3.4.2 COMPREHENSIVE TWO-DIMENSIONAL GAS CHROMATOGRAPHY– TIME OF FLIGHT MASS SPECTROMETRY (GC×GC–TOFMS)

Comprehensive two-dimensional gas chromatography (GC×GC) has been proven to be a powerful tool for the separation and analysis of complex systems because the information of GC×GC can be used to reveal the chemical composition of the components. However, qualitative analysis of complex chemical systems requires structural information provided by spectroscopic detectors such as MS. The two-dimensional column of GC×GC is equivalent to a fast chromatography, and the typical peak width of the column is 0.1–0.6 s. This requires that the frequency of the connected detector is fast enough. The acquisition frequency of conventional quadrupoles is too slow to meet the requirements of coupling with GC×GC. The time of flight mass spectrometry (TOFMS) enabled GC×GC–TOFMS to be applied to the identification of complex systems.

GC×GC–TOFMS is characterized by rapid response, good sensitivity, and high acquisition frequency (spectral collection rate is up to 1–500 full-range mass spectra/ second) without analyte mass-range limit. TOFMS also offers sufficient MS data to address the requirements of any comprehensive GC×GC separation for qualitative analysis of an unknown compound. Therefore, GC×GC–TOFMS is suitable for qualitative and quantitative analyses of complex mixture systems.

In the last number of years, analytical approaches employing gas chromatography coupled to time of flight mass spectrometry (GC–TOF MS) have proved to be a useful tool in the assessment of quality and safety of food and environmental matrices. Two-dimensional gas chromatograph-time of flight mass spectrometers have been used in the determination of pesticides in recent years. The application of GC×GC–TOFMS for the multi-residue analysis of pesticides is listed in Table 3.1.

TABLE 3.1

Multi-Residue Pesticide Analytical Methods Developed by GC×GC–TOFMS Technology

Pesticide	Sample	Recovery (%)	LOD (μg/kg)	LOQ (μg/kg)	Reference
34 organochlorine and pyrethroids	Fruits	60.1–110.5			[7]
29 organophosphorus, 27 organochlorine and eight carbamate pesticides	Brassica stalks, cucumbers, eggplant, and chili	68.3–117.8	0.010 – 6.032	0.035–20.107	[8]
423 pesticides, isomers, and pesticide metabolites	Tea	81.6–113			[11]
Cypermethrin, permethrin, chlorpyriphos, metalaxyl, etophenprox, and up to 160 pesticides	Grape and wine	70–120			[12]
Chlorpyrifos-methyl, vinclozoline, parathion-methyl, heptachlor, myclobutanil, buprofezin, flusilazole, and oxyfluorfen	Grape	70–110			[13]

3.5 APPLICATION OF CHROMATOGRAPHY–MASS SPECTROMETRY IN PESTICIDE RESIDUE

Pesticides have played an important role in the prevention and control of diseases, in the control of insects and weeds in agricultural production, and in promoting agricultural production and ensuring food security. However, due to the long-term abuse of pesticide application, the presence of their residues, especially in food and environmental samples, has caused serious problems to human health and the safety of the environment. Therefore, the great importance of the development of pesticide residue analysis technology is a given.

More than 1400 active substances are used in the world as pesticides with complex chemical structures, present at trace levels and dynamic-concentration fluctuation range in complex matrices, which indicates that pesticide residues analyses are complex and diverse. Chromatography is the main technology for pesticide residue analysis. This technology can separate many kinds of pesticides needing further complete qualitative and quantitative analysis. However, these methods need standard compounds and

TABLE 3.2

Application of Chromatography–Mass Spectrometry in Pesticide Residue Analysis

Compound	Matrix	Method	Recovery (%)	LOD (µg/kg)	LOQ (µg/kg)	Ref.
200 pesticides and pesticide metabolites	Honeybee	QuEChERS-GC-MS/MS & LC-MS/MS	70–120	—	1–100	[14]
27 pesticides	Wine	DLLME-GC-MS	66.7–126.1	0.025–0.88	0.082–2.94	[15]
216 pesticides	Edible vegetable oils	QuEChERS-GC-MS/MS	70–120	—	10	[16]
Eight pesticides	Tea	SPE-GC-MS	72.5–109.1	—	20–80	[17]
Seven pesticides	Water	SPE/SPME-GC-MS	63–104	—	0.2–3.5	[18]
47 pesticides	Tomato	QuEChERS-GC-MS	—	—	20	[19]
19 pesticides	Water, milk, honey, and fruit juice	SPE-DLLME-GC-MS	78.1–105	0.5–1.0	—	[20]
Neonicotinoids	Soil	LC-MS	9.41–100.2	0.01–0.84	0.05–2.79	[21]
Eight pesticides	Fruits and vegetables	DLLME-LC-MS/MS	82–137	0.02–0.32	0.07–1.06	[22]
117 pesticides	Coffee	LC-ESI-MS/MS	70–120	10–50	—	[23]
167 pesticides	Milk	LC-MS/MS	80.4–117.3	0.4–3.9	1.1–13.1	[24]
18 pesticides	Sediments	LC-Orbitrap MS	70.8–106.2	0.3–4	0.8–13	[25]
Neonicotinoid	Sugarcane juice	LC-MS/MS	62.06–129.9	0.7–2	2–5	[26]
199 pesticides	Spice matrices	UHPLC-Orbitrap-MS	70–120	—	10–20	[27]
Ten pesticides	Sewage sludge	LC-MS/MS	71–120	—	0.4–10	[28]
Flubendiamide	Honey	LC-TOF-MS	94–104	0.1–0.2	0.4–0.6	[29]
Thiacloprid	Soil	SFC-MS/MS	78.8–107.1	—	5	[30]
25 pesticides	Flour and pizza	SFC-MS/MS	≥87	0.3–1	0.75–15	[31]
Organochlorine pesticides	Tomatoes, eggplants, and cucumbers	GC-MS	70–116	0.02–0.26	0.06–0.87	[32]
Fenbuconazole and its chiral metabolites	Fruits, vegetables, cereals, and soil	SFC-MS/MS	76.3–104.6	—	0.13–3.31	[33]
423 pesticides	Green tea	GC×GC-TOFMS	81.6–113.0	0.04–4.15	0.07–6.92	[34]
68 pesticides	Oil seeds	GC×GC-TOFMS	—	—	—	[35]
160 pesticides	Grape and wine	GC×GC-TOFMS	70–120	—	≤ 10 (most analytes)	[36]
51 pesticides	Grapes	GC×GC-TOFMS	70–110	—	10	[37]

repeated verification under different conditions. In addition, chromatography tends to produce false positive and false negative results and cannot meet the requirements of the increasing number of pesticides, the expanding detection range, as well as the continuous improvement requests made of pesticide residue analysis in food safety and environmental monitoring. With the constant improvement of new ion sources and mass analyzers, the highly sensitive and accurate quantitative analysis of pesticide residues in complex matrices is possible. At present, the simultaneous detection of several hundreds of pesticides and the screening of unknown pesticides and metabolites can be achieved (Table 3.2). In general, chromatography–mass spectrometry has become an indispensable and important technology for the analysis of pesticide residues because of its wide application, high separation efficiency, rapid analysis, high sensitivity, and easy automation.

At present, there are hundreds of chromatography–mass spectrometry methods for the detection of pesticides residues, including Codex Alimentarius Commission (CAC) standards, national standards of different countries, and some laboratory pesticide residues detection methods. Those methods can include the detection of a single pesticide, the detection of residues of multiple pesticides according to diverse pesticide classes or in different matrices, the detection with different sample treatments, and detection based on different ion sources and mass analyzers. Among them, there are a number of multiresidue detection methods for pesticides, such as the California Department of Food and Agriculture (CDFA) method that can detect more than 360 pesticides and the DFG S19 method of Germany (more than 320 pesticides), which have been widely used in the quantitative determination of organochlorine, organophosphorus, and nitrogen-containing pesticides. In China, more than 500 pesticide residues can be simultaneously detected [38–40].

Tandem modes of chromatography–mass spectrometry for pesticide residue detection have been employed in food (agricultural products) and environmental samples. The main characteristics of food and environmental samples are as follows: (1) the amount of pesticide residues in each kilogram of sample is only in the range of milligrams, micrograms, or even picograms; (2) hundreds of pesticides are frequently used, all with different physicochemical properties. Some degradation products, metabolites, or impurities with toxicological significance should be detected. The detection methods should be applied according to the characteristics of various kinds of pesticides; (3) there are many kinds of complex matrices, such as agricultural and livestock products, food, soil, air, water, and so on. The water, fat, and sugar contents in these samples are very different; and (4) qualitative screening and component analysis of pesticides (including known and unknown pesticide analysis) require specific analyzers. In recent years, the advantages of supercritical fluid chromatography-liquid chromatography–mass spectrometry (SFC-HPLC–MS/MS) and multidimensional chromatography–mass spectrometry have made the separation and detection of structure analogues, isomers, enantiomers, and non-enantiomers of pesticides easier.

In conclusion, the combination of modern sample pretreatment technology and chromatography–mass spectrometry technology plays an important role in the detection and confirmation of pesticide residues. With the continuous optimization and improvement of modern technology, the detection methods for pesticide residues have been in the direction of lower detection limits, and faster analysis speed.

Some application examples are listed below [41].

3.5.1 Determination of Tebuconazole Enantiomers in Water and Zebrafish by SFC–MS/MS

This method is suitable for the determination of Tebuconazole enantiomers (Figure 3.2) in water and zebrafish samples. The limits of quantification (LOQs) range was 0.24–1.20 µg/kg and the average recovery was 79.8–108.4%, RSD ≤ 7.0%.

Sample preparation

Approximately 2.0 g of the blank zebrafish samples (2.0 mL water) was weighed into a 10 mL PTFE centrifuge tube. 2 mL of acetonitrile was added, and the tube was vortexed for 3 min. Then, 1 g of NaCl was added and the tube was vortexed again for 1 min followed by centrifugation for 5 min. Next, 1.0 mL of the upper layer was transferred into a single-use 2 mL centrifuge tube containing 150 mg of Florisil and 150 mg of anhydrous $MgSO_4$ (for the water samples, 1.0 mL of the upper layer was filtered directly using a 0.22 µm nylon syringe filter for SFC–MS/MS injection). The tubes were vortexed for 30 s and centrifuged for 5 min. Finally, the resulting supernatant was filtered using a 0.22 µm nylon syringe filter for SFC–MS/MS injection.

Chromatographic conditions

Chromatographic column: Chiralpak IA-3 (150 mm×4.6 mm, 3 µm), which was coated with amylose tris-3,5-dimethylphenylcarbamate. Mobile phase: solvent A (CO_2) and solvent B (methanol) at 87:13 (v/v) ratio and flow rate of 2.0 mL/min. A compensation pump flow rate of 0.15 mL/min, which used 0.1% FA-MeOH (v/v) as a post-column additive. The temperature of the column was 30°C and the injection volume 1 µL. The chromatogram of the analysis of racemic tebuconazole is shown in Figure 3.3.

MS/MS parameters

Multiple reaction monitoring (MRM) mode was used for MS detection. A dwell time of 130 ms per ion pair was used to maintain the high sensitivity of the analysis. The typical conditions were as follows: the cone voltage was 39 V; m/z 308.1 was selected as the precursor ion, m/z 70.02 was selected as the quantitative product ion

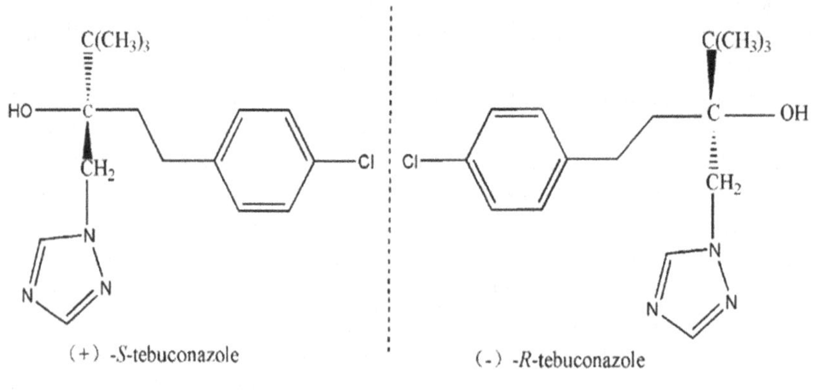

(+) -S-tebuconazole   (-) -R-tebuconazole

FIGURE 3.2   Chemical structures of tebuconazole stereoisomers. (Reproduced with permission from Liu et al., 2015, Copyright American Chemical Society.)

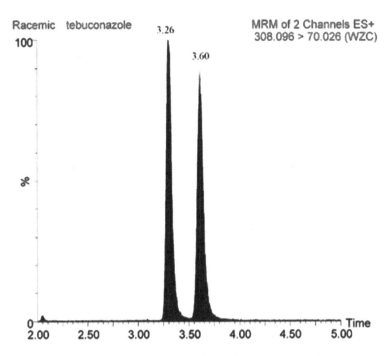

FIGURE 3.3   Chromatogram of racemic tebuconazole. (Reprinted with permission from Liu et al., 2015, Copyright American Chemical Society.)

and m/z 125.03 as the qualitative ion with the collision energy set to 23 and 37 V, respectively. Under these conditions, the retention times of (+)-S-tebuconazole and (−)-R-tebuconazole were approximately 3.26 and 3.60 min, respectively [42].

3.5.2   MULTI-RESIDUE DETERMINATION OF 334 PESTICIDES IN VEGETABLES BY GC–MS AND LC–MS

Sample preparation

Weigh 15 g of sample (accurate to 0.01 g) in a 100 mL centrifuge tube, add 30.0 mL acetonitrile and homogenize in a high-speed disperser for 2 min, add 7.5 g of anhydrous $MgSO_4$, then homogenize for 1 min, and centrifuge for 5 min. Take 10 mL of the supernatant into a 50 mL bottle, and concentrate to 1 mL on a rotary evaporator at a temperature below 50°C. Add 10 mL of acetonitrile to remove a small amount of water by forming an azeotrope. Finally, the volume of the solution is concentrated to approximately 1 mL. A graphite carbon black solid phase extraction column and a propylamino solid phase extraction column are connected in series from top to bottom, firstly pre-rinsed with 5 mL of acetonitrile + toluene (3+1), and the above concentrated solution is immediately loaded. Pesticides were eluted using 25 mL of acetonitrile + toluene (3+1), concentrated on a rotary evaporator at approximately 50°C to dryness; then volume was set to 2.5 mL using acetone and 25 μL of 5 mg/L internal standard solution, and the solutions were mixed before GC–MS analysis.

Pipette 1 mL of the above acetonitrile extract into a 2 mL centrifuge tube, add 50 mg of PSA and 200 mg of anhydrous $MgSO_4$, vortex for 1 min and centrifuge for 5 min. Take 0.5 mL and blow it slowly with nitrogen at 50°C. After drying, 0.5 mL of methanol was added, mixed, and filtered through a 0.22 μm filter for liquid chromatography analysis.

Gas chromatography–mass spectrometry

Gas chromatographic conditions: column: HP-5MS (30 m × 0.25 mm × 0.25 μm) quartz capillary; flow rate: 50.0 mL/min; injection volume: 20 μL.

Mass spectrometric parameters: EI source: 70 eV; ion source temperature: 230°C; quadrupole temperature: 150°C; GC–MS interface temperature: 280°C.

Liquid chromatography–mass spectrometry

HPLC conditions: column: C18 column, (30 μm, 150 mm × 2.1 mm); flow rate: 0.3 mL/min; injection volume: 10 μL; column temperature: 40°C.

Mass spectrometric parameters: ESI source: 3000 V; ion source temperature: 350°C.

The total ion chromatogram is shown in Figure 3.4 [43].

3.5.3   COMPREHENSIVE SUSPECT SCREENING OF 185 PESTICIDES AND THEIR TRANSFORMATION PRODUCTS (TPS) IN SURFACE WATER SAMPLES USING LIQUID CHROMATOGRAPHY–HIGH RESOLUTION MASS SPECTROMETRY (LC–HRMS)

This method was reported for the suspect screening of 185 water-soluble and readily ionizable pesticides registered in Switzerland and their TPs in surface water samples. Seventy percent of these target substances have a limit of quantitation (LOQ) < 5 ng L^{-1}, and the average recoveries are 75–125% for most compounds.

Sample preparation

A solid phase extraction method was employed to enrich the analytes from all water samples. Briefly, the pH was set to 6.5–6.7 (using formic acid or ammonia solutions); the sample was filtered, and 1 L of surface water was measured into a pre-rinsed glass bottle. Then, 200 ng of internal standard mix was added. The samples were passed over a multilayered cartridge containing Oasis HLB, Strata XAW, Strata XCW, and Isolute® ENV+. The elution of the analytes was achieved with 6 mL of ethyl acetate–methanol 50:50 v/v with 0.5% ammonia and 3 mL of ethyl acetate–methanol 50:50 v/v with 1.7% formic acid. The sample extracts were evaporated to 100 μL using a gentle nitrogen stream and reconstituted to 1 mL using nanopure water to give a final water–methanol ratio of 90:10 in the aliquot. The aliquots were stored at 4°C in a glass vial until analysis.

LC–HRMS

Chromatographic conditions: XBridge C18 column, 3.5 μm, 2.1×50 mm; flow rate: 200 μL/min; solvent A: 0.1% formic acid aqueous solution; solvent B: 0.1% formic

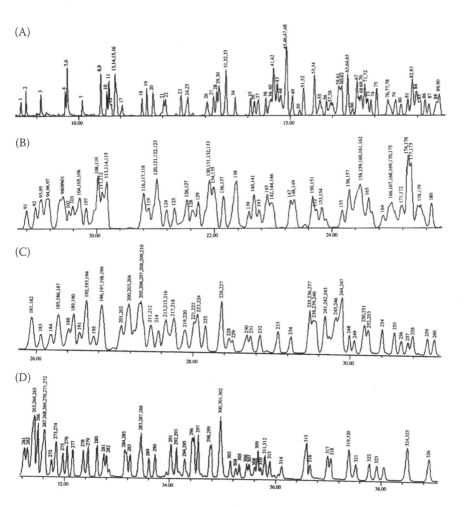

FIGURE 3.4 Number 1–90 total ion chromatogram of 305 standard substance (A) Number 91–180 total ion chromatogram of 305 standard substance (B) Number 181–260 total ion chromatogram of 305 standard substance (C) Number 260–327 total ion chromatogram of 305 standard substance (D).

acid in methanol; linear gradient procedure: 0 min 10% B, 0–4 min linear gradient to 50% B, 4–17 min linear gradient to 95% B, 17–25 min kept at 95% B, 25–25.1 min switch to 10% B, and 25.1–29 min kept at 10% B.

HRMS parameters: monitoring both positive and negative mode; ion source: HESI-II; spray voltage: 4000 V (+)/3000 V (−); sheath gas flow: 40 AU; capillary temperature: 350°C; heater temperature: 40°C; Orbitrap full scan; mass range: 100–1,000 m/z; mass resolution: 140,000; AGC target: 500,000; maximal injection time: 250 ms.

Data dependent MS/MS. Mass resolution: 17,500; microscan: 1; underfill ratio: 0.1%; isolation window: 1 m/z; dynamic exclusion: 8 s. Figure 3.5 shows an example for a detected substance in the suspect screening.

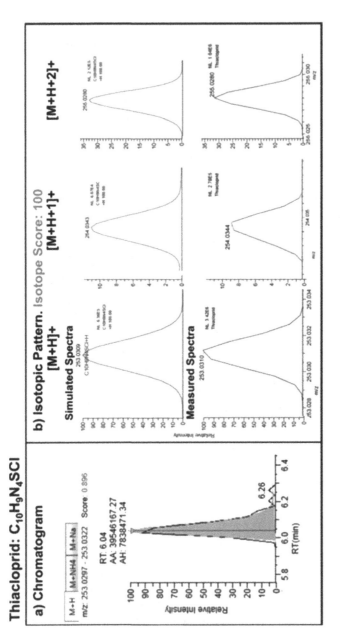

FIGURE 3.5 Example for a detected substance in the suspect screening: thiacloprid (15 ng/L) in the river Limpach: (a) measured chromatogram with peak fit by the software ExactFinder V 2.0 (gray shaded area). Score: peak score; RT: retention time; AA: peak area; AH: peak intensity; (b) isotopic pattern. (Reprinted with permission from Moschet et al, 2013, Copyright American Chemical Society.)

3.6 CONCLUSION AND FUTURE TRENDS

After decades of development, chromatography–mass spectrometry is widely used in the determination of pesticide residues, screening of unknown compounds, metabolite detection, and chiral pesticide separation in food and environmental samples, becoming a tool for qualitative and quantitative analysis of pesticide residues from a single pesticide to hundreds of pesticides. The developed methods of UPLC–MS/MS, multidimensional chromatography–mass spectrometry, and supercritical fluid chromatography–mass spectrometry provide important detection technologies to carry out the supervision and control of pesticide residues in food and environmental samples by the government and enterprises.

With the stricter management of pesticides and the high attention to food safety and environmental problems, the requirement for analysis of pesticide residues is much higher. In order to improve the capacity of analyzing pesticide residues, new technologies of higher sensitivity, higher resolution, and wider range have been applied. The future trends in the analysis of pesticide residues by chromatography–mass spectrometry will consider the following accomplishments: (1) modular detection systems and technical platforms for pesticide residues meeting the requirements of relevant regulations and standards; (2) detection methods for pesticide residues covering all maximum residue limits (MRL) of approved pesticides; (3) high-throughput and high sensitive detection database and platform for non-target pesticide analysis; (4) new materials and sample pretreatments for the detection of pesticide residues, in order to achieve higher fluxes, lower costs, and green purposes; and (5) on-line integration of chromatography–mass spectrometry with sample preparation equipment.

Pesticide residue analysis is a comprehensive and extensive analytical discipline. With the continuous development of science and technology, as well as the diversity and complexity of pesticide itself, chromatography–mass spectrometry analysis of pesticide residues will continue to be well innovated and developed.

REFERENCES

1. Mol, H. G., Zomer, P., Koning, M. 2012. Qualitative aspects and validation of a screening method for pesticides in vegetables and fruits based on liquid chromatography coupled to full scan high resolution (Orbitrap) mass spectrometry. *Analytical and Bioanalytical Chemistry* 403: 2891–2908.
2. Wang, X., Li, P., Zhang, W., et al. 2012. Screening for pesticide residues in oil seeds using solid-phase dispersion extraction and comprehensive two-dimensional gas chromatography time-of-flight mass spectrometry. *Journal of Separation Science* 35(13): 1634–1643.
3. Hird, S. J., Lau, B. P. Y., Schuhmacher, R., et al. 2014. Liquid chromatography–mass spectrometry for the determination of chemical contaminants in food. *TrAC Trends in Analytical Chemistry* 59: 59–72.
4. Yeboah, S. O., Mitei, Y. C., Ngila, J. C., et al. 2011. Compositional and structural studies of the major and minor components in three Cameroonian seed oils by GC–MS, ESI-FTICR-MS and HPLC. *Journal of Agricultural and Food Chemistry* 88(10): 1539–1549.
5. Senyuva, H. Z., Gökmen, V., Sarikaya, E. A. 2015. Future perspectives in Orbitrap™-high-resolution mass spectrometry in food analysis: A review. *Food Additives and Contaminants* 32(10): 1568–1606.

6. Desfontaine, V., Veuthey, J. L., Guillarme, D. 2017. Hyphenated detectors: Mass spectrometry. In *Supercritical Fluid Chromatography* Poole, C. (Ed.), Elsevier: Amsterdam, the Netherlands: 213–244.
7. Perrenoud, G. G. A., Veuthey, J. L., Guillarme, D. 2014. Coupling state-of-the-art supercritical fluid chromatography and mass spectrometry: From hyphenation interface optimization to high sensitivity analysis of pharmaceutical compounds. *Journal of Chromatography A* 1339: 174–184.
8. Dunkle, M., Vanhoenacker, G., David, F., Sandra, P. 2011. Agilent 1260 infinity SFC/MS solution superior sensitivity by seamlessly interfacing to the Agilent 6100 series LC/MS system. *Agilent Technologies*. Publication No. 5990-7972EN.
9. Zhang F. 2010. Study on the Determination Method of Pesticide Residues in Fruits. Central South University of Forestry and Technology, Changsha, Hunan, China.
10. Jiang, J., Peiwu, L., Xie, L. 2011. Solid phase extraction and comprehensive two-dimensional gas chromatography/time of flight mass spectrometry for the rapid detection of 64 pesticide residues in vegetables. *Analysis of Chemical Studies* 1: 72–76.
11. Jia, W., Chu, X., Zhang, F. 2015. Multiresidue pesticide analysis in nutraceuticals from green tea extracts by comprehensive two-dimensional gas chromatography with time-of-flight mass spectrometry, *Journal of Chromatography A* 22: 160–166.
12. Dasgupta, S., Banerjee, K., Patil, S.H., et al. 2010. Optimization of two-dimensional gas chromatography time-of-flight mass spectrometry for separation and estimation of the residues of 160 pesticides and 25 persistent organic pollutants in grape and wine. *Journal of Chromatography A* 1217: 3881–3889.
13. Banerjee, K., Patil, S.H., Dasgupta, S., et al. 2008. Optimization of separation and detection conditions for the multiresidue analysis of pesticides in grapes by comprehensive two-dimensional gas chromatography–time-of-flight mass spectrometry. *Journal of Chromatography A* 1190: 350–357.
14. Kiljanek, T., Niewiadowska, A., Semeniuk, S., et al. 2016. Multi-residue method for the determination of pesticides and pesticide metabolites in honeybees by liquid and gas chromatography coupled with tandem mass spectrometry—Honeybee poisoning incidents. *Journal of Chromatography A* 1435: 100–114.
15. Chen, B., Wu, F., Wu, W., et al. 2016. Determination of 27 pesticides in wine by dispersive liquid–liquid microextraction and gas chromatography–mass spectrometry. *Microchemical Journal* 126: 415–422.
16. Vázquez, P. P., Hakme, E., Uclés, S., et al. 2016. Large multiresidue analysis of pesticides in edible vegetable oils by using efficient solid-phase extraction sorbents based on quick, easy, cheap, effective, rugged and safe methodology followed by gas chromatography–tandem mass spectrometry. *Journal of Chromatography A* 1463: 20–31.
17. Deng, X., Guo, Q., Chen, X., et al. 2014. Rapid and effective sample clean-up based on magnetic multiwalled carbon nanotubes for the determination of pesticide residues in tea by gas chromatography–mass spectrometry. *Food Chemistry* 145: 853–858.
18. Bonansea, R. I., Améa, M. V., Wunderlin, D. A. 2013. Determination of priority pesticides in water samples combining SPE and SPME coupled to GC–MS. A case study: Suquía River basin (Argentina). *Chemosphere* 90: 1860–1869.
19. Restrepo, A. R., Ortiz, A. F. G., Ossa, D. E. H., et al. 2014. QuEChERS GC–MS validation and monitoring of pesticide residues in different foods in the tomato classification group. *Food Chemistry* 158: 153–161.
20. Shamsipur, M., Yazdanfar, N., Ghambarian, M. 2016. Combination of solid-phase extraction with dispersive liquid–liquid microextraction followed by GC–MS for determination of pesticide residues from water, milk, honey and fruit juice. *Food Chemistry* 204: 289–297.
21. Zhou, Y., Lu, X., Fu, X., et al. 2018. Development of a fast and sensitive method for measuring multiple neonicotinoid insecticide residues in soil and the application in parks and residential areas. *Analytica Chimica Acta* 1016: 19–28.

22. Lawal, A., Wong, R.C.S., Tan, G.H., et al. 2018. Multi-pesticide residues determination in samples of fruits and vegetables using chemometrics approach to QuEChERS-dSPE coupled with ionic liquid-based DLLME and LC–MS/MS. *Chromatographia* 81: 759–768.

23. Reichert, B., de Kok, A., Pizzutti, I. R., et al. 2018. Simultaneous determination of 117 pesticides and 30 mycotoxins in raw coffee, without clean-up, by LC-ESI-MS/MS analysis. *Analytica Chimica Acta* 1004: 40–50.

24. Golge, O., Koluman, A., Kabak, B. 2018. Validation of a modified QuEChERS method for the determination of 167 pesticides in milk and milk products by LC–MS/MS. *Food Anal. Methods* 11(4):1122–1148.

25. Nannou, C. I., Boti, V. I., Albanis, T. A. 2018. Trace analysis of pesticide residues in sediments using liquid chromatography–high-resolution Orbitrap mass spectrometry. *Analytical and Bioanalytical Chemistry* 410(7): 1977–1989.

26. Suganthi, A., Bhuvaneswari, K., Ramya, M. 2018. Determination of neonicotinoid insecticide residues in sugarcane juice using LCMS/MS. *Food Chemistry* 241: 275–280.

27. Goon, A., Khan, Z., Oulkar, D., et al. 2018. A simultaneous screening and quantitative method for the multiresidue analysis of pesticides in spices using ultra-high performance liquid chromatography–high resolution (Orbitrap) mass spectrometry. *Journal of Chromatography A* 1532: 105–111.

28. Ponce-Robles, L., Rivas, G., Esteban, B., et al. 2017. Determination of pesticides in sewage sludge from an agro-food industry using QuEChERS extraction followed by analysis with liquid chromatography-tandem mass spectrometry. *Analytical and Bioanalytical Chemistry* 409(26): 6181–6193.

29. Ares, A. M., Valverde, S., Bernal, J. L., et al. 2017. Determination of flubendiamide in honey at trace levels by using solid phase extraction and liquid chromatography coupled to quadrupole time-of-flight mass spectrometry. *Food Chemistry* 232: 169–176.

30. Li, R., Chen, Z., Dong, F., et al. 2018. Supercritical fluid chromatographic–tandem mass spectrometry method for monitoring dissipation of thiacloprid in greenhouse vegetables and soil under different application modes. *Journal of Chromatography B* 1081: 25–32.

31. Di Ottavio, F., Della Pelle, F., Montesano, C., et al. 2017. Determination of pesticides in wheat flour using microextraction on packed sorbent coupled to ultra-high performance liquid chromatography and tandem mass spectrometry. *Food Analytical Methods* 10(6): 1699–1708.

32. Ozcan, C., Balkan, S. 2017. Multi-residue determination of organochlorine pesticides in vegetables in Kirklareli, Turkey by gas chromatography–mass spectrometry. *Journal of Analytical Chemistry* 2017: 761–769.

33. Tao, Y., Zheng, Z. T., Yu, Y., et al. 2018. Supercritical fluid chromatography–tandem mass spectrometry-assisted methodology for rapid enantiomeric analysis of fenbuconazole and its chiral metabolites in fruits, vegetables, cereals, and soil. *Food Chemistry* 241: 32–39.

34. Jia, W., Chu, X., Zhang, F. 2015. Multiresidue pesticide analysis in nutraceuticals from green tea extracts by comprehensive two-dimensional gas chromatography with time-of-flight mass spectrometry. *Journal of Chromatography A* 1395: 160–166.

35. Wang, X., Wang, S., Cai, Z. 2013. The latest developments and applications of mass spectrometry in food-safety and quality analysis. *Trac-Trends in Analytical Chemistry* 52: 170–185.

36. Dasgupta, S., Banerjee, K., Patil, S. H., et al. 2010. Optimization of two-dimensional gas chromatography time-of-flight mass spectrometry for separation and estimation of the residues of 160 pesticides and 25 persistent organic pollutants in grape and wine. *Journal of Chromatography A* 1217(24): 3881–3889.

37. Banerjee, K., Patil, S. H., Dasgupta, S., et al. 2008. Optimization of separation and detection conditions for the multiresidue analysis of pesticides in grapes by comprehensive two-dimensional gas chromatography–time-of-flight mass spectrometry. *Journal of Chromatography A* 1190(1–2): 350–357.

38. Specht, W., Tillkes, M. 1985. Gas-chromatographische Bestimmung von Rückständen an Pflanzenbehandlungsmitteln nach Clean-up über Gel-Chromatographic und Mini-Kieselgel-Säulen-Chromatographi. *Analytical Chemistry* 322: 443–455.

39. GB/T 19426-2006. 2006Method for the determination of 497 pesticides and related chemicals residues in honey, fruit juice, and wine—GC–MS method. National Standards of the People's Republic of China.

40. GB/T 19649-2006. 2006Method for the determination of residues of 475 pesticides and related chemicals in grain valley—GC–MS. National Standards of the People's Republic of China.

41. Liu, N., Dong, F., Xu, J., et al. 2015. Stereoselective determination of tebuconazole in water and zebrafish by supercritical fluid chromatography tandem mass spectrometry. *Journal of Analytical Chemistry* 63: 6297–6303.

42. NY-T 1379-2007. 2007. Multi-residue determination of 334 pesticides in vegetable by GC–MS and LC–MS. Agricultural Industry Standards of the People's Republic of China.

43. Moschet, C., Piazzoli, A., Singer, H., et al. 2013. Alleviating the reference standard dilemma using a systematic exact mass suspect screening approach with liquid chromatography–high resolution mass spectrometry. *Analytical Chemistry* 85(21): 10312–10320.

4 Immunoassays and Biosensors

Jeanette M. Van Emon, Tanner Drum, and Kilian Dill

CONTENTS

4.1 Introduction 105
4.2 Immunoassays... 107
    4.2.1 General Overview for Immunoassays 107
    4.2.2 Method Development.. 109
    4.2.3 ELISA Methods for Pesticides.. 110
    4.2.4 Data Analysis.. 115
4.3 Biosensors... 117
    4.3.1 General Descriptions 117
    4.3.2 Microarrays... 120
    4.3.3 Biosensors Methods for Pesticides 122
        4.3.3.1 Potentiometric, Light Addressable Potentiometric
                 Sensor (LAPS), and Amperometric Detection 122
        4.3.3.2 Piezoelectric Measurements 123
        4.3.3.3 Surface Plasmon Resonance 124
        4.3.3.4 Conductive Polymers 124
4.4 Ongoing Developments... 125
4.5 Future Trends.. 126
References.. 127

4.1 INTRODUCTION

Monitoring and exposure data are critical to accurately determine the impact of pesticides and environmental contaminants on human health [1]. This is especially true for infants and young children, as well as the elderly and those with compromised immune systems. Uncertainties in the assessment of human exposures to exogenous compounds may be reduced using data obtained from dietary, epidemiological, and environmental monitoring measurement studies. Faster and more cost-effective analytical methods can facilitate the collection of data concerning particular target analytes that may impact human health and the environment. Methods are needed to determine which chemicals enter the food chain ending up in the human body, and to determine the body burden of various populations. Individuals and populations that have levels of pesticides above the concentrations associated with adverse health impacts must be identified for better risk management to minimize exposures. These data are needed to direct research to safeguard human and environmental health. Immunoassays and biosensors can provide fast, reliable, and cost-effective monitoring and measurement methods to be used in these endeavors [2].

In 1993, the United States National Academy of Sciences (NAS) issued a major report on pesticides in the diets of children. The report, "Pesticides in the Diets of Infants and Children" [3] recommended that US pesticide laws be revised to make foods safer for children. The Food Quality Protection Act [4] of 1996 (Public Law 104–170, 1996) was passed in response to the Academy's report. The FQPA is predicated on the need to reduce exposure to pesticides in foods, particularly for vulnerable groups. The purpose of the FQPA is to eliminate high-risk pesticide uses, not to eliminate pesticide use entirely. The Academy report recommended that pesticide residue monitoring programs target foods often consumed by children, and that analytical testing methods be standardized, validated, and subject to strict quality control and quality assurance programs [3]. These requirements often pose major analytical challenges to fully implement the FQPA.

The FQPA requires the US Environmental Protection Agency (EPA) to look at all routes and sources (i.e., food, air, water, pets, indoor environments) when setting limits on the amount of pesticides that can remain in food. Dietary and non-dietary exposures must now be considered in an integrated manner. This aggregate exposure approach clearly requires cost-effective analytical methods for a variety of analytes in different matrices.

Immunoassay detection methods were initially developed for clinical applications where their sensitivity and selectivity provided improvements in diagnostic capabilities. Clinical chemists developed highly successful methods for medical and health-care applications by leveraging the sensitivity and selectivity of the specific antibody interaction with large target analytes such as drugs, hormones, bacteria, and toxins. Pesticide residue chemists recognized the potential of immunochemical technology for small molecule detection in the 1970s [5]. Since that time, immunoassays have been successfully adapted for the analysis of a wide range of pesticides [6] and other potential environmental contaminants including PCBs, PAHs, dioxins, and metals [7–10].

Immunoassay detection methods range from cost-effective, high sample throughput methods for large-scale monitoring studies [11], to self-contained rapid testing formats. Immunoassays can provide rapid screening information or quantitative data to fulfill stringent data quality requirements. These methods have been used for the selective analyses of many compounds of environmental and human health concern. For water-soluble pesticides or compounds with low volatility, immunoassays can be faster, less expensive, and significantly more sensitive and reproducible than many other analytical procedures.

Biosensor technology also had its genesis in clinical applications. Medical diagnostic sensors designed for point-of-care use are small, portable devices, easy-to-use, and give rapid, quantitative results. These attributes are also important for unattended remote sensing of environmental contaminants and for monitoring pesticides and pesticide biomarkers [12]. Different biosensor formats have been reported for several classes of pesticides, microbiologicals, and chemicals in food including a genetic-based whole cell biosensor for chlorpyrifos and a portable biosensor for ethion [13–21].

4.2 IMMUNOASSAYS

All immunochemical methods are based on selective antibodies combining with a particular target analyte or analyte group. The selective binding between an antibody and a pesticide analyte has been used to analyze a variety of sample matrices for pesticide residues. Methods range from the determination of pesticide dislodgeable foliar residues on crops to monitoring dietary consumption, dust and soil exposures, and determining pesticide biomarkers in urine and serum [22, 23].

4.2.1 GENERAL OVERVIEW FOR IMMUNOASSAYS

Immunoassays have been routinely used in medical and clinical settings for the quantitative determination of proteins, hormones, and drugs with a molecular mass of several thousand Daltons. Immunoassay techniques including the enzyme-linked immunosorbent assay (ELISA) have also proven useful for environmental monitoring and human observational monitoring studies [6, 23]. Common environmental pollutants (i.e., pesticides) are typically small molecules with a molecular mass less than one thousand Daltons. This small size will not elicit antibody production. Small molecules can be used for antibody production when conjugated to carrier molecules such as proteins. The small molecule of interest is usually modified to introduce a chemical moiety capable of covalent binding. The small molecule, or hapten, is then converted to an immunogenic substance through conjugation to the carrier molecule for antibody production. The design of a hapten greatly affects the selectivity and sensitivity of the resulting antibody. The distinguishing features of the small molecule must be preserved while introducing the additional chemical group (i.e., -COOH, -OH, -SH, -NH$_2$,) for binding to the large carrier that is in part responsible for stimulating the immune system for antibody production. A linker chain or spacer arm (i.e., -CH$_2$CH$_2$CH$_2$CH$_3$) to distance the small molecule of interest from the carrier is typically used to facilitate immune recognition [5]. Hapten design, hapten synthesis, and antibody production are among the critical initial steps in developing immunoassays and antibodies with the desired selectivity for environmental pollutants.

A stepwise diagram for an ELISA is shown in Figure 4.1. This format is based on the immobilization of an antigen (i.e., the target analyte hapten conjugated to a protein) to a solid phase support such as a test tube or a 96-well microtiter plate [24]. The sample extract for a microplate format (in a water-soluble solvent), and a solution of specific antibody (typically in phosphate-buffered saline pH 7.4 containing 0.5% Tween 20) are added to the antigen-sensitized wells. The target analyte in solution and the immobilized antigen compete for binding sites on the specific antibody. The wells

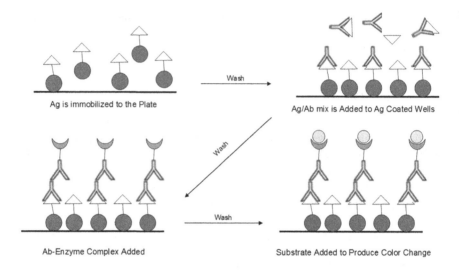

FIGURE 4.1   Indirect competitive ELISA.

are rinsed with buffer to remove antibody not bound to the solid phase antigen. The amount of antibody that can bind to the immobilized antigen on the plate is inversely related to the amount of analyte in the sample. The presence of bound primary antibody is determined with an enzyme-labeled species-specific secondary antibody that binds to the Fc region of the primary antibody. Alkaline phosphatase and horseradish peroxidase are two commonly used enzyme labels. Another buffer rinse removes unbound excess enzyme-labeled secondary antibody. The addition of a chromogenic substrate produces a colored end product that can be measured spectrophotometrically or kinetically for quantitation of analyte. This indirect competitive format is useful to support large observational studies due to its high level of performance, high sample throughput, adaptation to automation, and availability of commercial reagents. For extremely high sample throughput capability, microtiter plates containing 384 microwells can be used. In-depth details on how to develop antibodies and immunoassays, as well as data analysis are presented in Van Emon [2].

There are several permutations to the basic indirect competitive ELISA. Figure 4.2 depicts an immunoassay format utilizing immobilized antibody and an enzyme-labeled tracer [25]. Analyte in the sample competes with a known amount of enzyme-labeled analyte for binding sites on the immobilized antibody. In the initial step, the anti-analyte antibody is adsorbed to the side of a test tube or microtiter plate well. The analyte and an enzyme-labeled analyte are next added to the antibody-coated wells and competition for antibody binding occurs. After an incubation step, all unbound reagents are rinsed from the wells. Substrate is added for color development that is inversely related to the concentration of analyte present in the sample. This particular format is commonly used in immunoassay testing kits as it has minimal procedural steps. However, this format does not have the convenience of readily commercially available reagents (i.e., enzyme-labeled secondary antibody) and

Antibodies are immobilized to the Plate

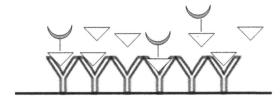

Analyte and Enzyme Labeled Hapten Compete for Antibody Sites

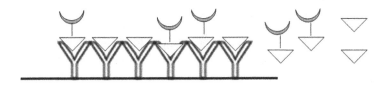

Wash Removes Unbound Analyte and Labeled Hapten

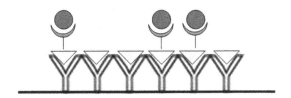

Substrate is Added for Color Detection

FIGURE 4.2 Direct competitive ELISA.

requires the synthesis or labeling of either the analyte or hapten which may not be straightforward and must be standardized and validated between batches.

4.2.2 METHOD DEVELOPMENT

The development of an immunoassay closely parallels the steps necessary for an instrumental analysis. A critical step in either type of analysis is presenting the analyte to the detector (e.g., antibody, mass spectrometer, electron capture, flame ionization) in a form that the detector can recognize. A derivatization step is often used in instrumental analysis enabling the analyte to pass through a GC, that is comparable to developing a hapten by "derivatizing" the molecule of interest for antibody production. A major difference is typically the extent of sample preparation required for immunoassays compared to instrumental methods. Frequently immunoassays do not require the same amount of sample cleanup as an instrumental method, providing savings in time and costs. Many methods have reported simply using a dilution series to remove interfering matrix substances [26, 27]. Solid phase extraction (SPE) can be used for either unprocessed samples or in tandem with enhanced solvent extraction (ESE) methods [28–32]. QuEChERS, which comes from the words "quick, cheap, effective, rugged, and safe", is a simple extraction procedure applicable to immunoassays as well as instrumental methods. Briefly, food samples are homogenized with solvent, agitated, then centrifuged followed by removal of the upper layer which is put through a dispersive solid phase extraction cleanup. This type of approach was used for the determination of O,O-diethyl and O,O-dimethyl organophosphorus pesticides in vegetable and fruit samples. The QuEChERS procedure was effective in minimizing matrix effects yielding recoveries of 89.4 to 135.5% as determined by ELISA [33].

Key to successful methods development is presenting the analyte to the antibody in a manner that is compatible with antibody function. As antibodies prefer an aqueous medium, the sample extract must be soluble in the buffer in which the immunoassay is performed. Organic solvents not soluble or miscible in water can be used for the initial extraction provided extracts are exchanged into a compatible solvent such as methanol prior to an immunoassay. Other organic solvents such as acetone, acetonitrile, dichloromethane (DCM), or hexane can be used as an extraction solvent; however, a solvent exchange step into an assay friendly solvent is necessary. The tolerance of organic solvents must be determined in each specific method as it is dependent upon the immunoreagents employed. For complex sample matrices such as soil, sediment, and fatty foods, extraction techniques and cleanup procedures may be required prior to immunoassay detection. The extraction techniques employed in instrumental methods, including shaking, sonication, supercritical fluid extraction (SFE), ESE, SPE, and now QuEChERS, can be used for immunochemical methods. The shaking method is common for field applications due to its simplicity. However, the shaking method may not provide adequate extraction efficiency depending upon the shaking time, temperature, analyte, and sample matrix [34]. The efficiency and reproducibility should be evaluated and documented for any extraction techniques before application to field samples. This can be accomplished through recoveries of target analytes from spiked and/or fortified samples as is typically done for instrumental methods.

4.2.3 ELISA METHODS FOR PESTICIDES

ELISA is a common immunoassay format that has been reported in the literature for determining pesticides and their metabolites in foods, as well as environmental and biological sample matrices [5, 27, 32, 35–54]. These pesticides include organophosphorus (OP) and organochlorine (OC) compounds, carbamates, neonicotinoids, sulfonylureas, pyrethroids, organochlorine, and many herbicides. Depending upon the specificity of the antibody and the hapten design, ELISA methods can be very selective for a specific target pesticide, providing quantitative measurements. Other methods employing less selective antibodies, having a high cross-reactivity for structurally similar pesticides, can be used as qualitative monitoring tools. Table 4.1 summarizes some of the ELISA methods developed for foods as well as environmental and biological samples.

TABLE 4.1
Pesticide Immunoassays

Analyte	Food Matrix	Reference	Analyte	Food Matrix	Reference
2,4-D	apple, grape, potato, orange, peach, urine	[39]	difenzoquat	beer, cereal, bread	[40]
3,5,6-TCP	urine, dust, soil	[43]	fenazaquin	apple and pear	[45]
4-nitrophenol parathion	soil	[29]	fenitrothion	apple and peach	[52]
acephate	analyte-fortified tap water, mulberry leaves, lettuce	[44]	fenthion	vegetable samples	[116]
acetamiprid	fruits, vegetables	[51]	glycine conjugate of cis/trans-DCCA	urine	[31]
alachlor, carbofuran, atrazine, benomyl, 2,4-D	beef liver, beef	[38]	glyphosate, atrazine, metolachlor mercapturate	water urine	[123]
atrazine	extra virgin olive oil	[114]	imidacloprid	fortified water samples	[117]
atrazine mercapturic acid	urine	[26, 32]	imidacloprid	fruit juices	[54]
azoxystrobin	grape extract	[115]	iprodione	apple, cucumber, eggplant	[53]
carbaryl (1-naphthyl methylcarbamate)	apple, Chinese cabbage, rice, barley	[15]	isofenphos	fortified rice and lettuce	[118]
carbaryl, endosulfan	rice, oat, carrot, green pepper	[72]	methyl-parathion and parathion	water and several food matrices	[119]
chlorpyrifos	fruits and vegetables	[49]	methyl-parathion	vegetable, fruit, soil	[46]
chlorpyrifos	Olive oil	[47]	pirimiphos-methyl	spiked grains	[120]
DDE	soil	[122]	tebufenozide	red and white wine	[121]
DDT and metabolites	drinking water, various foods	[42]	triazine herbicides	surface water, ground water	[28]

Assay performance must be demonstrated prior to applying the ELISA method to field or laboratory studies. For many reported laboratory-based ELISA methods, specific antibodies and coating antigens may only be available from the source laboratories while enzyme conjugates and substrates are commercial commodities. Generally, the protocols provided by the source laboratories should be used as starting points for determining optimal concentrations of immunoreagents for the particular analysis. Checkerboard titrations can be performed to determine the optimal concentrations of the antibodies and coating antigens. Whenever new lots of immunoreagents are used, they should be examined for their performance with previously used reagents. Protocols provided with commercial testing kits should be followed in the specified manner and reagents used within the expiration date. Most ELISA methods can offer comparable or better analytical precision (e.g., within ±20%) and accuracy (e.g., greater than ±80% of expected value) as conventional instrumental methods for analyzing pesticides. Calibration curves for quantification should reflect the composition of the sample extract. Standards should be prepared in the same buffer/solvent solution as the samples. Ideally, the standards should also include the same amount of matrix as the samples. This is particularly important when a simple sample preparation is used such as sample dilution without a cleanup. For example, if a food extract contains 20% orange juice the standards should also contain 20% orange juice (analyte-free prior to spiking). When assay performance is extremely well-documented as to the extent of the matrix effect, the matrix may be omitted and a conversion factor applied to the buffer standard curve to account for the matrix in the sample.

A laboratory-based ELISA method adapted to determine 3-phenoxy benzoic acid (3-PBA) in human urine samples was applied to sample subsets from two observational field studies. 3-PBA is a common urinary metabolite for several pyrethroid pesticides (cypermethrin, cyfluthrin, deltamethrin, esfenvalerate, permethrin) that contain the phenoxybenzyl group. The anti-PBA antibody had negligible cross-reactivity toward the parent pyrethroids but did recognize and react with the structurally similar 4-fluoro-3-PBA (FPBA) with a cross-reactivity of 72% [55]. This high cross-reactivity is advantageous as the 3-PBA ELISA could be used as a monitoring tool for determining a broad exposure to pyrethroids as FPBA is the metabolite for cyfluthrin (a pyrethroid pesticide containing a fluorophenoxybenzyl group). However, FPBA is typically detected at a much lower rate than 3-PBA in human urine samples. This phenomenon was seen in samples collected from residential settings as reported in the third CDC National Report on Human Exposure to Environmental Chemicals [56]. The assay protocol provided by the source laboratory was optimized using checkerboard titration experiments to determine the optimal concentrations of reagents for our application. The assay procedures were modified by preparing the standard solutions in a 10% methanol extract of 10% hydrolyzed drug-free urine in phosphate buffered saline (PBS). Calibration curves (Figure 4.3) for 3-PBA were generated based on ten concentration levels ranging from 0.00256 to 500 ng/mL (1:5 dilution series). The % relative standard deviation (%RSD) values of the triplicate analyses were less than 20% for the standard solutions. Day-to-day variation for the quality control (QC) standard solution (1.0 ng/mL) was within 13.1% (1.2±0.16 ng/mL) over a period of four months. The estimated assay detection limit was 0.2 ng/mL.

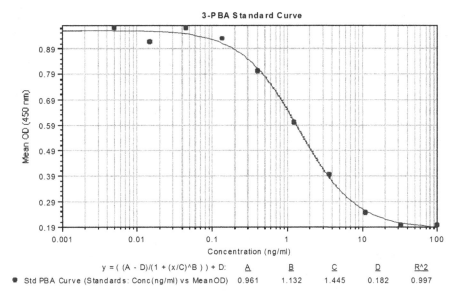

FIGURE 4.3 Calibration curve for 3-PBA immunoassay.

Quantitative recoveries of 3-PBA were achieved by ELISA (92±18%) in fortified urine samples. Approximately 100 human urine samples were prepared and analyzed by the ELISA method. The ELISA-derived 3-PBA concentrations correlated well with the GC/MS results for split samples. The Pearson correlation coefficient between the 3-PBA concentrations of the two methods was 0.952, which was statistically significant (p<0.0001). A non-significance outcome (p = 0.756) was also observed from the paired t-test indicating that there was no significant difference in measurements between the two analytical methods (ELISA and GC–MS) for a given sample. These data indicate that the ELISA could be applied to the analysis of the urinary biomarker, 3-PBA, for assessing human exposure to pyrethroids.

As most fruit and vegetable baby food preparations generally contain a significant amount (>80%) of water, ELISA methods have the advantage over instrumental methods in determining pesticides in this aqueous sample matrix. Various sample preparation methods were investigated for determining pesticides in baby foods using either GC–MS or ELISA methods [30]. A streamlined direct ELISA method consisting of dilution and filtration, prior to ELISA was evaluated on spiked baby foods at 1, 2, 5, 10, or 20 ppb. Quantitative recoveries (90–140 %) were achieved for atrazine in nonfat baby foods (i.e., pear, apple sauce, carrot, banana/tapioca, green bean). The performance of other ELISA testing kits was not as good as the atrazine-ELISA testing kit. Over-recoveries were observed for carbofuran and metolachlor testing kits in banana/tapioca and green bean. This was probably due to a sample matrix interference that was not completely removed by dilution. The off-line coupling of an enhanced solvent extraction procedure (150°C and 2000 psi with water) with an ELISA could determine atrazine in a more complex matrix of fatty baby foods. It was observed that the extraction temperature was an important factor to recover atrazine.

Different sample preparation procedures influenced the outcome of a magnetic particle ELISA for permethrin. Quantitative recoveries (>90%) were obtained when fortified soil samples were extracted with sonication using DCM, methyl-t-butyl ether (MTBE), or 10% ethyl ether (EE) in hexane. Recoveries were less than 50% from the fortified soil samples when the shaking method was employed (shaking with methanol for 1 hour). A longer shaking time (16 hours, overnight) was evaluated, using methanol, yielding recoveries of over 200% by ELISA. The longer shaking time extracted substances that interfered with the ELISA detection. This interference was also detected in the GC–MS analysis and persisted even after the SPE cleanup.

Interferences caused by sample matrix components are a concern for both conventional instrument methods and ELISA methods. In immunoassays, sample matrix effects may result from denaturation of the antibody or enzyme, or nonspecific binding between analyte, matrix, antibody, or enzyme. The matrix interferences can often be removed by a series of dilutions if a practical detection limit can still be achieved [27]. As shown above, cleanup methods for instrumental methods (e.g., SPE or column chromatographic separation) can also be performed prior to ELISA detection. Another effective cleanup method is immunoaffinity column chromatography that can be applied for the purification of sample extracts for either instrumental or ELISA detection [2, 57].

Atrazine was detected in complex sample media (soil, sediment, and duplicate diet food samples) using a combined method of ESE, followed by immunoaffinity column cleanup and a magnetic particle ELISA. Quantitative recoveries were achieved in fortified soil and sediment (93±17%) and food (100±15%) samples [58]. The ELISA data were in good agreement with the GC–MS data for these samples (the Pearson correlation coefficient was 0.994 for soil and sediment and 0.948 for food). However, the ELISA values were slightly higher than those obtained by GC–MS. This could be the result of sample loss during the solvent exchange step required for the GC–MS but not the ELISA. This bioanalytical approach is more streamlined than the GC–MS analysis lending itself to large-scale environmental monitoring and human exposure studies.

Oxidative stress and DNA damage have been proposed as mechanisms linking pesticide exposure to a variety of health effects including cancer. Exposure studies are finding that levels of oxidative stress markers correlate with levels of urinary pesticide metabolites, especially from the OP class. Isoprostane is a biomarker strongly indicative of oxidative stress in the human body. Levels of this biomarker in wastewater may provide a method to quantify the amount of oxidative stress experienced by a population serviced by the treatment facility. Samples were collected from a wastewater treatment facility. Both free and total isoprostane was measured using two immunoassays with different methods of preparation. An immunoassay method from Cayman Chemical was chosen due to its low detection level of 3 pg/mL and range of 0.8–500 pg/mL (Figure 4.4). The kit is very sensitive and requires an SPE cleanup. An Oxford Biomedical Research immunoassay kit was used despite higher detection levels, approximately 0.1–5 ng/mL, as it required no cleanup of the sample aside from dilution. This community urinalysis study is ongoing and seeks to develop a method to quantify the presence of isoprostane in wastewater to work toward a means to measure the overall health of a community serviced by a particular wastewater reclamation facility.

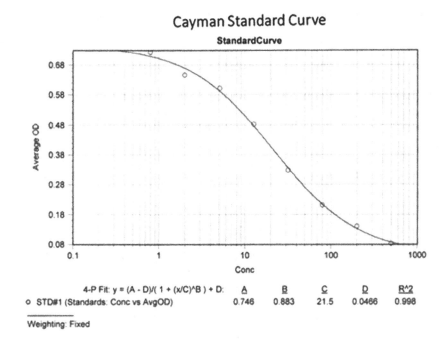

Cayman Standard Curve

StandardCurve

4-P Fit: y = (A - D)/(1 + (x/C)^B) + D:	A	B	C	D	R^2
○ STD#1 (Standards: Conc vs AvgOD)	0.746	0.883	21.5	0.0466	0.998

Weighting: Fixed

FIGURE 4.4   Calibration curve for an isoprostane immunoassay.

4.2.4 DATA ANALYSIS

Calculations of sample analyte concentrations in immunoassay methods are similar to those used in instrumental methods. A set of standard solutions covering the
working range of the method is used to generate the calibration curve, and the concentration of target analyte is calculated according to the calibration data. For the
96-microwell format, it is easy to include a standard curve on each plate along with
the samples. Thus, a calibration curve can be generated in the same 96-microwell
plate along with the samples. For test tube formats a standard curve series can be
interspersed among the samples. Many mathematical models have been used to con-
struct ELISA calibration curves including four-parameter logistic-log, log-log transforms, logistic-log transforms, and other models. The four-parameter logistic-log
model is commonly used for 96-microwell plate assays and is built into commercial
data analysis software [59]. The four-parameter logistic-log model is described as
follows: $y = (A - D)/1 + (X/C)^B + D$ where X is the concentration of the analyte, and
y is the absorbance for colorimetric endpoint determinations. For a more in-depth
analysis, the restricted cubic spline regression can also be used along with the four-
parameter logistic-log model.

Specifications are determined from each calibration curve for an expected mid-
point on the curve at 50% inhibition (IC_{50}), a maximum absorbance for the lower
asymptote (A), and a minimum absorbance for the upper asymptote (D). An established ELISA method usually has well-documented historical data for the specifications of the curve fit constants, such as the slope of the curve (B), and central point

of the linear portion of the curve (C). The specific curve fit constants may vary from day to day and the accepted ranges of such variations must be determined and documented. Triplicate analyses of each standard, control, and sample are generally performed for 96-microwell plate assays. The percent relative standard deviation (%RSD) of measured concentrations from triplicate analysis is usually within ±30% and can be as low as ±10%, depending upon the specific assay and required data quality objectives. Recoveries of positive controls and back-calculated standard solutions typically range from 70 to 130% or better. If the results of the samples are outside the calibration range, the sample is diluted and reanalyzed. Effects of the sample matrix can be determined by analyzing a number of samples at different dilutions. Typically, results from different dilutions should be within ±30%. Larger variations in the data suggest a matrix interference problem, indicating that a more effective extraction or cleanup procedure may be necessary.

When sample results are outside of the calibration range samples need to be diluted and reanalyzed. However, some of the commercial ELISA testing kits have a small dynamic optical density range (i.e., 1.0 OD to 0.35 OD) and small changes in OD correlate to large changes in derived concentrations. The differences between absorbance values and duplicate assays are generally small, and are well within the acceptance requirement (<10%) for the calibration standard solutions. However, the percent difference (%D) of the derived concentrations of the standard solution from duplicate assays sometimes may exceed 30%. The greater %D values obtained for some of the measured concentrations for the standards and samples may be due to a small volume of standard or sample retained in the pipette tip during the transfer step [8]. If the ELISA testing kit is to be used as a quantitative method, extreme care should be taken when transferring each aliquot of standard or sample. A trace amount of aliquot not delivered may result in a large variation in the data from duplicate analyses. The analyst should be alert in following the prescribed protocol when performing the assay.

To ensure the quality of the ELISA data, analytical quality control (QC) measures need to be integrated into the overall ELISA method. The QC samples may include: (1) negative and positive control standard solutions; (2) calibration standard solutions; (3) laboratory and field method blank; (4) fortified matrix samples; and (5) duplicate field samples. The assay performance can be monitored by characterization of the calibration curve and the data generated from the QC samples. The QC results will provide critical information such as assay precision, accuracy, detection limit, as well as overall method precision (including sample preparation and/or cleanup), accuracy, and detection limit when evaluating and interpreting the ELISA data.

Prior to applying an ELISA method for field application, the method needs to be evaluated and validated for its performance. The data generated from the ELISA method are usually compared with the data generated by a conventional instrument method (e.g., GC–MS). Various types of statistical analyses have been employed to compare the results between ELISA and GC–MS. For example, the Pearson correlation coefficient, commonly used, measures the extent of a general linear association between the ELISA and GC–MS data, and a parametric statistical test is performed to determine whether the calculated value of this correlation coefficient was significantly positive [60]. The slope of the established linear regression equation also can be used as guidance to determine if a 1:1 relationship exists for the ELISA and GC–MS data. The paired t-test [61] can be used to determine whether the measured ELISA and GC–MS concentrations differ significantly for a given sample at a 0.05 or 0.01 level of significance. Other nonparametric tests, namely, the Wilcoxon signed-rank test and the sign test, can also be performed on the sample-specific differences between ELISA and GC–MS data. These nonparametric tests can be used to determine if the median difference between the ELISA and GC–MS measurements among the samples is significantly different from zero [62]. The Wilcoxon signed-rank test is applied to differences between log-transformed measurements, as this test assumes that the differences have a symmetric distribution. In contrast, the sign test does not make this assumption and therefore does not require log transformations of the data. The McNemar's test of association can also be performed to determine whether there is any significant difference between the two methods in the proportion of samples having measurable levels that were at or above a specified threshold. The false negative and false positive rates can then be obtained at the specified concentration level. Chemometric data analysis for multi-analyte detection has also been applied.

4.3 BIOSENSORS

Biosensors are analytical probes composed of two components: a biological recognition element such as a selective antibody, enzyme, aptamer, receptor, DNA, microorganism, or cell, and a transducer that converts the biological recognition event into a measurable physical signal to quantitate the amount of analyte present. Biosensors must rapidly regenerate to provide continuous monitoring data yielding a response in real time. Analytical considerations such as sample preparation, matrix effects, and quality control measures must also be addressed in biosensor development. Matrix effects and the effect of sample on the recognition element are key issues for unattended sensors. Sensors that are easily fouled have limited reliability and application for environmental monitoring. Since biosensors utilize a biological recognition element, they may provide information on the effects of toxic substances as well as analytical measurements. Sensors for biochemical responses may assist in toxicity studies or human exposure assessments. A history of biosensors and a commentary on future potential development and uses can be found in [19]. Several pesticide biosensors have been reported for detecting various pesticides. Table 4.2 illustrates the application of biosensor technology to pesticide monitoring.

4.3.1 GENERAL DESCRIPTIONS

Biosensors can provide rapid and continuous *in situ* measurements for on-site or remote monitoring. Several different transducer types such as optical, electrochemical, piezoelectric, and thermometric can be employed. Acetylcholinesterase inhibition-based biosensors are based on the inhibition caused by the binding of an OP pesticide. The biosensor can be reactivated with the addition of various reagents for reuse. A review of these sensor types for pesticide (paraoxon, aldicarb, maneb) determination can be found in [63]. Immunosensors are based on specific antibodies for

TABLE 4.2

Examples of Biosensors for Determining Pesticides and Metabolites in Biological and Environmental Samples

Analyte	Sensor Type	Matrix	Range or LOD	Reference
atrazine	electrochemical immunosensor	orange juice	0.03 nmol/L	[17]
atrazine	electrochemical magnetoimmuno-sensor	orange juice	0.027 nmol/L	[71]
carbaryl, paraoxon	disposable screen-printed thick-film electrode	milk	20 µg/L (carbaryl) 1 µg/L (paraoxon)	[124]
carbofuran	flow injection electrochemical biosensor	fruits, vegetables, dairy products	1–100 nmole	[70]
dichlorvos	flow injection calorimetric biosensor	water	1 mg/L	[73]
dichlorvos	electrochemical biosensor	wheat	0.02 µg/g	[69]
fenthion	dipstick electrochemical immunosensor	water	0.01–1000 µg/L	[68]
malathion, dimethoate	amperometric biosensor	vegetables	malathion: 0.01–0.59 µM dimethoate: 8.6–520 µM	[14]
OP pesticides	fluorescence-based fiber-optic sensor	buffer	1–800 µM (paraoxon) 2–400 µM (DFP*)	[125]
OP pesticides and nerve agents	electrochem sensor using nanoparticles (ZrO_2) as selective sorbents	water	1–3 ng/mL	[105]
OP pesticides and nerve agents	flow injection amperometric biosensor using carbon nanotube modified glassy carbon electrode	water	0.4 pM	[106]
thiabendazole	fluorescence-based optical sensor	citrus fruits	0.09 mg/kg	[16]

*diisopropyl phosphorofluoridate (a nerve agent)

biological recognition and a transducer that converts the binding event of antibody to antigen to a physical signal.

Antibodies may be immobilized on membranes, magnetic beads, optical fibers, or embedded in polymers, or placed on metallic surfaces. In some types of sensors, such as those employing surface Plasmon resonance (SPR), evanescent waves, or piezoelectric crystals, the binding of antigen and antibody can be detected directly. With other transducers, an indicator molecule (either a labeled antigen or labeled secondary antibody) is required. An indicator may be fluorescent or it may be an enzyme that alters a colorimetric or fluorescent signal or produces a change in pH affecting the electrochemical parameters. Enzymes that are applicable to biosensors and approaches to improve selectivity and sensitivity of enzyme-based biosensors have been reviewed [64, 65].

Optical biosensors may measure fluorescence, fluorescence transfer, fluorescence lifetime, time-resolved fluorescence, color (either by absorbance or reflectance), evanescent waves, or an SPR response. Optical immunosensors are very rapid as they detect the antigen/antibody binding directly without requiring labeled reagents. Data can be generated in real time with devices applied to continuous monitoring situations such as effluent or runoff measurements from hazardous or agricultural waste streams. Optical immunosensors based on SPR employ an immobilized specific antibody on a metal layer. When the antigen binds, there is a minute change in the refractive index that is measured as a shift in the angle of total absorption of light incident on the metal layer. This technique was used to develop an SPR sensor to detect atrazine at 0.05 ppb in drinking water [66].

Fiber-optic biosensors are based on the transmission of light along silica glass or plastic fibers. The advantages of fiber-optic sensors are numerous: they are not subject to electrical interference, a reference electrode is not needed, immobilized reagent does not have to be in contact with the optical fiber, they can be miniaturized, and they are highly stable. A major advantage of these sensors is that they can respond simultaneously to more than one analyte and are useful for remotely monitoring hazardous environments or municipal water supplies.

Electrochemical biosensors offer the advantages of being effective with colored or opaque matrices and do not contain light-sensitive components. In an immunosensor format, the binding of antigen to antibody is visualized as an electrical signal. The response may be coupled to signal amplification systems such as an enzyme-conjugated secondary antibody, conferring very low detection limits. Amperometric sensors measure current when an electroactive species is oxidized or reduced at the electrode. Potentiometric sensors detect the change in charge of an antibody when it binds to an antigen.

Piezoelectric crystals are nonmetallic minerals (usually quartz), which conduct electricity and which develop a surface charge when stretched or compressed along an axis. The crystals vibrate when placed in an alternating electric field. The frequency of the vibration is a function of the mass of the crystal. Antibodies can be immobilized to the surface of piezoelectric crystals and the new vibrational frequency determined as a baseline measurement. The binding of analyte to the immobilized antibody alters the mass and vibrational frequency of the antibody–crystal system. This change in vibration can be measured to determine the amount of analyte detected.

Electroconductive polymer sensors have a specific antibody embedded in a conducting polymer matrix such as polypyrrole. When an analyte binds to the antibody, the ions in the matrix are less free to move, which decreases the ability of the polymer to conduct current. A reagent-less electrochemical DNA biosensor has been reported using an Au–Ag nanocomposite material adsorbed to a conducting polymeric polypyrrole [67]. The detection limit was 5.0×10^{-10} M of target oligonucleotides with a response time of 3 s. The integration of nanotechnology and sensor development will provide new analytical platforms and formats. Although new designs may first appear for clinical applications, these advancements will favorably impact the development of sensors for environmental measurements. Table 4.2 summarizes several pesticide biosensors that have been reported for various monitoring situations [14–17, 63, 68–73].

Redox enzyme-labeled antibodies have been employed in sensors for electrochemical detection. Combimatix developed the use of antibodies labeled with redox enzymes, such as horseradish peroxidase or laccase [74]. In these sensors, the enzymatic reaction gives a product, which can be reduced at the electrode, and electron charge is stored at the capacitor next to the electrode. Discharge of the capacitor then provides a current. This provides a format whereby a number of electrodes on the chip (10,000) may be labeled with unique capture antibodies. Numerous analytes may be tested at one time. This sensor format has been used for the detection of numerous chemical and biological agents and well as a host of biological antigens detected in the human body which could have applications for personalized medicine and treatments. Sensitivities can be achieved down to the picogram level. Systems are commercially available through Custom Array, Inc. for further applications.

A fluorescence-based immunosensor for the neonicotinoid insecticide imidacloprid was reported using a reagent-less biosensor called a quenchbody [75]. A quenchbody encompasses an antibody Fab fragment that is site-specifically labeled with a fluorescent dye. The Fab fragment quenches the dye and the addition of the target analyte stabilizes the antibody structure, displacing the quenched dye, resulting in an increase in fluorescence. The quenchbody method for imidacloprid had a detection limit of 10 ng/mL, a wide dynamic range, and showed little cross-reactivity. This procedure has potential for monitoring neonicotinoid residues on foods. Other quench-based antibody probes have been reported with applications for a wide range of analytes [76, 77].

4.3.2 Microarrays

Microarrays contain minute amounts of material (DNA, proteins, aptamers, antibodies, etc.) that are placed onto a matrix in an array format. The matrix is a solid support onto which a biological or organic material is placed. The solid support material can be plastic, glass, complementary metal–oxide–semiconductor (CMOS), gold, platinum, membranes, or other substances on which the reagents can be attached and still maintain function. The method of attachment can be covalent, hydrophobic, or through some tight affinity reagent, such as a biotin–streptavidin couple [78].

A microarray can be defined in terms of the number of spots (or electrodes) per chip/slide. By this definition, a low-density array may contain as little as 16 spots or as many as 96 spots. High-density arrays may have >500,000 spots. Lower density arrays are considered to be sensors, as microsensor detection is typically at the lower end of array density.

There are numerous methods used for array production. Arrays may result from "spotting" onto activated surfaces using robots to produce high-density arrays. Proteins or DNA are spotted onto activated surfaces (aldehydes, amines, etc.) so that either a chemical bond is formed or proteins can adhere through hydrophobic interaction. Another means of producing arrays is by photolithography using masks or lasers. This method has been used to produce *in situ* DNA or peptide-based arrays. In this specific case, a photolabile group is used on the 5' nucleotide end or photolabile groups are used as amino protection groups (peptides). The use of lasers or masks removes the labile group from a specific electrode or spot, promoting peptide bond or oligonucleotide bond formation. Conversely, this can also be accomplished using acid that is generated at a specific electrode. DNA and peptides can also be synthesized in this manner. The protecting groups are removed only at specific electrodes that generate acid resulting in an elongated nucleotide or peptide. The oligomers or peptides can be used as aptamers to capture specific molecules, such as pesticides, heavy metals, or other environmental contaminants. The method can also be extended to any synthesis procedure, providing an acid or base-labile group is present.

In the early developmental stages of either a microarray or a large sensor technique, the starting point is typically one or two electrodes. Much of the recorded electrochemical sensor data are based on just a few electrodes, as a particular technique may or may not be converted to a microarray. The decision to convert to a high-density array is dependent on many parameters such as reading times and hardwire issues. Detection methods in microarrays employ various techniques, including: fluorescence, luminescence, visible, electrochemical, Raman scattering, SPR, and electro-chemiluminescence, among others. The detection method used depends on the matrix and if the chip is hardwired. Typically, the light-based method can accommodate almost any matrix and production method. However, a laser scanner or CCD camera is required, which tends to be very expensive, increasing start-up costs. Electrochemical methods require chip hardwire in tandem with various detection methods. Amperometric detection, cyclic voltammetry, or the evaluation of a charge build-up on the electrode surface have all been employed.

The development of new microarray formats is keeping pace with new developments in nanotechnology, binding reagents, and detection systems. The pesticide metabolite 2,6-dichlorobenzamide (BAM) was detected in water samples using a microarray employing gold nanoparticle labels and a desktop scanner [79]. The signal from the scanner was proportional to the surface density of the bound antibody–gold conjugates which could be detected at a level of six attomoles. The microarray could qualitatively determine if BAM was present below or above the 100 ng/L target concentration for drinking water.

Another new unique detection technology being used in small microarrays is "cantilevers". Originally developed at Oak Ridge National Labs by Dr. Thundat [80], the principle is very simple. The samples are at the tip of a cantilever, which acts as a springboard. The amount of material on the cantilever changes the oscillation frequency. Thus, the attached antibody capturing the analyte assay changes the oscillation frequency at that cantilever. A laser is usually required to scan the cantilever.

Aptamers are single-stranded oligonucleotide sequences that can bind to many types of molecules with high affinity and specificity, similar to antibody–antigen binding. An aptamer-based microcantilever-array to detect profenofos was developed and applied to vegetable extracts [81]. The microcantilevers were functionalized with an aptamer specific for profenofos. The binding of profenofos to the aptamer induced a deflection of the microcantilever which could be detected in real-time using an optical system. Deflection of the microcantilever was positively correlated with the concentration of profenofos and could detect 1.3 ng/mL of the compound.

4.3.3 BIOSENSORS METHODS FOR PESTICIDES

Several types of biosensors have been developed for measuring pesticides in various sample media. However, the use of biosensors for obtaining environmental measurements is not as common as for immunoassay. This section presents the application of biosensor techniques for detecting pesticides and illustrates the potential of various sensor designs for environmental monitoring.

4.3.3.1 Potentiometric, Light Addressable Potentiometric Sensor (LAPS), and Amperometric Detection

Organophosphorus pesticides may be detected in a number of ways including potentiometric or amperometric methods. In both of these cases, enzymes such as organophosphorus hydrolase or urease may be employed. Dependent upon the structure of the analyte, the release of hydrogen ions can either be measured via a pH change or a p-nitrophenol group may be produced to give a redox compound for an electron shuttle.

Molecular Devices employs the use of a "Light Addressable Potentiometric Sensor" (LAPS) for detection. The samples are captured on membranes via vacuum filtration into discreet spots on a membrane [82]. The detection is pH-based, utilizing a sensitive LAPS method that can detect the urease enzyme conversion of urea in a pH-sensitive manner (potentiometric readings). For example, fluorescein-labeled anti-atrazine antibodies and atrazine covalently linked to biotin-DNP were used as reagents for atrazine detection. When the fluorescein-labeled antibody is bound to the biotinylated atrazine, the complex will bind to the streptavidin-coated membrane. If non-biotinylated atrazine (from the sample) is added to the mix, any antibody bound to this species will be washed away. Thus, in this competitive assay format, the fluorescein-labeled anti-atrazine antibody can either bind to the non-labeled or biotin-labeled atrazine. A species-specific secondary antibody labeled with urease reacts with the bound anti-atrazine antibody to generate a pH flux, providing the signal for the LAPS sensor. In this mode, there is an inverse relationship between signal and amount of non-labeled analyte found in solution. The largest signal output is seen when there is only atrazine present and the lowest signal is observed when a large quantity of non-labeled atrazine is present. Thus, if there is a large amount of environmental atrazine being measured the signal will be low. The result is a sigmoidal curve similar to the one shown in Figure 4.3 for the ELISA to detect 3-PBA. It should be noted that the detection range tends to be narrow with this format (due to the sigmoidal curve) and the sensitivity can be limited.

This assay is typically classified as a biosensor, as eight simultaneous assays can be performed using this system. In addition to using a fluorogenic substrate for assessing the effects of OP compounds, other means may be used to obtain information about these compounds in environmental samples. One of the simplest techniques is a potentiometric sensor based on pH changes. In this case, a simple biosensor that is sensitive to changes in pH would be adequate. The enzyme organophosphorus hydrolase needs only to be attached to the electrode, encompassed in a polymer, and attached to a bioresin over the electrode. Organophosphorus hydrolase catalyzes the hydrolysis of a wide range of OP pesticides (e.g., coumaphos, diazinon, dursban, ethyl parathion, methyl parathion, and paraoxon). The attached or trapped hydrolase then acts upon the OP compound to produce an alcohol and an acid. The resulting acid compound is monitored as a pH change at the electrode. This is a very simple system to use and is similar to LAPS detection.

Mulchandani and co-workers developed an assay where organophosphorus hydrolase was placed onto an electrode [83]. The phosphate hydrolysis product was monitored by measuring the current produced at the electrode. The output of the amperometric sensor could be correlated to the concentration of pesticide in sample solutions of soil and vegetation.

Another biosensor method is applicable to other OP compounds that produce PNP (p-nitrophenol) as a releasing compound. These compounds include ethyl parathion, methyl parathion, paraoxon, fenithrothion, and O-ethyl O-(4-nitrophenyl) phenylphosphonotioate (EPN). The released PNP is oxidized at the anode to insert a hydroxyl group that is ortho to the nitro group. In this case, the oxidation current is measured amperometrically at a fixed potential. The signal is linear to the concentration of PNP present. The analysis relies on the OP compound to be trapped or conjugated to material on the electrode.

The use of vibrational spectroscopy for sensor development has also been reported. The technique can be utilized in the form of Raman spectroscopy or simply infrared spectroscopy [84]. Although it is not the most sensitive detection, it can be useful and perhaps less expensive for use in some applications.

4.3.3.2 Piezoelectric Measurements

Many pesticides (e.g., organophosphates and carbamates) or their metabolites are cholinesterase inhibitors. This phenomenon can be used to develop sensors for the detection of these types of compounds. Using a piezoelectric sensor format, paraoxon was bound to an electrode (gold on a piezo/quartz surface) as the recognition element [85]. The analysis was performed by allowing a cholinesterase to interact with the modified electrode surface and with free paraoxon in a standard or sample. An oscillation change can be observed in terms of Hz or an electronic occurrence. A competitive assay was developed that allowed competition for cholinesterase between a cholinesterase inhibiting pesticide in solution and the inhibitor bound to the electrode surface. The ability of cholinesterase to bind to the paraoxon immobilized on the electrode is minimized or prevented in the presence of free inhibitor (analyte) in solution. In this case, the cholinesterase remains in solution bound to the pesticide in the sample. The sensing surface can be regenerated for reuse. The format can be used to develop better inhibitors and to quantitate OP compounds in solutions of environmental samples.

4.3.3.3 Surface Plasmon Resonance

SPR technology has been used in the biosensor field for some time and many sensors of this type are commercially available. The technique depends on the change in the reflectance angle (Plasmon) due to mass changes at the surface. Binding of proteins and small materials change the mass number at the surface and the reflectance angle is altered [86, 87]. SPR detection has demonstrated utility for many types of compounds. Initially, the technique was applied only to large molecules but as the technology has matured so has its potential for monitoring various pesticides, including photosynthetic inhibitors.

The crux of the system is a gold film on a glass surface. Attached to the gold film are self-assembled monolayers (SAMs) and capture reagents. These capture reagents may be antibodies, receptors, enzymes, ssDNA, streptavidin, and protein A or G (dependent upon the type of antibody used), as well as other reagents. As the specific species is captured, the mass on the chip surface increases and changes the specific reflection angle. In this technique, a herbicide such as atrazine may be detected in several modes. The simplest mode would be to attach an anti-atrazine antibody (as a whole or in parts) to the chip surface. If the solution being tested shows atrazine present, a signal response on the chip would be detected.

Another option would be to attach the photosynthetic reaction center (RC) from a purple bacterium to the sensing chip. This can be accomplished in a number of ways, but literature evidence suggests that histidine (His) tags can be conveniently used. The system can easily be reused as the RC can be removed and the chip regenerated once the assay is completed. Samples of atrazine are introduced and the signal is monitored. A positive response can be quantitated and the chip can be reactivated for the next sample.

4.3.3.4 Conductive Polymers

Conductive polymers may lead to an increase in the use of electrochemical detection methods for pesticides [88]. Many conductive polymers can be formed *in situ* directly over the electrode. Another added benefit is that interference from sample components is often reduced. Most of the polymers that have been used are electrochemically derived (synthesized *in situ*), formed by a host of starting materials. Additionally, many can be tethered to electrochemical conducting wires or even be encapsulated in a biopolymer matrix such as microgels [88–93]. A sensor using an electrodeposited conductive layer to detect the herbicide diuron [94] could also be applied to other substituted urea compounds.

For this technique to function, an enzymatic system is often used, such as glucose oxidase. Other enzymes may be employed, depending on the nature of the biosensor being developed and the anticipated monitoring applications. One application that appears to dominate for commercial development is that of a glucose sensor. Glucose is converted to gluconic acid and amperometric signals are observed based upon the production of hydrogen peroxide. The polymer may encapsulate the electrode or be placed on the electrode using microparticle slurries.

Another polymer that can be used is a water-soluble Os-poly(vinyl imidazole) redox hydrogel. Again, the electron transfer is very efficient and necessitates a redox enzyme being placed in the gel. A polypyrrole film has also been used in conjunction with NADH+ ferro/ferri cyanide redox chemistries. An enzyme is required whose function is to use NADP+ in conjunction with an enzymatic substrate to release a product and the co-factor, NADPH. The ferricyanide is present to efficiently shuttle the electrons.

There are also reports on the use of PVPOs(bpy) polymer and poly(maercapto-*p*-benzoquinone) on gold electrodes or within conducting hydrogels. For these systems, the redox enzyme horseradish peroxidase is used, or the cyclic voltammetry (CV) of the substrate, sulpho-*p*-benzoquinone (SBQ) is monitored. The types of solid supports and electrochemical methods are almost limitless.

4.4 ONGOING DEVELOPMENTS

Immunochemical methods can either be performed independently or coupled with other analytical techniques to produce powerful tandem methods for pesticide analysis. There are many opportunities to utilize immunoaffinity separation techniques coupled with immunoassay and/or instrumental methods to support environmental monitoring studies including:

- Immunoaffinity chromatographic separation of a group of structurally similar pesticides. This may be accomplished by utilizing either the high cross-reactivity of an antibody to a certain group of pesticides or using mixed antibodies that possess a combined affinity with a pesticide group.
- Hybrid affinity separation of multiple pesticides based on the integration of immunoaffinity chromatography and surface imprinting techniques. Hybrid affinity columns can be prepared by mixing one or more antibodies with one or more types of molecularly imprinted polymers.

A combination of on-line immunoaffinity separation with liquid chromatography–mass spectrometry (LC–MS) can provide rapid separation and detection of pesticides with a high degree of selectivity and sensitivity. Similar combinations can also be performed between immunoaffinity separation and flow injection analysis. The combination of immunoassay and sample preparation techniques such as ESE, QuEChERS, or SPE, or the on-line integration of an extraction procedure, immunoaffinity cleanup with an immunochemical or instrumental detection procedure can provide efficient analytical methods.

Molecularly imprinted polymers (MIPs) and aptamers have emerged as reagents (i.e., artificial antibodies) for pesticide immunoassays and immunosensors. These reagents have the potential to provide large amounts of reagents for methods development, alleviating the concern regarding a robust supply of key reagents. Some MIP-based affinity separation methods and biosensors have been developed for the extraction and determination of pesticides in aqueous samples. The continued development of aptamers, artificial nucleic acid ligands, to detect small and large molecules will increase the reagent supply for development of cost-effective methods.

The continued development of new formats, detection systems, and multi-analyte immunoassays will all continue to drive the use of immunoassays for pesticide and exposure monitoring [95–100].

4.5 FUTURE TRENDS

Immunoassay is a mature analytical technology with broad application to pesticide analysis. Extensive fundamental investigations, as well as technical improvements will make immunoassay methods more powerful tools for the identification and determination of a variety of pesticides. New breakthroughs in the development and application of immunoassays will result from the integration of future state-of-the-art research in several key areas including production of antibodies and antibody mimics, nanotechnology, new platforms, and detection systems.

Future research that may enhance the use of immunoassays and immunosensors for pesticide analysis is the development of novel antibodies for individual pesticide compounds, as well as reagents for more multi-analyte analyses. Improvements in nanobody production could provide selective and sensitive binding reagents that are more thermally stable and have a higher resistance to solvents than traditional antibodies [101]. Welcomed improvements also include the design and synthesis of new haptens using the latest concepts and techniques, better understanding and control of the combination of hapten molecules and macromolecular carriers through molecular modeling and 3-D visualization, and improving the efficiency of existing laboratory procedures. New platforms may also be integrated with new labels such as more robust enzymes, or highly sensitive visualization techniques such as laser-induced fluorescence detection (LIF) to produce even lower limits of detection.

Competitive immunoassay formats are typically employed for small molecules due to the single epitope available for antibody binding. Non-competitive sandwich assay formats for larger molecules tend to be more sensitive with good linearity. However, the target molecules in this format must have at least two sites for antibody binding. Research in this format for small molecules includes the development of antibodies that do not recognize the target analyte or its specific antibody. Rather, the reagent antibody recognizes the immunocomplex of target analyte-antibody [102]. Detection is proportional to the concentration of analyte present.

Nanotechnology is a rapidly growing discipline of scientific research and is being applied to a wide variety of fields including immunoassay and biosensor development. Nanomaterials with dimensions of less than 100 nanometers have physical and chemical properties that make them attractive for many applications requiring high strength, conductivity, durability, and reactivity. The application of nanotechniques in immunoassays has been of great interest to researchers [103]. New detection strategies based on gold and silver particles for immunoassay labeling have been successfully demonstrated to meet the needs of diverse detection methods. These particles have been used for various techniques such as scanning and transmission electron microscopy, Raman spectroscopy, and sight visualization due to their easily controlled size distribution, and long-term stability and compatibility with bio-macromolecules.

Nanoparticle-labeled microfluidic immunoassays have shown their unique advantages over conventional immunoassay formats for the detection of small molecules, macromolecules, and microorganisms. Sub-micron-sized striped metallic rods intrinsically encoded through differences in reflectivity of adjacent metal stripes have been used in autoantibody immunoassays. These bar-coded particles act as supports with antigens attached to the surface providing a permanent tag for the tracking of analyte [104].

Nanomaterials including gold, zirconia (ZrO_2), and carbon nanotubes have been applied as biosensors for monitoring OP pesticides [105–107]. An optical sensor based on fumed silica gel functionalized with gold nanoparticles has also been reported for OP pesticides [107]. Nanoparticles possess extraordinary optical properties that may offer alternative strategies for the development of optical sensors. An electrochemical sensor for detection of OP pesticides has been developed using ZrO_2 nanoparticles as selective sorbents possessing a strong affinity for the phosphoric group. The nitroaromatic OPs strongly bind to the ZrO_2 surface. A square-wave voltammetric analysis was used to monitor the amount of bound OP pesticide. Another sensitive flow injection amperometric biosensor for OP pesticides and nerve agents was developed using self-assembled acetylcholinesterase (AchE) on a carbon nanotube (CNT)-modified glassy carbon electrode [106]. The CNTs have two main functions for the biosensor; first, as platforms for AchE immobilization by providing a microenvironment that can maintain the bioactivity of AchE, and second, as a transducer for amplifying the electrochemical signal of the product of the enzymatic reaction. The integration of nano- and biomaterials could be extended to other biological molecules for future biosensor or immunoassay research.

Advancements in biosensor technology will also continue with expansion of multi-analyte detection and more rapid analytical capability. For example, a chip containing 92,000 electrodes with a 30-microsecond read has been investigated. With a 30-microsecond read time, enzymatic kinetic reads could be performed directly on the chip. However, the capability of 92,000 electrodes × 1000 reads presents storage, data acquisition, and conversion issues. The limiting factor at this time is computer capability. Other technologies such as a 40-second kinetic read of 12,000 electrodes with four or eight electrodes discharged at one time in microsecond intervals are near realization. A wealth of research into nanomaterials and biosensors including sample preparation strategies to analyze food samples can be found in [108–113]. The ultimate biosensor format would allow rapid analyte detection with real-time confirmatory analysis such as by direct coupling with instrumental detection.

Through future research, immunoassays and biosensors for pesticides and pesticide metabolites may enable additional applications for *in vitro* and *in vivo* studies in the interdependent fields of food safety and human exposure.

REFERENCES

1. Baker, S.R., and Wilkinson, C.F. The effects of pesticides on human health in *Advances in Modern Environmental Toxicology*, Vol. XVIII, Princeton Scientific Publishing, Princeton, NJ, p. 438, 1990.
2. Van Emon, J.M. (Ed.) *Immunoassay and Other Bioanalytical Techniques*, CRC Press, Taylor & Francis Group, Boca Raton, FL, 2007.
3. NRC, National Research Council. *Pesticides in the Diets of Infants and Children*, National Academy Press, Washington, DC, p. 386, 1993.

4. FQPA, Food Quality Protection Act of 1996, Public Law, 104–170, 1996.
5. Hammock, B.D., and Mumma, R.O. Potential of immunochemical technology for pesticide analysis in *Pesticide Analytical Methodology*, Harvey, J.J., and Zweig, G., (Eds.), American Chemical Society, Washington, D.C., pp. 321–352, 1980.
6. Van Emon, J.M., and Lopez-Avila, V. Immunochemical methods for environmental analysis, *Anal. Chem.*, 64 (2), 79A–88A, 1992.
7. Chuang, J.C., Miller, L.S., Davis, D.B., Peven, C.S., Johnson, J.C., and Van Emon, J.M. Analysis of soil and dust samples for polychlorinated biphenyls by enzyme-linked immunosorbent assay (ELISA), *Anal. Chim. Acta*, 376, 67–75, 1998.
8. Chuang, J.C., Van Emon, J.M., Chou, Y.-L., Junod, N., Finegold, J.K., and Wilson, N.K. Comparison of immunoassay and gas chromatography-mass spectrometry for measurement of polycyclic aromatic hydrocarbons in contaminated soil, *Anal. Chim. Acta*, 486, 31–39, 2003.
9. Khosraviani, M., Pavlov, A.R., Flowers, G.C., Blake, D.A. Detection of heavy metals by immunoassay: optimization and validation of a rapid, portable assay for ionic cadmium, *Environ. Sci. Technol.*, 32, 137–142, 1998.
10. Nichkova, M., Park, E., Koivunen, M.E., Kamita, S.G., Gee, S.J., Chuang, J.C., Van Emon, J.M., and Hammock, B.D. Immunochemical determination of dioxins in sediment and serum samples, *Talanta*, 63, 1213–1223, 2004.
11. Van Emon, J.M., and Gerlach, C.L. A status report on field-portable immunoassay, *Environ. Sci. Technol.*, 29 (7), 312A–317A, 1995.
12. Rodriguez-Mozaz, S., Lopez de Alda, M.J., and Barcelo, D. Fast and simultaneous monitoring of organic pollutants in a drinking water treatment plant by a multi-analyte biosensor followed by LC–MS validation, *Talanta*, 69 (2), 384, 2006.
13. Marco, M., and Barcelo, D. Environmental applications of analytical biosensors, *Meas. Sci. Technol.*, 7, 1547–1562, 1996.
14. Yang, Y., Guo, M., Yang, M., Wang, Z., Shen, G., and Yu, R. Determination of pesticides in vegetable samples using an acetylcholinesterase biosensor based on nanoparticles ZrO_2/chitosan composite film, *Int. J. Environ. Anal. Chem.*, 85 (3), 163–175, 2005.
15. Zhang, J. Tube-immunoassay for rapid detection of carbaryl residues in agricultural products, *J. Environ. Sci. Health Part B*, 41 (5), 693–704, 2006.
16. Garcia-Reyes, J.F., Llorent-Martinez, E.J., Ortega-Barrales, P., and Molina-Diaz, A. Determination of thiabendazole residues in citrus fruits using a multicommuted fluorescence-based optosensor, *Anal. Chim. Acta*, 557 (1–2), 95–100, 2006.
17. Zacco, E., Galve, R., Marco, M.P., Alegret, S., and Pividori, M.I. Electrochemical biosensing of pesticide residues based on affinity biocomposite platforms, *Biosens. Bioelectron.*, 22 (8), 1707–1715, 2007.
18. Verma, N., and Bhardwaj, A. Biosensor technology for pesticides – a review, *Appl. Biochem. Biotechnol.*, 175 (6), 3093–3119, 2015.
19. McGrath, T.F., Elliott, C.T., and Fodey, T.L. Biosensors for the analysis of microbiological and chemical contaminants in food, *Anal. Bioanal. Chem.*, 403, 75–92, 2012.
20. Whangsuk, W., Thiengmag, S., Dubbs, J., Mongkolsuk, S., and Loprasert, S. Specific detection of the pesticide chlorpyrifos by a sensitive genetic-based whole cell biosensor. *Anal. Biochem.*, 493, 11–13, 2016.
21. Khaled, E., Kamel, M.S., Hassan, H.N.A., Abdel-Gawad, H., and Aboul-Enein, H.Y. Performance of a portable biosensor for the analysis of ethion residues, *Talanta*, 119, 467–472, 2014.
22. Van Emon, J.M., and Gerlach, C.L. Environmental monitoring and human exposure assessment using immunochemical techniques, *J. Microbiol. Methods*, 32, 121–131, 1998a.
23. Van Emon, J.M. Immunochemical Applications in Environmental Science, *J. AOAC Int.*, 84 (1), 125, 2001.

24. Voller, A., Bidwell, D.E., and Bartlett, A. Microplate enzyme immunoassays for the immunodiagnosis of virus infections in *Manual of Clinical Immunolory*, Rose, N., and Friedman, H., (Eds.), American Society for Microbiology, Washington, DC, pp. 506–512, 1976.

25. Gee, S.J., Hammock, B.D., and Van Emon, J.M. *A User's Guide to Environmental Immunochemical Analysis, EPA/540/R-94/509*, March 1994, United States Environmental Protection Agency, Office of Research and Development, Washington, DC.

26. Jaeger, L.L., Jones, A.D., Hammock, B.D. Development of an enzyme-linked immunosorbent assay for atrazine mercapturic acid in human urine, *Chem. Res. Toxicol.*, 11, 342–352, 1998.

27. Chuang, J.C., Van Emon, J.M., Durnford, J., and Thomas, K. Development and evaluation for an enzyme-linked immunosorbent assay (ELISA) method for the measurement of 2,4-dichlorophenoxyacetic acid in human urine, *Talanta*, 67, 658–666, 2005.

28. Thurman, E.M., Meyer, M., Pomes, M., Perry C.A., and Schwab, A.P. Enzyme-linked immunosorbent assay compared with gas chromatography/mass spectrometry for the determination of triazine herbicides in water, *Anal. Chem.*, 62, 2043–2048, 1990.

29. Wong, J.M., Li, Q.X., Hammock, B.D., and Seiber, J.N. Method for the analysis of 4-nitrophenol and parathion in soil using supercritical fluid extraction and immunoassay, *J. Agric. Food Chem.*, 39, 1802–1807, 1991.

30. Chuang, J.C., Pollard, M.A., Misita, M., and Van Emon, J.M. Evaluation of analytical methods for determining pesticides in baby food. *Anal. Chim. Acta*, 399, 135–142, 1999.

31. Ahn, K.C., Ma, S., and Tsai, H. An immunoassay for a urinary metabolite as a biomarker of human exposure to the pyrethroid insecticide permethrin, *Anal. Bioanal. Chem.*, 384, 713–722, 2006.

32. Koivunen, M.E., Dettmer, K., Vermeulen, R., Bakke, B., Gee, S.J., and Hammock, B.D. Improved methods for urinary atrazine mercapturate analysis – assessment of an enzyme-linked immunosorbent assay (ELISA) and a novel liquid chromatography – mass spectrometry (LC–MS) method utilizing online solid phase extraction (SPE), *Anal. Chim. Acta*, 572, 180–189, 2006.

33. Zhao, F., Hu, C., Wang, H., Zhao, L., Yang, Z. Development of a MAb-based immunoassay for the simultaneous determination of *O,O*-diethyl and *O,O*-dimethyl organophosphorus pesticides in vegetable and fruit samples pretreated with QuEChERS. *Anal. Bioanal. Chem.*, 407, 8959–8970, 2015.

34. Chuang, J.C., Van Emon, J.M., Finegold, K., Chou, Y.-L., Rubio, F. Immunoassay method for the determination of pentachlorophenol in soil and sediment. *Bull. Environ. Contam. Toxicol.*, 76 (3), 381–388, 2006.

35. Van Emon, J.M., Hammock, B., and Seiber, J.N. Enzyme-linked immunosorbent assay for paraquat and its application to exposure analysis, *Anal. Chem.*, 58, 1866, 1986.

36. Van Emon, J.M., Seiber, J.N., and Hammock, B.D. Application of an enzyme-linked immunosorbent assay to determine paraquat residues in milk, beef, and potatoes, *Bull. Environ. Contam. Toxicol.*, 39, 490, 1987.

37. Van Emon, J.M., Seiber, J.N., and Hammock, B.D. Immunoassay techniques for pesticide analysis in *Analytical Methods for Pesticides and Plant Growth Regulators, Advanced Analytical Techniques*, Vol. XVII, Sherma, J. (Ed.), Academic Press, San Diego, CA, 1989, 217–263.

38. Nam, K., and King, J.W. Supercritical fluid extraction and enzyme immunoassay for pesticide detection in meat products, *J. Agric. Food Chem.*, 42, 1469–1474, 1994.

39. Richman, S.J., Karthikeyan, S., Bennett, D.A., Chung, A.C., and Lee, S.M. Low-level immunoassay screen for 2,4-dichlorophenoxyacetic acid in apples, grapes, potatoes, and oranges: circumventing matrix effects, *J. Agric. Food Chem.*, 44, 2924–2929, 1996.

40. Yeung, J.M., Mortimer, R.D., and Collins, P.G. Development and application of a rapid immunoassay for difenzoquat in wheat and barley products, *J. Agric. Food Chem.*, 44, 376–380, 1996.

41. Bashour, I.I., Dagher, S.M., Chammas, G.I., Kawar, N.S. Comparison of gas chromatography and immunoassay methods for analysis of total DDT in calcareous soils, *J. Environ. Sci. Health, Part B Pestic. Food Contam. Agric. Wastes*, B38 (2), 111–119, 2003.

42. Botchkareva, A.E., Eremin, S.A., Montoya, A., Marcius, J.J., Mickova, B., Rauch, P., Fini, F., and Girotte, S. Development of chemiluminescent ELISAs to DDT and its metabolites in food and environmental samples, *J. Immuno. Methods*, 283 (1–2), 45–57, 2003.

43. Chuang, J.C., Van Emon, J.M., Reed, A.W., and Junod, N. Comparison of immunoassay and gas chromatography–mass spectrometry methods for measuring 3,5,6-trichloro-2-pyridinol in multiple sample media, *Anal. Chim. Acta*, 517 (1–2), 177–185, 2004.

44. Lee, J.K., Ahn, K.C., Stoutamire, D.W., Gee, S.J., and Hammock, B.D. Development of an enzyme-linked immunosorbent assay for the detection of the organophosphorus insecticide acephate, *J. Agric. Food Chem.*, 51, 3695–3703, 2003.

45. Lee, J.K., Kim, Y.J., Lee, E.Y., Kim, D.K., and Kyung, K.S. Development of an ELISA for the detection of fenazaquin residues in fruits, *Agric. Chem. Biotech.*, 48 (1), 16–25, 2005.

46. Kolosova, A.Y., Park, J., Eremin, S.A., Park, S., Kang, S., Shim, W., Lee, H., Lee, Y., and Chung, D. Comparative study of three immunoassays based on monoclonal antibodies for detection of the pesticide parathion-methyl in real samples, *Anal. Chim. Acta*, 511 (2), 323–331, 2004.

47. Brun, E.M., Marta, G., Puchades, R., and Maquieira, A. Highly sensitive enzyme-linked immunosorbent assay for chlorpyrifos, application to olive oil analysis, *J. Agric. Food Chem.*, 53 (24), 9352–9360, 2005.

48. Morozova, V.S., Levashova, A.I., and Eremin, S.A. Determination of pesticides by enzyme immunoassay, *J. Anal. Chem.*, 60 (3), 202–217, 2005.

49. Koivunen, M.E., Gee, S.J., Park, E.-K., Lee, K., Schenker, M.B., Hammock, B.D. Application of an enzyme-linked immunosorbent assay for the analysis of paraquat in human-exposure samples, *Arch. Environ. Contam. Toxicol.*, 48, 184–190, 2005.

50. Gabaldon, J.A., Maquieria, A., and Puchades, R. Development of a simple extraction procedure for chlorpyrifos determination in food samples by immunoassay, *Talanta*, 71 (3), 1001–1010, 2007.

51. Watanabe, E., Miyake, S., Baba, K., Eun, H., and Endo, S. Immunoassay for acetamiprid detection: application to residue analysis and comparison with liquid chromatography, *Anal. Bioanal. Chem.*, 386 (5), 1441–1448, 2006a.

52. Watanabe, E., Baba, K., Eun, H., Arao, T., Ishii, Y., Ueji, M., and Endo, S. Evaluation of performance of a commercial monoclonal antibody-based fenitrothion immunoassay and application to residual analysis in fruit samples, *J. Food Prot.*, 69 (1), 191–198, 2006b.

53. Watanabe, E., and Miyake, S. Immunoassay for iprodione: key estimation for residue analysis and method validation with chromatographic technique, *Anal. Chim. Acta*, 583 (2), 370–376, 2007a.

54. Watanabe, E., Baba, K., Eun, H., and Miyake, S. Application of a commercial immunoassay to the direct determination of insecticide imidacloprid in fruit juices, *Food Chem.*, 102 (3), 745–750, 2007b.

55. Shan, G., Huang, H., Stoutamire, W., Gee, J., Leng, G., and Hammock, B.D. A sensitive class specific immunoassay for the detection of pyrethroid metabolites in human urine, *Chem. Res. Toxicol.*, 17, 218–225, 2004.

56. CDC. *Third National Report on Human Exposure to Environmental Chemicals.* Census for Diseases Control and Prenention, Altanta, GA, 2005. http://www.cdc.gov/exposurereport/

57. Van Emon, J.M., Gerlach, C.L., and Bowman, K. Bioseparation and bioanalytical techniques in environmental monitoring, *J. Chromatogr. B.*, 715, 211, 1998b.

58. Chuang, J.C., Van Emon, J.M., Jones, R., Durnford, J., and Lordo, R. Development and application of immunoaffinity column chromatography for atrazine in complex sample media, *Anal. Chim. Acta*, 583, 32–39, 2007.

59. Molecular Devices SOFTmax® PRO User Manual. Molecular Devices, Sunnyvale, CA, 1998.

60. Sendecor, G.W., and Cochran, W.G. *Statistical Methods*, Eighth Edition. Iowa State University Press, Ames, IA, 1989.

61. Hollander, M., and Wolfe, D.A. *Nonparametric Statistical Methods*. John Wiley and Sons, Inc., New York, 1973.

62. Rosner, B. *Fundamentals of Biostatistics*, Fifth Edition. Duxbury Press, Inc., North Scituate, MA, 2000.

63. Pundir, C.S., and Chauhan, N. Acetylcholinesterase inhibition-based biosensors for pesticide determination: a review. *Anal. Biochem.*, 429, 19–31, 2012.

64. Songa, E.A., and Okonkwo, J.O. Recent approaches to improving selectivity and sensitivity of enzyme-based biosensors for organophosphorus pesticides: a review. *Talanta*, 155, 289–304, 2016.

65. Van Dyk, J.S., and Pletschke, B. Review on the use of enzymes for the detection of organochlorine, organophosphate and carbamate pesticides in the environment. *Chemosphere*, 82, 291–307, 2011.

66. Minunni, M., and Mascini, M. Detection of pesticide in drinking water using real-time biospecific interaction analysis (BIA), *Anal. Lett.*, 26, 1441–1460, 1993.

67. Fu, Y., Yuan, R., Chai, Y., Zhou, L., Zhang, Y. Coupling of a reagentless electrochemical DNA biosensor with conducting polymer film and nanocomposite as matrices for the detection of the HIV DNA sequences, *Anal. Lett.* 39 (3), 1532–1236, 2006.

68. Cho, Y.A., Cha, G.S., Lee, Y.T., and Lee, H. A dipstick-type electrochemical immunosensor for the detection of the organophosphorus insecticide fenthion. *Food Sci. Biotech.*, 14 (6), 743–746, 2005.

69. Longobardi, F., Solfrizzo, M., Compagnone, D., Del Carlo, M., and Visconti, A. Use of electrochemical biosensor and gas chromatography for determination of dichlorvos in wheat, *J. Agric. Food Chem.*, 53 (24), 9389–9394, 2005.

70. Nikolelis, D.P., Simantiraki, M.G., Siontorou, C.G., and Toth, K. Flow injection analysis of carbofuran in foods using air stable lipid film based acetylcholinesterase biosensor, *Anal. Chim. Acta*, 537 (1–2), 169–177, 2005.

71. Zacco, E., Pividori, M.I., Alegret, S., Galve, R., and Marco, M.P. Electrochemical magnetoimmunosensing strategy for the detection of pesticides residues, *Anal. Chem.*, 78 (6), 1780–1788, 2006.

72. Zhang, C., Shang, Y., and Wang, S. Development of multianalyte flow-through and lateral-flow assays using gold particles and horseradish peroxidase as tracers for the rapid determination of carbaryl and endosulfan in agricultural products, *J. Agric. Food Chem.*, 54, 2502–2507, 2006.

73. Zheng, Y., Hua, T., Sun, D., Xiao, J., Xu, F., and Wang, F. Detection of dichlorvos residue by flow injection calorimetric biosensor on immobilized chicken liver esterase, *J. Food Eng.* 74 (1), 24–29, 2006.

74. Dill, K., Lui, R., and Grodzinski, P. (Eds.) *Microarrays: Preparation, Microfluidics, Detection Methods, and Biological Applications*, Springer Verlag, Berlin, Germany, 2009.

75. Zhao, S., Dong, J., Jeong, H., Okumura, K., and Ueda, H. Rapid detection of the neonicotinoid insecticide imidacloprid using a quenchbody assay, *Anal. Bioanal. Chem.*, 410, 4219–4226, 2018.

76. Abe, R., Jeong, H.-J., Arakawa, D., Dong, J., Ohashi, H., Kaigome, R., Saiki, F., Yamane, K., Takagi, H., and Ueda, H. Ultra Q-bodies: quench-based antibody probes that utilize dye-dye interactions with enhanced antigen-dependent fluorescence. *Sci. Rep.*, 4, 4640, 1–9, 2014.

77. Jeong, H.-J., Itayama, S., and Ueda, H. A signal-on fluorosensor based on quench-release principle for sensitive detection of antibiotic rapamycin, *Biosensors*, 5, 131–140, 2015.

78. Dill, K., Grodzinsky, P., and Liu, R. (Eds.) *Recent Advances in Microarray Technology*, Springer Verlag, Berlin, Germany, 2007.

79. Han, A., Dufva, M., Belleville, E., and Christensen, C.B.V. Detection of analyte binding to microarrays using gold nanoparticle labels and a desktop scanner, *LAB Chip.*, 3, 329–332, 2003.

80. Yue, M., Majumdar, A., and Thundat, T. Cantilever arrays: a universal platform for multiplexed label-free bioassays. Chapter 2 in *BioMEMS and Biomedical Nanotechnology*, Bashir, R. and Werely, S.T. (eds.), Springer–Verlag, Berlin, Germany, pp. 21–33, 2007.

81. Li, C., Zhang, G., Wu, S., and Zhang, Q. Aptamer-based microcantilever-array biosensor for profenofos detection, *Anal. Chim. Acta*, 1020, 116–122, 2018.

82. Dill, K. Sensitive analyte detection and quantitation using the threshold immunoassay system (chapter 9) in *Environmental Immunochemical Methods*, Van Emon, J.M., Gerlach, C.L., and Johnson, J.C., (Eds.), ACS symposium series 646, American Chemical Society, Washington, DC, 89–102, 1996.

83. Mulchandani, A., Chen, W., Mulchandani, P., Wang, J., and Rogers, K.R. Biosensors for direct determination of organophosphate pesticides, *Biosens. Bioelectron.*, 16, 225–230, 2001.

84. Neugebauer, U., Kurz, C., Bocklitz, T., Berger, T., Velten, T., Clement, J., Krafft, C., and Popp, J. Raman-spectroscopy based cell identification on a microhole array chip, *Micromachines*, 5, 204–215, 2014.

85. Makower, A., Hlamek, J., Skladal, P., Kerchen, F., and Scheller, F.W. New principle of direct real-time monitoring of the interaction of cholinesterase and its inhibitors by piezoelectric biosensor, *Biosens. Bioelectron.*, 18, 1329–1337, 2003.

86. Nakamura, C., Hasegawa, M., Nakamura, N., and Miyake, J. Rapid and specific detection of herbicides using a self-assembled photosynthetic reaction center from purple bacterium on a SPR chip, *Biosen. Bioelectron.*, 18, 599–603, 2003.

87. Strachan, G., Grant, S.D., Learmonth, D., Longstaff, M., Porter, A.J., and Harris, W.J. Binding characteristics of anti-atrazine monoclonal antibodies and their fragments sythesized in bacteria and plants, *Biosens. Bioelectron.*, 13, 665–673, 1998.

88. Liu, X., Neoh, K.G., and Kang, E.T. Enzymatic activity of glucose oxidase covalently wired via viologen to electrically conductive polypyrrole films, *Biosens. Bioelectron.*, 19, 823–834, 2004.

89. Retama, J.R., Cabarcos, E.L., Mecerreyes, D., and Lopez-Ruiz, B. Design of an amperometric biosensor using polypyrrole-microgel composites containing glucose oxidase, *Biosens. Bioelectron.*, 20, 1111–1117, 2004.

90. Chen, C., Jiang, Y., and Kan, J. A noninterference polypyrrole glucose biosensor, *Biosen. Bioelectron.*, 22, 639–643, 2006.

91. Vilkanauskyte, A., Erichsen, T., Marcinkeviciene, L., Laurinavicius, V., and Schuhmann, W. Reagentless biosensors based on co-entrapment of a soluble redox polymer and an enzyme within an electrochemically deposited polymer film, *Biosens. Bioelectron.*, 17, 1025–1031, 2002.

92. Lopez, M.A., Ortega, F., Dominguez, E., and Katakis, I. Electrochemical immunosensor for the detection of atrazine, *J. Mol. Recognit.*, 11, 178–181, 1998.

93. Gross, P., and Comtat, M. A bioelectrochemical polypyrrole-containing Fe(CN)63-interface for the design of a NAD-dependent reagentless biosensor, *Biosens. Bioelectron.*, 20, 204–210, 2004.

94. Maly, J., Masojidek, J., Masci, A., Ilie, M., Cianci, E., Foglietti, V., Vastarella, W., and Pilloton, R. Direct mediatorless electron transport between monolayer of photosystem II and poly(mercapto-p-benzoquinone) modified gold electrode---new design of biosensor for herbicide detection, *Biosens. Bioelectron.*, 21, 923–932, 2005.

95. Jiang, H., and Fan, M.-T. Multi-analyte immunoassay for pesticides: a review, *Anal. Lett.*, 45 (11), 1347–1364, 2012.

96. Li, Y.-F., Sun, Y.-M., Beier, R.C., Lei, H.-T., Gee, S., Hammock, B.D., Wang, H., Wang, Z., Sun, X., Shen, Y.-D., Yang, J.-Y., and Xu, Z.-L. Immunochemical techniques for multianalyte analysis of chemical residues in food and the environment: a review. *Trends Anal. Chem.*, 88, 25–40, 2017.

97. Boroduleva, A.Y., Manclús, J.J., Montoya, A., and Eremin, S.A. Fluorescence polarization immunoassay for rapid screening of the pesticides thiabendazole and tetraconazole in wheat, *Anal. Bioanal. Chem.*, 410, 6923–6934, 2018.

98. Ding, Y., Hua, X., Sun, N., Yang, J., Deng, J., Shi, H., and Wang, M. Development of a phage chemiluminescent enzyme immunoassay with high sensitivity for the determination of imidaclothiz in agricultural and environmental samples, *Sci. Total Environ.*, 609, 854–860, 2017.

99. Zhao, F., Tian, Y., Wang, H., Liu, J., Han, X., and Yang, Z. Development of a biotinylated broad-specificity single-chain variable fragment antibody and a sensitive immunoassay for detection of organophosphorus pesticides, *Anal. Bioanal. Chem.*, 408, 6423–6430, 2016.

100. Guo, Y., Tian, J., Liang, C., Zhu, G., and Gui, W. Multiplex bead-array competitive immunoassay for simultaneous detection of three pesticides in vegetables, *Microchim. Acta*, 180, 387–395, 2013.

101. Qiu, Y.-L., He, Q.-H., Xu, Y., Bhunia, A.K., Tu, Z., Chen, B., and Liu, Y.-Y. Deoxynivalenol-mimic nanobody isolated from a naïve phage display nanobody library and its application in immunoassay. *Anal. Chim. Acta*, 887, 201–208, 2015.

102. Liu, A., Anfossi, L., Shen, L., Li, C., and Wang, X. Non-competitive immunoassay for low-molecular-weight contaminant detection in food, feed and agricultural products: a mini-review, *Trends Food Sci. Technol.*, 71, 181–187, 2018.

103. Zhang, W., Guo, Z., Chen, Y., and Cao, Y. Nanomaterial based biosensors for detection of biomarkers of exposure to OP pesticides and nerve agents: a review, *Electroanalysis*, 29, 1206–1213, 2017.

104. Gonzalez-Buitrago, J.J. Multiplexed testing in the autoimmunity laboratory, *Clin. Chem. Lab. Med.*, 44 (10), 1169–1174, 2006.

105. Liu, G., and Lin, Y. Electrochemical sensor for organophosphate pesticides and nerve agents using zirconia nanoparticles as selective sorbents, *Anal. Chem.*, 77 (18), 5894–5901, 2005.

106. Liu, G., and Lin, Y. Biosensor based on self-assembling acetylcholinesterase on carbon nanotubes for flow injection/amperometric detection of organophosphate pesticides and nerve agents, *Anal. Chem.*, 78, 835–843, 2006.

107. Newman, J.D.S., Roberts, J.M., and Blanchard, G.J. Optical organophosphate sensor based upon gold nanoparticle functionalized fumed silica gel, *Anal. Chem.*, 79 (9), 3448–3454, 2007.

108. Weiying, Z., Yang, Y., Dan, D., Smith, J., Timchalk, C., Liu, D., and Lin, Y. Direct analysis of trichloropyridinol in human saliva using an Au nanoparticles-based immunochromatographic test strip for biomonitoring of exposure to chlorpyrifos, *Talanta*, 114, 261–267, 2013.

109. Arduini, F., Cinti, S., Scognamiglio, V., and Moscone, D. Nanomaterials in electrochemical biosensors for pesticide detection: advances and challenges in food analysis, *Microchimi. Acta*, 183 (7), 2063–2083, 2016.

110. Lee, K.L., You, M.L., Tsai, C.H., Lin, E.H., Hsieh, S.Y., and Ho, M.H. Nanoplasmonic biochips for rapid label-free detection of imidacloprid pesticides with a smartphone, *Biosens. Bioelectron.*, 75, 88–95, 2016.

111. Rawtani, D., Khatri, N., Tyagi, S., and Pandey, G. Nanotechnology-based recent approaches for sensing and remediation of pesticides, *J. Environ. Manage.*, 206, 749–762, 2018.

112. Zhang, W.Y., Asiri, A.M., Liu, D.L., Du, D., and Lin, Y.H. Nanomaterial-based biosensors for environmental and biological monitoring of organophosphorus pesticides and nerve agents, *Trac-Trends Anal. Chem.*, 54, 1–10, 2014.

113. Verma, M.L. Nanobiotechnology advances in enzymatic biosensors for the agri-food industry, *Environ. Chem. Lett.*, 15, 555–560, 2017.

114. Garces-Garcia, M., Morais, S., Gonzaliz-Martinez, F.A., Puchades, R., and Maquieira, A. Rapid immunoanalytical method for the determination of atrazine residues in olive oil, *Anal. Bioanal. Chem.*, 378 (2), 484–489, 2004.

115. Furzer, G.S., Veldhuis, L., and Hall, J.C. Development and comparison of three diagnostic immunoassay formats for the detection of azoxystrobin, *J. Agric. Food Chem.*, 54 (3), 688–693, 2006.

116. Cho, Y.A., Kim, Y.J., Hammock, B.D., Lee, Y.T., and Lee, H. Development of a microtiter plate ELISA and a dipstick ELISA for the determination of the organophosphorus insecticide fenthion, *J. Agric. Food Chem.*, 57, 7854–7860, 2003.

117. Lee, J.K., Ahn, K.A., Park, O.S., Kang, S.Y., and Hammock, B.D. Development of an ELISA for the detection of the residues of the insecticide imidacloprid in agricultural and environmental samples, *J. Agric. Food Chem.*, 49, 2159–2167, 2001.

118. Lee, W.Y., Lee, E.K., Kim, Y.J., Park, W.C., Chung, T., and Lee, Y.T. Monoclonal antibody-based enzyme-linked immunosorbent assays for the detection of the organophosphorus insecticide isofenphos, *Anal. Chim. Acta*, 557 (1–2), 169–178, 2006.

119. Skerritt, J.H., Guihot, S.L., Asha, M.B., Rani, B.E.A., and Karanth, N.G.K. Sensitive immunoassays for methyl-parathion and parathion and their application to residues in foodstuffs, *Food Agric. Immunol.*, 15 (1), 1–15, 2003.

120. Yang, Z., Kolosova, W.S., and Chung, D. Development of monoclonal antibodies against pirimiphos-methyl and their application to IC-ELISA, *J. Agric. Food Chem.*, 54, 4551–4556, 2006.

121. Irwin, J.A., Tolhurst, R., Jackson, P., and Gale, K.R. Development of an enzyme-linked immunosorbent assay for the detection and quantification of the insecticide tebufenozide in wine, *Food Agric. Immunol.*, 15 (2), 93–104, 2003.

122. Shivaramaiah, H.M., Odeh, I.O.A., Kennedy, I.R., and Skerritt, J.H. Mapping the distribution of DDT residues as DDE in the soils of the irrigated regions of northern new South Wales, Australia using ELISA and GIS, *J. Agric. Food Chem.*, 50 (19), 5360–5367, 2002.

123. Biagini, R.E., Smith, J.P., Sammons, D.L., MacKenzie, B.A., Striley, C.A.F., Robertson, S.K., and Snawder, J.E. Development of a sensitivity enhanced multiplexed fluorescence covalent microbead immunosorbent assay (FCMIA) for the measurement of glyphosate, atrazine and metolachlor mercapturate in water and urine, *Anal. Bioanal. Chem.*, 379 (3), 368–374, 2004.

124. Zhang, Y., Muench, S.B., Schulze, H., Perz, R., Yang, B., Schmid, R.D., and Bachmann, T.T. Disposable biosensor test for organophosphate and carbamate insecticides in milk, *J. Agric. Food Chem.*, 53 (13), 5110–5115, 2005.

125. Viveros, L., Paliwal, S., McCrae, D., Wild, J., and Simonian, A. A fluorescence-based biosensor for the detection of organophosphate pesticides and chemical warfare agents, *Sens. Actuators, B Chem.*, B115 (1), 150–157, 2006.

5 Quality Control and Quality Assurance

Árpád Ambrus and Gabriella Suszter

CONTENTS

5.1 Introduction 136
    5.1.1 Quality Systems 136
5.2 Factors Affecting the Reliability and Quality of Residue Data 138
    5.2.1 Samples and Sampling Operations 138
        5.2.1.1 Quality of Samples 139
        5.2.1.2 Sampling of Commodities of Plant and
                 Animal Origin 141
        5.2.1.3 Sampling Surface or Groundwater 142
        5.2.1.4 Sampling of Soil 142
        5.2.1.5 Packing of Samples 143
        5.2.1.6 Transport, Shipping, and Receiving of Samples 144
    5.2.2 Sample Preparation and Processing 144
    5.2.3 Stability of Residues 146
        5.2.3.1 Stability During Storage of Samples 146
        5.2.3.2 Stability of Residues During Sample Processing 148
        5.2.3.3 Stability of Analytical Standards 149
5.3 Method Validation 152
5.4 Internal Quality Control 154
5.5 Random Errors in the Measurement Results 155
    5.5.1 Characterization of the Uncertainty of the Results 155
    5.5.2 Calculation of Combined Uncertainty 156
    5.5.3 Determination of Uncertainties of Individual Steps of
        Analysis 159
        5.5.3.1 Sampling 159
        5.5.3.2 Subsampling 161
        5.5.3.3 Sample Processing 163
        5.5.3.4 Analysis 165
5.6 Systematic Errors—Bias of the Measurements 166
5.7 Interlaboratory Studies 168
References 169

5.1 INTRODUCTION

The results of measurements should provide reliable information and the laboratory should be able to prove the correctness of measurements with documented evidence. Analysts have serious responsibilities to produce correct and timely analytical results and are fully accountable for the quality of their work. The expanding national and international trade, and the responsibility of national registration authorities to permit the use of various chemicals long ago required reliable test methods, which were acceptable to all parties concerned. The accuracy and precision of the analytical results may be assured by proficient analysts applying properly validated methods, which are fit for the purpose, in a laboratory accredited according to the relevant standards or guidelines [1,2]. Several documents and guidelines have been developed to assist analysts to apply the relevant analytical quality control* (AQC) quality assurance[†] [3] (QA) principles in their diverse daily work, and to provide guidance for method validation and accreditation. One of the most comprehensive guidance documents on analytical quality control and method validation procedures for pesticide residue analysis in food and feed (EUAQCC), updated regularly, is issued by the European Commission [4].

The Codex Committee on Pesticide Residues (CCPR) continuously updates the Guidelines on Good Laboratory Practice [5], which were complemented with the guidelines on performance criteria and information on the minimum criteria for validation of methods published by the Codex Alimentarius Commission (CAC) as Codex Standard [6]. The EURACEM[‡]/CITAC[§] published additional guidelines on the application of quality assurance in non-routine laboratories' [7] interpretation of proficiency test results [8], and traceability of measurements [9]. These documents and guidelines (GLs) are complementary to the requirements of the ISO[¶]/IEC** 17025 and OECD[††] Good Laboratory Practice (GLP) Principles and provide detailed description of the general procedures and method performance criteria to demonstrate the reliability and "fit for the purpose" status of the results obtained.

This chapter aims to provide guidance on some critical quality control actions related to pesticide residue analysis which are often overlooked by the practicing analysts.

5.1.1 QUALITY SYSTEMS

The GLP [2] is a quality system concerned with the organizational processes and the conditions under which non-clinical health and environmental safety studies are planned, performed, monitored, archived, and reported. The ISO/IEC 17025: 2017

---

* *Quality control*: A system of procedures, checks, audits, and corrective actions to ensure that all technical, operational, monitoring, and reporting activities are of the highest achievable quality.

† *Quality assurance*: A guarantee that the quality of a product (analytical data set, etc.) is actually what is claimed on the basis of the *quality control* applied in creating that product. It includes all the planned and systematic actions implemented within the quality system, and demonstrated as needed, to provide adequate confidence that an entity will fulfill requirements for quality.

‡ EURACHEM: acronim for a network of organizations in Europe

§ Co-operation on International Traceability in Analytical Chemistry

¶ International Organization for Standardization

** International Electrotechnical Commission

†† Organization for Economic Co-operation and Development

Standard [1], replacing the ISO/IEC 17025:2005, contains all general requirements for the technical competence to carry out specific tests (but not complex studies), including sampling, that laboratories have to meet if they wish to demonstrate that they operate a quality system, and are able to generate technically valid results. It covers analytical tasks performed using standard methods, non-standard methods, and laboratory-developed methods, and incorporates all those requirements of ISO 9001 and ISO 9002 that are relevant to the scope of the services that are covered by the laboratory's quality system. The OECD GLP GLs and the ISO/IEC Standard focus on different fields of activities, but they have been developed simultaneously, and they specify the same basic requirements in terms of AQC.

Measurements of any type contain a certain amount of error. This error component may be introduced when samples are collected, transported, stored, or analyzed or when data are evaluated, reported, stored, or transferred electronically. It is the responsibility of the quality assurance programs to provide a framework for determining and minimizing these errors through each step of the sample collection, analysis, and data management processes. The process must ensure that we do the right experiment as well as doing the experiment right [10]. Systems alone cannot deliver quality. Staff must be trained, involved with the tasks in such a way that they can contribute their skills and ideas, and must be provided with the necessary resources. Accreditation of the laboratory by the appropriate national accreditation scheme, which itself should conform to accepted standards, indicates that the laboratory is applying sound quality assurance principles [1].

Internal quality control (QC) and proficiency testing are important parts of the quality assurance program which must also include the staff training, administrative procedures, management structure, auditing, etc. The laboratory shall document its policies, systems, programs, procedures, and instructions to the extent necessary to assure the quality of the results. The system's documentation shall be communicated to, understood by, available to, and implemented by the appropriate personnel.

The laboratory shall have *quality control procedures** for monitoring the batch to batch validity, accuracy, and precision of the analyses undertaken. Measurement and recording requirements are intended to demonstrate the performance of the analytical method in routine practice. The resulting data shall be recorded in such a way that trends are detectable and, where practicable, statistical techniques shall be applied to evaluate the results. This monitoring shall be planned and reviewed and may include, but not be limited to, the regular use of certified reference materials and/or internal quality control using secondary reference materials; participation in inter-laboratory comparison or proficiency-testing programs; performing replicate tests using the same or different methods; and retesting of retained items [1].

The analytical methods must be thoroughly validated before use according to recognized protocol. The methods must be carefully and fully documented, staff adequately trained in their use, and control charts should be established to ensure the procedures are under proper statistical control. Successful participation in proficiency test programs does not replace the establishment of within-laboratory performance of the method and its regular verification as part of internal quality control.

---

* Synonymous with the terms: analytical quality control (AQC) and performance verification.

The performance of the method should be fit for the purpose and fulfill the quality requirements in terms of accuracy, precision, sensitivity, and specificity [7]. Where possible, all reported data should be traceable to international standards by applying calibrated equipment and analytical standards with known purity certified by ISO accredited supplier [9].

Presently it is definitely more economical to contract out a few samples requiring tests with special methodology and expertise to well-established and experienced (preferably accredited) laboratories, than to invest a lot of time, instruments, etc. to set up and maintain a validated method (and the experience to apply it) for incidental samples in a laboratory.

As an external quality control, participating in proficiency testing schemes provides laboratories with an objective means to demonstrate their capability of producing reliable results.

5.2 FACTORS AFFECTING THE RELIABILITY AND QUALITY OF RESIDUE DATA

5.2.1 SAMPLES AND SAMPLING OPERATIONS

It is generally accepted that analytical results cannot be better than the sample which is analyzed. Even though the importance of reliable sampling has been long recognized, most regulatory laboratories concentrated only on the validation and establishment of performance characteristics of the methods. Very little attention was paid to the quality of the sample as the results of measurements were related only to the sample "as received" and not to the sampled commodity. The ISO/IEC 17025:2005 Standard has changed the situation, requiring appropriate methods and procedures for all tests including sampling, handling, transport, storage, and preparation of items to be tested, and the incorporation of sampling uncertainty in the combined uncertainty of the results when relevant [1]. The 2017 edition of the standard provides more detailed and specific requirements including the selection of samples or sampling sites, sampling plan, withdrawal, and preparation of a sample(s) from a substance, material, or product to yield the required information of the test to be performed.

Though ISO/IEC 17025 standards stated that the validation may include procedures for sampling, handling, and transportation, in case of the determination of pesticide residues the procedures for collecting, handling, and preparing the samples cannot be validated. Obtaining a representative sample that reflects the residue content of the sampled commodity or object can only be assured by careful planning of the sampling program, providing clear instructions for the actual sampling operation including packing and shipping of samples, allocating sufficient time for performing the related actions, and encouraging the sampling officer to record any deviation from the written protocol (e.g., truly random selection of sampling positions, condition of the materials to be sampled, etc.).

The sampling method depends on the objectives of the analyses; hence the sampling plan and protocol should be prepared jointly by the managers making decision based on the results, the analysts, and the representative of the sampling officers. The objectives of the investigation and the corresponding acceptable uncertainty of the measurement results, expressed with Equation 5.12, will determine the size, frequency (time or distance), spacing, mixing, dividing of samples, and consequently the time required for sampling and the cost of sampling, shipping, and analysis of samples. Careful balancing of cost and benefit is a key component in designing sampling plans.

The information on the uncertainty of sampling, sub-sampling, and sample processing is equally as important as the information on the uncertainty of analyses.

5.2.1.1 Quality of Samples

The purpose of sampling is to provide for a specific aim (determine one or some of the characteristic properties) a part of the object that is representative and suitable for analysis. The part of the object taken for further examination is the sample which is usually a very small portion (10^{-5}–10^{-6}) of the sampled object (e.g. 1–2 kg of apples taken from an orchard of 2 ha yielding 50000–60000 kg fruits, taking 20 soil cores from a 5 ha field, collecting few liters of water representing the composition of the stream of a large river, etc.). The sample may be a single unit or an increment, or it may contain several primary samples* defined by the sample size in case of a composite bulk sample, which the laboratory sample may be prepared from. The test portion (usually 1–50 g) is a representative part of the laboratory sample, which is extracted.

To prepare such a small fraction of the sampled object, providing unbiased information with quantifiable uncertainty requires well-defined procedures performed by responsible and properly trained technical staff. The samples and the test portions analyzed should satisfy some basic quality requirements:

- Represent the properties of the object under investigation (composition, particle size distribution)
- Be of a size that can be handled by the sampler and the analyst; keep the properties the object had at the time of sampling; be suitable to give the required information (e.g., mean composition, composition as a function of time or place); and keep its identity throughout the whole procedure of transport and analysis [11]

To develop a quality sampling plan the following actions should be taken and the following points should be considered:

- The purpose of the study (different sampling procedure would be required if we wanted to obtain information on the average residue in a commodity or the distribution of residues in crop units, within one field (or lot) or between fields)
- Clear definition of the object, which can usually be properly defined by the location of the stored material, spatial coordinates, and the time of an open area or agricultural field
- Collection of information on the properties of the objects prior to sampling (it may be necessary to inspect the site to determine the conditions and equipment required)

---

* One or more units taken from one position in a lot.

- Selection of suitable sampling method and tools; testing the suitability of containers to be used to collect, pack, and ship the samples, taking also into account health, safety, and security precautions
- Determination of the time required for reaching the sampling site and taking and handling the samples
- Provisions for prevention of contamination and deterioration of samples at all stages, including size reduction of bulk sample
- Arrangement for sealing, labeling, and delivering the samples and the sampling record to the laboratory in unchanged condition, and assuring integrity of the whole operation
- Preparation of pre-printed sampling record sheet which guides the operator to collect and record all essential information including deviations from the sampling protocol
- Training of sampling personnel to assure that they are aware of the purpose of the operation and the provisions to be taken for obtaining reliable samples (e.g., permitted flexibility to adapt the sampling method for the particular conditions, recording requirements, legal actions, etc.)

It is generally accepted that sample increments or primary samples should be taken randomly, which assumes that each of them making up the sampled lot has equal chance to be chosen. Obtaining random samples is usually a very difficult task and sometimes requires special equipment, for instance, to withdraw single increments of:

- Corn from a large barge or wagon.
- Peanuts packed in 80 kg bags transported in a fully packed camion (without breaking individual beans).
- Orange juice from a 50 L container.
  (Note that the heterogenic distribution of its components forms a dynamic system changing with time and location [pulp, which tends to fall to the bottom, peel oils, which tend to rise to the surface, foam, which sits on the top and gradually dissolves, important flavor volatiles that might evaporate, and juice which is a slurry of suspended particles].) [12]
- Heterogenic solids like granola.
  (The crunchy, sweet chunks of oats and honey form large groups and the crumbs fall to the bottom, while nuts and raisins hold their own position throughout the mix. Mixing the material by turning would increase the heterogeneity instead of reducing it [12]. The only solution is to take a depth-integrated sample from the top to the bottom using a probe of sufficiently large internal diameter and closable bottom to collect the large particles and keep also the crumbs when it is withdrawn from the sampled material.)

Further guidance for the scientific and systematic approach needed to develop or evaluate sampling protocols for defensible decisions can be found in the document titled 'GOOD Samples' including, among others, theories of sampling, sampling quality criteria, management, and health and safety considerations [13].

5.2.1.2 Sampling of Commodities of Plant and Animal Origin

For testing compliance with maximum residue limits (MRL) the CCPR has elaborated a sampling guideline [14] which has become widely accepted and used in many countries as well as in the European Union [15]. The Codex MRL for a plant, egg, or dairy product refers to the maximum residue level permitted to occur in a composite bulk sample,* which has been derived from multiple units of the treated product, while the MRLs for mammalian and poultry meat and edible offal refer to the maximum residue concentration in the tissues of individual animals or birds. It should be noted that the MRL includes the residues defined for enforcement purposes and which are present in the specified portion of the bulk/laboratory sample [16].

Each identifiable lot† to be checked for compliance must be sampled separately. The minimum number of primary samples to be taken depends on the size of the natural units of crops and the lot. For instance, a minimum of ten pieces and/or 1 kg shall be collected from medium- and small-sized units, and five pieces and/or 2 kg from units larger than 250 g. Each primary sample should be taken from a randomly chosen position as far as practicable. The primary samples should be combined and mixed well, if possible without damaging the individual units, to form the bulk sample. Where the bulk sample is larger than what is required for a laboratory sample‡, it should be divided to provide a representative portion. A sampling device, quartering, or another appropriate size reduction process may be used but units of fresh plant products or whole eggs should not be cut or broken. Where units may be damaged (and thus residues may be affected) by the processes of mixing or sub-division of the bulk sample, or where large units cannot be mixed to produce a more uniform residue distribution, replicate laboratory samples should be withdrawn, or the units collected should be allocated randomly to replicate laboratory samples at the time of taking the primary samples. In this case, the result to be used should be the mean of valid results obtained from the laboratory samples analyzed.

Further details for the minimum mass and the number of primary samples to be taken depending on the size of the sampled lot or the targeted (acceptable) violation rate for meat and edible offal are given in the guidelines [14,15].

Samples taken for residue analysis in supervised trials are usually larger than the sample size specified in the Codex GLs, as the main objective is to obtain the best estimate for the average residues [17]. The whole raw agricultural commodity (RAC), as it moves in commerce, should be taken as a sample. In addition to the treated sample(s), one sample of each matrix should be collected from the control plot and analyzed for each field trial site. Samples may be taken from the experimental site randomly, according to a systematic pattern or following some stratified random sampling design taking into account, for instance, the typical spatial distribution of

---

* For products other than meat and poultry, the combined and well mixed aggregate of the primary samples is taken from a lot. For meat and poultry, the primary sample is considered to be equivalent to the bulk sample.

† A quantity of a food material delivered at one time and known, or presumed, by the sampling officer to have uniform characteristics such as origin, producer, variety, packer, type of packing, markings, consignor, etc.

‡ The sample sent to, or received by, the laboratory. A representative quantity of material removed from the bulk sample.

residues in fruit trees [18]. It was shown that, where samples should be taken at different time intervals after the application of the pesticide for establishing decline curves, the least variation can be obtained if the primary sampling positions are selected randomly and marked before the first sampling, and the primary samples are collected from the close vicinity of the marked positions at the various sampling times [19].

Detailed guidance for collection of samples in supervised trials is given in Appendix V of the JMPR Manual [17].

Though the OECD Guidelines for crop field trials [20] permit cutting large crops at the supervised trial site and shipping only a portion of the whole crops, the JMPR accepts results from such trials only if it is shown that cutting the large crops does not affect their residue content. The recommended verification procedure to be applied before the trial is described in the JMPR Manual [17].

5.2.1.3 Sampling Surface or Groundwater

Sampling water is probably more challenging than sampling food. Defining the sampling objectives will help to determine the number of samples required, and the location and type of samples needed. The wide variety of sampling targets including, for instance dugouts, ponds, lakes, streams, large rivers, or the ocean, groundwater, drinking water, raw water distribution lines, or agricultural water require different sampling plans, methods, and equipment. There are numerous documents to provide general guidance or methods covering specific areas [21–26].

The following ISO Standards deal with various aspects of sampling water and handling water samples: ISO 5667-1:2006, ISO 5667-3:2012, ISO 5667-4:2016, SO 5667-5:2006, ISO 5667-6:2014.

5.2.1.4 Sampling of Soil

As for all types of studies, the identification of the objectives, and the object (decision unit, DU) of sampling are the first step in the preparation of sampling plans [12,27]. It is especially important for sampling of soil and may require prior survey of the target area [28]. The DU is the area or volume of soil targeted for sample collection and characterization. It may encompass an entire field, part of a field, or multiple fields.

The ISO 18400-101:2017 specifies the procedural elements to be taken in the preparation and application of a sampling plan and refers to the corresponding standards giving guidance for performing the key elements of the investigation [29].

The theoretical basis, complemented with some practical examples, of sampling particular materials is described by Gy [30] and further elaborated by Pitard [31]. According to Gy the sampling of a heterogeneous material containing a small quantity of a sought-for substance depends on particle shape, particle size distribution, the composition of the phases comprising the particulates, and the degree to which the substance sought is liberated from the remainder of the material (gangue) during particle size reduction by grinding or crushing. Gy's book is an excellent resource for planning sampling plans providing accurate results.

The objectives of the analysis of pesticide residue in soil are often to determine:

* The average concentration or its changes in time in various depth of soil
* The horizontal distribution of residues within the targeted area

- Vertical distribution of residues
- Identification of potentially contaminated sites

Stratified random sampling is recommended for the first two cases, taking 3–6 replicate composite samples, consisting of a minimum of 10 to 25 soil cores depending on the area, from each segment possessing similar characteristics (e.g., soil type, slope of the land, groundwater level, etc.). The consecutive sampling at various time intervals should be conducted from the same primary sampling positions identified preferably by global positioning system (GPS) coordinates, which can be easily determined with portable devices providing 0.1 m accuracy.

The microvariation of soil constituents can be reduced by taking cluster samples with a smaller probe along a circle of about 15 cm radius around the sampling point. Using soil auger of 2, or 2.5 cm internal diameter, six or four cores would provide the same surface area as one core of 5 cm diameter. Cluster sampling is especially useful for residue concentration–time relation studies.

The depth of soil cores depends on the objectives of the study. For pesticide residues applied on the surface of the soil, sampling the upper 5 cm layer is sufficient. However, for mobile substances sampling down to 25–50 cm may be required [32].

A systematic sampling plan, covering the sampled area with rectangular grids, is recommended when the spatial distribution of residues should be determined. Systematic sampling may also lead to the desired results when the patchy contamination of soil in orchards is to be determined. It was found that the sites of spray runoff (trunk of trees, dipping from large branches) are the most contaminated. Though, theoretically, the results of random sampling provide unbiased estimates for the population mean and standard deviation, there was no practical difference in the results obtained with random and systematic sampling [33].

There are many practical guidelines and standard operating procedures (SOPs) for taking soil samples, mainly for determining the available nutrients in soil. It should be noted that existing standard sampling protocols used for the evaluation of soil nutrients are likely insufficient for contaminants. The main reasons are the very low analyte levels and the differences in heterogeneity between nutrients and contaminants [27].

5.2.1.5 Packing of Samples

Individual samples should be placed in suitable containers, e.g., heavy polyethylene bags, and then put inside additional heavy paper bags. Polyethylene bags alone may become brittle in contact with dry ice and therefore there is a risk of breakage and subsequent loss of the sample. Glass or Teflon® containers should preferably be used for liquid samples and should be thoroughly cleaned and rinsed with one or more suitable pesticide-free solvents such as acetone, isopropyl alcohol, or hexane, and dried before use. Pesticides can migrate to the walls of a container and be adsorbed; hence even a glass container, after the sample is poured out, should be rinsed with solvent if the extraction is not made in the container itself.

Avoid other type of packing materials, especially polyvinylchloride (PVC), or plastic-lined caps of glass containers.

In summary, any type of container or wrapping material should be checked before use for possible interference with the analytical method at the limit of detection of the targeted analytes.

Each sample shall be uniquely identified with a proper label which cannot be removed from the packing material. Do not use felt pens for marking the bags as they may cause interference. Fasten boxes securely with strong twine, rope, or tape.

5.2.1.6   Transport, Shipping, and Receiving of Samples

Non-perishable commodities containing residues that are known to be stable over the period required to reach the laboratory can be shipped in a non-frozen state, provided that the samples are delivered to the laboratory within 24 hours. Perishable materials or samples containing labile or volatile pesticide residues may be cooled to close to 0°C at the experimental station and then packed in a cool box containing blue ice to keep them cool during transport. All samples should be protected against any effects which might cause degradation or contamination.

Where samples need to be frozen, use shipping containers of polystyrene foam, if available, as they are excellent for this purpose. If not available, use two cardboard boxes of slightly different sizes, with insulation between. Proper insulation is essential to ensure samples arrive at the residue laboratory still frozen. Sufficient dry ice must be used to assure that some of it remains upon delivery to the laboratory. This usually requires a minimum of 1 kg of dry ice per kg of sample. For journeys lasting more than two days, 2 kg of dry ice or more per kg of sample may be required. Poorly insulated containers require more dry ice. Use caution in handling dry ice (gloves and ventilated work area). Packages must comply with transport regulations.

In case of supervised trials carried out in compliance with GLP principles, a temperature monitoring device should preferably be placed beside the sample bag to verify that the samples have arrived deep-frozen to the laboratory.

When samples have to be shipped across national boundaries, quarantine regulations must be observed, and appropriate permits obtained well in advance of dispatching the samples.

The consignee should be advised by facsimile (FAX) or email giving full details of shipment of samples, including shipping document and flight numbers, so that delay in delivery to the laboratory is avoided.

Frozen samples must never be allowed to thaw, either before or during shipment. They must be shipped under conditions that permit their arrival at the residue laboratory still solidly frozen. Samples received deep-frozen by the laboratory should be stored deep-frozen until their processing. Storing all samples under –20°C is recommended if they were processed later.

5.2.2   SAMPLE PREPARATION AND PROCESSING

For food commodities, the Codex MRLs refer to the specified portion of the commodity to be analyzed [16]. The preparation of the analytical sample* may require

---

* The material prepared for analysis from the laboratory sample, by separation of the portion of the product to be analyzed.

removal of foreign materials and certain parts of the sampled material (such as shell of nuts, stone of mango or peach, adhering soil with gentle rinsing or brushing, outer withered loose leaves in case of plant materials, and peddles and remains of plants from soil, etc.). The Codex Classification of Food and Feed [34], containing also the description of the portion of commodities to be analyzed, is under revision by the CCPR and published gradually for commodity groups in its reports [35].

The sample preparation* procedures may significantly affect the residue level (e.g., how many outer leaves are removed from cabbage or head lettuce). As they cannot be validated and their contribution to the combined uncertainty of the results cannot be estimated, the sample preparation procedure should be clearly written, preferably as an SOP, and consistently followed without any deviation to obtain comparable results.

The residues in individual crop units are not uniformly distributed. For instance, the residue concentrations in the peel of citrus fruits, tomato, etc., can be quite different from those in the pulp. This is called compositional or constitutional heterogeneity[†] in the theory of sampling [12,36]. In addition, the average residues in crop units show large variation referred to as distributional heterogeneity [12]. Therefore the whole laboratory sample or its representative portion must be prepared in order to obtain the analytical sample, and the entire analytical sample should be chopped, ground, or mixed to obtain a well-mixed material from which the representative test portions can be withdrawn for extraction. The large crops making up the laboratory sample (e.g., five watermelons) may not be processed together due to the limited capacity of the equipment. In these cases, representative segment portions (not slices) should be cut from the individual units in such a way that the ratio of the surface and inner part remains the same [37–39].

The efficiency of the comminuting procedure depends on the equipment, maturity, and variety of the crops, but it is independent from the concentration and nature of the analyte, and the extraction method [40]. It may vary from sample to sample. It can be simply and quickly checked by taking a small portion and spreading it on a Petri dish or filtering after dilution with water (Figure 5.1).

If the particle size is less than 2–3 mm, 1–2 g test portions can usually be withdrawn with a relative standard deviation of sample processing, CV_{Sp}, ≤ 15%. It is more difficult to obtain a well-mixed matrix from plant materials with hard peal and soft pulp (tomato), than from a soft fruit (orange) [41,42]. The efficiency of the comminution can be increased and the test portion mass reduced by performing it in the presence of dry ice [41,42], followed by a secondary fine-milling step of a subsample of size already known to be representative, in a special freezer mill with liquid nitrogen [43,44]. Further advantage of cryo-milling is that it reduces the decomposition of labile compounds which may be caused by the cell fluids of high enzyme activity during the first few minutes of comminution [41,45]. Observe the freshly cut

---

[*] The first of two processes which may be required to convert the laboratory sample into the test (analytical) sample. The removal of parts that are not to be analyzed, if required. It is synonymous but distinctly different from sample processing.

[†] Heterogeneity: A lot is heterogeneous relative to a given characteristic if the characteristic is not uniformly distributed throughout the lot, the condition of a lot under which all the elements are not strictly identical.

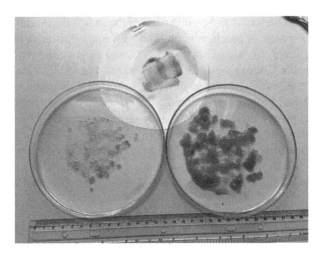

FIGURE 5.1   Visual testing of the efficiency of sample comminution. Note the sizes of pieces of tomato peels in a "well mixed" sample obtained with cryogenic processing and the result of an inefficient processing at room temperature. Photos of the homogenized material spread on a Petri dish or collected on filter paper can be used as records for quality control purposes.

surface of many fruits and vegetables such as, for instance, apple, mango, cucumber, or squash and note the excess of cell fluids, which may cause rapid degradation of pesticide residues present in the fruits or on their surface. When cryo-milling is used for sample processing, special measures are required to minimize the condensation of moisture.

Heterogeneous liquids, such as orange juice, should be subsampled following the principles described for sampling.

It was found that soil samples can be most effectively homogenized after adding sufficient amount of distilled water. Adding dry ice to soil samples did not improve the efficiency of homogenization [46].

Methods for determination of uncertainty of sample processing are described in Section 5.5.

5.2.3   STABILITY OF RESIDUES

The pesticide residues may be subject to different chemical reactions or evaporate after the samples are taken. The change of concentration of the residues should be avoided as far as possible to ensure the representativeness of the samples and the results.

5.2.3.1   Stability During Storage of Samples

The supervised trial samples are usually deep-frozen shortly after the sampling and shipped deep-frozen to the laboratory within the shortest possible time, where they are kept deep-frozen until analysis. During this storage period the concentration of residues of the pesticides and their metabolites may decline due to processes such as volatilization or reaction with enzymes. Storage stability tests are carried out with representative commodities to demonstrate the stability of residues during frozen storage prior to analysis. These studies are part of the data package submitted to support registration of a compound. The publication titled FAO/WHO Pesticide Residues – Evaluations [47] also includes information on the stability of residues during storage.

Where it is foreseen that the samples shall be stored in the laboratory over one month, and appropriate information on the stability of residues is not available on representative sample matrices under similar conditions as those under which the samples will be stored, a storage stability test should be carried out. The basic principles [17,48] to be considered for planning storage stability studies are briefly summarized below.

Stability data obtained on one commodity from a commodity group [4] (e.g., commodities with high water or oil content) can be extrapolated within the same group, provided that the storage conditions are comparable. The study can be performed with a sample containing field incurred residues, if the suitable homogeneity of the material had been verified before ($CV_{Sp} < 0.25$ or $0.3 \times CV_A$, see details in Section 5.4.3.3). Alternatively, the test portions withdrawn from the homogenized untreated sample matrix should be spiked and stored individually. Untreated sample material should be prepared and stored under the same conditions. The treated and blank test material should be sufficient for a minimum of 8×4 treated as well as untreated test portions for analyses with some extra material as reserve. The total number of test portions should be larger to accommodate an extended storage period if necessary. The active substance and its metabolites or degradation products included in the residue definitions for enforcement and dietary risk assessment should be tested separately if spiked test portions are used. The initial residue concentration should be sufficiently high (e.g., $>10 \times LOQ^*$) to enable the accurate determination of the residues if their concentration decreases during storage. Normally, analyses at five timepoints are sufficient. The first test should be performed at day 0 to verify the initial concentration, and the others selected according to the targeted storage period (e.g., 0, 1, 3, 6 ,12, 24, and 36 months or 0, 2, 4, 8, and 16 weeks if rapid decline of residues is suspected). At each time, two treated test portions and at least one freshly spiked untreated portion should be analyzed. If the difference between the duplicate analyses is greater than 20%, analyzing the third replicate test portion should be considered. The individual recoveries obtained should preferably be within the warning limits of the established method (see 5.3). If that is not the case, the residues measured in additional test portions of the stored samples should be repeated together with concurrent recoveries.

The results should be reported in the form of individual residue concentration (mg/kg) measured in the treated stored samples (called survived residues), the concurrent recoveries expressed in percentage of the spiked amount, and the standard uncertainty of the measurement determined independently as part of the validation of the analytical method.

---

* LOQ: limit of quantification

Where the storage stability study carried out with samples belonging to three representative commodity groups indicates that the residue is stable, then it can be assumed that the residues would be stable in other matrices stored under similar conditions.

5.2.3.2 Stability of Residues During Sample Processing

The laboratory sample processing received disproportionally little attention until recent years, though its contribution to the uncertainty and the bias of the results can be quite large. The analysts were aware of the rapid decomposition of dithiocarbamates or daminozide if they were in contact with the macerated samples and consequently eliminated the homogenization step from the method, but did not test the stability of other residues, or did not associated the loss of residues with their potential decomposition until some publications revealed the substantial decomposition (50–90%) of certain compounds (chlorothalonil, phthalimides, thiabendazole, dichlofluanid [49,50]). Further studies revealed that processing in the presence of dry ice (cryogenic milling) at or below −20°C reduced or practically eliminated the loss of pesticides which decomposed at ambient temperature [43,48]. Furthermore, cryogenic processing may provide more homogeneous sample matrix and reduce the uncertainty of sample processing.

The decomposition of analytes occurs during the first few minutes of comminution due to the high enzyme activity of the cell fluids. This is one of the reasons why large crops or whole units should not be cut or broken as part of the sampling procedure in order to reduce the sample mass (5.2.1.2).

The pesticides added to the test portion for recovery studies just before the extraction may not further decompose or its rate cannot be observed. It may probably be attributed to the inactivation of the enzymes by the extracting solvent and the different concentration of the chemicals in the diluted extract. It is emphasized that neither the recovery nor the proficiency tests reveal information on the effect of the sample processing on the potential decomposition of analytes and uncertainty of measured residues.

The decomposition of the residues depends on the composition of the sample material, and the homogenization process. When intensive and extended comminution in a high-speed blender is carried out to reduce the sample processing uncertainty, a significant bias can be introduced due to the decomposition of the residues. Because the rate of decomposition may depend on the laboratory equipment, the variety and maturity of the processed crop, and many other factors, there is not sufficient knowledge currently to extrapolate findings from one laboratory to another. Consequently, the laboratories, analyzing a wide range of pesticides in a large number of various commodities, should apply cryogenic processing as standard procedure to reduce the chance of producing biased results. Furthermore, they should verify the suitability of their procedures as part of the method validation with testing the stability of those compounds which are known to rapidly decompose under unfavorable conditions.

The cryogenic processing, performed in various ways, has become the standard procedure in many laboratories. One frequently applied version includes: preparation of the portion of sample to which the MRL applies (analytical sample) upon receiving the fresh sample in the laboratory; placing the analytical sample into the deep-freezer within the shortest possible time; chopping and/or grinding the sample in the presence of sufficient amount of dry ice (about 1:1 sample–dry ice ratio) to keep the temperature below -20°C (this process requires robust choppers with stainless steel bowl and lid); withdrawing the test portions needed for various extractions and confirmation of residues into appropriate unsealed containers and placing them in a deep-freezer for a minimum of 16 hours to allow the carbon dioxide to evaporate; and weighing the mass of the test portion, adding extraction solvent, and warming the test portion up to room temperature before proceeding with the extraction.

The stability of residues can be tested with a mixture of pesticides which contains a reference compound (R) known to be stable (e.g., buprofezin, chlorpyrifos,), at least one compound decomposing rapidly (chlorothalonil, dichlofluanid, captan), and the other compounds to be tested [40]. The test mixture should be carefully applied on the surface of the plant material, avoiding runoff. The treated sample should be kept in a fume cupboard until the solvent completely evaporates. The processing under ambient temperature can now be started, while the treated sample should be placed in a deep-freezer before cryogenic milling for a minimum of 16 hours. As a minimum, three test portions should be withdrawn from the comminuted material, and the extract should be analyzed in duplicate. The average recoveries obtained with independently spiked test portions should be taken into account when the residues measured in surface treated samples are compared to their nominal concentration adjusted with the recoveries of the reference compounds to compensate for the concentration differences in test portions of the "well-mixed" comminuted test matrix. Attention should be paid to perform the analyses of spiked test portions and those withdrawn from the surface treated plant matrix under the same conditions [43].

The stability of analytes primarily depends on its physicochemical properties, the sample material, and the method of comminution. Therefore, principally it should be tested for all residue matrix combinations, which is practically impossible. The stability tests should at least be performed with the residues included in the scope of the method and some representative matrices selected from the 11 commodity groups [4] considered to have similar properties relevant for the determination of pesticide residues. Special attention should be paid to those compounds which show degradation during extended storage of samples.

5.2.3.3 Stability of Analytical Standards

The purity of reference standards and accuracy of their concentration in stock and working solutions are the basis of obtaining unbiased results. Pure reference standards should be stored excluding light and moisture at low temperatures preferably in a freezer. Under such strictly controlled conditions, the pure reference materials may be stored without degradation for up to ten years, substantially exceeding the expiry date specified by the suppliers [4].

The stability of pesticides may depend on the solvent used. When the stock solutions prepared in dry toluene or acetone are stored in tightly closed glass containers in the freezer, most of pesticides are stable for at least five years, while the acetonitrile, methanol, or ethyl acetate solutions are usually stable for up to three years [4]. Very useful information can be obtained, among others, on the stability of pesticides for registered uses from the European Reference Laboratories (EURL) DataPool [51], but it may not be applicable under different laboratory conditions.

Regular testing of the actual concentration of the components of reference analytical standards shall be carried out by each laboratory because many of the varying factors affecting the stability of analytes are specific to a given laboratory.

After the purity of the old reference materials or concentration of stock solutions have been verified by comparing them to new ones, the original expiry date can be extended. The stability of the reference standard solutions should be checked by comparing the mean of a minimum of five replicate measurements made alternately with the old and new solutions. According to the EUAQCC [4], the mean values should not differ more than ±10% when the concentration (response) of the new standard solution is considered 100%.

$$\bar{C}_{\mathrm{diff}}\% = 100 \times \frac{\bar{C}_{\mathrm{new}} - \bar{C}_{\mathrm{old}}}{C_{\mathrm{new}}} \tag{5.1}$$

The compliance with this criterion should be checked with an appropriate statistical method. The usually applied two-sample t-test comparing the two mean values is not suitable for this purpose because it is designed to prove (alternative hypothesis) that the two mean values are different and not that the means of the two sets of measurements are equal or within a specified limit. For testing if the difference between the old and freshly prepared standard solutions is within the ±10% criterion, the two one sided t-test (TOST) is recommended [52]. The TOST requires the specification of acceptance criterion, ±Θ, which represents the largest difference in mean values that can be accepted for practical purposes. In case of pesticide residues, the acceptance criterion Θ is ±10% [4].

$$H_0 : \bar{C}_{\mathrm{new}} - \bar{C}_{\mathrm{old}} \leq \theta_{\mathrm{Lower}} \text{ or } \bar{C}_{\mathrm{new}} - \bar{C}_{\mathrm{old}} \geq \theta_{\mathrm{Upper}} \tag{5.2}$$

$$H_1 : \theta_{\mathrm{Lower}} \leq \bar{C}_{\mathrm{new}} - \bar{C}_{\mathrm{old}} \leq \theta_{\mathrm{Upper}} \tag{5.3}$$

When the difference in the mean values is within the boundaries (±Θ), the difference between the standard solutions is within the permissible deviation at the specified (usually 95%) confidence level.

The confidence interval, CI, for the difference in the mean concentrations is calculated as:

$$\mathrm{CI} = \bar{C}_{\mathrm{new}} - \bar{C}_{\mathrm{old}} \pm t_{[0.1;(2n-2)]} \times \sqrt{s_p^2 \left(\frac{1}{n_1} + \frac{1}{n_2} \right)} \tag{5.4}$$

where $\bar{C}_{\mathrm{new}}$ and $\bar{C}_{\mathrm{pld}}$ are the mean values from each sample set, $t_{0.1;(n_1+n_2-2)}$ is the t-value at 90% confidence level with $n_1 + n_2 - 2$ degrees of freedom, s_p is the estimate for the pooled standard deviation for the sample sets and n_1 and n_2 are the number of samples run in each set.

$$s_p^2 = \frac{(n-1) \times \left(S_{\mathrm{new}}^2 + S_{\mathrm{old}}^2 \right)}{2n-2} \tag{5.5}$$

Note that the 95% one-sided confidence limit is equivalent to the 90% two-sided confidence limit. The difference between the new and old solutions is acceptable if the CI is within $\pm\Theta = 0.1$ (10%). Using the identical input data (not included in the tables), the results obtained with the two-sample t-test and TOST are summarized in Tables 5.1 and 5.2.

The results included in the tables show that the TOST gave opposite outcome compared to the two-sample t-test in four out of five examples. In general, the two-sample t-test gives inconsistent results compared to TOST and should not be used for testing suitability of old standard solutions. The increase of replicate injections from five to seven did not influence the outcome of the tests.

Very good relationship was found [52] between the absolute difference of the average concentrations or responses of analytes (Δ) in the new (100%) and old standard solutions and the pooled relative standard deviations (CV_p) of the replicate injections (Figure 5.2).

TABLE 5.1
Comparison of the Two Sets of Measurements with Two-Sample t-Test

#	$n_1 = n_2$	CV_{new}	CV_{old}	Sp^2	$\bar{A}_{new}$	$\bar{A}_{old}$	t_{calc}	H_0
1	5	0.035355	0.022361	2.29E-04	1	0.950	2.71	Reject
2	5	0.035129	0.034775	1.17E-03	1	0.955	2.09	Reject
3	7	0.030615	0.030610	8.74E-04	1	0.930	4.43	Reject
4	7	0.011937	0.011937	1.32E-04	1	0.920	13.05	Reject
5	5	0.000406	0.000406	1.50E-07	1	0.905	387.48	Reject

Note: The table includes rounded calculated values. However, the calculations shall be performed with Excel without any rounding.

TABLE 5.2
Comparison of the Two Sets of Measurements with TOST

#	$n_1 = n_2$	CV_{new}	CV_{old}	CVp	$\bar{A}_{new}$	$\bar{A}_{old}$	CI^+	CI^-	H_0
1	5	0.035355	0.022361	0.030	1	0.954	0.067	0.026	Accept
2	5	0.035129	0.034775	0.035	1	0.955	0.091	-0.001	Accept
3	7	0.030615	0.03061	0.031	1	0.930	0.101	0.039	Reject
4	7	0.011937	0.011937	0.012	1	0.920	0.092	0.068	Accept
5	5	0.000406	0.000406	0.0004	1	0.905	0.096	0.094	Accept

Note: The table includes rounded calculated values. However, the calculations shall be performed with Excel without any rounding.

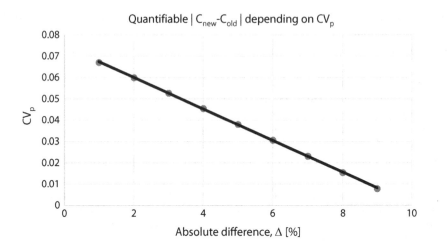

FIGURE 5.2   Relationship between the absolute difference [Δ%] of average concentrations or responses of the new (100%) and the old standard solutions and the pooled CV_p of the replicate measurements made with the two standard solutions. (Reproduced with permission from Ambrus, Á; Noonan, G; Németh, A.; et al. Testing the accuracy of analytical standard solutions used for quantitative determination of pesticide residues. *J. AOAC Int.* 2017; 100, 4, 1–4.)

$$CV_p = -0.0074\Delta + 0.0748 \left(R^2 = 0.9999 \right) \qquad (5.6)$$

The figure can be used to verify visually the "equivalency" of the two standard solutions by plotting CV_p as the function of Δ. If CV_p is above the acceptance criterion line, the concentrations of the tested compounds differ more than the ±Θ (±10%) at 95% probability level.

As Table 5.2 and Figure 5.2 indicate, very good reproducibility of injections (<1%) is required for verification of compliance with the acceptable difference above 8.7%, and a repeatability of 0.08% of the ten replicate injections would be required to confirm that the difference is exactly 10%.

5.3   METHOD VALIDATION

Method validation typically follows the development of a method and its performance requirements such as calibration, system suitability, stability of analytes, and limits of detection and quantification had been satisfactorily established. The validation is a process to demonstrate that the method is fit for the purpose.

The concepts of method validation have been developed and partly updated simultaneously by AOAC[*] International [53], EURACHEM [54], IUPAC[†] Working Party [55], and several national organizations. The general criteria set by the different guidelines are similar and provide the basis for assuring reliability of the methods validated for one or a few analyte–sample matrix combinations. However,

---

[*] AOAC: Association of Official Analytical Chemists
[†] IUPAC: International Union of Pure and Applied Chemistry

these general guidelines are not directly applicable to the methods used in pesticide residue analysis as they cannot address its specific requirements and limitations. To provide guidance on in-house method validation to analysts, national authorities, and accreditation bodies, a Guideline for Single-laboratory Method Validation was developed and discussed at an International Workshop [56]. The guidelines were included in the Good Laboratory Practice GLs of CCPR and its updated versions [5,6]. The guidelines also provide specific information for extension of the method to a new analyte and/or new sample matrix, and adaptation of a fully validated method in another laboratory.

According to the guidelines the method validation is not a one-time, but continuous operation including the performance verification during the use of the method. Information essential for the characterization of a method may be gathered during the development or adaptation of an analytical procedure; establishment of acceptable performance; regular performance verification of methods applied in the laboratory; and demonstration of acceptable performance by participation in proficiency test or inter-laboratory collaborative study.

Before validation of a method commences, the method must be optimized, an SOP describing the method in sufficient detail should be prepared, and the staff performing the validation should be experienced with the method. Parameters to be studied are: stability of residues during sample storage, sample processing, and in analytical standards; efficiency of extraction; homogeneity of analyte in processed samples; selectivity of separation; specificity of analyte detection; calibration function; matrix effect; analytical range, limit of detection, LOQ, and ruggedness of the method.

The validation should be performed in case of *individual methods* with the specified analyte(s) and sample materials or using sample matrices representative of those to be tested by the laboratory; *group specific methods* with representative commodity(ies)* and a minimum of two representative analytes† selected from the group; multi-residue methods (MRM) with representative commodities and a minimum of ten representative analytes. For method validation purposes, commodities should be differentiated sufficiently but not unnecessarily. The concentration of the analytes used to characterize a method should be selected to cover the analytical ranges of all analytes. Full method validation shall be performed in all matrices and for all compounds specified if required by the relevant legislation. The extent of validation is always a balance between costs, risks, and technical possibilities [1].

The EUAQCC [4] and CACGL 90-2017 [5] provide detailed information concerning the recommended procedures to carry out the validation and the generally acceptable performance criteria for single and multi-residue methods used for screening and quantitative determination.

The method is considered applicable for an analyte if its performance satisfies the basic requirements specified in these documents.

---

* Single food or feed used to represent a commodity group for method validation purposes. A commodity may be considered representative on the basis of proximate sample composition, such as water, fat/oil, acid, sugar, and chlorophyll contents, or biological similarities of tissues, etc.
† Analyte chosen to represent a group of analytes which are likely to be similar in their behavior through a multi-residue analytical method, as judged by their physicochemical properties e.g. structure, water solubility, Kow, polarity, volatility, hydrolytic stability, pKa, etc.

5.4 INTERNAL QUALITY CONTROL

Based on the validation and optimization data generated, a quality control scheme should be designed. The performance of the method should be regularly verified during its use as part of the internal quality control program of the laboratory. The internal quality control/performance verification is carried out to:

1. monitor the performance of the method under the actual conditions prevailing during its use, and take into account the effect of inevitable variations caused by, for instance, the composition of samples, performance of instruments, quality of chemicals, varying performance of analysts, and laboratory environmental conditions;
2. demonstrate that the performance characteristics of the method are similar to those obtained during method validation, the application of the method is under "statistical control", and the accuracy and uncertainty of the results are comparable to the performance characteristics established during method validation.

The results of internal quality control provide essential information for the confirmation and refinement of performance characteristics established during the initial validation, and extension of the scope of the method. Detailed instructions and acceptable performance parameters for the whole process of analysis of pesticide residues (sampling, sample preparation and processing, extraction, clean-up, concentration/reconstitution and storage of extracts, chromatographic separation and determination, calibration for quantification, on-going method performance verification during routine analysis, identification and confirmation, preparation of handling analytical standards, method performance criteria) are given in the EUAQCC [4].

Some key components of the QC scheme are reemphasized hereunder. The correct preparation of analytical standards should be verified by comparing its analyte content to the old standard or preparing the new standard in duplicate at the first time. A balance with 0.01 mg sensitivity should not be used to weigh less than 10 g standard material. The dilutions of standard solutions should be made independently based on weight measurement except the last step for which an A-grade volumetric flask should be used [57].

For the most effective internal quality control, analyze samples concurrently with quality control check samples (recovery, reanalyses of retained test portions, quality of calibration, etc.). The initial control charts can be constructed with the average recovery (Q) of representative analytes in representative matrices and the typical within-laboratory reproducibility coefficient of variation (CV_{typ}) of analysis for checking acceptability of individual recovery results. The warning and action limits are $Q \pm 2*CV_{Atyp}*Q$ and $Q \pm 3*CV_{Atyp}*Q$, respectively. At the time of the use of the method, the recoveries obtained for individual analyte–sample matrices are plotted in the chart. The long-term reproducibility of the MRM can be demonstrated by plotting on the control chart all recovery values of compounds that can be characterized with the same typical average recovery and CV_A obtained during the method validation.

Based on the results of internal quality control tests, refine the control charts at regular intervals if necessary. If the analyte content measured in the quality control check samples is outside the action limits, the analytical batch (or at least the analysis of critical samples in which residues found are ≥0.7 MRL and 0.5 MRL for regularly and occasionally detected analytes, respectively) may have to be repeated. When the results of quality control check samples fall repeatedly outside the warning limits (1 in 20 measurements outside the limit is acceptable), the application conditions of the method must be checked, the sources of error(s) must be identified, and the necessary corrective actions have to be taken before the use of the method is continued.

The differences of the replicate measurements of test portions of positive samples can be used to calculate the overall within-laboratory reproducibility of the method (CV_{Ltyp} calculated with Equation 5.22). The CV_{Ltyp} will also include the uncertainty of sample processing but will not indicate if the analyte is lost during the process.

The applicability of the method for the additional analytes and commodities shall be verified as part of the internal quality control program. All reported data for a specific pesticide matrix combination should be supported with either validation or performance verification performed on that particular combination.

5.5 RANDOM ERRORS IN THE MEASUREMENT RESULTS

The interpretation of the results and making of correct decisions require information on the accuracy and precision of the measurements. The measurement process is subject to several influencing factors which may contribute to random, systematic, and gross errors [5,58]. The quality control of the process aims to monitor the uncertainty (repeatability, reproducibility) and trueness of the measurement results.

The uncertainty of the measurements (u) is mainly due to some random effects. The uncertainty 'estimate' describes the range around a reported or experimental result [58] within which the true value can be expected to lie at a defined level of probability. This is a different concept to measurement error (or accuracy of the result) which can be defined as the difference between an individual result and the true value. It is worth noting that, while the overall random error cannot be smaller than any of its contributing sources, the systematic error can be zero even if each step of the determination of the residues provides biased results. Another important difference between the random and systematic errors is that once the systematic error is quantified the results measured can be corrected for the bias of the measurement, while the random error of a measurement cannot be compensated for, but its effects can be reduced by increasing the number of observations.

5.5.1 CHARACTERIZATION OF THE UNCERTAINTY OF THE RESULTS

The uncertainty components of a residue analytical result may be grouped into external operations and laboratory procedures according to the major phases of the determination [59]. The external operation, including sampling (S), packing, shipping, and storage of samples, results in the laboratory sample(s). The laboratory phase may comprise the subsampling (sample size reduction), preparation, processing, and analysis of the sample. The large samples, for instance five pieces of cabbage heads (3–6 kg each), watermelons (5–12 kg), or jackfruits (15–25 kg) required as a minimum for one laboratory sample [14,15] should be subsampled (SS) to obtain a representative portion which can be further processed with the available equipment.

Soil samples should also be prepared for instance by removing pebbles, remains of plant materials, etc.

Water samples may have to be filtered and the solid particles analyzed separately depending on the purpose of the analysis.

Usually the sample size reduction is carried out after the sample preparation separating the portion of commodity to be analyzed [16]. The sample processing (Sp), which has a different meaning from sample preparation in pesticide residue analysis, aims to obtain a well-mixed ("homogeneous") analytical sample by chopping, mincing, grinding, etc. from which the representative small test portions (typically 1–25 g) can be withdrawn for extraction. The analysis of sample (A) may require several operations depending on the analytes, sample material, and mode of detection: extraction, clean-up, derivatization, detection, and quantification of the analytes.

The uncertainty of sampling, packing, shipping, storage, and preparation of samples cannot be determined. To minimize the random error and to avoid the systematic errors, the laboratories should prepare detailed description of the procedures which should be implemented by properly trained staff applying appropriate tools and materials.

The major sources of the random and systematic errors [60] are summarized in Table 5.3. The random error reflecting the uncertainty of the results is usually expressed with the standard deviation (SD) or the relative standard deviation (CV) of the repeated measurements. Their nature and contribution to the combined uncertainty of the results will be discussed in the following sections.

5.5.2 Calculation of Combined Uncertainty

The combined uncertainty of the measured residues (u_c) can be calculated applying the basic rules of error propagation [58,61]. These basic equations can be used for continuous data populations following various distributions, such as normal, rectangular, or triangular. The method of calculation depends on how the result is obtained.

1. The result is the sum of separately measured values (P, Q, R), such as residues of parent compound and its metabolites:

$$Y = C_1 P \pm C_2 Q \pm C_3 R \ldots \tag{5.7}$$

Applying their standard deviations (SD_P, SD_Q, SD_R) and multiplying factors (C_1, C_2, C_3), for instance, for molecular weight correction, the combined uncertainty of the Y is calculated as:

$$S_{(y(xP,Q,R))} = \sqrt{\left(C_1 \times SD_P\right)^2 + \left(C_2 \times SD_Q\right)^2 + \left(C_3 \times SD_R\right)^2 \ldots} \tag{5.8}$$

TABLE 5.3

Major Sources of Random and Systematic Errors in Pesticide Residue Analysis[a]

Operation	Sources of Systematic Error	Sources of Random Error
Sampling	Wrong sampling design or operation; Degradation, evaporation of analytes during preparation, transport, and storage.	Inhomogeneity of analyte in sampled object; Varying ambient (sample material) temperature during transport and storage; Varying sample size.
Sample preparation	The portion of sample to be analyzed (analytical sample) may be incorrectly selected.	The analytical sample is in contact and contaminated by other portions of the sample; Rinsing/brushing is performed to varying extents, stalks and stones may be differentially removed; Inhomogeneity of the analyte in single units of the analytical sample; Heterogeneity of the analyte in the ground/chopped analytical sample.
Sample processing	Decomposition of analyte during sample processing, cross-contamination of the samples.	Variation of temperature during the homogenization process; Texture (maturity) of plant materials affecting the efficiency of homogenization process; Varying chopping time, particle size distribution.
Extraction/Clean-up	Incomplete recovery of analyte; Interference of co-extracted materials (load of the adsorbent).	Variation in the composition (e.g., water, fat, and sugar content) of sample materials taken from a commodity; Temperature and composition of sample/solvent matrix.
Quantitative determination	Interference of co-extracted compounds; Incorrect purity of analytical standard; Biased weight/volume measurements; Determination of substances which do not originate from the sample (e.g., contamination from the packing material); Determination of substance differing from the residue definition; Biased calibration.	Variation of nominal volume of devices within the permitted tolerance intervals; Precision and linearity of balances; Variable derivatization reactions; Varying injection, chromatographic, and detection conditions (matrix effect, system inertness, detector response, signal-to-noise variation, etc.); Operator effects (lack of attention); Calibration.

[a] Some processes and actions may cause both systematic and random error. They are listed where the contribution is larger.

2. The result is obtained with multiplication or division:

$$Y = \frac{k \times P}{Q \times R} \tag{5.9}$$

The relative standard deviation (coefficient of variation) of the P value is:

$$CV_P = \frac{SD_P}{P} \tag{5.10}$$

The relative uncertainty (random error) of the calculated Y value is obtained as:

$$CV_Y = \sqrt{\left(k \times CV_p\right)^2 + CV_Q^2 + CV_R^2} \tag{5.11}$$

The final result of the residue analysis is obtained by multiplying the combined effects of systematic and random errors of each step involved. Let's assume that the average residue content of the sampled object is X mg/kg. Due to the inherent variability of sampling (CV_S), subsampling (CV_{SS}), sample processing (CV_{Sp}), and analysis (CV_A), they alter the measured residue with factors of 1.3, 1.1, 1, and 0.8. Then the measured residue will be 1.79X mg/kg.

The combined relative standard uncertainty (CV_R) of the results can be calculated according to Equation 5.12:

$$CV_R = \sqrt{CV_S^2 + CV_{SS}^2 + CV_{Sp}^2 + CV_A^2} \tag{5.12}$$

The laboratory operations can be combined (CV_L') and separate the uncertainty of the field and laboratory phase of the determination of pesticide residues:

$$CV_R = \sqrt{CV_S^2 + CV_L'^2} \tag{5.13}$$

If subsampling is not required, the uncertainty of within-laboratory operations (CV_L) comprises the contribution of sample processing and analysis:

$$CV_L = \sqrt{CV_{Sp}^2 + CV_A^2} \tag{5.14}$$

If the CV_A is higher than the acceptable performance criteria [4,6]. the analytical phase can be further subdivided (e.g., extraction, evaporation, instrumental determination, etc.) to estimate their uncertainty separately. The individual contributions to the uncertainty of the analysis phase (CV_A) can only be conveniently determined by applying ^{14}C labeled compounds [46].

Where the combined uncertainty is calculated from the linear combination of the variances of its components (Equation 5.8), according to the Welch-Statterthwaite formulae, the effective degree of freedom (v_{eff}) of the estimated combined uncertainty (u_c) is:

$$v_{eff} = \frac{u_{c(y)}^4}{\sum_{i=1}^{N} \frac{u_{i(y)}^4}{v_i}}$$ (5.15)

where

$$u_{c(y)}^2 = \sum_{i=1}^{N} u_{i(y)}^2$$ (5.16)

and

$$v_{eff} \leq \sum_{i=1}^{N} v_i$$ (5.17)

The $SD_{c(y)} = u_{c(y)}$ values may be replaced with $SD_{c(y)}/y = CV$ values where the combined uncertainty is calculated from the relative standard deviations (Equation 5.12) [62]. If the value of v_{eff} obtained from Equation 5.15 is not an integer, which will usually be the case in practice, it can be truncated to the next lower integer. The effective degree of freedom shall be used in further statistical tests performed with the results of analysis.

5.5.3 DETERMINATION OF UNCERTAINTIES OF INDIVIDUAL STEPS OF ANALYSIS

5.5.3.1 Sampling

The deposit of pesticide residues on treated surface is very uneven (distributional heterogeneity) due to various factors such as application technique, canopy structure, microclimate, or crop growth. Wash off or irrigation can also cause substantial loss of residues [18,63]. Extensive studies involving the analysis of the residues of 46 pesticides in 100–320 individual units of 20 different fruits and vegetables revealed that the relative standard deviations of residues derived from a single field or lot ranged between 15–140% with a weighted average of about 80% in case of single fields of known treatment history and 110% for market samples [64]. Analysis of 12087 replicate composite samples taken from supervised field trials resulted in a weighted average CV_S of 22% for composite samples and 78% for unit crop residues assuming that 12 crop units made up a composite sample [63]. These results concord with the general findings of previous studies [65,66].

Based on the statistical analysis of replicate samples taken from supervised trials Farkas could estimate the crop specific uncertainty of sampling for 24 crop groups (Table 5.4) and 106 individual crops [39].

Furthermore, Farkas and co-workers concluded [67] that: (i) the relative 95% range of the average CV_S values does not depend on the variability of residues in crop units forming the parent population (CV_1: 0.11–1.44) and the physical-chemical properties of pesticides; (ii) the CV_S are substantially increasing up to eight replicate samples, but

TABLE 5.4

Estimated Sampling Uncertainties for Crop Groups [1] Recommended for Practical Use[a]

Crop Groups	#	N[a]	CV_S	CV_{Sprim}	$UCLCV_{prim}$
Small-sized fruits[b]	4	768	0.33	0.96	1.1
Medium-sized fruits	12	2139	0.27	0.52	0.57
Large-sized fruits	4	560	0.30	0.78	0.62
Medium-sized vegetables	4	1211	0.36	1.1	1.2
Bush berries	4	171	0.18	0.98	1.1
Legume vegetables	7	211	0.33	1.1	1.2
Brassica vegetables	4	698	0.32	1.5	1.1
Cucurbits	5	337	0.37	1.4	1.1
Leafy vegetables	11	1872	0.29	0.87	0.92
Root and tuber vegetables	6	256	0.30	0.89	1.0
Stalk and stem vegetables	2	276	0.20	0.59	0.69
Pulses	4	346	0.40	1.2	1.4
Cereal grains	6	340	0.21	0.62	0.71
Grasses, for sugar or syrup production	1	15	0.71	2.9	4.4
Tree nuts	2	101	0.19	0.57	0.72
Oilseeds	5	247	0.33	0.98	1.1
Seeds for beverages and sweets	1	22	0.55	1.6	2.6
Legume forage and fodder	4	288	0.28	0.83	0.96
Straw, hay (of legume feeds)	6	523	0.30	0.88	0.91
Cereal forage, fodder and straw	10	1176	0.29	0.86	0.97
Grass forage	2	19	0.22	0.65	1.1
Grass hay	1	18	0.15	0.46	0.75
Dried herbs	2	99	0.23	0.67	0.84
By-products for animal feed	3	391	0.23	0.66	0.75

[a] modified from reference 67.
[1] crop groups according to the Codex commodity classification [34]
number of commodities belonging to the group;
N number of replicate sample pairs used for estimation of CV_S;
CV_{Sprim} relative standard deviation of primary samples. It can be used for calculation of sampling uncertainty of composite samples (CV_n) of different sizes (n) based on the general equation of sampling: $CV_n = \dfrac{CV_{prim}}{\sqrt{n}}$;
UCL CV_{prim} upper 95% confidence limit of estimated CV_{Sprim}.

the gain gradually decreases at larger numbers of samples. The latter finding concords with the conclusion of Lyn [68]. The EURACHEM Guide also recommends duplicate sampling of a minimum of eight independent targets and analyses of each sample in duplicate [69]. The ISO standard for sampling bulk materials recommends taking a minimum of 2 replicate samples from a minimum of 20 independent lots [70].

All these results indicate that a reliable estimate for the uncertainty of sampling cannot be obtained from a few samples, and unless independent estimates for individual crops based on a sufficient level of data are available, the typical CV_S values recommended for commodity groups (Table 5.4) provide the practical solution. The table includes estimates obtained from $N \geq 8$ sample pairs as a lower level of data was considered insufficient for calculation. Furthermore, the authors emphasized that estimates obtained from less than 20 data pairs should be used with caution and, if possible, they should be refined with additional new data.

Concerning the sampling uncertainty, one should always remember that the MRLs refer to the residues in the bulk sample. Hence, for testing compliance with an MRL any amount of material satisfying the minimum sample size [14,15] is sufficient and the sampling uncertainty need not be taken into account for the evaluation of the uncertainty of the results. On the other hand, where the compliance of a lot before shipment is to be verified, the sampling uncertainty must be included in the combined expanded uncertainty of the measured residue value [39].

Gy's theory [30] warns us to remember that when all other errors of sampling are eliminated the fundamental sampling error (FSE) still remains as it is an intrinsic characteristic of the material. The relationship of characteristics of sampled material and CV_{FSE} can be described as:

$$CV_{FSE}^2 = Cd^3 \left(\frac{1}{M_s} - \frac{1}{M_L} \right)$$ (5.18)

where C is the sampling constant (specific to analyte and commodity); d is the diameter of the largest elements (cm) (95th percentile of the size distribution); M_s is the mass of multiple increment composite sample collected (g); and M_L is the mass of the lot from which the composite sample is collected (g). Usually M_L is much larger than M_s and Equation 5.18 can be simplified:

$$CV_{FSE} = \sqrt{\frac{Cd^3}{M_s}}$$ (5.19)

indicating that, for a particular material, the relative uncertainty of sampling can be efficiently decreased by grinding to fine particles a large amount of the material collected from random positions to be sampled, mixing it well, and then withdrawing as large a portion as practical (M_s) from randomly selected positions of the ground material to obtain the laboratory sample. Note that CV_{FSE} is inversely proportional to the square root of the mass of composite sample.

5.5.3.2 Subsampling

The mass of a minimum of five pieces of some large-sized commodities (e.g., watermelon, cabbage, winter squash, jackfruit) may be too large to be comminuted in the usual blenders. Their mass should be reduced by cutting appropriate number of segments representing approximately similar surface–mass ratio in longitudinal direction from stem to the top. From each crop unit an equal number of segments shall be randomly chosen and combined for comminution. The number of segments depends on the size of the crop.

TABLE 5.5
Example for Making Up Subsamples from Randomly Selected Segments of Five Large Fruits

			Fruits and Segments Selected		
#	A	B	C	D	E
1	1	1	1	1	6
2	6	4	6	3	2
3	6	4	6	6	1
4	3	6	2	4	5
5	2	1	4	1	2
6	5	3	4	3	6
7	2	2	5	1	5
8	3	6	3	1	5
9	1	5	6	3	3
10	6	2	3	3	5

serial number of generated subsamples.

In a specific study [38], five jackfruits* (16.5–18 kg each) treated with thiophanate-methyl according to normal farming practice and harvested on the following day were cut into six segments of similar size. The residues were determined in each segment separately. The average residue of the kth fruit ($\bar{c}_k$) was calculated from the masses (m_{ki}) of the 'i' segments and the corresponding residue concentration (c_i):

$$\bar{c}_k = \frac{\sum_{i=}^{6} m_i \times c_i}{\sum_{i=1}^{6} m_i} \qquad (5.20)$$

The grand average of residues in the five fruits was calculated inserting $\bar{c}_k$ and $\sum m_i$ in Equation 5.20. The ranges of residue contents ($mg\,kg^{-1}$) of individual segments of the five fruits and the calculated averages given in brackets were: first, 0.11–0.55 (0.29); second, 0.14–0.46 (0.26); third, 0.1–0.42 (0.26); fourth, 0.36–0.61 (0.44); and fifth, 0.18–0.35 (0.31). From the residue concentrations and masses of individual segments, 50 subsamples consisting of one segment from each fruit were generated with random sampling without replacement applying an Excel macro. In this way only one segment was selected from each fruit for one subsample. The composition of the first ten virtual subsamples is shown in Table 5.5.

The modeling of possible composition of subsamples from individual fruit segments revealed that the relative uncertainty of residues derived from subsampling (CV_{SS}) without the contribution of sample processing (CV_{Sp}) and analysis (CV_A) (see Section 5.4.3.4) was 17%.

* Courtesy of Mr. Kit Chan, Malaysia.

The results highlight the importance of carrying out subsampling by paying attention to the constitutional heterogeneity of the residues within large fruits and vegetables, especially where most of the residues are expected to be on the surface. In such cases, the independent preparation and analysis of two subsamples, containing one segment from each crop, is recommended and the weighted average of the two residue values be reported.

The distributional and compositional heterogeneity of residues in large crops shows large variation. Application of the basic principle of replicate design [69] for eight subsamples would provide information on the uncertainty of the intermediate steps and the final results only for that particular sample. It may be the solution if only one sample is to be analyzed by the laboratory. Where similar samples are expected regularly, the initial uncertainty of the measured residues (CV'_L) can be calculated from the results of the analyses of two to four replicate subsamples prepared as described above:

$$CV'_L = \frac{R_{max} - R_{min}}{d_2 \times \bar{R}}$$ (5.21)

Where the d_2 factors, obtained from range statistics for replicate samples [71]. are 1.128, 1.693, and 2.059 for two, three, and four replicate results, respectively. When detectable residues from further samples (n) are available, the initial uncertainty estimate shall be refined by analyzing duplicate subsamples and calculate the refined within-laboratory reproducibility with Equation 5.22:

$$CV'_L = \sqrt{\frac{\sum_{i=1}^{n} \Delta^2}{2n}}$$ (5.22)

Where $\Delta = (R_1 - R_2)/\bar{R}$, the relative difference of replicate results. To obtain reliable estimate, the CV'_L should be calculated from ≥ 15 independent samples.

In cases where the CV'_L is larger than 25%, the acceptance criterion for replicate measurements [4] and the contribution of the sample processing and analyses steps should be separately determined and corrective actions, if possible, should be taken.

5.5.3.3 Sample Processing

The homogeneity (well-mixed status) of the processed analytical sample cannot be verified with the usual recovery studies, but it should be tested either with samples treated with pesticides according to the normal agricultural practice [46,72] or a small part of the surface of the crops should be treated with suitable test compounds [41,42]. A third alternative is to treat a small portion of the sample matrix with the test compound and then mix it with the rest of the sample [40,46,73].

The uncertainty of sample processing can be quantified as part of the method validation by applying fully nested or staggered nested [74] experimental design and evaluating the results with ANOVA. In this case, uncertainty information can be obtained only for the size of the test portion. If the expectable uncertainty should be determined for a given range of test portion sizes in order to optimize the analytical procedure, the concept of the sampling constant [75] can be used. The sampling constant for sample processing, K_{Sp}, is defined as:

$$K_{Sp} = m \times CV_{Sp}^2 \tag{5.23}$$

Where m is the mass of the test portion taken from the analytical sample as a single increment, and CV_{Sp} is the relative standard deviation of the concentration of the analyte in the test portions of size m.

If the analytical sample is well mixed the sampling constant should be the same for small (Sm) and large (Lg) portions, and Equation 5.23 can be written as:

$$m_{Lg} \times CV_{Lg}^2 = m_{Sm} \times CV_{Sm}^2 \tag{5.24}$$

$$s_{Lg}^2 = s_{Sm}^2 \times \frac{m_{Sm}}{m_{Lg}} \tag{5.25}$$

The well-mixed condition of the comminuted laboratory sample can be verified by analyzing large and small test portions ($m_{lg} \geq 10\ m_{sm}$) repeatedly, ≥ 7 times. Since the standard deviation calculated from a few replicate test portions is very unprecise, according to Wallace and Kratochvil the ratio of:

$$\frac{S_{Sm}^2 \times m_{Sm}}{S_{Lg}^2 \times m_{Lg}} \tag{5.26}$$

should be tested with F-test at 90% or lower confidence level. If the F-test indicates that the difference is not significant the processed sample can be considered statistically well mixed. In such cases the K_{Sp} should be calculated from the CV_{Lg} values as it is more precise than the CV_{Sm}. Then the expected sample processing uncertainty can be calculated for any test portion size $\geq m_{Sm}$. This is a great advantage compared to the duplicate method.

Equation 5.23 can be written as:

$$CV_{Sp} = \sqrt{\frac{K_{Sp}^2}{m}} \tag{5.27}$$

which indicates that the K_{Sp} incorporates the Gy's sampling constant [30] (Equation 5.19) and the third power of particle size of comminuted materials ($K_{Sp} = Cd^3$). In practice, it means that if we take 1 g test portion from a comminuted matrix shown to be well-mixed at 10 g test portion level, the $CV_{Sp(1\ g)}$ will be at least 3.2 times higher than $CV_{Sp(10g)}$, and the comminuted material should not be considered well-mixed at 1 g test portion level. Therefore, the test portion size should not be reduced without testing again the well-mixed condition of the homogenized matrix.

The acceptable variability of sample processing depends on the variability of the other steps of the determination. As far as possible it should not contribute substantially to the combined uncertainty of the reported results. When the combined uncertainty of the measurement results should include the sampling uncertainty, then the CV_{Sp} should be ≤ 8–10% ($<0.3CV_S$) depending on the crop analyzed. Where only the CV_L is taken into account the CV_{Sp} should preferably be less than $0.3 \times CV_A$. The efficiency of sample processing can be substantially improved in two steps. First the whole analytical sample or subsample is comminuted in a large blender which can accommodate the whole sample, and from the well-mixed matrix with a known K_{Sp1} value, a large portion (m_2) of the well-mixed matrix is further homogenized with a Warring blender or freezer mill. The process can be carried out with added dry ice [41] or liquid nitrogen [44]. The K_{Sp2} should be determined as described above. The combined sample processing relative standard deviation CV_{Spc} obtained with the two-step procedure analyzing a small test portion (m_A) is calculated with Equation 5.28:

$$CV_{Spc} = \sqrt{\frac{K_{Sp1}}{m_2} + \frac{K_{Sp2}}{m_A}} \tag{5.28}$$

The efficiency of sample processing depends on, among other factors, the type and variety of the sample and the implementation of the process, but it is independent from the analyte. It should be determined as part of the method validation with the representatives of commodity groups [4] having different physical properties (e.g., tomato, celery, sunflower seed, wheat grain and straw, etc.). As sample processing can be a significant contributor to the combined uncertainty, it should be tested regularly as part of the performance verification of the laboratory phase of the determination of pesticide residues. It can be most economically carried out with the reanalysis of the retained test portions of samples (containing the analyte in detectable concentration) included in different analytical batches. The reanalysis of retained test portions should include all samples and analytes within the scope of a (multi-residue) method to verify or refine its performance characteristics established during method validation. The within-laboratory reproducibility of the procedure (CV_L) can be calculated with Equation 5.22.

If the order of analyses of test portions is recorded, useful information can be obtained from the tendency of their differences. For instance, if the results of the second analysis are generally lower than the first one it may indicate, for instance, that the residues decomposed during storage or the nominal concentration of analytical standard solutions used for calibration changed.

5.5.3.4 Analysis

Five-five test portions taken from five different types of soils were spiked with acetone solution of ^{14}C-labeled atrazine at 0.05 mg/kg level. After letting the acetone evaporate, the test portions were extracted on different days by three analysts. The within-laboratory reproducibility CV was 2% [73]. The results indicate that extraction does not practically affect the reproducibility of analysis (CV_A). The major part of the variability of results of residue analysis is derived from the further steps (evaporation, clean-up, and instrumental analyses), which may require special attention if the combined relative reproducibility uncertainty of the results (CV_L, or CV'_L) is getting close to the upper acceptable limit of 25% [4].

The uncertainty of the analyses (CV_A) is determined from replicate test portions spiked before extraction. The test portions should be withdrawn from the homogenized analytical samples which preferably do not contain the analytes included in the spiking solution. In some cases, certified reference materials or a remaining portion of properly stored proficiency test materials with known analyte concentration are available which can also be used. Where untreated samples are not available, or the final extract of the blank sample gives detectable response, the analyte equivalent of the average instrument signal obtained from the test portion before spiking shall be considered. Recovery tests performed according to the above-mentioned options will not provide information on CV_L or CV'_L.

EUQAQC provides detailed guidance [4] on minimum requirements regarding the selection of representative matrices and analytes and frequency of recovery tests together with the acceptable performance criteria.

5.6 SYSTEMATIC ERRORS—BIAS OF THE MEASUREMENTS

Systematic errors can occur in all phases of the measurement process. However, it practically cannot be quantified during sampling and other external or field phases of the process. The most accurate and precise determination of the systematic error including that caused by the efficiency of extraction and dispersion of residues in the treated material can be carried out with radio labeled compounds [73]. Unfortunately, routine pesticide residue laboratories very rarely have access to facilities suitable for working with radioisotopes. As an alternative solution, extraction solvents of different polarity may be used to extract the residues from samples containing incurred residues.* Generally the efficiency of extraction can be improved by processing the sample to fine particles applying cryogenic processing or freezer mill.

Very useful information on the efficiency of extraction, stability of residues during storage of samples, and proportion of residues in edible portions can be found in the FAO/WHO series of "Pesticide Residues–Evaluations", which are published annually by FAO, and can be freely downloaded from the Website of the Pesticide Management Group [47]. Another source of information is the data submitted to national registration authorities for supporting the claim for registration of the pesticides. Though the whole package is confidential, that part relating to the analysis of residues could be made accessible for laboratories analyzing pesticide residues.

The laboratories should test the bias of their measurement results from the extraction step onward by performing recovery studies which can provide information on the systematic error and precision of the procedure only from the point of spiking. Thus, the usual procedure will not indicate the efficiency of extraction, or the loss of residues during storage and sample processing.

Different areas of analytical chemistry have different practices and legal requirements concerning the use of recovery factors. The IUPAC Guidelines for the use of recovery information aid the preparation of the "best estimate of the true result" and

---

* Residue in a commodity resulting from specific use of a pesticide, consumption by an animal, or environmental contamination in the field, as opposed to residues from laboratory fortification of samples.

to contribute to the comparability of the analytical results reported [76]. When the average recovery is statistically significantly different from 100%, based on t-test, the results should generally be corrected for the average recovery. Generally average recoveries between 70–120% are acceptable without adjusting the measured residues [4]. In exceptional cases, consistently lower recovery with reproducibility relative standard deviation of ≤20% would also be acceptable [6].

It should be noted however, that in case of pre-marketing or export control it is advisable to adjust the measured result with the average recovery, if that is significantly different from 100%, to avoid dispute situation. The shipment may be simply rejected due to the lower recoveries of analytical method used in the exporting country compared to that applied in the importing country. Another area where reporting the most accurate result is necessary, is providing data for the estimation of exposure to pesticide residues. In order to avoid any ambiguity in reporting the results, the analyst should give the uncorrected as well as the corrected value, and the reason for and the method of the correction [17].

The uncertainty of the mean recovery, $\bar{Q}$, obtained from n replicate measurements:

$$CV_{\bar{Q}} = \frac{CV_A}{\sqrt{n}} \qquad (5.29)$$

affects the uncertainty of the reported results CV_{Acor}:

$$CV_{Acor} = \sqrt{CV_A^2 + CV_{\bar{Q}}^2} \qquad (5.30)$$

The increase of the uncertainty of the residue values adjusted for the recovery can be practically eliminated if the mean recovery is determined from ≥15 measurements ($CV_{Acor} \leq 1.03 \times CV_A$). On the other hand, if corrections would be made with a single procedural recovery, the uncertainty of the corrected result would be $1.41 \times CV_A$. Therefore, such correction should be avoided as far as practical.

The recovery values obtained from performance verification usually symmetrically fluctuate around their mean, which indicates that the measured values are subject to random variation. If the procedural recovery performed with an analytical batch is within the expected range, based on the mean recovery and within-laboratory reproducibility of the method, the analyst demonstrated that the method was applied with expected performance. Therefore, the correct approach is to use the typical recovery established from the method validation and the long-term performance verification (within-laboratory reproducibility studies) for correction of the measured residue values, if necessary.

Under certain circumstances, such as extraction of soil samples, the extraction conditions might not be fully reproduced from one batch of samples to the next, leading to much higher within-laboratory reproducibility than repeatability. In this case, the use of concurrent recovery for adjusting the measured residues may provide more accurate results. Where correlation between the results is quantifiable, it may be necessary to perform at least two recovery tests in one analytical batch covering the expected residue range and use their average value for correction to reduce the uncertainty of the results.

5.7 INTERLABORATORY STUDIES

Regular participation in interlaboratory studies and proficiency tests is an important part of quality assurance programs and a basic requirement for accreditation, as it provides the opportunity for the laboratories to prove the suitability of their methods, comparability of their results, and proficiency of the staff in their application. Good results obtained in these studies only show the capability of the laboratory, but they do not guarantee similar performance during the daily work, which must be shown with the results of internal quality control.

The harmonized criteria for testing the proficiency of laboratories had been jointly elaborated by ISO, AOAC, and IUPAC [77]. The current proficiency test programs [78,79] organized by several organizations are based on the revised versions of the harmonized criteria.

Within the EURL proficiency test program, the participants are normally informed in advance about the lists of targeted pesticides that must be analyzed and those may be analyzed on a voluntary basis. The carefully homogenized and tested material contains field incurred residues or is spiked with mixtures of analytes, which are known to be stable during the expected duration of the study. The reported results are first screened for false positive and false negative results and obviously erroneous data, then statistically evaluated for analytical outliers with robust statistics. The assigned value for the mean ($\hat{x}$) is established by calculation of the robust mean, the median, or the mode depending on the distribution of the results after removal of spurious and outlier values. The target standard deviation, σ, should reflect the best practice or 'fit for purpose' for the given analyte, which may be derived from the results of collaborative studies, previous proficiency tests, or predicted based on the Horwitz equation or those suggested by Thompson [80]. The z-score is calculated from the assigned, ($\hat{x}$) and reported (x_i) value and targeted standard deviation (σ_t): $z = (x_i - \hat{x})/\sigma_t$. The interpretation of the z-score is based on normal statistics. Laboratories performing as expected should produce results within the $2 \times z$ in most of the cases, but 1 result (questionable) out of 20 may be between $2–3 \times z$. Results above $3 \times z$ are unacceptable, should occur very rarely (the probability is 0.3%), and "require action".

In order to evaluate each laboratory's overall performance, and taking into account all pesticides analyzed, the EU proficiency test program applies the average of squared z-scores [81]:

$$AZ^2 = \frac{\sum_{i=1}^{n} z^2}{n} \tag{5.31}$$

where n is the number of quantified analytes reported.

The application of squared z-score widens the scale of describing the performance of the laboratories. For instance, a just acceptable result $z = 2$ is considered unsatisfactory $z^2 = 4 > 3$. A z-score of 0.5 would be counted with 0.25 in Equation 5.31. Obtaining an AZ^2 around 0.02–0.03 for 19 residues shows the excellent general performance of the reporting laboratories.

There are some variations in the evaluation of proficiency test results. The reports provide detailed explanation on their statistical evaluation. Although participation in proficiency test is obligatory for certain laboratories in the EU, and one of the requirements of accreditation according to ISO/IEC 17025 Standard, proficiency scheme providers, participants, and end users should avoid judging the performance and ranking of laboratories on the basis of their z-scores alone.

REFERENCES

1. International Standard Organization. ISO/IEC 17025: 2017 General requirements for the competence of testing and calibration laboratories, 2017; 30 pp.
2. OECD Series on 'Principles of Good Laboratory Practice and compliance monitoring' Nos. 1–17, OECD Environmental Directorate, Paris, 1998–2016. http://www.oecd.org/chemicalsafety/testing/oecdseriesonprinciplesofgoodlaboratorypracticeglpandcompliancemonitoring.htm
3. Stephenson, G.R.; Ferris, I.G.; Holland, P.T.; Nordberg, M. Glossary of terms relating to pesticides. (IUPAC Recommendations 2006). *Pure Appl. Chem.*, 2006, 78 11, 2075–154.
4. European Commission. Guidance document on analytical quality control and method validation procedures for pesticide residues and analysis in food and feed, SANTE /18813/2017, 2017; 46 pp.
5. Codex Alimentarius. Guidelines on good laboratory practice in residue analysis, CAC/ GL 40-1993, Rev.1-2003, Amended 2010; 36 pp. www.fao.org/input/download/standards/378/cxg_040e.pdf
6. Codex Alimentarius. Guidelines on performance criteria for methods of analysis for the determination of pesticide residues in food and feed, CAC/GL 90-2017, 2017; 13 pp. http://www.fao.org/fao-who-codexalimentarius/sh-proxy/fr/?lnk=1&url=https%25 3A%252F%252Fworkspace.fao.org%252Fsites%252Fcodex%252FStandards%252FCA C%2BGL%2B90-2017%252FCXG_090e.pdf
7. Holcombe, D. Quality assurance for research and development and non-routine analysis, EURACHEM/CITAC, 1998; 67 pp. https://www.eurachem.org/images/stories/ Guides/pdf/rdguide.pdf.
8. Mann, I.; Brokman, B. *Selection, Use and Interpretation of Proficiency Testing (PT) Schemes*, 2nd ed. EURACHEM, 2011; 52 pp. https://www.eurachem.org/index.php/ publications/guides/usingpt
9. Ellison, S.L.R.; King, B.; Rösslein, M.; Salit, M.; Williams, A. (eds.) Traceability in chemical measurement – A guide to achieving comparable results in chemical measurement, EURACHEM/CITAC, 2003; 44 pp. http://extras.springer.com/2007/978-3-540-71271-8/MISC/content.pdf
10. King, B. Quality in the analytical laboratory. In *6th International Symposium on Quality Assurance and TQM for Analytical Laboratories*; Parkany, M. (ed.); The Royal Society of Chemistry: Oxford, UK, 1995; 8–18.
11. Kateman, G.; Buydens, L. *Quality Control in Analytical Chemistry*, 2nd ed. John Wiley & Sons, Inc. New York, NY, 1993; Chapter 2.
12. Cook, J.M.; Ambrus, Á. Theory and practice of sampling for pesticide residue analysis. In *Food Safety Assessment of Pesticide Residues*; Ambrus, Á.; Hamilton, D. (eds.); World Scientific Publishing Europe Ltd.: London; 2017; 327–403.
13. Association of American Feed Control Officials. GOOD samples, 2015; 81 pp. https:// www.aafco.org/Portals/0/SiteContent/Publications/GOODSamples.pdf

14. Codex Alimentarius. Recommended methods of sampling for the determination of pesticide residues for compliance with MRLs, CAC GL 33/1999, 1999; 18 pp. http://www.fao.org/fao-who-codexalimentarius/codex-texts/guidelines/en/

15. European Commission. Directive. 2002/63/EC on establishing community methods of sampling for the official control of pesticide residues in and on products of plant and animal origin and repealing Directive 79/700/EEC. *Off. J. Eur. Communities*, 2002, 187, 30–43.

16. Codex Alimentarius. Portion of commodities to which codex maximum residue limits apply and which is analyzed, CAC GL 41-1993, Amendment 2010; 9 pp. http://www.fao.org/fao-who-codexalimentarius/codex-texts/guidelines/en/

17. Ambrus, Á. *Submission and Evaluation of Pesticide Residues Data for the Estimation of Maximum Residue Levels in Food and Feed*, 3rd ed. FAO Plant Production and Protection Paper, 225. FAO: Rome, 2016; 1–298. http://www.fao.org/3/a-i5452e.pdf

18. Ambrus, Á. The influence of sampling methods and other field techniques on the results of residue analysis. In *Pesticide Residues*; Frehse, H.; Geissbühler, H. (eds.); Pergamon Press: Oxford; 1979; 6–18.

19. Ambrus, Á.; Lantos, J. Evaluation of the studies on decline of pesticide residues. *J. Agric. Food Chem.*, 2002; 50, 4846–51.

20. OECD. OECD Guidelines for the testing of chemicals test no. 509: crop field trials, 2009; 44 pp. http://www.oecd-ilibrary.org/environment/test-no-509-crop-field-trial _9789264076457-en

21. U.S. Geological Survey. Collection of water samples, Chapter A4. In *National Field Manual for the Collection of Water-Quality Data*. USGS: Reston, VA, 2006; 231 pp. https://water.usgs.gov/owq/FieldManual/chapter4/pdf/Chap4_v2.pdf

22. WHO. *Guidelines for Drinking-Water Quality: Fourth Edition Incorporating the First Addendum*. WHO, Geneva, 2017; 564 pp. www.who.int/water_sanitation_health/ publications/2011/dwq_guidelines/en

23. Alberta Agriculture and Forestry: Water sampling: Methods based on source. Includes 9 separate documents. Government of Alberta, 2017; http://www1.agric.gov. ab.ca/$department/deptdocs.nsf/all/wqe11070

24. Government of Western Australia. Surface water sampling methods and analysis – technical appendices, 2009; 69 pp. https://www.water.wa.gov.au/__data/assets/pdf_ file/0019/2935/87152.pdf

25. US EPA. *Surface Water Sampling*. SOP, 2013; 22 pp. https://www.epa.gov/sites/produc-tion/files/2015-06/documents/Surfacewater-Sampling.pdf

26. Duncan, D.; Harvey, F.; Walker, M. *EPA Guidelines: Regulatory Monitoring and Testing, Water and Wastewater Sampling*. ISBN 978-1-921125-47-8, 2007; 58 pp.

27. Ramsey, C.A. Considerations for sampling contaminants in agricultural soils. *J. AOAC Int.*, 2015; 98, 309–15.

28. Esbensen, K.H.; Ramsey, C.A. QC of sampling processes—a first overview: from field to test portion. *J. AOAC Int.*, 2015; 98, 282–7.

29. International Standard Organization. ISO 18400-101:2017. Soil quality – Sampling – Part 101: Framework for the preparation and application of a sampling plan, 2017. https://www.iso.org/standard/62842.html

30. Gy, P. *Sampling for Analytical Procedures*. John Wiley & Sons Ltd.: Chichester, UK; 1999; 153 pp.

31. Pitard, F.P. *Pierre Gy's Sampling Theory and Sampling Practice*. CRC Press LLC: Boca Raton, FL; 1993; 483 pp.

32. USDA Natural Resources Conservation Service. *Sampling Soils for Nutrient Management*. 2 pp. https://www.nrcs.usda.ov/Internet/FSE_DOCUMENTS/nrcs144 p2_051273.pdf

33. Kratochvil, B.; Taylor, J.N. Sampling for chemical analysis. *Anal. Chem.*, 1981, 53, 924–938A.
34. FAO. Codex Classification of food and animal feed. In *Joint FAO/WHO Food Standards Programme*, Codex Alimentarius V. 2, Pesticide Residues in Food 2nd ed. FAO: Rome; 1993, last revised in 2006; 199 pp. http://www.fao.org/tempref/codex/Meetings/CCPR/ccpr38/pr38CxCl.pdf.
35. Codex Alimentarius. Reports of the Codex Committee on Pesticide residues. Published annually on the CAC Website. http://www.fao.org/fao-who-codexalimentarius/meetings/archives/en/?y=2001&s=1999
36. Codex Alimentarius. General guidelines on sampling, CAC/GL 50-2004, 2004; 69 pp. http://www.fao.org/fao-who-codexalimentarius/codex-texts/guidelines/en/
37. Ambrus, Á. (Ed.) *Pesticide Residue Analytical Methods*, Vol. I (Növényvédőszermaradékok meghatározási módszerei). Ministry of Agriculture: Budapest; 1976; 258 pp.
38. Omeroglua, P.Y.; Ambrus, Á.; Boyacioglu, D.; Solymosné, M.E. Uncertainty of the sample size reduction step in pesticide residue analysis of large-sized crops. *Food Addit. Contam. Part Part A*, 2013; 30 (1), 116–26.
39. Farkas, Zs.; Cook, J.M.; Ambrus, Á. Estimation of uncertainty of measured residues and testing compliance with MRLs. In *Food Safety Assessment of Pesticide Residues*; Ambrus, Á.; Hamilton, D. (eds.); World Scientific Publishing Europe Ltd.: London; 2017; 404–66.
40. Ambrus, Á.; Buczkó, J.; Hamow, K.Á. et al. Contribution of sample processing to variability and accuracy of the results of pesticide residue analysis in plant commodities. *J. Agric. Food Chem.*, 2016; 64, 6071–81.
41. Maestroni, B.; Ghods, A.; El-Bidaoui, M. et al. Testing the efficiency and uncertainty of sample processing using ^{14}C labelled chlorpyrifos Part I. II. In *Principles of Method Validation*; Fajgelj, A.; Ambrus, Á. (eds.); The Royal Society of Chemistry: Cambridge, UK; 2000; 49–74.
42. Fussell, R.J.; Hetmanski, M.T.; Macarthur, R. et al. Measurement uncertainty associated with sample processing of oranges and tomatoes for pesticide residue analysis. *J. Agric. Food. Chem.*, 2007; 55, 1062–70.
43. Riter, L.S.; Lynos, K.J.; Wujcikt, C.E.; Buchholz, L.M. Interlaboratory assessment of cryomilling sample preparation for residue analysis. *J. Agric. Food. Chem.*, 2015; 63, 4405–8.
44. Riter, L.S.; Chad, E.; Wujcik, C.E. Novel two-stage fine milling enables high-throughput determination of glyphosate residues in raw agricultural commodities. *J. AOAC Int.*, 2017; 101, 1–9.
45. Fussell, R.J.; Addie, K.J.; Reynolds, S.L.; Wilson, M.F. Assessment of the stability of pesticides during cryogenic sample processing. 1. Apples. *J. Agric. Food. Chem.*, 2002; 50, 441–8.
46. Suszter, G.; Ambrus, Á.; Schweikert-Turcu, M.; Klaus, P.M. Estimation of efficiency of processing soil samples. *J. Environ. Sci. Health Part B*, 2006; 41, 531–52.
47. FAO. *Pesticide Residues in Food - Evaluations*. Published annually in the series of Plant Production and Protection Papers. http://www.fao.org/agriculture/crops/core-themes/theme/pests/jmpr/jmpr-rep/en/
48. OECD. OECD Guidelines for the testing of chemicals, Test No. 506: Stability of pesticide residues in stored commodities, 2007; 12 pp. http://www.oecd-ilibrary.org/environment/test-no-506-stability-of-pesticide-residues-in-stored-commodities_9789264061927-en
49. El-Bidaoui, M.; Jarju, O.P.; Maestroni, M.; Phakaeiw, Y.; Ambrus, Á. Testing the effect of sample processing and storage on stability of residues. In *Principles of Method Validation*; Fajgelj, A.; Ambrus, Á. (eds.); The Royal Society of Chemistry: Cambridge, UK; 2000; 75–88.

50. Hill, A.R.C.; Harris, C.A.; Warburton, A.G. Effects of sample processing on pesticide residues in fruits and vegetables. In *Principles and Practices of Method Validation*; Fajgelj, A.; Ambrus, A. (eds.); The Royal Society of Chemistry: Cambridge, UK; 2000; 41–8.
51. EURL-DataPool. http://www.eurl-pesticides-test.eu/ (accessed March 2018).
52. Ambrus, Á.; Noonan, G.; Németh, A. et al. Testing the accuracy of analytical standard solutions used for quantitative determination of pesticide residues. *J. AOAC Int.*, 2017; 100 4, 1058–61.
53. AOAC. Appendix D: Guidelines for collaborative study procedures to validate characteristics of a method of analysis. *J. Assoc. Off. Anal. Chem.*, 1989; 72, 694–704. http://www.aoac.org/aoac_prod_imis/AOAC_Docs/StandardsDevelopment/Collaborative_Study_Validation_Guidelines.pdf
54. Magnussson, B.; Örnemark, U. (eds.) *The Fitness for Purpose of Analytical Methods: A Laboratory Guide to Method Validation and Releted Topics*, 2nd ed. EURACHEM, 2014; 70 pp. https://www.eurachem.org/images/stories/Guides/pdf/MV_guide_2nd_ed_EN.pdf
55. Thompson, M.; Ellison, S.L.R.; Wood, R. Harmonized guidelines for single laboratory validation of methods of analysis. *Pure Appl. Chem.*, 2002; 74, 835–55.
56. Fajgelj, A.; Ambrus, A. (eds.) *Principles and Practices of Method Validation*. The Royal Society of Chemistry: Cambridge UK, 2000; 305 pp.
57. Ambrus, Á., Hamow, K.Á.; Kötelesné, S.G.; Németh, A.; Solymosné, M.E. Accuracy of analytical standard solutions and uncertainty in their nominal concentrations. *J. Food Invest.*, 2017; LXIII 1, 1398–421.
58. Miller, J.N.; Miller, J.C. *Statistics and Chemometrics for Analytical Chemistry*, 6th ed. Pearson Education Limited: Harlow, UK; 2010; 297 pp.
59. Ambrus, A. Reliability of measurement of pesticide residues in food. *Accred. Qual. Assur.*, 2004; 9, 288–304.
60. Codex Alimentarius. Guidelines on estimation of uncertainty of results, CAC/GL 59-2006; 16 pp. http://www.fao.org/input/download/standards/10692/cxg_059e.pdf
61. Ellison, S.L.R.; Williams, A. *EURACHEM/CITAC Guide: Quantifying Uncertainty in Analytical Measurement*, 3rd ed. ISBN 978-0-948926-30-3. 2012; 141 pp. https://www.eurachem.org/index.php/publications/guides/quam
62. BIPM-JCGM. Evaluation of measurement data – Guide to the expression of uncertainty in measurements, JCGM 100:2008, Annex G4. JCGM, 2008; 73–4. https://www.bipm.org/en/publications/guides/gum.html
63. Farkas, Zs. Optimization of sampling procedures for verifying compliance of commodities with maximum residue limits of pesticides. PhD Dissertation in Hungarian, 2017. https://szie.hu/node/976
64. Horváth, Zs.; Ambrus, Á.; Mészáros, L.; Braun, S. Characterization of distribution of pesticide residues in crop units. *J. Environ. Sci. Health Part B*, 2013; 48, 615–25.
65. Ambrus, Á.; Soboleva, E. Contribution of sampling to the variability of pesticide residue data. *J. AOAC Int.*, 2004; 87, 1368–79.
66. Ambrus, Á. Estimation of sampling uncertainty for determination of pesticide residues in plant commodities. *J. Environ. Sci. Health Part B*, 2009; 44, 1–13.
67. Farkas, Zs.; Horváth, Zs.; Szabó, I.J.; Ambrus, Á. Estimation of sampling uncertainty of pesticide residues based on supervised residue trial data. *J. Agric. Food Chem.* 2015; 63, 4409–17.
68. Lyn, J.A.; Ramsey, M.H.; Coad, D.S.; Damant, A.P.; Wood, R.; Boon, K.A. The duplicate method of uncertainty estimation: are eight targets enough? *Analyst*, 2007; 132, 1147–52.
69. Ramsey, H.M.; Ellison, S.L.R. *Measurement Uncertainty Arising from Sampling: A Guide to Methods and Approaches*. EURACHEM, 2007; 111 pp. https://www.eurachem.org/images/stories/Guides/pdf/UfS_2007.pdf

70. International Standard Organization. ISO 11648-1. *Statistical Aspects of Sampling from Bulk Materials –Part 1: General Principles.* 2003; 92 pp.

71. Anderson, L.R. *Practical Statistics for Analytical Chemists.* Appendix D12. Van Nostrand Reinhold Company: New York; 1987.

72. Ambrus, Á.; Solymosné-Majzik, E.; Korsós, I. Estimation of uncertainty of sample preparation for the analysis of pesticide residues. *J. Environ. Sci. Health Part B*, 1996; 31, 443–50.

73. Suszter, G.K.; Ambrus, Á. Testing the efficiency of extraction of incurred residues from soil with optimized multi-residue method. *J. Environ. Sci. Health Part B*, 2017; 52, 547–56.

74. Lyn, J.A.; Ramsey, M.H.; Fussell, R.J.; Wood, R. Measurement uncertainty from physical sample preparation: estimation including systematic error. *Analyst*, 2003; 128, 1391–8.

75. Wallace, D.; Kratochvil, B. Visman equations in the design of sampling plans for chemical analysis of segregated bulk materials. *Anal. Chem.*, 1987; 59, 226–32.

76. Thompson, M.; Ellison, S.L.R.; Fajgelj, A.; Willets, P.; Wood, R. Harmonised guidelines for the use of recovery information in analytical measurement. *Pure Appl. Chem.*, 1999; 71, 337–48.

77. Thompson, M.; Ellison, S.L.R.; Wood, R. International harmonised protocol for the proficiency testing of analytical laboratories, IUPAC Technical Report. *Pure Appl. Chem.*, 2006: 78 145 .

78. FERA Science Ltd. Protocol for proficiency testing schemes V.6, 2017; 19 pp. https://fapas.com/sites/default/files/2017-05/FeraPTSprotocol_pt1_common.pdf

79. EU Reference Laboratories for Residues of Pesticides. http://www.eurl-pesticides.eu/docs/public/home.asp?LabID=100&Lang=EN

80. Thompson, M. Recent trends in inter-laboratory precision at ppb and sub-ppb concentrations in relations to fitness for purpose criteria in proficiency testing. *Analyst*, 2000; 125, 385–6.

81. Medina-Pastor, M.; Mezcua, C.; Rodríguez-Torreblanca, A.; Fernández-Alba, R. Laboratory assessment by combined z-score values in proficiency tests: experience gained through the European Union proficiency tests for pesticide residues in fruits and vegetables. *Anal. Bioanal. Chem.*, 2010; 397, 3061–70.

6 Determination of Pesticides in Food of Vegetal Origin

Jon W. Wong, Kai Zhang, Douglas G. Hayward,
Alexander J. Krynitsky, Frank J. Schenck, James B.
Wittenberg, Jian Wang, and Paul Yang

CONTENTS

List of Abbreviations.. 175
6.1 Extraction Solvents Used in Multiresidue Pesticides Methods for
  Vegetal-Based Foods 176
6.2 Cleanup Procedures for Sample Extracts 182
6.3 QuEChERS.. 184
6.4 Gas Chromatography–Mass Spectrometry....................................... 187
6.5 Liquid Chromatography–Mass Spectrometry 189
6.6 Liquid and Gas Chromatography–High Resolution Mass Spectrometry 190
6.7 Future Challenges and Considerations 193
References... 195

LIST OF ABBREVIATIONS

APCI	Atmospheric pressure chemical ionization
CAC	Column adsorption chromatography
CDFA	California Department of Food and Agriculture
DDA	Data dependent acquisition
DIA	Data independent acquisition
dSPE	dispersive solid phase extraction
ECD	Electron-capture detector
ElCD	Electrolytic conductivity detector
EMR	Enhanced matrix removal-lipid
ESI	Electrospray ionization
FDA	U.S. Food and Drug Administration
FID	Flame ionization detector
FLD	Fluorescence detector
FPD	Flame photometric detector
GC–MS	Gas chromatography–mass spectrometry
German DFG	Deutsche Forschungsgemeinschaft (German Research Foundation)

GPC	Gel permeation chromatography
HRMS	High resolution mass spectrometry
IT–MS	Iontrap mass spectrometry
K-D	Kuderna-Danish evaporator
LC	High performance liquid chromatography
LC–MS	Liquid chromatography–mass spectrometry
MRM	Multiple reaction monitoring
MS/MS	Tandem mass spectrometry
MS^E	Alternate low-energy and high-energy MS/MS experiment
NFA	National Food Administration of Sweden
NPD	Nitrogen phosphorus detector
Orbitrap	Orbital trap mass analyzer
PAM	Pesticides analytical manual
PCD	Post column derivatization
PRM	Parallel reaction monitoring
PSA	Primary-secondary amine linked to silica particles
Q-HRMS	Quadrupole high resolution mass spectrometry
Q-Orbtrap	Quadrupole orbital trap mass analyzer
QqQ	Triple quadrupole mass spectrometry
Q-TOF-MS	Quadrupole time-of-flight mass analyzer
QuEChERS	Quick, Effective, Cheap, Easy. Rugged, and Safe
QuPPe	Quick polar pesticide
RSD	Relative standard deviation
SIM	Single ion monitoring
SIM	Single ion monitoring
SNR	Signal-to-noise ratio
SPE	Solid phase extraction
SRM	Selected reaction monitoring
TOF-MS	Time-of-flight mass analyzer
UHPLC	Ultrahigh performance liquid chromatography
vDIA	Variable DIA
XSD	Halogen selective detector

6.1 EXTRACTION SOLVENTS USED IN MULTIRESIDUE PESTICIDES METHODS FOR VEGETAL-BASED FOODS

Extraction of pesticides from plant-based foods with an organic solvent is the common approach for routine isolation of pesticides for analysis and detection. Fresh plant-based food is sufficiently and properly homogenized and comminuted to form a liquified or fluid-like state so that the pesticides are isolated and extracted with organic solvent. Water is typically added to hydrate low moisture foods such as nuts, dried fruits, cereal grains, and botanicals to further increase surface area and improve the extraction of the nonpolar pesticides into the organic solvent. Acetonitrile, acetone, ethyl acetate, and methanol have been commonly used as extraction solvents for isolating pesticides (Krynitsky and Lehotay, 2002). Acetonitrile, acetone, and methanol are water-miscible solvents and require an additional non-miscible solvent such as petroleum ether or dichloromethane to back-extract the pesticides from the aqueous phase. Ethyl acetate is essentially immiscible in water and can both extract non-polar pesticides and partition pesticides from the aqueous phase in a single step. Acetonitrile and, to some extent, acetone, have a unique character compared to the other solvents because both can partition from the aqueous phase when a salt such as sodium chloride is added to the organic solvent–aqueous mixture. The use of these organic solvents to extract pesticides from plant matrices for sample preparation and analysis is described below.

Acetonitrile. Mills et al. (1963) from the U.S. Food and Drug Administration (FDA) developed one of the earliest multiresidue procedures that involved "stripping" or extracting pesticides from plant foods using acetonitrile, partitioning the acetonitrile extract with sodium chloride and petroleum ether, cleanup with Florisil column chromatography, and analysis by gas chromatography–microcoulometric detection. The Mills procedure, as illustrated in Figure 6.1, has been successfully studied, collaborated, and used for the analysis of organochlorine and some nonpolar organophosphorus pesticides in a variety of foods (Smart, 1976). The Mills procedure was further expanded to detect diverse classes of organochlorine, organophosphorus, and organonitrogen pesticides. The multiresidue procedures used by the California Department of Food and Agriculture (CDFA) prior to 2010 represent the types of modifications and changes associated with the original

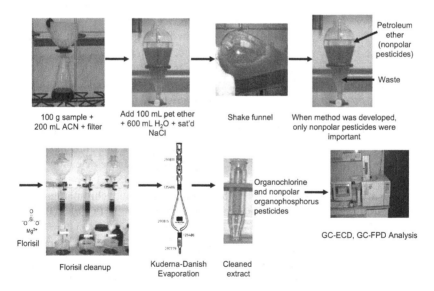

FIGURE 6.1   Steps involved in the Mills Procedure (Mills et al., 1963) utilizing acetonitrile as an extraction solvent (blending with the plant sample), followed by partitioning with sodium chloride and petroleum ether, sample cleanup using Florisil column chromatography, and gas chromatographic analysis. Excessive solvent is removed by Kuderna-Danish evaporation. (Photo provided by F. Schenck.)

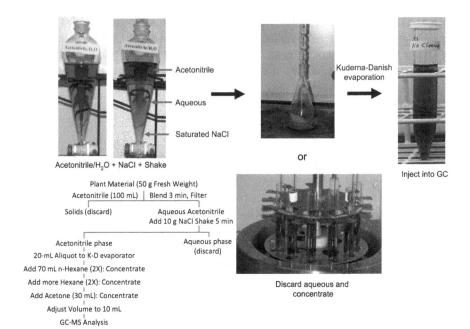

FIGURE 6.2   The California Department of Food and Agriculture utilizes acetonitrile salt-out extraction and several evaporation steps for analysis by gas chromatography. (Photo provided by F. Schenck.)

Mills method (Lee et al., 1991). As illustrated in Figure 6.2, food samples (50 g) are blended with acetonitrile, filtered, and transferred into a separatory funnel where the acetonitrile phase is partitioned with a saturated sodium chloride solution. n-Hexane is added to the acetonitrile, the organic extract is concentrated using a Kuderna-Danish (K-D) evaporator, and the reduced extracts are analyzed for non-polar and thermally stable organochlorine and organophosphorus pesticides using gas chromatography (GC) equipped with electrolytic conductivity detection (ECD) or flame photometric detection (FPD), respectively. A third aliquot undergoes cleanup using an aminopropyl-bonded silica solid phase extraction (SPE) for the analysis of carbamate pesticides by high-performance liquid chromatography-post column derivatization-fluorescence detection (HPLC-PCD/FLD). The FDA Mills procedure has been accepted by various pesticide laboratories and is part of the FDA Pesticide Analytical Manual (PAM; U.S. Food and Drug Administration, 1999). Acetonitrile is used for multiresidue pesticide procedures by various government agencies such as Agriculture and Agri-Food Canada (Fillion et al., 1995), Florida Department of Agriculture and Consumer Services (Cook et al., 1999), and Japan Department of Food Safety (Japan Ministry of Health, Labour, and Welfare, 2006). It is also the basis for the Quick, Easy, Cheap, Effective, Rugged, and Safe (QuEChERS) procedure (Anastassiades et al., 2003). Mol et al. (2008) evaluated different organic solvents, acetone, acetonitrile, and methanol, for a generic extraction procedure and found 1% formic acid–acetonitrile–water was the most favorable extraction solvent based on method performance (i.e., accuracy and precision) and minimization of matrix effects.

Acetone. An alternative approach using acetone as the extraction solvent was developed to address the diversity of the chemical and physical properties of different pesticides, such as the highly polar organophosphorus insecticide, methamidophos. The lower boiling point of acetone (57°C) versus acetonitrile (82°C) was beneficial to easily remove the solvent and concentrate the extract for analysis. The German multiresidue S8 (DFG) and the FDA PAM procedures are examples of acetone extraction procedures (Becker, 1971; Luke et al., 1975). The S8 procedure involves blending 100 g of comminuted sample with 200 mL of acetone and filtering. The filtrate is partitioned twice with a sodium chloride solution and dichloromethane and the organic extract is evaporated with a rotary evaporator. The concentrated extract is cleaned up using an activated carbon-silica column, re-concentrated using a rotary evaporator, and solvent exchanged with hexane for GC equipped with either an electron-capture detector (ECD), alkali flame ionization detector (AFID), or nitrogen-specific detector (e.g., nitrogen-phosphorous detector, NPD) for the analysis of organochlorine and organophosphorus pesticides and triazine herbicides, respectively. The original Luke method (1975) similarly involves the blending of 100 g of sample with 200 mL acetone and filtering. The filtered extract is shaken and partitioned with petroleum ether and dichloromethane. The upper layer is dried with sodium sulfate and transferred directly to a K-D concentrator. The lower layer is mixed with sodium chloride and the mixture is extracted twice with dichloromethane and filtered to the K-D concentrator. The filtered extract is evaporated with steam and the concentrate is solvent exchanged into acetone. An aliquot of the acetone extract was analyzed for organophosphorus and organonitrogen pesticides using GC-KCl thermionic detection. The second aliquot was subject to Florisil cleanup followed by GC–ECD or GC–flame ionization detection (FID) for the analysis of organochlorine and hydrocarbon-based pesticides, respectively. The Luke method as illustrated in Figure 6.3 was further modified (Luke et al., 1981) to eliminate the Florisil cleanup step and replace GC–ECD with the GC–electrolytic conductivity (GC–ELCD) or GC–FPD, expanding the number of pesticides analyzed and detected to include organonitrogen and polar organophosphorus pesticides. The procedure has undergone many intra- and inter-laboratory studies (Sawyer, 1985; Pang et al., 1995 a,b) and demonstrated successful reproducibility. The Luke method, along with the Mills method, also became part of the FDA PAM (U.S. Food and Drug Administration, 1999). The Dutch mini-Luke (NL) method is currently used in the Netherlands as a multiresidue procedure to analyze pesticides in fruits and vegetables by both GC–mass spectrometry (GC–MS) and liquid chromatography–mass spectrometry (LC–MS) (NVWA, Lehotay et al., 2005; Hiemstra and de Kok, 2007; Lozano et al., 2016).

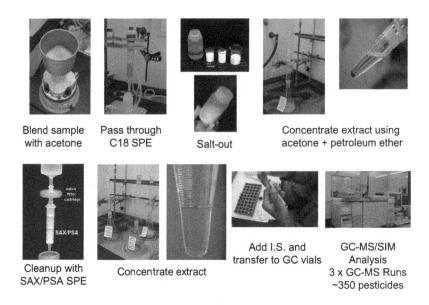

Blend sample with acetone	Pass through C18 SPE	Salt-out	Concentrate extract using acetone + petroleum ether
Cleanup with SAX/PSA SPE	Concentrate extract		Add I.S. and transfer to GC vials GC-MS/SIM Analysis 3 x GC-MS Runs ~350 pesticides

FIGURE 6.3 Modified version of the FDA Luke acetone method (Luke et al., 1975, 1981). Food sample is blended with acetone, partitioned with salts and petroleum ether, followed by cleanup using solid-phase extraction (strong anion exchange/primary-secondary amine SPE cartridge). Excessive solvent is removed by Kuderna-Danish evaporation and the concentrated extract is prepared for analysis by various gas chromatography detectors (flame photometric and halogen selective detection and mass spectrometry). (Photo provided and used by permission by I. Cassias.)

Ethyl Acetate. Ethyl acetate extraction of pesticides from foods has its basis in the acetone extraction procedure followed by dichloromethane/n-hexane partitioning and cleanup using gel permeation chromatography (GPC) of plant and animal food extracts containing a high fat content (DFG S 9, Specht and Tillkes, 1980; Roos et al., 1987). The GPC step removes the number of interfering plant co-extractives and high molecular weight fats and lipids that may interfere with the detection of the pesticides in fatty foods. The use of the dichloromethane–n-hexane partitioning step was later replaced by ethyl acetate–cyclohexane to eliminate the toxic exposure to dichloromethane, in which the implementation of these modifications became the basis of the DFG S19 method (Specht et al., 1980, 1995). The National Food Administration (NFA) of Sweden further modified the acetone-based extraction procedure by replacing acetone with ethyl acetate as the extraction solvent and eliminating the hexane–dichloromethane partitioning step (Andersson and Pålshedo, 1991). The ethyl acetate extract is compatible with GPC cleanup and the finished extract is subjected to analysis by GC–ECD, GC–FPD, or GC–NPD. Later, Pihlström and Österdahl (1999) replaced the GPC cleanup step by loading the ethyl acetate extract through polystyrene-divinylbenzene SPE cartridges, and eluting with ethyl acetate for the analysis of pesticides in low-fat content fruits and vegetables. The NFA of Sweden has performed the most work with ethyl acetate as an extraction solvent for multiresidue pesticide analysis using GC–MS and LC–MS, and the procedure is known as the Swedish Ethyl Acetate (SweEt) method (Jansson et al., 2004; Pihlström et al., 2007; Swedish National Food Administration). Several laboratories have used these ethyl acetate extraction procedures and GC–MS and LC–MS for pesticide analysis. (Mol et al., 2007; Banerjee et al., 2007, 2012; Aysal et al., 2007).

Methanol. Methanol was used as a solvent coupled with ultrasonic homogenization to extract thermally labile carbamate pesticides from plant crops (Krause, 1980, 1992). High moisture plant foods were chopped (100 g) and mixed with 300 mL of methanol and homogenized with a Polytron® blender. The extract was filtered, the volume adjusted, and the filtrate concentrated to 75 mL using a vacuum rotary evaporator. The extract is transferred to a separatory funnel, a 20% sodium chloride solution is mixed, and acetonitrile is added to partition the aqueous and organic layers. The aqueous layers are then discarded. Petroleum ether is added to the acetonitrile extract, shaken, and allowed to partition, and the combined extract collected in a second separatory funnel. Plant pigments such as chlorophyll and carotenoids are removed by passing the extracts through a diatomaceous Earth (Celite) cartridge. The acetonitrile is extracted with dichloromethane, the lower dichloromethane–acetonitrile layer is evaporated and reconstituted. The reconstituted extract is loaded onto a column consisting of silanized Celite and activated charcoal. The column is eluted with 1:3 toluene:acetonitrile, evaporated to dryness using a vacuum rotary evaporator, and reconstituted with methanol. The extract is analyzed for carbamate insecticides by HPLC-PCD/FLD (Moye and Scherer, 1977). The procedure has been collaboratively studied in six laboratories (Krause, 1985). The applicability of the methanol extraction procedure for carbamates was successfully evaluated for additional pesticide types such as fluorescent pesticides, organophosphorus insecticides, a synergist, and herbicides in produce. One of the weaknesses of the procedure was the presence of matrix interferants with some of the pesticides detected by fluorescence. Chaput (1988) modified the FDA–Krause procedure by incorporating GPC with the Celite/activated charcoal column cleanup for HPLC-PCD/FLD analysis of carbamates in several produce samples (apple, broccoli, cabbage, cauliflower, potato). Other variations of the FDA–Krause procedure replaced methanol as the extracting solvent with acetonitrile (Ting et al., 1984) and acetone (Podhorniak et al., 2004). The European Reference Laboratory (EURL) method EN 14185-2 uses acetone extraction and a diatomaceous Earth or ChemElut® column for cleanup (Sannino, 2008). Another EURL method, EN 15637, also uses methanol extraction with ChemElut® cartridges and LC–MS to analyze a variety of pesticides (Klein and Alder, 2003). Granby et al. (2004) and Hanot et al. (2015) developed simplified methods to analyze fresh produce and cereals utilizing methanolic ammonium–acetate/acetic acid and buffered methanol–water extraction, respectively, followed by centrifugation and filtration before LC–MS/MS analysis. The Quick Polar Pesticide (QuPPe) method uses acidified methanol and stable isotopes for the quantitation of highly polar pesticides in various plant crops by LC–MS (Anastassiades et al., 2017).

6.2   CLEANUP PROCEDURES FOR SAMPLE EXTRACTS

After extraction, the solvent extract typically undergoes cleanup procedures to separate or remove co-extractives that can interfere with the instrumental detection of
pesticides. There are risks that the cleanup steps can also adsorb, degrade, or remove
some of the pesticide residues from the extract and care must be taken in evaluating the cleanup steps in a multiresidue pesticide procedure. The most commonly
used cleanup procedures for multiresidue pesticide procedures for vegetal foods of
high water content are column adsorption chromatography and SPE (Krynitsky and
Lehotay, 2002).

Column adsorption chromatography is a sample preparation method that concentrates and purifies chemical analytes from solution by sorption onto a packed column,
followed by elution of the analyte with an appropriate solvent system (Thurman and
Mills, 1998). Column chromatography cleanup typically involves four steps (Hopper,
1992): 1) conditioning the sorbent, usually with the eluting solvent; 2) applying the
sample solution containing the pesticides so that the pesticide is adsorbed or retained
on the sorbent; 3) eluting the analyte from the column with eluting solvents of different polarity or ionic strengths; and 4) collecting the concentrated and purified extract
solution. The inorganic sorbents, silica, alumina, Florisil, diatomaceous earth, and
carbon, originally used for normal phase chromatography, would be used for normal
phase cleanup of organic extracts from various food types.

Florisil is a synthetic magnesium silicate produced from the thermal treatment
of magnesium sulfate and sodium silicate and is a highly selective adsorbent to isolate pesticides (Lee et al., 1982; Afifi et al., 2010). The FDA–Mills acetonitrile and
Luke acetone methods use Florisil to clean up a wide range of low-to-moderate polar
pesticides and it is effective to remove lipids and pigments in plant-based foods.
Increasing the polarity of the eluant by the addition and successive increase of diethyl
ether (6, 15, and 50%) in petroleum ether can increase the recovery of a wide range
of non-polar and polar pesticides, ranging from DDT to azinphos-methyl (US FDA,
1999; Sherma and Zweig, 1972). Activation of Florisil and other related sorbents
such as silica and alumina is achieved by heating the sorbent to remove any residual
water, while deactivation is achieved by a small addition of water (1–10% solution)
which could improve the recovery of the relatively polar pesticides (Langlois et al.,
1964). Similar to Florisil, silica gel, a form of amorphous silica with weak acidic
properties, is also a normal-phase sorbent used to cleanup plant-based food extracts.
Miyahara et al. (1993) used acetone extraction followed by silica column chromatography cleanup with 10% water deactivation for organochlorine pesticide analysis.
The sorbents can also be used to remove fatty acids and lipids from various food
types. Alumina (Al_2O_3), along with the other inorganic-based sorbents, are commonly used to clean up fatty sample extracts as illustrated by the FDA–Mills procedure (US FDA, 1999). Extracts of plant and animal foods with a high fat content are
dispersed on deactivated alumina and the pesticides are removed with acetonitrile
or acetone, while most of the fat is retained on the alumina sorbent. The residues
in the acetonitrile eluant are partitioned back into petroleum ether when water is
added to the eluant, which reduces the pesticide solubility in water. The residues are
then further purified on Florisil using the three-ethyl ether/petroleum ether eluants.

Diatomaceous earth is derived from the siliceous skeletons of minute aquatic plants known as diatoms (Hopper, 1992). This sorbent, discussed in the previous sections to cleanup aqueous methanol extracts in the EN 15637 procedure, is also known as Celite, Chromosorb W, Chromosorb P, Hyflo Super-Cel, and Chem Elut and is approximately 65–87% SiO_2, 2–12% Al_2O_3, and up to 3% Fe_2O_3. These are part of supported liquid extraction procedures in which the diatomaceous earth sorbent is packed in a cartridge, the extract (aqueous or nonaqueous) is loaded onto the cartridge, and an immiscible or typically organic solvent is used to perform the extraction. Supported liquid extraction can reduce emulsions and simultaneously provide cleanup. In the EN 15637 method, the aqueous methanol extract is loaded onto a diatomaceous earth cartridge to form an aqueous film coating on the sorbent (Klein and Alder, 2003). The pesticides are extracted from the aqueous methanol coating by partitioning/eluting the cartridge with dichloromethane. The extracts are then concentrated and re-dissolved in methanol/water and analyzed by LC–MS/MS. Packing these inorganic sorbents into glass columns must be meticulously performed to prevent air bubble formation and maintain column uniformity (Afifi et al., 2010). Carbon has been used as a sorbent to cleanup extracts by adsorption of non-polar, lipophilic, chlorophyll, and high molecular weight chemical components. Nakamura et al. (1994) developed a comprehensive multiresidue pesticide method to optimize the analysis of organochlorine, organonitrogen, and organophosphorus pesticides using acetone, acetonitrile, and ethyl acetate extraction solvents in high-moisture and high-fat foods. Florisil, silica, and Celite sorbents obtained from commercial sources were used for column cleanup, and the pesticides were detected by GC–FPD, GC–NPD, and GC–ECD. For column adsorption chromatography cleanup, carbon is usually mixed with diatomaceous earth or silica because, by itself, bulk carbon is a fine powder and has poor flow characteristics (Ambrus and Thier, 1986). Another problem with most of these sorbents is that they are susceptible to water exposure and manufacturing variabilities in production.

Column cleanup has historical roots in chromatography as described by Thurman and Mills (1998). Developments in sorbent technology for chromatographic separations such as ion exchange, reversed phase, and other different mixed-mode chemistries bonded to silica or polymeric particles provided new applications for SPE cleanup of pesticides in foods. The particle size of silica gels used in SPE sorbents (40–60 µm) is larger than the particles used in HPLC (5 µm) in order to promote flow with gravity or a manifold with in-house vacuum. Free accessible silanol groups on the surface of the silica gel are chemically modified with bonded phases containing functional groups that have reversed or normal-phase, mixed-mode, or ion exchange properties. The most common bonded phases are reversed phase, ranging from methyl (CH_3) to octadecyl (C18) or phenyl- groups; mixed-mode stationary phases containing alkyl substituted ligands such as pentafluorphenyl, cyano (CN) or cyano-alkyl, and amino (NH_2); and ion exchange groups or ligands. Ion exchange sorbents are characterized as either cation- or anion-exchange and as weak or strong. The latter describes the functional group of the ligand as a strong or weak acid or base, indicating the strength of ionization to displace and "exchange" the anion or cation. A weak ion exchange resin indicates that ion exchange is dependent on the pH environment, whereas a strong ion exchange resin indicates that the ligand is always charged independently of the pH. Other sorbents specifically designed for SPE applications are polymers composed of crosslinked hydrophilic N-vinylpyrrolidone and hydrophobic divinylbenzene monomers, polystyrene and divinylbenzene, or a surface modified polymer consisting of styrene and divinylbenzene.

After the introduction of SPE in the late 1970s and early 1980s, the large columns used for column adsorption chromatography were replaced with SPE cartridges (Picó et al., 2007). The SPE sorbents were either used for analyte isolation or matrix removal (Schenck and Wong, 2008). Analyte isolation typically involves adsorption of the pesticide analytes to the column, followed by desorption of the analyte through an eluting solvent or a series of eluting solvents. In matrix removal, SPE is used as a filter that traps or retains primarily the chemical interferants from the matrix as the pesticides of interest elute with the same extract solution without much interaction of the sorbent. For most multiresidue pesticide procedures, SPE cleanup procedures involve matrix removal (Schenck and Wong, 2008). Reversed-phase sorbents such as C8-C18 and styrene-divinylbenzene polymers are often used to remove hydrophobic and lipophilic interferants from the matrix (Cairns et al., 1993; Cook et al., 1999; PAM, 1999). Ion exchange sorbents such as aminopropyl and primary-secondary amine sorbents retain polar organic sugars, humic acids, and fatty acids. Graphitized carbon is very effective in removing chlorophyll and flat aromatic heterocyclic compounds that have strong interaction with carbon (Chun et al., 2003; Tong et al., 2006; Song et al., 2007). For complicated matrices, a combination of tandem SPE columns have been utilized to remove the chemical interferants from sample matrices (Fillion et al., 1995; Sheridan and Meola, 1999).

6.3   QuEChERS

Anastassides et al. (2003) developed the Quick, Easy, Cheap, Effective, Rugged, and Safe (QuEChERS) procedure for analyzing pesticides in plant-based foods, which has also been applied to the analysis of other compounds of importance such as drugs, toxins, pollutants, other chemical residues and contaminants in various matrices such as animal tissue, environmental matrices, and consumer-based products (Majors, 2007; Botitsi et al., 2011). The original purpose of QuEChERS was to develop a simple, rapid, and inexpensive multiresidue procedure that streamlined the number of preparation steps and reduced the amount of solvent and reagent use. The QuEChERS procedure is illustrated in Figure 6.4. The original work by Anastassiades et al. (2003) involved 10 g of a comminuted and homogenized fresh produce sample of high water content and 10 mL of acetonitrile containing a quantitative internal standard, triphenylphosphate, which adjusts for water content difference and volume fluctuations. Extraction and liquid–liquid partitioning occur when the sample, acetonitrile, anhydrous magnesium sulfate (4 g), and sodium chloride (1 g) are added to the centrifuge tube. The centrifuge tube is vigorously shaken and centrifuged and the supernatant is collected for dispersive solid-phase extraction (dSPE) cleanup. dSPE is performed by mixing the acetonitrile extract with a bulk SPE sorbent containing 25 mg primary-secondary amine linked to silica particles (PSA) and anhydrous magnesium sulfate (150 mg) in a test tube. This process removes chemical interferants in the food matrix, such as organic acids and some types of color

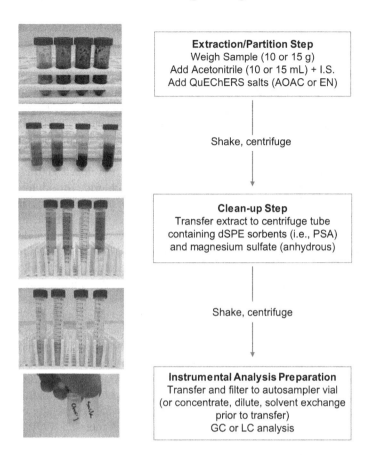

Extraction/Partition Step
Weigh Sample (10 or 15 g)
Add Acetonitrile (10 or 15 mL) + I.S.
Add QuEChERS salts (AOAC or EN)

Shake, centrifuge

Clean-up Step
Transfer extract to centrifuge tube
containing dSPE sorbents (i.e., PSA)
and magnesium sulfate (anhydrous)

Shake, centrifuge

Instrumental Analysis Preparation
Transfer and filter to autosampler vial
(or concentrate, dilute, solvent exchange
prior to transfer)
GC or LC analysis

FIGURE 6.4 Schematic of the Quick, Easy, Cheap, Effective, Rugged, and Safe (QuEChERS) procedure. The procedure consists of extraction and cleanup phases. Extraction is achieved by acetonitrile salt-out extraction and cleanup consists of mixing the acetonitrile with a mixture of anhydrous magnesium sulfate and SPE sorbents. The extraction can be further processed to incorporate other cleanup processes or injected into a liquid or gas chromatograph. (Photo provided by J. Wang.)

pigments. The centrifuge tube is shaken and centrifuged and the cleaned-up extract is analyzed by GC–MS/single ion monitoring (SIM) analysis. Performance of the QuEChERS procedure revealed acceptable recoveries > 95% (81–101%) and repeatabilities < 5% for a wide range of commonly used polar and basic pesticides such as methamidophos, acephate, omethoate, imazalil, and thiabendazole.

Further modifications and fine-tuning of QuEChERS were investigated by both Lehotay et al. (2005, 2007) and Anastassiades et al. (2007). This work included employing LC–MS analysis and improving the performance of difficult pesticides such as captan, chlorothalonil, dichlofluanid, and pymetrozine. Lehotay et al. (2005) modified the procedure by adding 1% acetic acid to the acetonitrile extraction solvent, using the same amount of anhydrous magnesium sulfate, and replacing sodium chloride with sodium acetate to achieve a pH ~ 6. The number of pesticides analyzed in their study included 229 pesticides by GC–MS and LC–MS platforms. Recoveries for most of the pesticides ranged between 70–120% (90–110% for 206 pesticides) and repeatabilities < 10%, problematic pesticides (recoveries < 50%) were asulam, pyridate, dicofol, thiram, and chlorothalonil. The QuEChERS results were comparable to those obtained from the established Dutch (NL or Mini-Luke) acetone procedure. The acetate-buffered QuEChERS procedure was used in an AOAC collaborative study involving 13 laboratories from 7 countries to evaluate 18 blind duplicates spiked at 9 spiking levels in 3 plant matrix commodities (grape, lettuce, orange). The successful completion of the study became an official AOAC method, 2007.01 (Lehotay, 2007). Meanwhile, Anastassides et al. (2007) developed a buffered system consisting of anhydrous magnesium sulfate, sodium chloride, disodium hydrogen citrate sesquihydrate, and trisodium citrate dihydrate to maintain a buffering pH of 5 to 5.5 to account for the stability of base-labile pesticides such as tolylfluanid, dichlofluanid, captan, folpet, dicofol, and pyridate and acid-sensitive pesticides such as sulfonylurea herbicides, pymetrozine, carbosulfan, and dioxacarb. The QuEChERS procedure utilizing a citrate buffered system was approved as an official procedure, Standard Method EN 5662, by the European Committee for Standardisation (CEN) (EURL-FV, 2010a).

Other existing procedures using ethyl acetate and acetone, mentioned in the previous sections, also underwent similar modification as acetonitrile, to improve operation efficiency. The Luke method (Luke et al., 1975, 1981), underwent modification using smaller sample sizes and less solvent by Dutch researchers to become more resource- and cost-efficient (Lehotay et al., 2005; Hiemstra and de Kok, 2007; Lozano et al., 2016). The acetone-based procedure DFG S19 (Specht et al., 1980,1995, 2000) was modified by replacing acetone with ethyl acetate which eventually became the basis of the Swedish Ethyl acetate (SweEt) procedure (Andersson, 1991; Andersson and Pålshede, 1991; Pihlström et al., 2007). The Dutch and Swedish procedures are validated multiresidue procedures for the European Reference Laboratories for Residues of Pesticides (EURL-FV 2014, EURL-FV, 2010b). The popularity of QuEChERS is due to its flexibility and modularity. The extraction can be modified and adjusted for the specific properties of the pesticides of interest by modifying the solvent choice, adjusting the pH and sample size, and implementing different cleanup sorbents and techniques. Although most pesticides are non-polar in nature, there are classes of pesticides such as chlorphenoxy acid herbicides, diquat herbicides, and organophosphorus acid herbicides, that will not work with the conventional or buffered (AOAC or EN) QuEChERS procedures. QuEChERS was modified to analyze acidic herbicides such as chlorphenoxy acid herbicides by acidifying the extracts to a lower pH value to extract the pesticides into the acetonitrile phase (Sack et al., 2015). Steinborn et al. (2017) have also modified QuEChERS to address hydrolysis of pesticides that have been conjugated or are in their ester form such as chlorphenoxy acid herbicides. This extraction procedure typically utilizes 10 g sample with acetonitrile and sodium hydroxide solution to simultaneously hydrolyze and extract the esterified and conjugated pesticides. The hydrolysis is terminated by the addition of acid, followed by the addition of salts to induce phase separation of the acetonitrile phase from the aqueous matrix. The QuPPe method was developed by Anastassiades et al. (2017) to analyze highly polar pesticides (log $K_{ow} < -2$) that are not amenable to established multiresidue procedures and are more conducive to non-polar chemicals. QuPPe is best optimized using stable isotope dilution assay/LC–MS to compensate for matrix effects. This allows for the native and isotope compounds to chromatographically co-elute and to compensate for matrix effects.

A tremendous amount of commercial investment in QuEChERS is in the cleanup step. Cleanup materials such as PSA, graphitized carbon, and C-18 have specific applications to remove organic acids, pigments, fats, and other matrix co-extractives. The goal of improved cleanup with no or minimal loss of the pesticides has impacted the development of new sorbents with specific applications to pigment removal in pigmented foods and lipid and fat removal in plant foods such as nuts, legumes, and avocado. New sorbents such as Chlorofiltr® and zirconium oxide bonded to silica, zirconium oxide and C18 bonded to silica, materials known as enhanced matrix removal-lipid (EMR®) and LipiFiltr®, and cross-linked polymers (Oasis Prime®) have been evaluated with promising results to remove pigments and fats (Walorczyk et al., 2015; Dias et al., 2016; He et al., 2017). López-Blanco et al. (2016) evaluated the performance of different sorbents for the cleanup of vegetables of high fat content for LC–QqQ–MS/MS analysis. Olives, olive oil, and avocado were extracted with the acetate-buffered QuEChERS (AOAC **2007.01**, Lehotay, 2007) procedure and the extracts were subjected to different cleanup procedures: 1) 250 mg PSA + 250 mg C18; 2) 250 mg Z-Sep⁺; and 3) 1 g activated EMR sorbent for the analysis of 67 pesticides by LC–MS/MS. The optimal extraction efficiencies were obtained using EMR in all three matrices, although the PSA + C18 and Z-Sep⁺ performed comparably with the olive and avocado matrices. Recoveries were 70–120%, RSD < 10% for 80% of the pesticides studied. Han et al. (2016) effectively evaluated avocado, kale, pork, and salmon using QuEChERS extraction followed by EMR-lipid cleanup to analyze 65 pesticides and 52 environmental contaminants. Walorczyk et al. (2015) developed a two-step dispersive solid phase extraction strategy in removing chlorophyll and lipids in soybeans extracts derived from EN QuEChERS extraction procedure by employing a combination of PSA, Chlorofiltr®, and magnesium sulfate in the first cleanup step and GCB and C18 in the second cleanup step for analysis by GC–MS/MS.

6.4 GAS CHROMATOGRAPHY–MASS SPECTROMETRY

The introduction of GC–MS in the 1980s made an impact in multiresidue pesticide analysis. This led to the elimination of multiple injections in various GCs equipped with element selective detectors for the analysis of the different classes of pesticides that consisted of different heteroatoms or conjugated moieties. GC–MS provides the retention time and structural information required for identification. Stan (2000) evaluated a commercial GC–MS consisting of a quadrupole mass filter by evaluating pesticides in food extracts prepared from the acetone-based DFG 19 procedure. GC–MS analysis was performed in full scan mode, matching the experimental full MS spectrum to the spectra provided in the National Bureau Standards (National Institute for Standards and Technology) probability library. A second analysis using GC–MS in SIM mode was also performed for detection and quantitation of pesticides at the 10 parts per billion (ppb) level. Cairns et al. (1993) demonstrated that

GC equipped with quadrupole ion trap mass spectrometry (GC–IT-MS) was able to detect and quantify 245 pesticides in full scan MS from a Luke method extract from different produce samples. The study showed that GC–MS was acceptable for trace level quantitation at 0.25 and 1.00 ppm, sensitive for most of the pesticides studied, but that further sample cleanup is required to improve on detection and quantitation.

The improved selectivity and sensitivity of GC–MS has affected the development of multiresidue pesticide procedures. GC–MS was initially developed to complement and confirm the presence of the pesticides with the existing GC methods. The Swedish procedure was modified by replacing the GPC cleanup step with a cross-linked polystyrene polymer solid-phase extraction sorbet for analysis by GC–thermionic sensitive detection (TSD), GC–ECD, and GC–IT-MS (Pihlström and Ősterdahl, 1999). Electronic pressure control (EPC) devices allowed GC and GC–MS systems to synchronize their retention times (i.e., retention time locking, RTL) to operate under the same chromatographic conditions that led the CDFA to develop and implement a retention time library for over 500 pesticides (Lee and Richman, 2002). Mercer et al. (2004, 2006) adopted the Luke acetone procedures and RTL for GC–MS/SIM analysis. By then, the Luke method was modified by implementing SPE cleanup using octadecyl-, strong anion exchange-, and aminopropyl-linked silica sorbents instead of Florisil columns, resulting in the analysis of a wider range of pesticides as illustrated in Figure 6.3. Pang et al. (1999, 2000) conducted studies by performing collaborative studies using the Luke method on pyrethroids using GC–MS/SIM. The work by Fillion et al. (1995, 2000) illustrated that a wider range of pesticide types could be analyzed in a single run using GC–MS/SIM without the need for element selected detectors. Their work also included the analysis of carbamate pesticides by evaporating and reconstituting an aliquot of the extract for analysis by HPLC–PCD/FLD. Fillion's work utilized the available tools to screen for pesticides, and involved acetonitrile salt-out extraction, followed by SPE cleanup using a dual column cleanup involving graphitized carbon black and primary-secondary amine, and instrumental determination by GC–MS/SIM or HPLC–PCD/FLD. Many other laboratories adopted Fillion's GC–MS/SIM procedure for the analysis of fresh vegetal-based foods for both GC–MS and LC–MS analysis, and these are official pesticide methods for Canada and Japan (Chun, 2004; Kakimoto et al., 2003; Ueno et al., 2004; Okihashi et al., 2005, 2007; Pang et al., 2006; Takatori et al., 2008).

Gas chromatography equipped with a triple quadrupole mass analyzer began to replace the single quadrupole analyzers, offering improved selectivity and sensitivity (Garrido Frenich et al., 2005, 2008; Fernández Moreno et al., 2008; Hernández et al., 2013). Garrido Frenich et al. (2005) demonstrated one of the first uses of gas chromatography–triple quadrupole mass spectrometry (GC–QqQ-MS/MS) operating in selected reaction monitoring mode (SRM) for pesticide analysis. They extracted 5 g of a homogenized vegetable sample with 10 mL ethyl acetate using a Polytron® homogenizer. The extract was filtered through an anhydrous sodium sulfate cartridge to remove residual water, evaporated to dryness with nitrogen, and redissolved in 1.0 mL cyclohexane containing 0.5 μg/mL caffeine as an internal standard. Analysis was achieved by two injections. The first injection was used to screen for the presence of 130 pesticides based on retention time and instrumental response of a precursor-to-product ion transition characteristic of the pesticide. If the screening indicated the presence of one or more pesticides, then a second injection was performed for identification and quantitation based on the retention time, the presence of two or three selected precursor-to-product transition ions, and the relative ratio intensities within a tolerance ± 20% of the transition ions. As more laboratories adopted GC–QqQ-MS technology, procedures were developed so that a single injection could be implemented to quantitate and identify pesticides.

Many laboratories continue to use the sample preparation procedure or simplified established procedures that were developed for GC–MS/SIM and GC–QqQ-MS/MS analysis. Other laboratories adopted or modified QuEChERS for GC–QqQ-MS/MS analysis. These modifications include changes in extraction solvents such as ethyl acetate, or cleanup sorbents such as the inclusion of C-18 and/or graphitized carbon black to remove plant sterols and pigments (Okihashi et al., 2007; Walorczyk, 2008, 2015; Wong et al., 2010; Ma et al., 2013; Norli et al., 2016; Lee et al., 2017; Ahammed Shabeer et al., 2018).

6.5 LIQUID CHROMATOGRAPHY–MASS SPECTROMETRY

Gas chromatographic analysis of insecticides and herbicides such as carbamates and phenylureas, respectively, were limited due to the pesticide lability and decomposition from the elevated temperature conditions in the GC inlet and column. The introduction of liquid chromatography in the 1980s provided accurate and stable analysis of these pesticide types by ultraviolet detection (UV) of characteristic chromophores (Lawrence, 1977; Krause, 1980) or derivatization methods to produce fluorophores that could easily be detected by fluorescence detection (FLD) (Krause, 1985; Podhorniak et al., 2004). The pesticides of the 1960s and 1970s were becoming unpopular and less used due to bioaccumulation, harmful health effects to mammal species, persistence in the environment, and reduced acceptance. The next generation of pesticides were more pest-targeted, less toxic to humans and other species, and less persistent in the environment. Procedures were needed and required to address the analysis of these upcoming pesticides which tend to be thermally labile and not amenable for GC or GC–MS analysis and lacked suitable chromophores for detection.

LC–MS, particularly liquid chromatography triple quadrupole mass spectrometry (LC–QqQ-MS/MS) in single (or multiple) reaction monitoring (SRM, MRM) mode initiated the change from the detailed multi-step procedures of the 1980s to early 2000s to the more simplified approaches for sample preparation used today such as QuEChERS (Anastassiades et al., 2007). LC–QqQ-MS/MS is selective and sensitive because SRM scans for user-defined precursor and product ions, eliminating the background of other compounds present in the sample matrix at the m/z that may coelute with the compound analyte of interest. This enhances the signal-to-noise ratio (SNR) for the pesticide so that detection at the levels of interest (typically low parts-per-billion concentration levels) can be easily determined. The breakthrough of LC–QqQ-MS/MS for residue analysis is mainly contributed to the development of the triple quadrupole mass analyzers coupled to HPLC or ultra-high performance liquid chromatography (UHPLC), improved electronics for faster scanning of more precursor-to-product ion transitions, and advanced computer technologies enabling data acquisition, storage, and processing effective and efficient data analysis for a large number of pesticides. The pioneering work of Klein and Alder (2003) demonstrated that LC–QqQ-MS/MS was capable of analyzing 100 pesticides in a single run. Other laboratories (Taylor et al., 2002; Jansson et al., 2004; Akiyama et al., 2007; Hiemstra et al., 2007) soon followed and applied their LC and GC pesticide procedures once the technology became accessible. Pesticides that normally required LC–UV, LC–FLD, or GC and GC–MS procedures, such as benzimidazoles, carbamates, N-methyl carbamates, and organophosphorus compounds, could now be analyzed by LC–MS, since these different chemical classes were susceptible to ionization by electrospray (ESI) or atmospheric pressure chemical ionization (APCI). Testing of various food commodity groups were also evaluated to study the influence of different matrices on recoveries and ion suppression effects.

QuEChERS has been the preferred multiresidue pesticide procedure for LC–MS/MS analysis using acetonitrile salt-out extraction followed by dispersive solid-phase extraction cleanup using PSA (Wang and Leung, 2009a,b; Kmellár et al., 2010; Sack et al., 2011; Zhang et al., 2011; Rajski et al., 2013; Namikawa et al., 2014). Although QuEChERS has been the method of choice for multi-pesticide methods, other solvents have been compared and evaluated. Pizzuti et al. (2009) validated and compared acetonitrile (QuEChERS using 1% acetic acid in acetonitrile without dispersive SPE cleanup) and acetone (Dutch mini-Luke procedure) extraction procedures of 169 pesticides in soy grains by LC–MS/MS. The investigators found that QuEChERS performed slightly better than the Dutch mini-Luke procedure based on method detection limit, average recovery, and relative standard deviation. The investigators showed that the differences in performance results between the two methods were due to the pesticide extractability into the solvent. Mol et al. (2008) investigated and evaluated three extraction systems: acetonitrile, methanol, and acetone containing 20% water and 1% formic acid on maize, milk, meat, eggs, and honey. The samples were extracted by shaking, centrifuged, and filtered for LC–QqQ-MS/MS analysis of 172 pesticides. The results showed among the three procedures, the acetone extraction procedure provided the best recoveries (70–120%, < 20% RSD) while acetonitrile and methanol exhibited the most and least favorable performances in response to ion suppression and compensating for matrix effects, respectively.

6.6 LIQUID AND GAS CHROMATOGRAPHY–HIGH RESOLUTION MASS SPECTROMETRY

Recently, liquid and gas chromatography coupled to high resolution mass spectrometry (LC–HRMS, GC–HRMS) have become of interest to pesticide laboratories because of their ability to operate in full scan (MS) and MS/MS modes with high resolving power that can discriminate between the pesticide of interest and isobaric compounds that may be present in the food matrix. Time-of-flight (TOF) and Orbitrap mass analyzers have been evaluated for both quantitative and qualitative analyses. Detailed descriptions of the design and how these two HRMS platforms operate are described elsewhere (Boesl, 2017; Botitsi et al., 2011; Eliuk and Makarov, 2015; Kaufmann and Teale, 2017; Kaufmann, 2018). Prior to 2013, much of the work on HRMS focused on standalone TOF and Orbitrap MS. Early and pioneering work on LC–TOF-MS by Ferrer and Thurman (2005a,b, 2007a,b; Ferrer et al., 2011; Thurman et al., 2005a,b, 2011) demonstrated the potential of HRMS systems to screen food samples using software procedures and databases. In combination with a quadrupole filter, hybrid instruments (Q-HRMS) such as QTOF-MS and Q-Orbitrap-MS, provide additional selectivity and identification when operated in MS/MS modes such as data-dependent and data-independent acquisition (DDA, DIA) and parallel reaction monitoring (PRM) (Agüera et al., 2017; Kaufmann, 2018).

The terms "targeted" and "non-targeted" are often used to differentiate analytical procedures associated with LC–MS platforms. Targeted approaches are used to analyze the known chemicals of interest which require prior knowledge of the analyte (i.e., retention time, precursor and product ions, collision energies) and require analytical standards for the instrumentation setup and data analysis. Non-targeted approaches typically involve procedures that analyze the chemicals that are unknown or the chemicals are determined without the assistance of a suspect list (Sobus et al., 2017). Non-targeted approaches are very challenging and instead are modified as a suspect screening approach by comparing the experimental result against using a database of chemical compounds. Examples of Q-HRMS targeted procedures include DDA (with an inclusion list) and PRM. Wang et al. (2014) and Jia et al. (2014) performed LC–Full scan MS/DDA-MS/MS for the analysis of pesticides and veterinary drugs using the Q-Orbitrap platform, while Zhu et al. (2017) analyzed 581 pesticides, mycotoxins, and rodenticides on a QTOF platform.

Non-targeted acquisition requires little or no prior knowledge of the pesticide, such as the retention time. This can be achieved by LC–HRMS in the absence of a quadrupole, acquiring data in full scan mode and/or all-ion fragmentation (AIF) mode and LC–Q-HRMS in DIA mode (Veneable et al., 2004; Silva et al., 2006; Gillet et al., 2012; Arnhard et al., 2015). Data analysis for DIA and AIF experiments requires comparison, analysis, and retrieval of the experimental results of other chemicals stored in a compound database. The compound database typically consists of information such as name, CAS number, molecular formula, chromatographic retention time, and mass spectral information (precursor and product ions) for each compound. Polgár et al. (2012) developed an accurate mass database of 1396 compounds consisting of 850 parent compounds, 447 fragment ions, and 99 metabolites. They prepared extracts from fresh produce using the AOAC acetate-buffered QuEChERS procedure and analysis by UHPLC-QTOF-MS. Saito-Shida et al. (2016) applied an acetonitrile salt-out extraction and C18 and GCB/PSA SPE cleanup for the quantitative analysis of pesticide residues in fruits and vegetables using an LC–QTOF-MS. The QTOF was operated in both full scan MS and MS/MS modes using DIA or MS^E which incorporates an alternating low collision energy at 4 eV and a high collision energy ramp of 10–40 eV to provide the characteristic precursor and product ions, respectively. The LC–QTOF-MS procedure was found to detect pesticides at 10 µg/kg as well as provide unambiguous identification. The first demonstration of DIA on a quadrupole-Orbitrap MS was performed by Zomer and Mol (2015) using a non-targeted approach of data acquisition with fragmentation. Their non-targeted approach used a combination of a full scan acquisition event followed with variable DIA (vDIA) MS/MS scan for screening applications. Other laboratories (Rajski et al., 2017; Goon et al., 2018; Wang et al., 2017) have also used developed vDIA acquisition procedures for pesticide screening. In one study, Wang et al. (2017) developed a target screening of 448 pesticide residues in fruits and vegetables using UHPLC–ESI-Q-Orbitrap and a compound database (CDB). Their work, for the first time, described a procedure for the development of a pesticide CDB and its applications for target screening using LC–ESI-Q-Orbitrap Full MS/ dd-MS/MS and Full MS/DIA. They optimized screening applications by addressing in-spectrum mass correction, retention time alignment, and response threshold adjustment when building the CDB, which are critical parameters to minimize false negative/positive rates for routine screening. The validated target screening method is capable of screening at least 94% and 99% of 448 pesticides at 10 and 100 µg/kg, respectively.

Earlier work prior to 2015 using GC–HRMS focused on GC coupled to a time-of-flight mass analyzer and potential applications to pesticide analysis (Cervera et al., 2012). Using QuEChERS extracts and a GC–TOF-MS platform, only half of the 55 pesticides studied could be validated at the 10 µg/kg level. These platforms typically had resolution less than 10,000 and the cost, resolution limitation, and inadequate sensitivity did not provide much impact or incentive for multiresidue pesticide analysis. However, recent developments in the improvement of HRMS has made significant advances in GC–HRMS. The review by Špánik and Machyňáková (2018) summarizes various approaches and applications of GC–HRMS for non-targeted analysis to identify contaminants in food and environmental and biological samples. Cao et al. (2015) produced extracts prepared from salt-out acetonitrile extraction, followed by SPE cleanup using primary secondary amine and graphitized carbon followed by GC–QTOF-MS/MS and GC–QqQ-MS/MS analysis. They found that QqQ analysis provided better sensitivity but QTOF was preferred for screening the plant food samples. The work was continued by Li et al. (2018) by expanding the pesticide database to screen 439 pesticide residues in fruits and vegetables. Mol et al. (2016) investigated GC-electron impact ionization (EI)-HRMS on an Orbitrap mass analyzer in full scan mode for the analysis of 54 pesticides in tomato, leek, and orange matrices. They validated the acetate-buffered QuEChERS method (AOAC 2007.1) with a final solvent exchange into iso-octane and achieved recoveries 70–120% and repeatability < 10%. Pesticides were identified based on the existence of precursor and product ions with mass accuracies < 5 ppm and their ion ratios not deviating the ± 30% relative from the reference ion ratio. The experimental accurate mass ions and mass spectra obtained from GC–HRMS were compared and matched to a NIST library database of pesticides and other potential chemical contaminants. Uclés et al. (2017) validated a method using the citrate QuEChERS procedure for 210 pesticide residues in fresh fruits and vegetables and analyzed the extracts using GC–Orbitrap MS in Full Scan/SIM mode with a resolution setting of 60,000. The method was applied to the routine analysis of 102 fresh produce samples. The method demonstrated selectivity due to the determination of high accurate masses and the performance of the GC–HRMS in full scan was similar to GC–QqQ-MS/MS analysis. Portolés et al. (2010) evaluated an APCI source for GC–QTOF-MS analysis of 100 pesticide residues in nectarine, orange, and spinach. The use of APCI generated the molecular ions which are lacking in highly fragmented generated EI spectra. GC–HRMS quantitation and identification can be based on the molecular ion and using the quadrupole mass filter to conduct DIA (or MS^E)-MS/MS experiments that are commonly used in LC–HRMS. Cervera et al. (2014) developed a screening method for 130 pesticides in apples, carrots, courgettes, lettuce, oranges, red peppers, strawberries, and tomatoes using the acetate-based QuEChERS procedure and solvent exchanged to toluene, followed by analysis by GC-APCI-QTOF-MS. The screening strategy consisted of a rapid searching and detection and a refined identification step based on DIA (or MS^E). Identification was based on the presence of one precursor ion one product ion with mass accuracy < 5 ppm and the product to precursor ratio in reference to the calibration standard. Cheng et al. (2017) also demonstrated and confirmed the practicality of using DIA (or MS^E) for the GC–APCI-QTOF-MS analysis of 15 organophosphorus pesticides. They used the non-buffered QuEChERS method with sodium chloride and magnesium sulfate for acetonitrile salt-out extraction, followed by cleanup with C18 for apple and cucumber or with PSA cleanup for cabbage and solvent transferred to hexane.

6.7 FUTURE CHALLENGES AND CONSIDERATIONS

The workflow for multiresidue pesticide procedures for plant foods discussed in this chapter is defined by extraction, cleanup, and instrumentation. These multiresidue pesticide workflows have evolved over the years primarily due to improvements in technologies. Although there have been other extraction technologies available, solid–liquid extraction between a homogenized plant food sample and an organic solvent is still the most common approach. Minimal sample preparation approaches such as QuEChERS or dilute-or-shoot procedures have become popular due to the availability and technical advancements in chromatographic separations (UHPLC, LC, GC) and mass spectrometry. Figure 6.5 shows a method genealogy of how the original methods utilizing acetonitrile, acetone, ethyl acetate, and methanol have evolved over the last sixty years. A common observation is that the sample preparation is becoming simpler and utilizing smaller sample sizes for analysis by LC–MS and GC–MS. Future advancement in multiresidue pesticide procedures could involve robotic automation in sample preparation, instrumental analysis, and data processing.

Much of the work presented in this chapter is focused on the analysis of pesticides in high moisture plant foods. However, another area of plant-based materials that is of interest due to its economic importance is dried plant products such as botanical dietary supplements, spices, tea, tobacco, and cannabis. These plant-based materials are difficult to analyze because they are concentrated and dehydrated so their sample sizes are typically smaller than the sizes analyzed for fresh produce. Matrix interferants, especially in dried botanicals, can affect chromatographic separation and suppress signal intensities of the pesticide analytes (Chen et al., 2016; Hayward et al., 2013, 2015). The matrices of these products are also very complex because they contain essential oils, usually terpenes that have a similar molecular weight range and chemical properties to pesticides. This makes cleanup very difficult, limiting the effect of SPE and GPC. Many pesticides and related compounds have their origins in natural products such as in the case of synthetic pyrethroid pesticides that have structural and physical features that resemble naturally occurring pyrethrins that

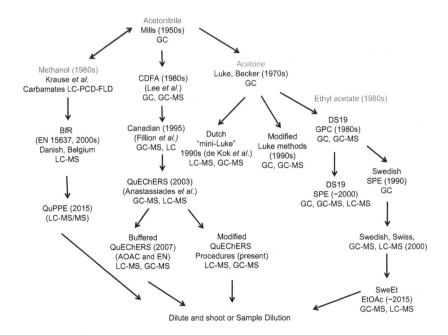

FIGURE 6.5   Pesticide genealogy beginning with the Mills Method (1963) for acetonitrile, Luke/Becker methods using acetone as the extraction solvent of the 1970s, ethyl acetate (~1980s), and methanol (mid-1980s). The procedures underwent modifications due to the introduction of new technologies, especially in improvements or technological advances in chromatography–mass spectrometry. Resulting procedures used in today's multiresidue pesticide procedures are QuEChERS (acetonitrile), the Dutch mini-Luke method (acetone), and the Swedish Ethyl Acetate (SweET) method (ethyl acetate). Dilute-and-shoot LC–MS methods involve the use of acetonitrile and methanol. Today's multiresidue pesticide methods have evolved from the earlier procedures due to smaller analytical sample sizes and solvent volume, minimal sample preparation, and advances in instrumentation, particularly the use of mass spectrometry.

plants synthesize to ward off plant-eating pests and animals (Ware, 1989). This adds further difficulty in developing sorbents with specific molecular recognition features that can isolate the pesticide from the complex botanical matrices. The challenge is to develop effective and innovative ways to separate and clean up pesticides from the matrix or to develop or improve new or existing technologies in separations and mass spectrometry (or any new detection technology) that are not affected by matrix interferants.

Targeted GC–QqQ-MS/MS and LC–QqQ-MS/MS approaches for multiresidue pesticide methods for plant foods are the current and established procedures because they have been well studied, validated, and documented. Established criteria for quantitation and identification of pesticides are well-established for GC–MS and LC–MS analysis (US FDA, 2012, 2015; Codex Alimentarius, 2017; European Commission, 2017). There is still no universal method to detect all pesticides and other residues and contaminants that may be present in the food. This will be a challenging task since the chemicals of interest may have varying chemical and physical properties that would make sample preparation, chromatographic separation, MS ionization, and detection difficult to accomplish using one generic approach. Recently, research on chromatography-based HRMS has received much attention because of its potential to screen for pesticides beyond a small targeted list. HRMS analysis relies on software tools that can screen, retrieve, and archive experimental results with compound databases, which can comprise thousands of potential pesticide candidates. The implementation of APCI in GC–HRMS would allow GC analysis to adopt some of the LC–HRMS approaches such as DIA that could be used to screen additional chemicals beyond pesticides. The creation and expansion of databases for HRMS analysis would cover not only more pesticides but also other chemical residues and contaminants of interest, using a generic sample preparation and chromatography–HRMS approaches.

REFERENCES

Afifi FU, Hajjo RM, Battah AH. Medicinal plants, pesticide residues, and analysis. In: *Handbook of Pesticides: Methods of Pesticide Residues Analysis.* L.M.L. Nollet and H.S. Rathore (eds.), CRC Press, Boca Raton, FL, 2010, pp. 401–34.

Agüera A, Martínez-Piernas AB, Campos-Mañas MC. Analytical strategies used in HRMS. In: *Applications in High Resolution Mass Spectrometry Food Safety and Pesticide Residue Analysis.* R. Romero-González and A.G. Frenich (eds.), Elsevier, Amsterdam, Netherlands, 2017, pp. 59–82.

Ahammed Shabeer TP, Girame R, Utture S, Oulkar D, Banerjee K, Ajay D, Arimboor R, Menon KRK. Optimization of multi-residue method for targeted screening and quantitation of 243 pesticide residues in cardamom (*Elettaria cardamomum*) by gas chromatography tandem mass spectrometry (GC-MS/MS) analysis. *Chemosphere.* 2018, *193*, 447–453.

Akiyama Y, Yoshioka N, Matsuoka T. Multiresidue analysis of 500 pesticide residues in agricultural products using GC/MS and LC/MS. In: *Pesticide Chemistry: Crop Protection, Public Health, Environmental Safety.* H. Ohkawa, H. Miyagawa, PW Lee (eds)., Wiley-VCH Verlag GmbH & Co, Weinheim, Germany, 2007, pp. 395–400.

Ambrus A, Thier H-P. Application of multiresidue procedures in pesticides residues analysis. *Pure & Appl. Chem.* 1986, *58*(7), 1035–62.

Anastassiades M, Kolberg DI, Benkenstein A, Eichhorn E, Zechmann S, Mack D, Wildgrube, C Sigalov I, Dörk D, Barth A. Quick Method for the Analysis of numerous Highly Polar Pesticides in Foods of Plant Origin via LC–MS/MS involving Simultaneous Extraction with Methanol (QuPPe-Method). European Reference Laboratories for Residues and Pesticides. 2017. http://www.eurl-pesticides.eu/userfiles/file/EurlSRM/meth_QuPPe-PO_EurlSRM.pdf

Anastassiades M, Lehotay SJ, Štajnbaher D, Schenck FJ. Fast and easy multiresidue method employing acetonitrile extraction/partitioning and "dispersive solid-phase extraction" for the determination of pesticide residues in produce. *J. AOAC Int.* 2003, *86*(2), 412–31.

Anastassiades M, Tasdalen B, Scherbaum E, Štajnbaher D. Recent developments in QuEChERS methodology for pesticide multiresidue analysis. In: *Pesticide Chemistry: Crop Protection, Public Health, Environmental Safety.* H. Ohkawa, H. Miyagawa, P.W. Lee (ed)., Weinheim, Germany, 2007, pp. 439–58.

Andersson A. The Swedish control of pesticide residues in fruits and vegetables. In: *Residuos de plaguicidas: actas del II Seminario Internacional sobre Residuos de plaguicidas.* A. Valverde-Garcia, E. González Pradas. Almería, Spain, 1991, pp. 43–58.

Andersson A, Pålsheden H. Comparison of the efficiency of different GLC multi-residue methods on crops containing pesticide residues. *Fresenius J. Anal. Chem.* 1991, *339*, 365–7.

Arnhard K, Gottschall A, Pitteri F, Oberacher H. Applying 'sequential windowed acquisition of all theoretical fragment ion mass spectra' (SWATH) for systematic toxicological analysis with liquid chromatography-high-resolution tandem mass spectrometry. *Anal. Bioanal. Chem.* 2015, *407*, 405–14.

Aysal P, Ambrus Á, Lehotay SJ, Cannavan A. Validation of an efficient method for the determination of pesticide residues in fruits and vegetables using ethyl acetate for extraction. *J. Environ. Sci. Health, Part B.* 2007, *42*(5), 481–90.

Banerjee K, Oulkar DP, Dasgupta S, Patil SB, Patil SH, Savant R, Adsule PG. Validation and uncertainty analysis of a multi-residue method for pesticides in grapes using ethyl acetate extraction and liquid chromatography-tandem mass spectrometry. *J. Chromatogr. A.* 2007, *1173* (1–2), 98–109.

Banerjee K, Utture S, Dasgupta S, Kandaswamy C, Pradhan S, Kulkami S, Adsule P. Multiresidue determination of 375 organic contaminants including pesticides, polychlorinated biphenyls, and polyaromatic hydrocarbons in fruits and vegetables by gas chromatography-triple quadrupole mass spectrometry with introduction of semi-quantification approach. *J. Chromatogr. A.* 2012, *1270*, 283–95.

Becker VG. Gaschromatographische simultanbestimmung von chlorierten kohlen wasserstoffen und phosphorsäuresstern in pflanzlichem material. *Deutsche Lebensm.-Rundschau: Zeitsch. Lebensmittel. Lebensmittel.* 1971, *67*, 125–6.

Boesl U. Time-of-flight mass spectrometry: Introduction to the basics. *Mass Spectrom. Rev.* 2017, *36*, 86–109.

Botitsi HV, Garbis SD, Economou A, Tsipi DF. Current mass spectrometry strategies for the analysis of pesticides and their metabolites in food and water matrices. *Mass Spectrom. Rev.* 2011, *30*, 907–39.

Cairns T, Chiu KS, Navarro D, Siegmund E. Multiresidue pesticide analysis by ion-trap mass chromatography. *Rapid Commun. Mass Spectrom.* 1993, *7*, 971–88.

Cao X, Pang G, Jin L, Kang J, Hu X, Chang Q, Wang M, Fan C. Comparison of the performances of gas chromatography-quadrupole time of flight mass spectrometry and gas chromatography-tandem mass spectrometry in rapid screening and confirmation of 208 pesticide residues in fruits and vegetables. *Chinese J. Chromatogr.* 2015, *33*(4), 389–96.

Cervera MI, Portolés T, López FJ, Beltrán J, Hernández F. Screening and quantification of pesticide residues in fruits and vegetables making use of gas chromatography-quadrupole time-of-flight mass spectrometry with atmospheric pressure chemical ionization. *Anal. Bioanal. Chem.* 2014, *406*(27), 6843–55.

Cervera MI, Portolés T, Pitarch E, Beltrán J, Hernández F. Application of gas chromatography time-of-flight mass spectrometry for target and non-targett analysis of pesticide residues in fruits and vegetables. *J. Chromatogr. A.* 2012, *1244*, 168–77.

Chaput D. Simplified multiresidue method for liquid chromatographic determination of N-methyl carbamate insecticides in fruits and vegetables. *J. Assoc. Off. Anal. Chem.* 1988, *71*(3), 542–6.

Chen Y, Lopez S, Hayward DG, Park HY, Wong JW, Kim SS, Wan J, Reddy RM, Quinn DJ, Steiniger D. Determination of multiresidue pesticides in botanical dietary supplements using gas chromatography-triple-quadrupole mass spectrometry (GC–MS/MS). *J. Agric. Food Chem.* 2016, *64*(31), 6125–32.

Cheng Z, Dong F, Xu J, Liu X, Wu X, Chen Z, Pan Z, Gan J, Zheng Y. Simultaneous determination of organophosphorus pesticides in fruits and vegetables using atmospheric pressure gas chromatography quadrupole-time-of-flight mass spectrometry. *Food Chem.* 2017, *231*, 365–73.

Chun OK, Hee GK, Kim MH. Multiresidue method for the determination of pesticides in Korean domestic crops by gas chromatography/mass selective detection. *J. AOAC Int.* 2003, *86*(4), 823–31.

Codex Alimentarius. Guidelines on performance criteria for methods of analysis for the determination of pesticide residues in food and feed. CAC/GL 90-2017. Food and Agriculture Organization of the United Nations. World Health Organization. http://www.fao.org/fao-who-codexalimentarius/sh-proxy/fr/?lnk=1&url=https%253A%252F%252Fworkspace.fao.org%252Fsites%252Fcodex%252FStandards%252FCAC%2BGL%2B90-2017%252FCXG_090e.pdf

Cook J, Beckett MP, Reliford B, Hammack W, Engel M. Multiresidue analysis of pesticides in fresh fruits and vegetables using procedures developed by the Florida Department of Agriculture and Consumuer Services. *J. AOAC Int.* 1999, *82*(6), 1419–35.

Deutsche Forschungsgemeinschaft DFG. Manual of Pesticide Residue Analysis, volume I. H.-P. Their and H. Zeumer (eds.), VCH Verlagsgesllschaft mbH, D-6940 Weinheim Federal Republic of Germany, 1987.

Dias JV, Cutillas V, Lozano A, Pizzutti IR, Fernández-Alba AR. Determination of pesticides in edible oils by liquid chromatography-tandem mass spectrometry employing new generation materials for dispersive solid phase extraction cleanup. *J. Chromatogr. A.* 2016, *1462*, 8–18.

EURL-FV. Multiresidue method using QuEChERS followed by GC-QqQ/MS/MS and LC-QqQ/MS/MS for fruits and vegetables. 2010a. http://www.crl-pesticides.eu/library/docs/fv/CRLFV_Multiresidue_methods.pdf

EURL-FV. Analysis of pesticide residues in fruit and vegetables with ethyl acetate extraction using gas and liquid chromatography with tandem mass spectrometric detection 0.1. European Reference Laboratories for Residues and Pesticides. 2010b. http://www.crl-pesticides.eu/library/docs/fv/ethyl_acetate_extraction.pdf

EURL-FV. Dutch Mini-Luke ("N-") extraction method followed by LC and GC–MS/MS for multi-residue analysis of pesticides in fruits and vegetables. European Reference Laboratories for Residues and Pesticides. 2014. http://www.eurl-pesticides.eu/userfiles/file/NL-miniLuke-extraction-method.pdf

Eliuk S, Makarov A. Evolution of orbitrap mass spectrometry instrumentation. *Ann. Rev. Anal. Chem.* 2015, *8*, 61–80.

European Commission Directorate General for Health and Food Safety. Guidance document on analytical quality control and method validation procedures for pesticide residues and analysis in food and feed. SANTE/11813/2017. https://ec.europa.eu/food/sites/food/files/plant/docs/pesticides_mrl_guidelines_wrkdoc_2017-11813.pdf

Fernández Moreno JL, Garrido Frenich A, Plaza Bolaños P, Martínez Vidal JL. Multiresidue method for the analysis of more than 140 pesticide residues in fruits and vegetables by gas chromatography coupled to triple quadrupole mass spectrometry. *J. Mass Spectrom.* 2008, *43*, 1235–54.

Ferrer I, García-Reyes JF, Mezcua M, Thurman EM, Fernández-Alba AR. Multi-residue pesticide analysis in fruits and vegetables by liquid chromatography–time-of-flight mass spectrometry. *J. Chromatogr. A.* 2005, *1082*(1), 81–90.

Ferrer I, Thurman EM, Fernández-Alba AR. Quantitation and accurate mass analysis of pesticides in vegetables by LC/TOF-MS. *Anal. Chem.* 2005, *77*(9), 2818–25.

Ferrer I, Thurman EM. Multi-residue method for the analysis of 101 pesticides and their degradates in food and water samples by liquid chromatography/time-of-flight mass spectrometry. *J. Chromatogr. A.* 2007, *1175*(1), 24–37.

Ferrer I, Thurman EM, Zweigenbaum J. LC/TOF-MS analysis of pesticides in fruits and vegetables: The emerging role of accurate mass in the unambiguous identification of pesticides in foods. *Methods Mol. Biol.* 2011, *747*, 193–218.

Fillion J, Hindle R, Lacroix M, Selwyn J. Multiresidue determination of pesticides in fruit and vegetables by gas chromatography–mass-selective detection and liquid chromatography with fluorescence detection. *J. AOAC Int.* 1995, *78*(5), 1252–66.

Fillion J, Sauvé F, Selwyn J. Multiresidue method for the determination of residues of 251 pesticides in fruits and vegetables by gas chromatography/mass spectrometry and liquid chromatography with fluorescence detection. *J. AOAC Int.* 2000, *83*(3), 698–713.

Garrido Frenich A, González-Rodríguez MJ, Arrebola FJ, Martínez Vidal JL. Potentiality of gas chromatography-triple quadrupole mass spectrometry in vanguard and rearguard methods of pesticide residues in vegetables. *Anal. Chem.* 2005, *77*, 4640–8.

Garrido-Frenich A, Plaza-Bolaños P, Martínez Vidal JL. Comparison of tandem-in-space and tandem-in-time mass spectrometry in gas chromatography determination of pesticides: Application to simple and complex food samples. *J. Chromatogr. A.* 2008, *1203*, 229–38.

Gillet LC, Navaroo P, Tate S, Röst H, Selevsek N, Reiter L, Bonner R, Aebersold R. Targeted data extraction of the MS/MS spectra generated by data-independent acquisition: A new concept for consistent and accurate proteome analysis. *Mol. Cell. Proteomics.* 2012, *11*, 1–17.

Goon A, Zhan K, Oulkar D, Shinde R, Gaikwad S, Banerrjee K. A simultaneous screening and quantitative method for the multiresidue analysis of pesticides in spices using ultra-high performance liquid chromatography-high resolution (Orbitrap) mass spectrometry. *J. Chromatogr. A.* 2018, *1532*, 105–11.

Granby K, Andersen JH, Christensen HB. Analysis of pesticides in fruit, vegetables and cereals using methanolic extraction and detection by liquid chromatography-tandem mass spectrometry. *Anal. Chim. Acta.* 2004, *520*, 165–76.

Han L, Matarrita J, Sapozhnikova Y, Lehotay SJ. Evaluation of a recent product to remove lipids and other matrix co-extractives in the analysis of pesticides and environmental contaminants in foods. *J. Chromatogr. A.* 2016, *1449*, 17–29.

Hanot V, Goscinny S, Deridder M. A simple multi-residue method for the determination of pesticides in fruits and vegetables using a methanolic extraction and ultra-high-performance liquid chromatography-tandem mass spectrometry: Optimization and extension of scope. *J. Chromatogr. A.* 2015, *1384*, 53–66.

Hayward DG, Wong JW, Park HY. Determinations for pesticides on black, green, oolong, and white teas by gas chromatography triple-quadrupole mass spectrometry. *J. Agric. Food Chem.* 2015, *63*(37), 8116–24.

He Z, Wang Y, Wang L, Peng Y, Wang W, Liu X. Determination of 255 pesticides in edible vegetable oils using QuEChERS method and gas chromatography tandem mass spectrometry. *Anal. Bioanal. Chem.* 2017, *409*, 1017–30.

Hernández F, Cervera MI, Portolés T, Beltrán J, Pitarch E. The role of GC–MS/MS with triple quadrupole in pesticide residue analysis in food and the environment. *Anal. Methods.* 2013, *5*, 5875–94.

Hiemstra M, de Kok A. Comprehensive multi-residue method for the target analysis of pesticides in crops using liquid chromatography-tandem mass spectrometry. *J. Chromatogr. A.* 2007, *1154*, 3–25.

Hopper ML. Solid-phase partition column technology. In: *Emerging Strategies for Pesticide Analysis.* T Cairns and J Sherma (eds.), CRC Press, Boca Rotan, FL, 1992, pp. 39–50.

Jansson C, Pihlström T, Österdahl B-G, Markides KE. A new multi-residue method for analysis of pesticide residues in fruit and vegetables using liquid chromatography with tandem mass spectrometry. *J. Chromatogr. A.* 2004, *1023*, 93–104.

Japan Ministry of Health, Labour and Welfare, Department of Food Safety. Analytical methods for residual compositional substances of agricultural chemicals, feed additives, and veterinary drugs in food, 2006. http://www.mhlw.go.jp/english/topics/foodsafety/positivelist060228/dl/060526-1a.pdf

Jia W, Chu X, Ling Y, Huang J, Chang J. High-throughput screening of pesticide and veterinary drug residues in baby food by liquid chromatography coupled to quadrupole Orbitrap mass spectrometry. *J. Chromatogr. A.* 2014, *1347*, 122–8.

Kakimoto Y, Ohtani Y, Fuanki N, Joh T. Simultaneous determination of pesticide residues in fruits and vegetables by GC/MS (SCAN mode) and HPLC. *Shokuhin Eiseigaku Zasshi (J. Food Hyg. Soc. Japan),* 2003, *44*(5), 253–62.

Kaufmann A. Analytical performance of the various acquisition modes in Orbitrap MS and MS/MS. *J. Mass Spectrom.* 2018. DOI:10.1002/jms/4195.

Kaufmann A, Teale P. Capabilities and limitations of high-resolution mass spectrometry (HRMS): time-of-flight and Orbitrap. In: *Chemical Analysis of Non-Antimicrobial Veterinary Drug Residues in Food,* First edition. JF Kay, JD MacNeil, J Wang (eds.), John Wiley & Sons, Inc., Hoboken, NJ, 2017, pp. 93–139.

Klein J, Alder L. Applicability of gradient liquid chromatography with tandem mass spectrometry to the simultaneous screening for about 100 pesticides in crops. *J. AOAC Int.* 2003, *86*(5), 1015–37.

Kmellár B, Abrankó L, Fodor P, Lehotay SJ. Routine approach to qualitatively screening 300 pesticides and quantification of those frequently detected in fruit and vegetables using liquid chromatography tandem mass spectrometry (LC–MS/MS). *Food Add. Contam. Part A. Chem. Anal. Control. Expo. Risk Assess.* 2010, *27*(10), 1415–30.

Krause RT. Multiresidue method for determining N-methylcarbamate insecticides in crops using high performance liquid chromatography. *J. Assoc. Off. Anal. Chem.* 1980, *63*(5), 1114–24.

Krause RT. Liquid chromatographic determination of N-methylcarbamate insecticides and metabolites in crops. I. Collaborative study. *J. Assoc. Off. Anal. Chem.* 1985, *69*(4), 726–33.

Krause RT. Multiresidue methods for carbamate pesticides. In: *Emerging Strategies for Pesticide Analysis.* T Cairns, J Sherma (eds.), CRC Press, Boca Rotan, FL, 1992, pp. 99–126.

Krynitsky AJ, Lehotay SJ. Overview of analytical technologies available to regulatory laboratories for the determination of pesticide residues. In: *Handbook of Residue Analytical Methods for Agriculture.* PW Lee (ed.), John Wiley & Sons, Ltd, Chichester, United Kingdom, 2002, pp. 753–86.

Langlois BE, Stemp AR, Liska BJ. Rapid cleanup of dairy products for analysis of chlorinated insecticide residue by electron capture gas chromatography. *J. Agric. Food Chem.* 1964, *12*(3), 243–5.

Lawrence JF. Direct analysis of some carbamate pesticides in foods by high-performance liquid chromatography. *J. Agric. Food Chem.* 1977, *25*(1), 211–12.

Lee HB, Chau ASY, Kawahara F. Organochlorine Pesticides. In: *Analysis of Pesticides in Water. Volume II: Chlorine- and Phosphorus-Containing Pesticides.* ASY Chau, BK Afghan, JW Robinson (eds.), CRC Press, Boca Raton, FL, 1982, pp. 1–60.

Lee J, Kim L, Shin Y, Lee J, Lee J, Kim E, Moon J-K, Kim J-H. Rapid and simultaneous analysis of 360 pesticides in brown rice, spinach, orange, and potato using microbore GC–MS/MS. *J. Agric. Food Chem.* 2017, *65*, 3387–95.

Lee SM, Papathakis ML, Feng, H-MC, Hunter GF, Carr JE. Multipesticide residue method for fruits and vegetables: California Department of Food and Agriculture. *Fresenius J. Anal. Chem.* 1991, *339*, 376–83.

Lee SM, Richman SJ. Pesticide residue procedures for agricultural commodities: An international view. In: *Pesticides in Agriculture and the Environment.* WB Wheeler (ed.), Marcel Dekker Inc., New York, 2002, pp. 163–208.

Lehotay SJ. Determination of pesticide residues in foods by acetonitrile extraction and partitioning with magnesium sulfate: Collaborative study. *J. AOAC Int.* 2007, *90*(2), 485–520.

Lehotay SJ, de Kok A, Hiemstra M, van Bodegraven P. Validation of a fast and easy method for the determination of residues from 229 pesticides in fruits and vegetables using gas and liquid chromatography and mass spectrometric detection. *J. AOAC Int.* 2005, *88*(2), 595–614.

Lehotay SJ, Maštovská K, Lightfield AR. Use of buffering and other means to improve results of problematic pesticides in a fast and easy method for the analysis of fruits and vegetables. *J. AOAC Int.* 2005, *88*(2), 615–29.

Li J-X, Li X-Y, Chang Q-Y, Li Y, Jin L-H, Pang G-F, Fan C-L. Screening of 439 pesticide residues in fruits and vegetables by gas chromatography-quadrupole-time-of-flight mass spectrometry based on TOF accurate mass database and Q-TOF spectrum library. *J. AOAC Int.* 2018, *101*, 1–8.

López-Blanco R, Nortes-Méndez R, Robles-Molina J, Moreno-González D, Gilbert-López B, García-Reyes JF, Molina-Díaz A. Evaluation of different cleanup sorbents for multiresidue analysis in fatty vegetable matrices by liquid chromatography tandem mass spectrometry. *J. Chromatogr. A.* 2016, *1456*, 89–104.

Lozano A, Kiedrowska B, Scholten J, de Kroon M, de Kok A, Fernández-Alba AR. Miniaturisation and optimisation of the Dutch mini-Luke extraction method for implementation in the routine multi-residue analysis of pesticides in fruits and vegetables. *Food Chem.* 2016, *192*, 668–81.

Luke MA, Froberg JE, Doose GM, Masumoto HT. Improved multiresidue gas chromatographic determination of organophosphorus, organonitrogen, and organohalogen pesticides in produce using flame photometric and electrolytic conductivity detectors. *J. AOAC.* 1981, *64*(5), 1187–95.

Luke MA, Froberg JE, Masumoto HT. Extraction and cleanup of organochlorine, organophosphorus, organonitrogen, and hydrocarbon pesticides in produce for determination by gas-liquid chromatography. *J. AOAC.* 1975, *58*(5), 1020–6.

Ma Z, Zhao W, Li L, Zheng S, Lin H, Zhang Y, Gao Q, Liu S. Rapid determination of 129 pesticide residues in vegetables and fruits by gas chromatography-triple quadrupole mass spectrometry. *Se Pu.* 2013, *31*(3), 228–39.

Majors RE. QuEChERS - A new sample preparation technique for multiresidue analysis of pesticides in foods and agricultural samples. *LCGC North America.* 2007, *26*(5), 436–45.

Mills PA, Onley JH, Gaither RA. Rapid method for chlorinated pesticide residues in nonfatty foods. *J. AOAC.* 1963, *46*(2), 186–91.

Miyahara M, Murayama M, Suzuki T, Saito Y. Silica gel chromatographic cleanup procedure for organochlorine pesticide analysis with capillary gas chromatography. *J. Agric. Food Chem.* 1993, *41*, 221–6.

Mol HG, Rooseboom A, van Dam R, Roding M, Arondeus K, Sunarto S. Modification and re-validation of the ethyl acetate-based multiresidue method for pesticides in produce. *Anal. Bioanal. Chem.* 2007, *389*(6), 1715–54.

Mol HGJ, Plaza-Bolaños P, Zomer P, de Rijk TC, Stolker AAM, Mulder PPJ. Toward a generic extraction method for simultaneous determination of pesticides, mycotoxins, plant toxins, and veterinary drugs in feed and food matrixes. *Anal. Chem.* 2008, *80*, 9450–9.

Mol HGJ, Tienstra M, Zomer P. Evaluation of gas chromatography-electron ionization-full scan high resolution Orbitrap mass spectrometry for pesticide residue analysis. *Anal. Chim. Acta.* 2016, *935*, 161–72.

Moye HA, Scherer SJ. Dynamic fluorogenic labelling of pesticides for high performance liquid chromatography: Detection of N-methylcarbamates with o-phthalaldehyde. *Anal. Lett.* 1977, *10*(13), 1049–73.

Nakamura Y, Tonogai Y, Sekiguchi Y, Tsumura Y, Nishida N, Takakura K, Isechi M, Yuasa K, Nakamura M, Kifune N, Yamamoto K, Terasawa S, Oshima T, Miyata M, Kamakura K, Ito Y. Multiresidue analysis of 48 pesticides in agricultural products by capillary gas chromatography. *J. Agric. Food Chem.* 1994, *42*, 2508–18.

Namikawa M, Shibata S, Shiomi T, Nakagawa T, Ban S, Tomita Y, Semura S, Nakao Y, Banno Y, Kawakami M. Simultaneous analysis of residual pesticides in fruit, vegetables, brown rice, and oolong tea by LC–MS/MS. *Shokuhin Eiseigaku Zasshi.* 2014, *55*(6), 279–89.

Norli HR, Christiansen AL, Stuveseth K. Analysis of non-cleaned QuEChERS extracts for the determination of pesticide residues in fruit, vegetables and cereals by gas chromatography-tandem mass spectrometry. *Food Addit. Contam. Part A Chem. Anal. Control Expo. Risk Assess.* 2016, *33*(2), 300–12.

NVWA, Netherlands Food and Consumer Product Safety Authority. Dutch mini-Luke ("NL-") extraction method followed by LC and GC–MS/MS for multi-residue analysis of pesticides in fruits and vegetables. European Reference Laboratory EURL-FV.

Okihashi M, Kitagawa Y, Akutsu K, Obana H, Tanaka Y. Rapid method for the determination of 180 pesticide residues in foods by gas chromatography/mass spectrometry and flame photometric detection. *J. Pestic. Sci.* 2005, *30*(4), 368–77.

Okihashi M, Kitagawa Y, Obana H, Tanaka Y, Yamagishi Y, Sugitake K, Saito K, Kubota M, Kanai M, Ueda T, Harada S, Rapid multiresidue method for the determination of more than 300 pesticide residues in food. *Food.* 2007, *1*(1), 101–10.

Okihashi M, Takatori S, Kitagawa Y, Tanaka Y. Simultaneous analysis of 260 pesticide residues in agricultural products by gas chromatography/triple quadrupole mass spectrometry. *J. AOAC Int.* 2007, *90*(4), 1165–79.

Pang GF, Chao YZ, Fan CL, Zhang JJ, Li XM, Zhao TS. Modification of AOAC multiresidue method for determination of synthetic pyrethroid residues in fruits, vegetables, and grains. Part I: Acetonitrile extraction system and optimization of florisil cleanup and gas chromatography. *J. AOAC Int.* 1995, *78*(6), 1481–8.

Pang GF, Can YZ, Fan CL, Zhang JJ, Li XM, Mu J, Wang DN, Liu SM, Song WB, Li HP, Wong SS, Kubinec R, Tekel J, Tohotna S. Interlaboratory study of identification and quantitation of multiresidue pyrethroids in agricultural products by gas chromatography–mass spectrometry. *J. Chromatogr. A.* 2000, *882*, 231–8.

Pang GF, Fan CL, Liu YM, Cao YZ, Zhang JJ, Li XM, Li ZY, Wu YP, Guo TT. Determination of residues of 446 pesticides in fruits and vegetables by three-cartridge solid-phase extraction-gas chromatography–mass spectrometry and liquid chromatography-tandem mass spectrometry. *J. AOAC Int.* 2006, *89*, 740–71.

Pang GF, Fan CL, Chao YZ, Zhao TS. Modification of AOAC multiresidue method for determination of synthetic pyrethroid residues in fruits, vegetables, and grains. Part II: Acetone extraction system. *J. AOAC Int.* 1995, *78*(6), 1489–96.

Pang GF, Cao YZ, Fan CL, Zhang JJ, Li XM. Multiresidue gas chromatographic method for determining synthetic pyrethroid pesticides in agricultural products: Collaborative study. *J. AOAC Int.*, 1999, *82*(1), 186–212.

Picó Y, Fernández M, Ruiz MJ, Font G. Current trends in solid-phase-based extraction techniques for the determination of pesticides in food and environment. *J. Biochem. Biophys. Methods.* 2007, *70*, 117–31.

Pihlström T, Blomkvist G, Friman P, Pagad U, Österdahl B-G. Analysis of pesticide residues in fruits and vegetables with ethyl acetate extraction using gas and liquid chromatography with tandem mass spectrometric detection. *Anal. Bioanal. Chem.* 2007, *389*, 1773–89.

Pihlström T, Österdahl B-G. Analysis of pesticide residues in fruit and vegetables after cleanup with solid-phase extraction using ENV+ (polystyrene-divinylbenzene) cartridges. *J. Agric. Food Chem.* 1999, *47*, 2549–52.

Pizzuti IR, de Kok A, Hiemstra M, Wickert C, Prestes OD. Method validation and comparison of acetonitrile and acetone extraction for the analysis of 169 pesticides in soya grain by liquid chromatography–tandem mass spectrometry. *J. Chromatogr. A.* 2009, *1216*, 4539–52.

Podhorniak LV, Schenck FJ, Krynitsky A, Griffith F. Multiresidue method for N-methyl carbamates and metabolite pesticide residues at the parts-per-billion level in selected representative commodities of fruit and vegetable crop groups. *J. AOAC Int.* 2004, *87*(5), 1237–51.

Polgár L, García-Reyes JF, Fodor P, Gyepes A, Dernovics M, Abrankó L, Gilbert-Lópea B, Molica-Díaz A. Retrospective screening of relevant pesticide metabolites in food using liquid chromatography high resolution mass spectrometry and accurate-mass databases of parent molecules and diagnostic fragment ions. *J. Chromatogr. A.* 2012, *1249*, 83–91.

Portolés T, Sancho JV, Hernández F, Newton A, Hancock P. Potential of atmospheric pressure chemical ionization source in GC–QTOF MS for pesticide residue analysis. *J. Mass Spectrom.* 2010, *45*(8), 926–36.

Rajski Ł, Lozano A, Uclés A, Ferrer C, Fernández-Alba AR. Determination of pesticide residues in high oil vegetal commodities by using various multi-residue methods and cleanups followed by liquid chromatography tandem mass spectrometry. *J Chromatogr A.* 2013, *1304*, 109–20.

Roos AH, Van Munsteren J, Nab FM, Tuinstra LT Universal extraction/cleanup procedure for screening of pesticides by extraction with ethyl acetate and size exclusion chromatography. *Anal. Chim. Acta.* 1987, *196*, 95–102.

Sack C, Smoker M, Chamkasem N, Thompson R, Satterfield G, Masse C, Mercer G, Neuhaus B, Cassias I, Lin Y, Macmahon D, Wong J, Zhang K, Smith RE. Collaborative validation of the QuEChERS procedure for the determination of pesticides in food by LC–MS/MS. *J. Agric. Food Chem.*, 2011, *59*(12), 6383–411.

Sack C, Vonderbrink J, Smoker M, Smith RE. Determination of acid herbicides using modified QuEChERS with fast switching ESI+/ESI- LC–MS/MS. *J. Agric. Food Chem.* 2015, *63*, 9657–65.

Saito-Shida S, Nemoto S, Teshima R, Akiyama H. Quantitative analysis of pesticide residues in vegetables and fruits by liquid chromatography quadrupole time-of-flight mass spectrometry. *Food Addit. Contam. Part A. Chem. Anal. Anal. Control Expo. Risk Assess.* 2016, *33*(1), 119–27.

Sannino A. Pesticide residues. In: *Wilson & Wilson's Comprehensive Analytical Chemistry. Volume 51. Food Contaminants and Residue Analysis.* Y Picó (ed.), Elsevier, Amsterdam, the Netherlands, 2008, pp. 257–306.

Sawyer LD. The Luke et al. method for determining multipesticide residues in fruits and vegetables: Collaborative study. *J. AOAC* 1985, *68*(1), 64–71.

Schenck FJ, Wong JW. Determination of pesticides in food of vegetal origin. In: *Analysis of Pesticides in Food and Environmental Samples.* JL Tadeo (ed.), CRC Press, Boca Rotan, FL, 2008, pp. 151–76.

Sheridan RS, Meola JR. Analysis of pesticide residues in fruits, vegetables, and milk by gas chromatography/tandem mass spectrometry. *J. AOAC Int.* 1999, *82*, 982–90.

Sherma J, Zweig G. *Gas Chromatographic Analysis: Analytical Methods for Pesticides and Plant Growth Regulators.* G. Zweig (ed.), Academic Press Inc., New York, NY, 1972.

Silva JC, Gorenstein MV, Li GZ, Vissers JPC, Geromanos SJ. Absolute quantification of proteins by LCMSE. *Mol. Cell Proteomics.* 2006, *5*, 145–56.

Smart NA. Collaborative studies of methods for pesticides residues analysis. In: *Residue Reviews: Residues of Pesticides and Other Contaminants in the Total Environment*, vol. 64. F.A. Gunther, J.D. Gunther (eds.), Springer-Verlag, New York NY, 1976, pp. 1–16.

Sobus JR, Wambaugh JF, Isaacs KK, Williams AJ, McEachran AD, Richard AM, Grulke CM, Ulrich EM, Rager JE, Strynar MJ, Newton SR. Integrating tools for non-targeted analysis research and chemical safety evaluations at the US EPA. *J. Expo. Sci. Environ. Epidemiol.* 2017. DOI:10.1038/s41370-017-0012-y.

Song S, Ma X, Li C. Multi-residue determination method of pesticides in leek by gel permeation chromatography and solid phase extraction followed by gas chromatography with mass spectrometric detector. *Food Control.* 2007, *18*, 448–53.

Špánik I, Machyňáková A. Recent applications of gas chromatography with high-resolution mass spectrometry. *J. Sep. Sci.* 2018, *41*, 163–79.

Specht W, Pelz S, Gilsbach W. Gas-chromatographic determination of pesticide residues after cleanup by gel-permeation chromatography and mini-silica gel-column chromatography. *Fresenius J. Anal. Chem.* 1995, *353*, 183–90.

Specht W, Tillkes M. Gas-chromatographic determination of pesticide residues after cleanup by gel-permeation chromatography and mini-silica gel-column chromatography. *Fresenius Z. Anal. Chem.* 1980, *301*, 300–7.

Stan H-J. Pesticide residue analysis in foodstuffs applying capillary gas chromatography with mass spectrometric detection. State-of-the-art use of modified DFG-multimethod S19 and automated data evaluation. *J. Chromatogr. A.* 2000, *892*, 347–77.

Steinborn A, Alder L, Spitzke M, Dörk D, Anastassiades M. Development of a QuEChERS-based method for the simultaneous determination of acidic pesticides, their esters, and conjugates following alkaline hydrolysis. *J. Agric. Food Chem.* 2017, *65*, 1296–305.

Swedish National Food Administration. Analysis of pesticide residues in fruits and vegetables with ethyl acetate extraction using gas and liquid chromatography with tandem mass spectrometrid detection 0.1. European Reference Laboratory EURL-FV.

Takatori S, Okihashi M, Okamoto Y, Kitagawa Y, Kakimoto S, Murata H, Sumimoto T, Tanaka Y. A rapid and easy multiresidue method for the determination of pesticide residues in vegetables, fruits, and cereals using liquid chromatography/tandem mass spectrometry. *J. AOAC Int.* 2008, *91*(4), 871–83.

Thurman EM, Ferrer I, Fernández-Alba AR. Matching unknown empirical formulas to chemical structure using LC/MS TOF accurate mass and database searching: example of unknown pesticides on tomato skins. *J. Chromatogr. A.* 2005, *1067*(1–2), 127–34.

Thurman EM, Ferrer I, Zweigenbaum JA, García-Reyes JF, Woodman M, Fernández-Alba AR. Discovering metabolites of post-harvest fungicides in citrus with liquid chromatography/time-of-flight mass spectrometry and ion trap tandem mass spectrometry. *J. Chromatogr. A.* 2005, *1082*(1), 71–80.

Thurman EM, Mills MS. Overview of solid-phase extraction. In: *Solid-Phase Extraction: Principles and Practice.* E.M. Thurman, M.S. Mills (eds.), John Wiley & Sons, Inc., New York, NY, 1998, pp. 1–23.

Ting K-C, Kho PK, Musselman AS, Root GA, Tichelaar GR. High performance liquid chromatographic method for determination of six N-methylcarbamates in vegetables and fruits. *Bull. Environ. Contam. Toxicol.* 1984, *33*, 538–47.

Taylor MJ, Hunter K, Hunter KB, Lindsay D, LeBouhellec S. Multi-residue method for rapid screening and confirmation of pesticides in crude extracts of fruits and vegetables using isocratic liquid chromatography with electrospray tandem mass spectrometry. *J. Chromatogr. A.* 2002, *98*(2), 225–36.

Tong L, Ma X, Li C. Application of gas chromatography-tandem mass spectrometry (GC–MS–MS) with pulsed splitless injection for the determination of multiclass pesticides in vegetables. *Anal. Lett.* 2006, *39*(5), 985–96.

U.S. Food and Drug Administration. *Pesticide Analytical Manual*, 3rd edition, volumes I and II. 1999. U.S. Department of Health and Human Services. https://www.fda.gov/Food/FoodScienceResearch/LaboratoryMethods/ucm2006955.htm

U.S. Food and Drug Administration, FDA Office of Regulatory Affairs. ORA-LAB.10 Guidance for the Analysis and Documentation to Support Regulatory Action on Pesticide Residues, Version 1.2, revised 05/17/2012.

U.S. Food and Drug Administration. Acceptance criteria for confirmation of identity of chemical residues using exact mass data within the Office of Foods and Veterinary Medicine Program. 2015. https://www.fda.gov/downloads/ScienceResearch/Field Science/UCM491328.pdf.

Uclés S, Uclés A, Lozana A, Martinez-Bueno MJ, Fernandez-Alba AR. Shifting the paradigm in gas chromatography mass spectrometry pesticide analysis using high resolution accurate mass spectrometry. *J. Chromatogr. A*. 2017, *1501*, 107–16.

Ueno E, Oshima H, Saito I, Matsumoto H, Yoshimura Y, Nakazawa. Multiresidue analysis of pesticides in vegetables and fruits by gas chromatography/mass spectrometry after gel permeation chromatography and graphitized carbon column cleanup. *J. AOAC Int.* 2004, *87*(4), 1003–15.

Walorczyk S. Application of gas chromatography/tandem quadrupole mass spectrometry to the multi-residue analysis of pesticides in green leafy vegetables. *Rapid Commun. Mass Spectrom.* 2008, *22*(23), 3791–801.

Walorczyk S, Drożdżyński D, Kierzek R. Two-step dispersive-solid phase extraction strategy for pesticide multiresidue analysis in a chlorophyll-containing matrix by gas chromatography–tandem mass spectrometry. *J Chromatogr. A*. 2015, *1412*, 22–32.

Walorczyk S, Drożdżyński D, Kierzek R. Determination of pesticide residues in samples of green minor crops by gas chromatography and ultra performance liquid chromatography coupled to tandem quadrupole mass spectrometry. *Talanta*. 2015, *132*, 197–204.

Wang J, Leung D. Applications of ultra-performance liquid chromatography electrospray ionization quadrupole time-of-flight mass spectrometry on analysis of 138 pesticides in fruit- and vegetable-based infant foods. *J. Agric. Food Chem.* 2009, *57*(6), 2162–73.

Wang J, Leung D. Determination of 142 pesticides in fruit- and vegetable-based infant foods by liquid chromatography/electrospray ionization-tandem mass spectrometry and estimation of measurement uncertainty. *J. AOAC Int.* 2009, *92*(1), 279–301.

Wang J, Chow W, Chang J, Wong JW. Ultrahigh-performance liquid chromatography electrospray ionization Q-Orbitrap mass spectrometry for the analysis of 451 pesticide residues in fruits and vegetables: Method development and validation. *J. Agric. Food Chem.* 2014, *62*(42), 10375–91.

Wang J, Chow W, Chang J, Wong JW. Development and validation of a qualitative method for target screening of 448 pesticide residues in fruits and vegetables using UHPLC/ESI Q-Orbitrap based on data-independent acquisition and compound database. *J. Agric. Food Chem.* 2017, *65*(2), 473–93.

Ware GW. *The Pesticide Book*, 3rd edition. 1989. Thomson Publications, Fresno, CA. W.H. Freeman and Company.

Wong JW, Zhang K, Tech K, Hayward DG, Makovi CM, Krynitsky AJ, Schenck FJ, Banerjee K, Dasgupta S, Brown D. Multiresidue pesticide analysis in fresh produce by capillary gas chromatography–mass spectrometry/selective ion monitoring (GC–MS/SIM) and –tandem mass spectrometry (GC–MS/MS). *J. Agric. Food Chem.* 2010, *58*(10), 5868–83.

Zhang K, Wong JW, Yang P, Tech K, Dibenedetto AL, Lee NS, Hayward DG, Makovi CM, Krynitsky AJ, Banerjee K, Jao L, Dasgupta S, Smoker MS, Simonds R, Schreiber A. Multiresidue pesticide analysis of agricultural commodities using acetonitrile salt-out extraction, dispersive solid-phase sample cleanup, and high-performance liquid chromatography-tandem mass spectrometry. *J. Agric. Food Chem.* 2011, *59*(14), 7636–46.

Zhu F, Ji W, Liu H, Jia Y, Cai M, Zhang H. Rapid screening and identification of food poisonings by ultra high performance liquid chromatography coupled with quadrupole-time of flight mass spectrometry. *Chinese J. Chromatogr.* 2017, *35*(9), 957–62.

Zomer P, Mol HG. Simultaneous quantitative determination, identification and qualitative screening of pesticides in fruits and vegetables using LC–Q-Orbitrap™-MS. *Food Addit. Contam. Part A. Chem. Anal. Control Expo. Risk Assess.* 2015, *32*(10), 1628–36.

7 Determination of Pesticide Residues in Food of Animal Origin

Giulia Poma, Marina López-García,
Roberto Romero González,
Antonia Garrido Frenich, and Adrian Covaci

CONTENTS

List of Abbreviations 208
7.1   Introduction 209
7.2   Sample Preparation 211
    7.2.1   Sample Pretreatment 211
    7.2.2   Extraction Techniques 221
    7.2.3   Cleanup and Fractionation 223
7.3   Analytical Techniques 224
    7.3.1   Gas Chromatography Coupled to Mass Spectrometry 224
        7.3.1.1   Low Resolution Mass Spectrometry Applications 224
        7.3.1.2   High Resolution Mass Spectrometry Applications 226
        7.3.1.3   Other GC Applications 226
    7.3.2   Liquid Chromatography Coupled to Mass Spectrometry 226
        7.3.2.1   Low Resolution Mass Spectrometry Applications 227
        7.3.2.2   High Resolution Mass Spectrometry Applications 227
7.4   Internal Quality Control 228
    7.4.1   Basic Activities of IQC in Pesticide Residue Analysis 229
    7.4.2   IQC Measures 229
7.5   Application to Real Samples 230
    7.5.1   Multiresidue Analysis of Pesticides 230
    7.5.2   Organochlorine Pesticides 231
    7.5.3   Organophosphorus Pesticides 231
    7.5.4   Carbamates 231
    7.5.5   Pyrethroids 232
    7.5.6   Neonicotinoids 233
7.6   New Trends and Emerging Issues in Analytical Methods 233
Acknowledgments 235
References 235

LIST OF ABBREVIATIONS

AChE	Acetylcholinesterase
ACN	Acetonitrile
APCI	Atmospheric pressure chemical ionization
CBs	Carbamates
CME-UABE	Coacervation microextraction ultrasound-assisted back-extraction technique
CRMs	Certified Reference Materials
DAD	Diode array detector
DCPA	Dimethyl 2,3,5,6-tetrachloro-1,4-benzenedicarboxylate
DLLME	Dispersive liquid–liquid microextraction
d-SPE	Dispersive solid phase extraction
ECD	Electron capture detector
EFSA	European Food Safety Authority
EI	Electron impact
EQC	External quality control
ESI	Electrospray ionization
EtAc	Ethyl acetate
EU	European Union
GC	Gas chromatography
GCB	Graphitized carbon black
GC×GC	Two-dimensional or comprehensive GC
GPC	Gel permeation chromatography
HILIC	Hydrophilic interaction liquid chromatography
HRMS	High-resolution mass spectrometry
HS–SPME	Headspace–SPME
IQC	Internal quality control
IT	Ion trap
K_{ow}	Octanol–water partition coefficient
LC	Liquid chromatography
LLE	Liquid–liquid extraction
LLE-LTP	Liquid–liquid extraction with purification at low temperature
LOQ	Limit of quantification
LPGC	Low pressure gas chromatography
LRMS	Low resolution mass spectrometry
MeOH	Methanol
MEPS	Microextraction in packed sorbent
$MgSO_4$	Magnesium sulfate
MRLs	Maximum residue levels
MS	Mass spectrometry
MS/MS	Tandem mass spectrometry
MSPD	Matrix solid-phase dispersion
Na_2SO_4	Sodium sulfate
NaCl	Sodium chloride
NCI	Negative chemical ionization

n-Hex	*n*-hexane
NPD	Nitrogen–phosphorus detector
OCs	Organochlorines
OPs	Organophosphates
PAD	Photodiode array detector
PCBs	Polychlorinated biphenyls
PCI	Positive chemical ionization
PLE	Pressurized liquid extraction
PSA	Primary secondary amine
PT	Proficiency testing
Py	Pyrethroids
QqQ	Triple quadrupole
Q-TOF	Quadrupole/TOF
QuEChERS	Quick, Easy, Cheap, Effective, Rugged, and Safe
SBSE	Stir bar sorptive extraction
SFE	Supercritical fluid extraction
SIM	Selected ion monitoring
SLE	Solid-liquid extraction
SPE	Solid phase extraction
SPME	Solid-phase microextraction
TOF	Time-of-flight
UHPLC	Ultra-high-performance liquid chromatography
VWD	Variable wavelength detector

7.1 INTRODUCTION

Organic pollutants present in food of animal origin can be classified in two main categories: *contaminants* and *residues*. *Contaminants* (e.g., dioxins, polychlorinated biphenyls – PCBs) are substances that are not deliberately added to the food and that can enter food during its production process, transformation, storage, packaging, transport, or because of fraudulent practices. *Residues* are compounds that can occur in foodstuffs because of intentional usage of phytosanitary or veterinary products during plant or animal production. Pesticides (found incorporated in the food or on the surface of food) can thus be included in the latter category.

Pesticides are a group of chemicals used either to directly control pest populations or to prevent or reduce pest damage on crops, landscape, or animals. Consequently, they may be present in fresh or processed animal foodstuffs, if animals have been fed with contaminated feed or water, or if practices involving pesticides are performed in food-processing factories and/or where animals eat and live (e.g., stables, roosts, meadows and pastures, grazing lands, beehives, etc.). Pesticides can be classified by the target organisms they are designed to control. The main classes are represented by *insecticides* and *acaricides* (to control insects, acari, and mites), *fungicides* (to control fungi and oomycetes), *herbicides* and *algicides* (to control weeds and algae), and *rodenticides* (to control rodents). In food of animal origin, pesticide residues can be grouped into main chemical families of public and regulatory concern, such as *organochlorines* (OCs), *organophosphates* (OPs), *carbamates* (CBs), *pyrethroids* (Py), and *neonicotinoids*.

Knowledge of the physicochemical properties of pesticides is very important in environmental risk assessment, because they influence the distribution, persistence, and fate of parent compounds and their metabolites into the environment. The most important physicochemical properties are vapor pressure, water solubility, and the octanol–water partition coefficient (K_{ow}), which dictates how a pesticide will distribute among animal (fatty) tissues.

Chronologically, OCs were the first class of chemicals used to control pests in agriculture and public health. They are persistent, lipophilic (Log K_{ow} values of approximately 6), they can bioaccumulate and biomagnify through the food chain, and present potential for adverse effects in humans and the environment. Because of worldwide restrictions on their usage and production since the 1970s and 1980s (with exceptions for malaria control in some developing countries), other classes of pesticides, such as OPs, CBs, and Py, have been used as alternatives to OCs on a large scale. Despite their lower environmental persistence compared to OCs and their wider range of lipophilicity (Log K_{ow} ranging between −1 and 6), OPs and CBs can also accumulate in fatty matrices of animal origin. However, unlike most OCs, OPs and CBs are stored in animal fatty tissues for a shorter period of time (e.g., days). Nevertheless, these compounds present high acute toxicity. For these reasons, OPs and CBs have been gradually replaced by Py pesticides, known as non-persistent lipophilic compounds (Log K_{ow} ranging between 4.2 and 7) with low water solubility and lower toxicity to mammals and birds. Lastly, neonicotinoids are a more recent class of insecticides, currently among the most popular and widely used insecticides. They are much more toxic to invertebrates, like insects, than to mammals, birds, and other higher organisms. What has made neonicotinoid insecticides popular in pest control is their systemic nature and high water solubility, which allows them to be applied to soil and be taken up by plants. However, in recent years, serious concerns raised about the impact neonicotinoids may have on non-target organisms (e.g., honeybees).

Pesticides can also be classified according to their mode of action [1]. The most common mechanisms are listed below:

Enzymatic inhibition: This is, by far, the most important mechanism of toxicity. The toxicant reacts with an enzyme or a transport protein and inhibits its normal function. Typical pesticides included in this group are OPs and CBs which affect the nervous system by disrupting acetylcholinesterase (AChE) activity. However, some enzymatic inhibitors have little specificity and, therefore, different enzymes may be targeted.

Disturbance of the chemical signal transmission: The toxicant imitates the true chemical signal substances, by transmitting a signal too strongly, too long lasting, or at the wrong time. Typical examples are neonicotinoids, acting as agonists on AChE, inducing continuous excitation of the neuronal membranes, producing discharges leading to paralysis and cell energy exhaustion.

Generation of reactive molecules: The toxicant forms reactive radicals of intermediates which are very aggressive and can attack biomolecules non-selectively. The classical example of a free radical producing poison is the herbicide paraquat, which delivers an electron to molecular oxygen, further producing a reactive hydroxyl radical.

Disturbance of the chemical properties of cell membranes: For Py, the insecticidal action depends on the ability to bind to and disrupt voltage-gated sodium channels of insect nerves.

Despite undeniable benefits of pesticides in controlling pests and plant disease vectors and controlling nuisance organisms that harm human activities and structures, pesticide residues on foodstuffs, resulting from the use of plant protection products, may pose a risk to public health [2]. For this reason, a comprehensive legislative framework has been established in the European Union (EU) which defines maximum residue levels (MRLs) of pesticides in or on food and feed of plant and animal origin (Regulation EC 396/2005) [3]. These MRLs, derived after a comprehensive assessment of the properties of the active substance and the intended use of the pesticide, are based on good agricultural practice and the lowest consumer exposure necessary to protect vulnerable consumers.

This chapter critically reviews the most recent analytical methods (2010–18) for pesticide analysis (including practical aspects on sample preparation, analytical techniques, and quality control) in food of animal origin (e.g., milk and dairy products, meat, fish and seafood, honey, and eggs). As such, it updates the previous version edited in 2008 [4]. Current issues and future perspectives regarding pesticide analysis are also discussed.

7.2   SAMPLE PREPARATION

Despite the tremendous growth during recent years in the number of papers dealing with the determination of pesticide residues in various food of animal origin, no standard analytical procedures have been set yet for these compounds. This has resulted in a variety of analytical approaches for both sample preparation and instrumental analysis. Because of the low levels at which these compounds may be present in complex animal matrices, sample treatments often include several steps for exhaustive extraction and pre-concentration of the target compounds, usually followed by purification before final chromatographic separation, detection, and quantification. In most cases, the need for additional fractionation depends on the selected chromatographic approach, detection systems, and/or on the specific study goal. Tables 7.1 and 7.2 summarize relevant data on selected analytical procedures used for the determination of pesticide residues in a wide variety of food samples of animal origin. Furthermore, due to the specific physicochemical properties of various classes of pesticides, their determination may require specific analytical approaches, as has been presented below.

7.2.1   SAMPLE PRETREATMENT

Only freeze-drying and homogenization are usually carried out before extraction of biological samples. Alternatively, (semi)liquid (e.g., eggs or milk) samples may be freeze-dried and then treated as any other solid biological sample. Losses of volatile

TABLE 7.1

Overview of Typical Analytical Procedures Used for the Determination of Pesticides by GC–MS Techniques in Food of Animal Origin

Pesticide	Sample Type (g, mL)	Extraction Procedure	Clean-Up	Analytical Column	Instrumental Analysis	Recovery (%)	Ref.
22 OCs	Fish (5)	PLE	GPC	BPX-50 (30 m × 0.25 mm, 0.25 µm)	GC–EI-TOF-MS	71–113	[14]
11 OCs	Honey (0.5)	DLLME	–	Zebron ZB 5-MS (30 m × 0.25 mm, 0.25 µm)	GC–MS	35–83	[20]
18 OCs	Honey (20)	PLE	SPE aluminum–silica (methylene chloride:n-hex 1:1)	DB-5MS (30 m × 0.25 mm, 0.25 µm)	GC–IT-MS	60–85	[17]
15 OCs	Honey (5)	DLLME	–	Heliflex AT™-5MS (30 m × 0.25 mm, 0.25 µm)	GC–EI-IT-MS/MS (SIM)	81–117	[75]
OPs	Honey (10)	CME-UABE	–	VF-5MS (25 m × 0.25 mm, 0.25 µm)	GC–EI-IT-MS	90–107	[25]
OPs+Py	Beef, chicken, eggs, milk, fish (0.1–0.3)	SLE (n-Hex: DCM 2:1)	SPE basic Al+C18	DB-5ms (15 m × 0.25 mm × 0.1 µm)	GC–ECNI-MS/MS	27–128	[47]
Py	Bovine milk (5)	LLE-LTP (ACN)	–	OV-5 (15.0 m × 0.25 mm × 0.1 µm)	GC–ECD GC–MS/MS	87–100	[108]
50 Multiclass	Meat (2.5)	"dilute and shoot" (EtAc)	Florisil	DB-5ms (30 m × 0.25 mm × 0.25 µm)	GC–MS/MS GC–APCI-QTOF-MS	70–120	[6]
34 Multiclass	Milk and cream (4)	GPC	SPE PSA+GCB (acetone/toluene)	HP5 ms (30 m × 0.25 mm, 0.25 µm) coupled to BPX-50 (1.5 m × 0.15 mm, 0.15 µm)	GC × GC–TOF-MS	66–84	[44]

(Continued)

TABLE 7.1 (CONTINUED)

Overview of Typical Analytical Procedures Used for the Determination of Pesticides by GC–MS Techniques in Food of Animal Origin

Pesticide	Sample Type (g, mL)	Extraction Procedure	Clean-Up	Analytical Column	Instrumental Analysis	Recovery (%)	Ref.
174 Multiclass	Egg (5)	MSPD (Cyclohexane/EtAc)	GPC+SPE C18 and PSA (ACN)	HP5 ms (30 m × 0.25 mm, 0.25 μm)	GC–EI-QqQ-MS/MS	70–120	[9]
14 Multiclass	Milk (15)	QuEChERS	d-SPE MgSO$_4$, C18 and PSA	Rtx-5MS (30 m × 0.25 mm, 0.25 μm)	GC–MS	72–99	[57]
25 Multiclass	Beef, cattle liver, chicken, fish, and milk (5)	SLE (NaCl)	SPE AL2O3 (acetone-n-hex 1:1)	DB-1701 MS (30 m × 0.25 mm, 0.25 μm)	GC–EI-MS (SIM)	68–111	[8]
80 Multiclass	Egg and meat (5)	QuEChERS	n-Hex + d-SPE C18 (ACN)	HP-5MS UI (15 m × 0.25 mm, 0.25 μm) coupled to HP-5MS UI (15 m × 0.25 mm, 0.25 μm)	GC–QqQ-MS/MS	70–120	[40]
28 Multiclass	Honey (10)	SLE (Water)	SPE Florisil (methylene chloride:n-hex (80:20, v/v) and n-hex:acetone (60:40, v/v))	RT-5MS (30 m × 0.25 mm, 0.1 μm)	GC–IT-MS/MS	75–102	[48]
4 Multiclass	Honey (22)	LLE-LTP	Florisil	HP-5 (30 m × 0.25 mm, 0.1 μm)	GC–EI-MS (SIM)	93–101	[60]
53 Multiclass	Honey (1)	SLE (MeOH+citrate)	Florisil (EtAc:hexane 1:1.25)	HP-5MS (30 m × 0.25 mm, 0.25 μm)	GC–MS	30–128	[49]

(*Continued*)

TABLE 7.1 (CONTINUED)
Overview of Typical Analytical Procedures Used for the Determination of Pesticides by GC–MS Techniques in Food of Animal Origin

Pesticide	Sample Type (g, mL)	Extraction Procedure	Clean-Up	Analytical Column	Instrumental Analysis	Recovery (%)	Ref.
22 Multiclass	Honey (3)	MEPS	–	NST-05MS (30 m × 0.25 mm, 0.25 μm)	GC-EI-MS	82–114	[21]
80 Multiclass	Honey (5), honeybees (10), and pollen (2)	QuEChERS	d-SPE PSA and C18	DB-XLB (30 m × 0.25 mm, 0.25 μm)	GC-EI-TOF-MS	60–120	[76]
109 Multiclass	Meat and fish (10)	PLE	GPC	DB-5MS (30 m × 0.25mm, 0.25 μm)	GC-EI-MS (SIM)	63–108	[16]
9 Multiclass	Eggs (0.25)	MSPD with Florisil (ACN)	SPE-C18	DB-5ms (30 m × 0.25 mm × 0.25 μm)	GC-EI-MS	62–95	[118]
192 Multiclass	Meat of beef, pig, poultry (5)	QuEChERS	d-SPE	DB-5ms (15 m × 0.53 mm × 1 μm)	LPGC–MS/MS	70–120	[32]
Multiclass	Milk (20), honey (20)	SPE-C18+DLLME	–	CP-Sil 8 CB (50 m × 0.25 mm × 0.25 μm)	GC-MS	66–99	[101]

TABLE 7.2
Overview of Typical Analytical Procedures Used for the Determination of Pesticides by LC–MS Techniques in Food of Animal Origin

Pesticide	Sample Type (g, mL)	Extraction Procedure	Clean-Up	Analytical Column	Instrumental Analysis	Recovery (%)	Ref.
14 OPs	Honey, pollen, and honeybee (3)	QuEChERS	d-SPE C18 (ACN 1% acetic acid)	Kinetex C18 (100 × 4.6 mm, 5.0 µm)	HPLC–ESI–QqQ-MS/MS	–	[39]
6 OPs	Fish tissue (5)	"Dilute and shoot" (ACN)	SPE NH$_2$ cartridge	Atlantis T3 C18 (2.1 × 150 mm, 3.0 µm)	HPLC–ESI–QqQ-MS/MS	80–99	[84]
CBs+OPs	Meat, eggs, cheese, seafood (10)	QuEChERS	d-SPE	Synergi MAX-RP-80A (50 mm × 2 mm × 4 µm)	HPLC–MS/MS	81–111	[30]
CBs	High fat cheese (3)	QuEChERS	d-SPE	ZORBAX ECLYPSE plus RRHD (50 mm × 2.1 mm × 1.8 µm)	UHPLC–MS/MS	70–115	[119]
5 Neonicotinoids	Honey (2)	SPE-DLLME	–	Spherisorb ODS2 C18 (150 × 4 mm, 5.0 µm)	LC–APCI–IT-MS/MS	90–103	[23]
13 Neonicotinoids	Honey and honeybee (2)	Incubate (30 min) + SLE (ACN:EtAc (8:2, v/v)) (for honeybee) Incubate (30 min)+SLE (ACN:EtAc (8:2, v/v)) + ACN 20% of TEA) (for honey)	SPE Sep-Pak Alumina N Plus Long (for honeybee) SPE Strata X-CW (for honey)	Luna C18 (100 × 2.0 mm, 3.0 µm)	HPLC–ESI–QqQ-MS/MS	85–112	[85]

(Continued)

TABLE 7.2 (CONTINUED)

Overview of Typical Analytical Procedures Used for the Determination of Pesticides by LC–MS Techniques in Food of Animal Origin

Pesticide	Sample Type (g, mL)	Extraction Procedure	Clean-Up	Analytical Column	Instrumental Analysis	Recovery (%)	Ref.
7 Neonicotinoids	Honey (5)	DLLME	–	Zorbax Eclipse XDB-C18 (50 × 4.6 mm, 1.8 μm)	LC–multimode interface-QqQ-MS/MS	74–114	[18]
7 Neonicotinoids	Honey (5–15)	DLLME (Sample: 5 g) QuEChERS (Sample: 15 g)	– d-SPE PSA and MgSO$_4$	Zorbax Eclipse XDB-C18 (50 × 4.6 mm, 1.8 μm)	LC–multimode interface-QqQ-MS/MS	69–113 72–95	[19]
13 Neonicotinoids	Honeybees and bee products (15)	QuEChERS (ACN (2% triethylamine))	d-SPE PSA and MgSO$_4$ + SPE C18 (ACN (2% triethylamine))	Acquity HSS T3 column (10 cm × 2.1 mm, 1.8 μm)	UHPLC-ESI-QqQ-MS/MS	35–145	[86]
6 Neonicotinoids	Honey (5)	QuEChERS	d-SPE PSA and MgSO$_4$	Hypersilgold C18 (50 × 2.1 mm, 1.9 μm)	UHPLC–HESI-QqQ-MS/MS	75–114	[90]
4 Neonicotinoids	Honey (1)	QuEChERS	d-SPE PSA and MgSO$_4$	Zorbax Eclipse XDB-C18 (250 × 4.6 mm, 5.0 μm)	LC-ESI-QTRAP-MS/MS	–	[52]
10 Neonicotinoids	Honey (5)	QuEChERS	d-SPE PSA and MgSO$_4$	Synergi Fusion RP C18 (50 × 2 mm, 4.0 μm)	HPLC-ESI-QqQ-MS/MS	60–114	[51]
7 Neonicotinoids, 6 metabolites	Honey (2)	SLE (ACN:EtAc 4:1 + TEA)	SPE - Strata X-CW	LUNA C18 100A (100 mm × 2 mm × 3 μm)	LC-MS/MS	85–112	[85]
7 Neonicotinoids	Honey (15) Honey (0.5)	QuEChERS (ACN) DLLME (ACN: DCM 1:4)	d-SPE –	ZORBAX ECLYPSE XDB-C18 (50 mm × 4.6 mm ×1.8 μm)	HPLC-DAD	70–120	[69]

(Continued)

TABLE 7.2 (CONTINUED)
Overview of Typical Analytical Procedures Used for the Determination of Pesticides by LC–MS Techniques in Food of Animal Origin

Pesticide	Sample Type (g, mL)	Extraction Procedure	Clean-Up	Analytical Column	Instrumental Analysis	Recovery (%)	Ref.
7 Neonicotinoids	Muscle and liver (2.5)	PLE	SPE Oasis HLB (MeOH)	Symmetry Shield™ RP C18 (150 × 2.1 mm, 3 µm)	HPLC–ESI–QqQ-MS/MS	83–102	[15]
7 Neonicotinoids	Eggs, chicken, pork (1)	SLE (ACN)	SPE-HLB	HSS T3 (100 mm × 2.1 mm × 1.8 µm)	UHPLC–ESI-MS/MS	80–103	[110]
4 Neonicotinoids	Bovine milk (5)	SPE (DCM)	Diatomaceous earth – Chem Elute	Synergi Hydro-RP-C18 (250 mm × 4.6 mm × 4 µm)	HPLC–DAD	85–100	[114]
>350 Multiclass	Honey (2.5)	"Dilute and shoot" (ACN 1% formic acid, v/v)	–	Hypersil GOLD aQ C18 column (100 × 2.1 mm, 1.7 µm)	UHPLC–ESI-Exactive-Orbitrap-MS	60–120	[5]
83 Multiclass	Honey (5)	QuEChERS	–	Xterra C18 (150 × 2.1 mm, 5.0 µm)	HPLC–ESI-Q-Exactive-MS/MS; LC–Exactive-MS	53–114	[53]
115 Multiclass	Honeybees, honey, and bee pollen (1)	QuEChERS	d-SPE PSA and $MgSO_4$ + SPE Extra Bond C18 (ACN)	Zorbax Eclipse XDB-C18 (2.1 × 150 mm, 3.5 µm)	LC–ESI–QqQ-MS/MS	59–117	[87]
13 Multiclass	Honey (1)	LLE	–	Atlantis™ dC18 (100 × 2.1 mm, 3.0 µm)	HPLC–ESI-QTRAP-MS/MS	63–117	[94]
30 Multiclass	Honey (5)	QuEChERS	n-Hex	Kinetex C18 (50 × 2.1 mm,1.7 µm)	UHPLC–ESI-QqQ-MS/MS	34–86	[88]

(Continued)

TABLE 7.2 (CONTINUED)
Overview of Typical Analytical Procedures Used for the Determination of Pesticides by LC–MS Techniques in Food of Animal Origin

Pesticide	Sample Type (g, mL)	Extraction Procedure	Clean-Up	Analytical Column	Instrumental Analysis	Recovery (%)	Ref.
80 Multiclass	Honey (5), honeybees (10), and pollen (2)	QuEChERS	d-SPE PSA and C18	Nucleodur Sphinx RP-C18 (50 × 2 mm, 1.8 µm)	HPLC–ESI-QqQ-MS/MS	60–120	[76]
3 Multiclass	Honey (10)	QuEChERS	d-SPE PSA and MgSO$_4$	XTerra C18 (50 × 3 mm, 3.5 µm)	LC–ESI-QqQ-MS/MS	90–120	[70]
9 Multiclass	Honey (10)	QuEChERS	d-SPE PSA and MgSO$_4$	XTerra (50 × 3 mm, 3.5 µm)	LC–APCI-QqQ-MS/MS	70–112	[50]
30 Multiclass	Honey (1)	QuEChERS	d-SPE	Poroshell 120 EC-C18 (3 × 100 mm, 2.7 µm)	LC–ESI-QqQ-MS/MS	80–110	[33]
12 Multiclass	Honey (1.5)	QuEChERS	d-SPE PSA and MgSO$_4$	Luna C18 (250 × 4.60 mm, 5.0 µm)	LC–APCI-IT-MS/MS	87–98	[42]
7 Multiclass	Honey and pollen (1)	SPE Florisil (MeOH)	–	Symmetry Shield (150 × 2.1 mm, 3.5 µm)	HPLC–HESI-QqQ-MS/MS	90–102	[11]
54 Multiclass	Fish and honey (1)	QuEChERS	d-SPE PSA, anhydrous MgSO$_4$ and GCB	Kinetex XB-C18 (5 × 2.10 mm, 1.7 µm)	UHPLC–HESI-LTQ-Orbitrap-MS/MS	70–100	[45]
13 Multiclass	Fish muscle (3)	QuEChERS	n-Hex	Zorbax Eclipse C18 (50 × 2.1 mm, 1.8 µm)	LC–ESI-QTRAP-MS/MS	6–110	[95]
82 Multiclass	Fish (5)	SLE (ACN:water (80:20, v/v) 0.1% formic acid)	–	Acquity UHPLC BEH C18 (2.1 × 100 mm, 1.7 µm)	UHPLC–ESI-QTOF-MS/MS	–	[38]

(Continued)

TABLE 7.2 (CONTINUED)
Overview of Typical Analytical Procedures Used for the Determination of Pesticides by LC–MS Techniques in Food of Animal Origin

Pesticide	Sample Type (g, mL)	Extraction Procedure	Clean-Up	Analytical Column	Instrumental Analysis	Recovery (%)	Ref.
228 Multiclass	Fish (20)	QuEChERS	d-SPE	HSS T3 (50 mm × 2.1 mm × 1.8 µm)	UHPLC–MS/MS	70–120	[31]
18 Multiclass	Egg (5) and egg products (2.5)	QuEChERS	d-SPE PSA, C18, GCB and MgSO₄	Kinetex C18 (100 × 3 mm, 2.6 µm)	HPLC–ESI-QqQ-MS/MS	71–95	[82]
174 Multiclass	Egg (5)	MSPD (cyclohexane/EtAc)	GPC+SPE C18 (MeOH 0.5% acetic acid)	Zorbax Eclipse plus-C18 (2.1 × 100 mm, 3.5 µm)	HPLC–ESI-QqQ-MS/MS	70–120	[9]
80 Multiclass	Eggs and meat (5)	QuEChERS	n-Hex + d-SPE C18 (ACN)	Acquity UPLC BEH C18 (2.1 × 100 mm, 1.7 µm)	LC–Turbo spray-QTRAP-MS/MS	74–121 70–120	[40]
>350 Multiclass	Meat (2.5)	"Dilute and shoot" (ACN 1% formic acid, v/v)	–	Hypersil GOLD aQ C18 column (100 × 2.1 mm, 1.7 µm) Zorbax Eclipse Plus C18 (100 × 2.1 mm, 1.8 µm)	UHPLC–ESI-Exactive-Orbitrap-MS UHPLC–ESI-QqQ-MS/MS	70–120	[6]
128 Multiclass	Meat (2)	QuEChERS	d-SPE ODS and MgSO₄	Hypersil C18 (150 × 2.1 mm, 5.0 µm)	LC–ESI-QqQ-MS/MS	70–100	[91]
192 Multiclass	Beef, pig, and poultry meat (5)	QuEChERS	d-SPE	BEH C18 (100 mm × 2.1 mm×1.7 µm)	UHPLC–MS/MS	70–120	[32]
118 Multiclass	Milk (10), liver (1), and muscle tissue (5)	QuEChERS	d-SPE PSA and C18 (only for liver)	Hypersil Gold AQ C18 (2.1 × 50 mm, 1.9 µm)	UHPLC–HESI-Exactive-Orbitrap-MS	–	[43]

(Continued)

TABLE 7.2 (CONTINUED)

Overview of Typical Analytical Procedures Used for the Determination of Pesticides by LC–MS Techniques in Food of Animal Origin

Pesticide	Sample Type (g, mL)	Extraction Procedure	Clean-Up	Analytical Column	Instrumental Analysis	Recovery (%)	Ref.
48 Multiclass	Milk (10)	SPE C18 (MeOH)	–	Acquity UPLC™ BEH C18 (100 × 2.1 mm, 1.7 µm)	UHPLC-ESI-QqQ-MS/MS	<30–129	[10]
28 Multiclass	Milk, baby food, and meat (1)	"Dilute and shoot" (MeOH 1% formic acid, v/v)	–	ZIC-pHILIC (150 mm × 4.6 mm, 5 µm)	LC-ESI-QqQ-MS/MS	70–120	[83]
31 Multiclass	Milk (15)	LLE (ACN)	SPE C18 (ACN)	SB-C18 (2.1 × 150 mm, 5.0 µm)	HPLC-ESI-QTRAP-MS/MS	62–107	[41]
40 Multiclass	Milk, yogurt, and cheese (5)	SLE (ACN:EtAc:acetic acid (49.5:49.5:1, v/v/v))	Low temperature+SPE Oasis HLB (MeOH: ACN:ammonium hydroxide (47.5:47.5:5, v/v/v))	ZORBAX SB-C18 (100 × 2.1 mm, 3.5 µm)	LC-ESI-QqQ-MS/MS	67–107	[55]
36 Multiclass	Milk and other matrices (2–10)	QuEChERS	d-SPE PSA and C18	Acquity UPLC™ BEH C18 (100 × 2.1 mm, 1.7 µm)	UHPLC–HESI-Exactive-Orbitrap-MS	–	[97]
333 Multiclass	Baby food (meat-based food and milk-based formulae) (5)	QuEChERS (ACN:water (84:16, v/v))	–	Accucore C-18 aQ (100 × 2.1 mm, 2.6 µm)	UHPLC–HESI-Q-Orbitrap-MS/MS	80–111	[54]
Transformation products	Honey and meat (2.5)	"Dilute and shoot" (ACN 1% formic acid, v/v) or QuEChERS	d-SPE PSA and GCB	Hypersil GOLD aQ C18 column (100 × 2.1 mm, 1.7 µm)	UHPLC-ESI-Exactive-Orbitrap-MS	–	[46]

compounds might occur during freeze-drying. The samples should be stored in an inert material (e.g., glass containers) to avoid possible sorption of the pesticides into the storage medium or to prevent contamination of the sample from the storage medium (e.g., with phthalates). Most commonly, samples are kept in glass or Teflon containers. Screw caps should be lined with solvent-rinsed aluminum foil or with Teflon inserts. Polyethylene or other plastics should be avoided, unless a thorough validation has been previously carried out and contamination has been ruled out. Alternatively, solid samples can be wrapped in aluminum foil and then inserted into plastic bags. Preferably, samples should be frozen (−20 or −80°C) as soon as possible after sampling. Freezing and storage of multiple small samples suitable for analysis, rather than larger masses, is recommended to avoid multiple freezing/thawing of tissue and to reduce sample handling, which, in turn, reduces the potential for contamination.

In general, for pesticide residue analysis, maintaining sample tissues in their original wet state is regarded as the most appropriate approach for preparing samples. Instead, homogenized samples should be mixed with a desiccant such as sodium sulfate, Celite, or Hydromatrix to bind water. The desiccant must be free of analytes, e.g., by heating at a high temperature in the case of sodium sulfate or by pre-extraction (Celite or Hydromatrix).

7.2.2 EXTRACTION TECHNIQUES

For food samples of animal origin, the selection of the most suitable extraction technique mostly depends on the nature of the matrix investigated; different procedures are used for solid and liquid samples. The amount of sample to analyze largely depends on the anticipated sample contamination level and on the sensitivity provided by the detection technique.

Solid samples can be extracted by different techniques (Tables 7.1 and 7.2). The key-points are the use of adequate solvent mixtures (e.g., low boiling point to facilitate concentration), adequate exposure time between solvents and the sample matrix, and limitation of sample handing steps, e.g., avoid filtration steps by using semi-automated extraction systems (e.g., pressurized liquid extraction, PLE). Contamination by carry-over (e.g., high pesticide residues in the previous analyzed sample) should be avoided by thoroughly cleaning any lab equipment from batch to batch. Purity of extraction solvents should also be guaranteed by only using pesticide-grade or high-purity glass-distilled solvents, to avoid the presence of contaminant residues in the solvent volume concentration step.

Different extraction methods have been applied for the determination of pesticides extraction in food from animal origin. The conventional solid–liquid extraction (SLE) technique is still widely used, performed by homogenizing the solid sample with a variety of solvents, e.g., acetonitrile (ACN) 1% formic acid, v/v [5,6] or McIlvain solution [7], partitioning of analytes from the matrix to the solvent. Such techniques allow quantitative extraction of pesticides directly from dry matrices or after drying the original fresh sample with anhydrous Na_2SO_4 [8]. Alternatively, the use of matrix solid-phase dispersion (MSPD) in different variants is a suitable choice, which results in an intimate contact between the sample components and the sorbent particles and in a more efficient retention of impurities [9]. For liquid or viscous samples, solid phase extraction (SPE) can be employed for the extraction of pesticide residues from milk [10] and honey [11]. SPE is based on the elution of analytes with appropriate solvents, such as methanol (MeOH), which were previously retained in the cartridge [12]. However, high solvent consumption and laborious sample handling can be expected when using these methodologies. Therefore, an automated extraction technique, such as PLE, has been applied for the extraction of milk, egg and meat [13,14], liver and fish [15,16], and honey [17]. The use of elevated temperatures and pressure increases diffusion and desorption rates. Although PLE is an extraction technique that provides good recoveries and reduces the use of solvents, it has high costs and the preparation of the extraction cells is tedious and time-consuming. Moreover, the preparation of an homogeneous dry sample from wet tissue for PLE can be challenging due to the limited volume of PLE extraction vials.

Miniaturized techniques have also been applied for the extraction of pesticide residues from honey samples by dispersive liquid–liquid microextraction (DLLME) [18–20], microextraction in packed sorbent (MEPS) [21], and solid-phase microextraction (SPME) [22]. Moreover, Campillo et al. (2013) have combined SPE extraction with DLLME for the extraction of neonicotinoids from honey [23]. DLLME is a simple and rapid extraction technique that consumes a low volume of organic solvent. On the other hand, MEPS can be considered a green extraction method because the sorbent can be reused more than 40 times with minimum loss of extraction efficiency [21]. However, due to the complexity of the honey-matrix, is often necessary to perform a previous step, such as liquid–liquid extraction (LLE). Another environmentally friendly method has been developed by Campillo et al. (2012) for the determination of organotin compounds in honey samples using headspace–SPME (HS–SPME) [22]. Although conventional SPME does not require solvent or multi-stepped cleanup/concentration procedures, adsorption efficiency can be affected by complex matrices, especially when the fiber is directly immersed in the sample. Other non-classical techniques have been employed for the analysis of pesticides, such as purge and trap [24], and coacervation microextraction ultrasound-assisted back-extraction technique (CME–UABE) [25], in honey and animal tissues.

Despite the advantages of these extraction techniques, none of them have overcome critical flaws and practical limitations to enable their widespread implementation. For example, PLE requires time-consuming manual steps and specialized items that necessitate thorough cleaning after use, the sample throughput is not always optimal, and instruments and maintenance are often expensive. Other techniques, such as MSPD and SPME, do not provide a wide enough analytical scope within a single procedure, or enough selectivity. In 2003, Anastassiades et al. developed a method for the extraction of multi-class pesticide residues from fruits and vegetables, naming it QuEChERS (Quick, Easy, Cheap, Effective, Rugged, and Safe) [26]. The optimized original method consisted of the extraction of the homogenized food sample with ACN, followed by the addition of salts (e.g., anhydrous $MgSO_4$, NaCl). The extract was then cleaned up by dispersive solid phase extraction (d-SPE) with primary secondary amine (PSA), C18, and/or graphitized carbon black (GCB) sorbents, with the aim of eliminating possible interfering compounds from the food extract (e.g., organic acids, certain polar pigments, sugars, etc.). After centrifugation, the supernatant was finally ready for analysis in the GC–MS system [27]. Although the original QuEChERS method is not likely to be applicable for the extraction of more lipophilic pesticides from high fatty samples, it is considered optimal for their extraction from low fatty foods and for the extraction of polar and semi polar pesticides from a wide variety of fatty foods [28]. Subsequent adjustments/modifications to the original method were developed to improve its performance for some more complex analytes and foodstuffs (e.g., foods of animal origin which contain proteins, lipids, and saturated and monounsaturated fatty acids), to avoid pesticide degradation and to diminish the matrix effects [29–33]. For these reasons, the QuEChERS method and its following modifications, with its unique combination of extraction and cleanup procedures, solvents, salts, and sorbents, is currently considered the standard sample preparation procedure for pesticides analysis in foodstuffs [27].

Another approach for the determination of multiple classes of pesticides in different food matrices (such as honey [5], meat [6,34], baby food [35] and liver [36]) is the "dilute and shoot" method, by using ACN (1% formic acid, v/v) as extraction solvent. A critical factor is the extraction time for properly mixing the target compounds with the solvent, and 1 h is usually enough to allow an exhaustive extraction of the target compounds. Considering that more than ten samples could be simultaneously analyzed in the "dilute and shoot" extraction, the sample throughput can be considered suitable to apply this approach in routine analysis [37].

7.2.3 Cleanup and Fractionation

The non-selective nature of the exhaustive extraction procedures and the complexity of sample matrices result in complex extracts that often require further purification. SPE and d-SPE are currently the most commonly used purification steps [38]. Several sorbents have been used depending on the type of matrix, such as C_{18} [39–41] in egg, meat, honey, and milk samples, PSA [36,42,43] in honey, liver, milk, and meat samples, GCB [44–46] in fish, honey, milk, cream, and meat matrices, alumina [8,47] in several matrices such as meat, fish, egg, milk, and liver matrices, Florisil [35,48,49] in honey and baby food samples, and anhydrous $MgSO_4$ [50–52], mostly in honey matrices. In some cases, depending on the type of matrix and the determination technique, there is no need to use these clean-up stages [53,54]. Nowadays, prepacked columns for the SPE and sorbents for d-SPE extraction are commercially available, OASIS HLB being one of the most used cartridges [55]. Typical purification and fractionation procedures have been summarized in Tables 7.1 and 7.2.

For biological samples, including food of animal origin, lipid elimination should be accomplished by non-destructive methods, before chromatographic analysis. Traditionally, *n*-hexane saturated with ACN is used for the extraction of fat content. However, it is not effective during the extraction of weak and non-polar compounds. To overcome this problem, the low-temperature cleanup could be an alternative extraction to remove most of lipids co-extracted in ACN [55]. Because the fatty substances have lower melting points than the solvent, freezing centrifugation is a simple approach to remove the fatty co-extracted interferents [38]. Gel permeation chromatography (GPC) is a non-destructive treatment applied for lipid elimination. GPC is mainly performed either in automated systems [14,16] or by gravity flow columns [9]. The current use of pre-packed polystyrene-divinyl benzene based high performance GPC columns has resulted in higher separation efficiencies, improved reproducibility, and lower solvent consumption as compared to manually packed columns. Satisfactory isolation of the target compounds from the co-extracted organic material after single GPC analysis has been achieved for some samples that contain limited amounts of lipids like fish and meat (approx. 2% of lipid). However, for more complex matrices, such as milk, cream, and egg (with more than 40% of lipid), GPC followed by a further cleanup step by SPE may be required to remove remaining low molecular weight lipids, waxes, and pigments [9,44].

7.3 ANALYTICAL TECHNIQUES

The choice of the best analytical technique for the detection and quantification of pesticide residues strongly depends on the analyte's polarity. Compounds with high log K_{ow} (e.g., OCs, Py, and most OPs) are non-polar and are preferably analyzed by gas chromatography (GC), while polar compounds, such as CBs, neonicotinoids, and some OPs, are amenable by liquid chromatography (LC). GC and LC can be coupled to several detectors such as electron capture detector (ECD) [56–62], nitrogen–phosphorus detector (NPD) [24,49,63,64], photodiode array detector (PAD) [65–68], diode array detector (DAD) [23,69–71], and variable wavelength detector (VWD) [72], depending on the targeted compounds. However, EU requirements indicate that all confirmatory methods for pesticide residues in food must use mass spectrometry (MS) detection [73]. Nowadays, LC and GC coupled to MS [20] or to tandem mass spectrometry (MS/MS) [18] are the most widely used analytical techniques for the determination of pesticide residues in food of animal origin. Low-resolution mass spectrometry (LRMS) usually provides better robustness and lower limits than high-resolution mass spectrometry (HRMS). However, the number of pesticides analyzed by HRMS is commonly higher than those analyzed by LRMS. For instance, Gómez-Pérez et al. (2014) determined more than 350 compounds in meat by LC coupled to an Exactive-Orbitrap detector (HRMS) [6]; meanwhile, García-Chao et al. (2010) quantified seven pesticides in honey and pollen by LC coupled to a triple quadrupole (QqQ) analyzer (LRMS) [11]. Moreover, retrospective analysis can be performed when HRMS is applied.

7.3.1 GAS CHROMATOGRAPHY COUPLED TO MASS SPECTROMETRY

GC–MS is the most commonly applied technique for the determination of volatile and semi-volatile analytes of low-medium polarity [74]. Furthermore, it has become a powerful technique for the quantitative determination at low levels of contaminants in complex matrices, such as honey and meat [12].

7.3.1.1 Low Resolution Mass Spectrometry Applications

There are three modes of GC–MS available, electron impact (EI), positive chemical ionization (PCI), and negative chemical ionization (NCI), with the first one being the most widely used. Due to its adequate sensitivity and selectivity, GC–MS in selected ion monitoring (SIM) is commonly used in the determination of different classes of pesticides in animal tissues [8,16,62], milk [8,57], fish [8,16], and honey [20–22,24,49,60,75].

GC–MS/MS commonly provides higher sensitivity and selectivity, as well as a higher degree of certainty than GC–MS in SIM mode, because it involves at least two stages of mass analysis, separated by a fragmentation step. The most common tandem mass spectrometers for GC, namely ion trap (IT) [17,25,48,75] and QqQ [9,40,47], are important tools in food analysis. GC–IT-MS/MS has been employed in honey samples for the analysis of Ops [25], OCs [17,75] (Figure 7.1), or for the determination of multiclass pesticide as OCs and Ops [48]. However, GC–QqQ-MS/MS has been less frequently used to determine pesticide residues in meat [40,47], fish, egg [9,40,47], and milk samples [47].

In the aforementioned GC–MS applications, fused capillary columns with bonded phases of different polarities (non-polar NST-05 MS; low polar HP-5, DB-5MS, ZB-5, VF-5MS, DB-5MS, ZB-5MS, HP-5MS, RTX-5MS, AT-5MS; and low/mid polar DB-1701) have been used at various lengths (15–30 m), internal diameters (0.25–0.32 mm), and film thickness (0.10–0.50 μm) (Table 7.1).

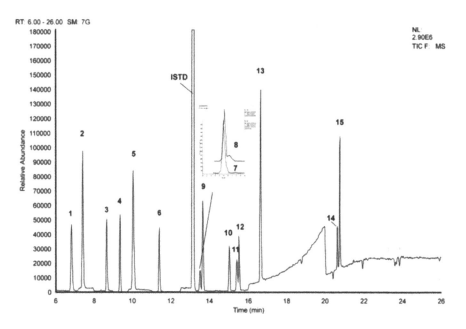

FIGURE 7.1 GC–IT/MS (SIM) chromatogram of organochlorine pesticides spiked at 1 μg/L (corresponding to 20 ng/g) in pooled honey sample after DLLME. Peak identification: (1) etridiazole; (2) chloroneb; (3) propachlor; (4) trifluralin; (5) hexachlorobenzene; (6) chlorothalonil; (7) chlorpyrifos; (8) cyanazine; (9) DCPA; (10) trans-chlordane; (11) cis-chlordane; (12) trans-nonachlor; (13) chlorobenzilate; (14) cis-permethrin; (15) trans-permethrin. (Reprinted with permission from Zacharis, C. K. et al. Dispersive liquid–liquid microextraction for the determination of organochlorine pesticides residues in honey by gas chromatography–electron capture and ion trap mass spectrometric detection. *Food Chem.* 2012, 134(3), 1665–1672.)

7.3.1.2 High Resolution Mass Spectrometry Applications

Even though LRMS has been used for the determination of pesticide residues in food of animal origin, it is sometimes not selective enough and only unit mass resolution is provided. Besides, in one single run, a "limited" number of compounds can be simultaneously determined [37]. Therefore, HRMS has been applied to improve the requirements mentioned above. This technique provides some advantages, such as non-target approach, post-target analysis, and the (theoretically) unlimited determination of compounds. A great variety of pesticides could be detected in food of animal origin. Therefore, HRMS is a powerful technique to cover a wide range of pesticides, reducing the cost of the analysis and improving the cost-effectiveness of the developed method [37].

The time-of-flight detector (TOF) has been applied in different matrices like fish [14] and honey, honeybee, and pollen [76] for the determination of multiple classes of pesticides. Both of them operated in the EI mode and employed a BPX-50 column [14] and a DB-XLB column [76]. On the other hand, to improve the drawbacks of the TOF detector, a hybrid system, consisting of a quadrupole/TOF (Q-TOF), has been applied for the determination of 50 pesticides in meat samples [34]. In this study, atmospheric pressure chemical ionization (APCI) has been used as an alternative to conventional ion sources applied in GC (EI or CI), minimizing the matrix effect observed in applying other ionization techniques. An overview of the recent literature for the application of GC coupled to TOF and Q-TOF in the determination of pesticides residues have been performed by Elbashir and Aboul-Enein (2017) [74].

7.3.1.3 Other GC Applications

Two-dimensional or comprehensive GC (GC × GC) involves the separation of target analytes by two orthogonal capillary columns in which the first column has normally larger diameter and longer length than the second column. Hayward et al. (2010) determined 34 compounds in milk and cream by GC × GC coupled to a TOF detector, the first column being a HP5-MS (30 m × 0.25 mm i.d., 0.25 µm film thickness) and the second one a BPX-50 (1.5 m × 0.15 mm i.d., 0.15 µm film thickness) [44]. The observed peaks were higher and sharper (smaller peak width) than those obtained by conventional GC approaches, offering sensitivity enhancement. Besides, the second column could separate the co-eluting peaks from the first column [77].

Several pesticides have optically active or chiral isomers (e.g., cis/trans-imiprothrin, resmethrin, or tetramethrin) [47]. As a consequence, biotransformation reactions in biological samples can result in non-racemic patterns in environmental samples. Dallegrave et al. (2016) analyzed the isomeric composition of some Py, cyfluthrin, and cypermethrin in chicken, fish, eggs, milk, and beef by GC coupled to a QqQ [47]. Crucial for chiral analysis is the availability of chiral capillary GC columns such as those with various cyclodextrins chemically bonded to a polysiloxane. While use of chiral GC separations is not part of routine pesticide analysis, it is a well-developed technology that is relatively easy to implement in existing GC–MS instruments.

7.3.2 LIQUID CHROMATOGRAPHY COUPLED TO MASS SPECTROMETRY

LC is the preferred method when dealing with thermally labile, ionic, and polar compounds (such as CBs, neonicotinoids, urea, and phenoxy herbicides or benzimidazoles). Some reviews focused on the applications of LC–MS (LRMS and HRMS) for the determination of pesticide residues in environmental and food analysis [78] or in food [79] have recently been published. This technique meets the EU legislation requirements [73,80] to ensure appropriate selectivity and sensitivity.

7.3.2.1   Low Resolution Mass Spectrometry Applications

LC has been coupled with different types of MS detector for the monitoring of pesticide residues in food of animal origin and QqQ is the most common analyzer [6,9–11,15,18,19,35,39,50,51,55,70,76,81–91]. Besides, other detectors have been employed in the quantification of pesticides such as single quadrupole (Q) [92], IT [13,23,42,93], and QTRAP [40,41,52,94,95]. Of the different LC–MS methodologies, electrospray ionization (ESI) and APCI are the most used. Both techniques allow the detection of pesticides in food of animal origin in positive ionization mode [23,90,94] or both ionization modes, negative and positive [11,42,91]. García-Chao et al. (2010) and Kasiotis et al. (2014) determined fipronil and its metabolites by LC coupled to a QqQ in negative ionization mode in honey [11,87]. In addition, Lichtmannegger et al. (2015) quantified 80 multiclass pesticides in egg and meat by LC coupled to QTRAP [40]. Most of these applications involve the use of the C18 reversed phase column at different sizes, e.g., (100 × 2.1 mm i.d., 1.7 µm particle size) [10] or (50 × 2.0 mm i.d., 4.0 µm particle size), although other reversed phase columns have been employed like C8 (Table 7.2) [93]. The elution most widely used is a gradient of methanol–water, methanol–aqueous solution of formic acid, or methanol–aqueous solution of ammonium formate. Danezis et al. (2016) applied a QqQ analyzer coupled with a hydrophilic interaction liquid chromatography (HILIC) column using a gradient acetonitrile–water-aqueous ammonium formate for the analysis of cyromazine, amitrole, and the metabolite of propylenethiourea simultaneously with plant hormones, veterinary drugs, and mycotoxins in milk, baby food, and meat [83].

7.3.2.2   High Resolution Mass Spectrometry Applications

The Orbitrap mass analyzer has been applied for the determination of several types of contaminants or residues, including pesticides. For instance, 118 alkaloids, CBs, OPs, and veterinary drugs have been analyzed in several matrices, such as milk, liver, and muscle tissue [43]; 36 pesticides, mycotoxins, antibiotics, and veterinary drugs in milk, and bakery products [7,97]; pesticides, antibiotics, and veterinary drugs in honey [5,53], meat [6], baby food [35], meat [46]; rodenticides and transformation products in liver [36].

Most of the literature reports the use of ESI in both ionization modes (positive and negative) [5–7,35,36,46,54] or acquired the data employing only the positive mode [38,45,53,97].

However, comparing to the LRMS analyzers, Orbitrap provides lower sensitivity, but better selectivity. Hybrid analyzers can even improve this last characteristic, bearing in mind that precursor ions can be selected. They are based on a hybrid system that couples two analyzers such as a quadrupole with an Orbitrap (Q–Exactive) or TOF [53,54]. Farré, Picó, and Barceló (2014) used a combination of a linear ion trap with an Orbitrap analyzer (LTQ Orbitrap) for the determination of 54 pesticide residues in fish and honey matrices [45]. Moreover, Nácher-Mestre et al. (2013) analyzed 82 compounds in fish by a hybrid system consisting of a Q–TOF [38].

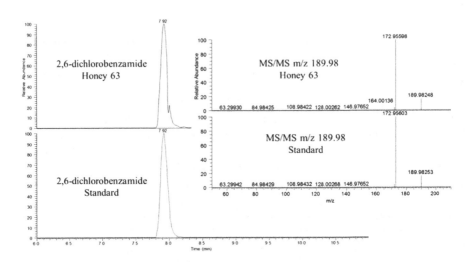

FIGURE 7.2 Identification of 2,6-dichlorobenzamide in honey ([M +H]⁺= 189.9821, *m/z* ±8 ppm). (Reprinted with permission from Cotton, J. et al. High-resolution mass spectrometry associated with data mining tools for the detection of pollutants and chemical characterization of honey samples. *J. Agric. Food Chem.* 2014, 62(46), 11335–11345.)

On the other hand, HRMS allows post-target analysis to search for additional compounds, avoiding additional analyses. In that sense, Nácher-Mestre et al. (2013) performed a retrospective approach in order to detect and tentatively identify other compounds [38]. In addition, the need to analyze transformation products has increased in the last few years because, sometimes, they are also toxic compounds. HRMS is a powerful tool for the determination of target and non-target compounds at low concentration levels in complex matrices. For instance, a retrospective analysis was applied to identify transformation products of chlorpyrifos (one of the most widely used OP insecticides) in different food matrices, such as honey and meat (United States, EPA, 2013) [46]. Furthermore, Cotton et al. (2014) detect a metabolite of dichlobenil (banned in France since 2010) in lavender honey samples, 2,6-dichlorobenzamide (Figure 7.2) [53].

7.4 INTERNAL QUALITY CONTROL

Quality control is a part of quality management focused on individual measures used to fulfill the quality requirements [98], including internal and external activities.

Internal quality control (IQC) is a set of procedures undertaken by laboratory staff for the continuous monitoring of an operation and the measurements to evaluate whether results are reliable enough to be released [99]. On the other hand, *external quality control* (EQC), also known as proficiency testing (PT), evaluates a laboratory's performance against both its own requirements and similar laboratories. However, both activities are strictly related, because laboratories that receive poor PT scores often encounter analytical problems and show a lack of attention to their internal quality control. In the framework of using validated analytical methods by internationally accepted criteria [80], IQC activities must be suitable to ensure the performance of the analytical method in routine practice. This goal is achieved when the quality level of the results allows the detection of unexpected or unwanted modifications of the method capabilities that have occurred during sample analysis.

7.4.1 BASIC ACTIVITIES OF IQC IN PESTICIDE RESIDUE ANALYSIS

1. The laboratory must be divided into well-defined working areas to avoid contamination of standards, samples, and extracts.
2. All equipment (measurement instruments, balances, flasks, pipets, etc.) must be regularly cleaned and calibrated.
3. Personnel must be qualified, motivated, and continuously trained.
4. Sampling procedures require well-trained staff. Samples must be transported to the laboratory and processed as soon as possible. If not possible, they must be stored in adequate conditions to assure their stability.
5. The quality of reagents must be appropriate for the performed test. Individual primary calibration solutions, generally 100–1000 mg/L, should be prepared by weighing not less than 10 mg of the pesticide using a five-decimal place balance.
6. The method validation process should include the estimation of uncertainty, essential in establishing the comparability of the measurements.

7.4.2 IQC MEASURES

IQC measures [100] must be included in each analytical batch to evaluate whether or not the processed batch satisfies the pre-set quality criteria and its results can be accepted.

1. *Solvent blanks* eliminate false positives by cross-contamination or carryover in the extraction process, used instruments, or chemicals. *Procedural blanks* should be analyzed to detect interferences of sample matrix and laboratory background contamination.
2. *Certified Reference Materials* (CRMs) or *home-made reference materials* (checked for stability and homogeneity), prepared by spiking a blank sample with 5–10% of the targeted pesticides, are used to evaluate the accuracy of the performed analysis and to investigate possible variation between batches of samples. CRMs should be analyzed in every batch of samples by applying the analytical method and providing additional information about instrument performance (instrument sensitivity, column performance, etc.). The variation in the data obtained from the analysis of the quality control sample is normally monitored on a quality control chart. However, CRMs related to pesticide determination in food from animal origin are still scarce, and they show limitations, such as a limited number of certified pesticides, wide uncertainty ranges, higher concentrations than the current range of interest, or a physical state not matching routine samples.

3. A *calibration curve* must be carried out for every batch of samples, although single-level calibration can be performed if the response in the sample is close to the response of the standard. A minimum of three standard concentrations, prepared in solvent or in extract of blank matrix, should be used for each pesticide [80], although more levels are desirable (usually five to eight points). The lowest calibration point should be equal to or lower than the limit of quantification (LOQ). The fit of the calibration function (linear, quadratic, etc.) must be checked to ensure that it is satisfactory. Individual points of the calibration curve should not differ more than ±20% (±10% if the MRL is exceeded or approached).

4. *Recoveries* and *accuracy* of spiked matrices or blanks (by adding all analytes to a matrix/blank at a concentration equal to the LOQ) should be used to check the extraction efficiency of each batch of samples. Acceptable recovery ranges have been set in relation to the amount of the analyte(s) added to the blank matrix. For instance, if a CRM is used at 10 µg/kg, a range from −20 to +10% has been set as acceptable by the Commission Decision 2002/657 [73], whereas, when spiked samples are evaluated, acceptable recoveries within the range 70–120% are accepted by current SANTE guidelines [80].

5. *Replicated samples* provide a less formal approach than quality control samples to check for drifts. The results obtained from the analysis of these samples must be comparable, considering the uncertainty of the method. They are included in the batch at intervals of a certain number of samples (e.g., every ten samples) and their presence is known by the analyst.

6. *Blind samples* are replicate samples placed in the analytical batch without being known by the analyst. They are complementary to replicated samples, providing information about the analyst's proficiency.

7.5 APPLICATION TO REAL SAMPLES

7.5.1 MULTIRESIDUE ANALYSIS OF PESTICIDES

Despite their numerous merits, pesticides are considered one of the most dangerous environmental contaminants. These chemicals can be transferred from plants to animals via the food chain, leading to their bioaccumulation in food of animal origin, such as meat, fish, eggs, milk, and honey [2]. Since diet is the main source of chronic exposure to low doses of pesticides, humans are mainly exposed to these chemicals through diet [2]. Consequently, there is an increasing demand worldwide for reliable and sensitive analytical methods for multi-residue surveillance of pesticides [101]. Fast, simple, efficient, and cost-effective analytical methods for pesticides are thus needed in food safety monitoring programs to provide high sample throughput and accurate results [32].

In the next paragraphs, the most recent analytical methods for the determination of pesticides in products of animal origin are reported for five main groups of pesticides, namely OC pesticides, OP pesticides, CBs, Py, and neonicotinoids.

7.5.2 ORGANOCHLORINE PESTICIDES

The determination of OC residues in biological matrices (including food of animal origin) has been a major subject because of their potential risk for human health, persistence, and tendency to bioaccumulate [102]. OC pesticides have been heavily used in the past, but they are now outdated and, with few exceptions, are banned in most industrialized countries. Since the environmental levels of various OCs are on a decreasing trend, the studies focusing on the determination of OCs in food matrices often deal with low analyte levels and many interfering substances. The extraction of OCs is usually performed by SPME, stir bar sorptive extraction (SBSE), and MSPD [102,103]. In addition, due to their good thermal stability, GC coupled to ECD or to MS or MS/MS is considered as the preferred method for the analysis of OC pesticides residue in food of animal origin.

7.5.3 ORGANOPHOSPHORUS PESTICIDES

OPs represent a group of compounds with a wide range of physicochemical properties, but all having the same mode of action, which has enabled the development of compounds against ecto- and endoparasites, active in the vapor phase or in the soil, and used against a wide range of crop and public health pests. Although most OPs are insecticides, some of them have also fungicidal or herbicidal activities. Many OPs present high acute toxicity and are suspected of carcinogenic, mutagenic, and endocrine disruptive effects. In the past decade, most published studies based their analysis on the use of GC coupled to various detectors. GC–ECD and GC–atomic emission detector (AED) have been applied to the determination of OPs in honey, and GC–NPD and GC–flame photometric detector (FPD) were applied to the analysis of OPs in dairy products, meat, and honey [4].

More recently, analytical methods for multi-class analysis of selected pesticides, including OPs, based on QuEChERS extraction and d-SPE cleanup, and on analysis using low pressure (LP)-GC–MS/MS and LC–MS/MS were developed and applied for several foods of animal origin (e.g., fish, seafood, eggs, dairy products, meat, and poultry) [30,32,104].

7.5.4 CARBAMATES

CB insecticides are commonly used as surface sprays or baits in the control of household pests. Their mode of action is similar to that of the OP insecticides, but their inhibitory effect on cholinesterase is shorter. CBs can vary in their spectrum of activity, mammalian toxicity, and persistence, but they are generally unstable compounds that break down in the environment within weeks or months. The presence of CB residues in food of animal origin, as mainly attributed to the contamination of feedstuff, is a matter of concern because of their possible adverse effects in humans. Since, in some cases, CB residues appear in samples at trace levels, an efficient sample preparation is a key aspect in the development of appropriate analytical methods. In addition, foodstuffs of animal origin are generally considered complex matrices, where lipid, carbohydrate, and protein components could often interfere in CB analytical determination [105,106]. For this reason, sample pre-treatments to remove the high molecular-mass lipid from the sample are often required.

A modified QuEChERS procedure using Supel™ QuE Z-Sep+ as sorbent, followed by ultra-high-performance liquid chromatography with MS/MS (UHPLC–MS/MS), has been used for the determination of CB residues in dairy products (i.e., high-fat cheese) [106]. This approach has also been successfully used in the determination of residues of seven CB pesticides in milk [107], where different d-SPE sorbents (e.g., PSA and C18) were tested. A modified version of the QuEChERS approach was used by Holmes et al. (2015) to quantify 185 pesticide residues, including CBs, in samples of fresh salmon. Their validated QuEChERS method used ethyl acetate (EtAc) as extraction solvent and involved two freezing steps and a C18 d-SPE for lipid removal. The quantification was performed using a combination of LC–MS/MS and GC–MS/MS. With their approach, 708 salmon samples collected as part of the U.S. Department of Agriculture's Pesticide Data Program were analyzed over 12 months [31]. A sample analysis method based on QuEChERS and liquid chromatography–tandem mass spectrometry was validated by Chung and Chan (2010) in the analysis of ultra-trace levels of 24 CB residues in a total diet study involving diversified food types (including, eggs, dairy products, meat, poultry, and seafood) [30].

7.5.5 PYRETHROIDS

Py are a class of synthetic organic insecticides derived from pyrethrins. They have been used worldwide since the 1980s because of their high level of effectiveness and low toxicity compared to other insecticides, such as OPs and CBs. They are currently among the major kinds of pesticides used worldwide and constitute the majority of commercial household insecticides. Py are hydrophobic molecules which tend to deposit in the lipid fraction of food and, because fatty food of animal origin (e.g., eggs, milk, meat, fish) is an essential component of the human diet, the possibility of bioaccumulation of these pesticides in fatty food can present a hazard to humans [47].

The complexity of these samples requires efficient extraction and cleanup procedure prior to analysis, in order to isolate the targeted pesticides and allow reproducible results. For this purpose, in the past decade, different sample extraction procedures have been used based on techniques such as LLE, supercritical fluid extraction (SFE), PLE, SPME, MSPD, etc. [4,108]. Py residues in different food matrices have mainly been analyzed either by gas chromatography with GC–ECD or GC–MS and GC–MS/MS and by LC–UV or LC–MS, and LC–MS/MS.

More recently, extraction of Py pesticide residues in fatty food samples of animal origin (including chicken, beef, fish, eggs, and milk) has been performed using DLLME [101], ultrasound solvent-assisted extraction followed by SPE cleanup [47], LLE with purification at low temperature (LLE–LTP) [108], and QuEChERS [109]. For the analysis of Py residues in food of animal origin, GC–MS and LC–MS or GC–MS/MS and LC–MS/MS are still among the most common quantification techniques. These combinations of extraction procedures and analytical methods claim to provide good precisions, wide linear ranges, and sensitive limits of detection for the analysis of Py pesticides in such samples.

7.5.6 NEONICOTINOIDS

Neonicotinoids are a class of neuro-active insecticides chemically similar to nicotine. They are mostly applied against soil pests, seed, and animal pests, as well as foliar treatments. Compared to organophosphate and CB insecticides, neonicotinoids cause less toxicity in birds and mammals than insects. It has been estimated that neonicotinoids are currently among the most widely used insecticides in the worldwide market. Analytical methods for the determination of neonicotinoid residues in food mainly include HPLC and LC–MS. More recently, for the detection of pesticides, chromatographic columns with sub 2 μm particle size at elevated pressure up to 15,000 psi (UHPLC) have been introduced coupled to mass spectrometry. This technique has been used for the quantification of neonicotinoid residues in food of animal origin (including eggs, chicken, and pork meat) by SPE and UHPLC–ESI–MS/MS [110].

When first introduced, neonicotinoids were thought to have low toxicity to many insects, but recent research has suggested a potential toxicity to honeybees and other beneficial insects even with low levels of contact. Honeybee exposure to neonicotinoids can lead to contamination of apiarian products, especially honey, which is the most commonly consumed bee product. Because of the potential threat to human health, the European Union established MRLs for acetamiprid, clothianidin, imidacloprid, thiacloprid, and thiamethoxam in the range of 50–200 μg/kg [85]. In addition, in response to the European Food Safety Authority (EFSA) study on the safety of three neonicotinoids (namely clothianidin, imidacloprid, and thiamethoxam) [111], the European Commission recommended a restriction of their use across the European Union in crops attractive to pollinators, emphasizing the awareness of the potential harmful impact of the neonicotinoids on honeybees and their products (EU Regulation 485/2013) [112,113]. Therefore, monitoring and determination of trace levels of neonicotinoids in honey are necessary and demand highly efficient, selective, and sensitive analytical techniques. Several sample preparation techniques, including LLE, SPE, QuEChERS, and combinations of them, DLLME, and diatomaceous earth material (Chem Elut cartridges) were reported for the sample preparation in honey (and in other food of animal origin) [69,85,114]. Neonicotinoids are usually determined by LC–DAD, ultraviolet, fluorescence, mass spectrometric (MS or MS/MS), and electrochemical detectors [69]. GC analysis is more complex due to neonicotinoids' low volatility and relatively high polarity [69].

7.6 NEW TRENDS AND EMERGING ISSUES IN ANALYTICAL METHODS

Although pesticide analysis is a mature area within environmental and food analytical chemistry, analytical methods are constantly evolving and improving and, undoubtedly, new technologies will emerge in the future. While in the 1990s GC–MS was mainly used to confirm identification of analytes after quantification by means of specific detection methods (like ECD, FPD, and NDP), in the 2000s, GC–MS/MS became widely used for identification and quantification of pesticide residues in food of animal origin [2]. Nowadays, LC techniques coupled to (tandem) mass spectrometry (LC–MS and LC–MS/MS) are considered the "gold standard" in routine laboratories to perform pesticide residue analyses, providing the sensitivity, selectivity, and specificity needed to meet EU legislation requirements for the analysis of pesticides in food samples [78]. They are generally preferred over GC because several currently used pesticides are quite polar, thermally labile, or not easily vaporized, and thus, are unsatisfactorily detected by GC. MS/MS detectors are also sensitive enough to quantify ultra-traces that are "blind" to "non-targeted" approaches, or to identify unknown compounds not included in pre-target lists.

In these regards, the recent introduction of HRMS analyzers (e.g., Orbitrap and time of flight – TOF) for screening pesticide residues is a powerful tool for the detection of target/non-target pesticides in food, confirming the identification of contaminants and elucidating the structure of their metabolites. This technique works in full scan mode (possibly identifying an "unlimited" number of compounds in the samples), it gives accurate masses for both parent and fragment ions, and enables the measurement of the elemental formula of a compound, achieving compound identification. Thus, in addition to target pesticide residue analysis, other related contaminants (i.e., other pesticides or their metabolites) could be detected and identified, increasing considerably the scope of the analyses and the analytical capabilities of the performing laboratories. However, several important limitations should be considered for the implementation of these HRMS detectors in routine laboratories, such as their higher price and lower sensitivity compared to target MS/MS techniques. However, in the last few years, the sensitivity provided by HRMS analyzers allows for the quantification of pesticide residues at concentrations below 10 µg/kg. In addition, systematic approaches for the detection of untargeted pesticides and their metabolites by HRMS analyzers are still scarce and, since the compound identification is a labor-intensive activity, automated strategies are needed.

To detect a wide range of compounds by HRMS analyzers, generic extraction methods, such as "dilute and shoot" and QuEChERS, are currently performed to ensure their identification during analysis. In addition, an interesting trend in sample pre-treatment is the miniaturization of this step, reducing the organic solvents consumption and achieving "greener" analytical performances. Also, semiautomatic procedures can improve the sample throughput and the accuracy of the extraction method, avoiding the unwanted co-extraction of interferences.

For GC amenable pesticides, comprehensive GC×GC has recently gained interest in the analysis of food of animal origin [115]. The applications of GC×GC in food analysis are both untargeted, when coupled to TOF-MS, and targeted, when coupled to MS. However, it is in the field of untargeted analysis (both qualitative profiling and fingerprinting of foodstuffs) that GC×GC shows its superiority over conventional GC [115].

Together with the recent miniaturized extraction and pre-concentration techniques (e.g., SDME, SBSE, DLLME, MSPE, etc.), there is currently a tendency to reduce GC separation times [116,117]. In this respect, the rapid GC separation of pesticides by "fast GC" is performed on short micro-bore (≤0.18 mm i.d.) or narrow-bore (≤0.25 mm i.d.) columns, which reduce peak widths and shorten total run times to a few minutes [117]. Another recently applied technology is the low-pressure vacuum outlet gas chromatography (LPGC), which speeds the analyte separation by applying a vacuum to the column in GC to generate a greater pressure drop than possible with the outlet at atmospheric pressure [116].

In conclusion, by combining suitable extraction methods to efficient chromatographic procedures and mass analyzers, wider scope analytical methods can be developed, allowing the most reliable determination of pesticide residues and identifying non-targeted compounds (e.g., contaminants and their metabolites) potentially present in the food samples.

ACKNOWLEDGMENTS

Giulia Poma acknowledges funding from the University of Antwerp for her postdoctoral fellowship.

REFERENCES

1. Stenersen, J. *Chemical Pesticides: Modes of Action and Toxicology.* CRC Press, Boca Raton, FL, 2004.
2. LeDoux, M. Analytical methods applied to the determination of pesticide residues in foods of animal origin. A review of the past two decades. *J. Chromatogr. A* 2011, *1218*(8), 1021–1036.
3. Regulation, E. Regulation (EU) No 396/2005 on maximum residue levels of pesticides in or on food and feed of plant and animal origin, and amending Council Directive 91/414/EEC. *Off. J.* 2005, *L70*, 28–39.
4. Tadeo, J. L. *Analysis of Pesticides in Food and Environmental Samples.* CRC Press: Boca Raton, FL, 2008, 384.
5. Gómez-Pérez, M. L.; Plaza-Bolaños, P.; Romero-González, R.; Martínez-Vidal, J. L.; Garrido-Frenich, A. Comprehensive qualitative and quantitative determination of pesticides and veterinary drugs in honey using liquid chromatography-Orbitrap high resolution mass spectrometry. *J. Chromatogr. A* 2012, *1248*, 130–138.
6. Gómez-Pérez, M. L.; Romero-González, R.; Plaza-Bolaños, P.; Génin, E.; Vidal, J. L. M.; Frenich, A. G. Wide-scope analysis of pesticide and veterinary drug residues in meat matrices by high resolution MS: Detection and identification using Exactive-Orbitrap. *J. Mass Spectrom.* 2014, *49*(1), 27–36.
7. De Dominicis, E.; Commissati, I.; Gritti, E.; Catellani, D.; Suman, M. Quantitative targeted and retrospective data analysis of relevant pesticides, antibiotics and mycotoxins in bakery products by liquid chromatography-single-stage Orbitrap mass spectrometry. *Food Addit. Contam. - Part A Chem. Anal. Control Expo. Risk Assess.* 2015, *32*(10), 1617–1627.
8. Li, Y.; Wang, M.; Yan, H.; Fu, S.; Dai, H. Simultaneous determination of multiresidual phenyl acetanilide pesticides in different food commodities by solid-phase cleanup and gas chromatography-mass spectrometry. *J. Sep. Sci.* 2013, *36*(6), 1061–1069.
9. Hildmann, F.; Gottert, C.; Frenzel, T.; Kempe, G.; Speer, K. Pesticide residues in chicken eggs - a sample preparation methodology for analysis by gas and liquid chromatography/tandem mass spectrometry. *J. Chromatogr. A* 2015, *1403*, 1–20.
10. Aguilera-Luiz, M. M.; Plaza-Bolaños, P.; Romero-González, R.; Vidal, J. L. M.; Frenich, A. G. Comparison of the efficiency of different extraction methods for the simultaneous determination of mycotoxins and pesticides in milk samples by ultra high-performance liquid chromatography–tandem mass spectrometry. *Anal. Bioanal. Chem.* 2011, *399*(8), 2863–2875.

11. García-Chao, M.; Agruña, M. J.; Calvete, G. F.; Sakkas, V.; Llompart, M.; Dagnac, T. Validation of an off line solid phase extraction liquid chromatography–tandem mass spectrometry method for the determination of systemic insecticide residues in honey and pollen samples collected in apiaries from NW Spain. *Anal. Chim. Acta* 2010, *672*(1–2), 107–113.

12. Souza Tette, P. A.; Guidi, L. R.; De Abreu Glória, M. B.; Fernandes, C. Pesticides in honey: A review on chromatographic analytical methods. *Talanta* 2016, *149*, 124–141.

13. Brutti, M.; Blasco, C.; Picó, Y. Determination of benzoylurea insecticides in food by pressurized liquid extraction and LC–MS. *J. Sep. Sci.* 2010, *33*(1), 1–10.

14. Drábová, L.; Pulkrabová, J.; Kalachová, K.; Hradecký, J.; Suchanová, M.; Tomaniová, M.; Kocourek, V.; Hajšlová, J. Novel approaches to determination of PAHs and halogenated POPs in canned fish. *Czech J. Food Sci.* 2011, *29*(5), 498–507.

15. Xiao, Z.; Li, X.; Wang, X.; Shen, J.; Ding, S. Determination of neonicotinoid insecticides residues in bovine tissues by pressurized solvent extraction and liquid chromatography–tandem mass spectrometry. *J. Chromatogr. B Anal. Technol. Biomed. Life Sci.* 2011, *879*(1), 117–122.

16. Wu, G.; Bao, X.; Zhao, S.; Wu, J.; Han, A.; Ye, Q. Analysis of multi-pesticide residues in the foods of animal origin by GC–MS coupled with accelerated solvent extraction and gel permeation chromatography cleanup. *Food Chem.* 2011, *126*(2), 646–654.

17. Wang, J.; Kliks, M. M.; Jun, S.; Li, Q. X. Residues of organochlorine pesticides in honeys from different geographic regions. *Food Res. Int.* 2010, *43*(9), 2329–2334.

18. Jovanov, P.; Guzsvány, V.; Franko, M.; Lazić, S.; Sakač, M.; Šarić, B.; Banjaca, V. Multi-residue method for determination of selected neonicotinoid insecticides in honey using optimized dispersive liquid–liquid microextraction combined with liquid chromatography–tandem mass spectrometry. *Talanta* 2013, *111*, 125–133.

19. Jovanov, P.; Guzsvány, V.; Franko, M.; Lazić, S.; Sakač, M.; Milovanović, I.; Nedeljković, N. Development of multiresidue DLLME and QuEChERS based LC–MS/MS method for determination of selected neonicotinoid insecticides in honey liqueur. *Food Res. Int.* 2014, *55*, 11–19.

20. Kujawski, M. W.; Pinteaux, E.; Namieśnik, J. Application of dispersive liquid–liquid microextraction for the determination of selected organochlorine pesticides in honey by gas chromatography–mass spectrometry. *Eur. Food Res. Technol.* 2012, *234*(2), 223–230.

21. Salami, F. H.; Queiroz, M. E. C. Microextraction in packed sorbent for the determination of pesticides in honey samples by gas chromatography coupled to mass spectrometry. *J. Chromatogr. Sci.* 2013, *51*(10), 899–904.

22. Campillo, N.; Viñas, P.; Peñalver, R.; Cacho, J. I.; Hernández-Córdoba, M. Solid-phase microextraction followed by gas chromatography for the speciation of organotin compounds in honey and wine samples: A comparison of atomic emission and mass spectrometry detectors. *J. Food Compos. Anal.* 2012, *25*(1), 66–73.

23. Campillo, N.; Viñas, P.; Férez-Melgarejo, G.; Hernández-Córdoba, M. Liquid chromatography with diode array detection and tandem mass spectrometry for the determination of neonicotinoid insecticides in honey samples using dispersive liquid–liquid microextraction. *J. Agric. Food Chem.* 2013, *61*, 4799–4805.

24. Chienthavorn, O.; Dararuang, K.; Sasook, A.; Ramnut, N. Purge and trap with monolithic sorbent for gas chromatographic analysis of pesticides in honey. *Anal. Bioanal. Chem.* 2012, *402*(2), 955–964.

25. Fontana, A. R.; Camargo, A. B.; Altamirano, J. C. Coacervative microextraction ultrasound-assisted back-extraction technique for determination of organophosphates pesticides in honey samples by gas chromatography–mass spectrometry. *J. Chromatogr. A* 2010, *1217*(41), 6334–6341.

26. Anastassiades, M.; Lehotay, S. J.; Štajnbaher, D.; Schenck, F. J. Fast and easy multiresidue method employing acetonitrile extraction/partitioning and "dispersive solid-phase extraction" for the determination of pesticide residues in produce. *J. AOAC Int.* 2003, *86*(2), 412–431.

27. González-Curbelo, M.; Socas-Rodríguez, B.; Herrera-Herrera, A. V.; González-Sálamo, J.; Hernández-Borges, J.; Rodríguez-Delgado, M. Evolution and applications of the QuEChERS method. *TrAC - Trends Anal. Chem.* 2015, *71*, 169–185.

28. Lehotay, S. J.; Mastovská, K.; Yun, S. J.; Regional, E.; Service, Q.; Food, T. U. S. Evaluation of two fast and easy methods for pesticide residue analysis in fatty food matrixes. *J. AOAC Int.* 2005, *88*(2), 630–638.

29. Zhang, H.; Wang, J.; Li, L.; Wang, Y. Determination of 103 pesticides and their main metabolites in animal origin food by QuEChERS and liquid chromatography–tandem mass spectrometry. *Food Anal. Methods* 2017, *10*(6), 1826–1843.

30. Chung, S. W. C.; Chan, B. T. P. Validation and use of a fast sample preparation method and liquid chromatography–tandem mass spectrometry in analysis of ultra-trace levels of 98 organophosphorus pesticide and carbamate residues in a total diet study involving diversified food types. *J. Chromatogr. A* 2010, *1217*(29), 4815–4824.

31. Holmes, B.; Dunkin, A.; Schoen, R.; Wiseman, C. Single-laboratory ruggedness testing and validation of a modified QuEChERS approach to quantify 185 pesticide residues in salmon by liquid chromatography– and gas chromatography–tandem mass spectrometry. *J. Agric. Food Chem.* 2015, *63*(21), 5100–5106.

32. Han, L.; Sapozhnikova, Y.; Lehotay, S. J. Method validation for 243 pesticides and environmental contaminants in meats and poultry by tandem mass spectrometry coupled to low-pressure gas chromatography and ultrahigh-performance liquid chromatography. *Food Control* 2016, *66*, 270–282.

33. Barganska, Z.; Slebioda, M.; Namiesnik, J. Determination of pesticide residues in honeybees using modified QUEChERS sample work-up and liquid chromatography–tandem mass spectrometry. *Molecules* 2014, *19*(3), 2911–2924.

34. Gómez-Pérez, M. L.; Plaza-Bolaños, P.; Romero-González, R.; Martínez Vidal, J. L.; Garrido Frenich, A. Evaluation of the potential of GC–APCI-MS for the analysis of pesticide residues in fatty matrices. *J. Am. Soc. Mass Spectrom.* 2014, *25*(5), 899–902.

35. Luz Gómez-Pérez, M.; Romero-González, R.; José Luis Martínez, V.; Garrido Frenich, A. Analysis of pesticide and veterinary drug residues in baby food by liquid chromatography coupled to Orbitrap high resolution mass spectrometry. *Talanta* 2015, *131*, 1–7.

36. López-García, M.; Romero-González, R.; Frenich, A. G. Determination of rodenticides and related metabolites in rabbit liver and biological matrices by liquid chromatography coupled to Orbitrap high resolution mass spectrometry. *J. Pharm. Biomed. Anal.* 2017, *137*, 235–242.

37. Romero-González, R.; Garrido Frenich, A. Application of HRMS in pesticide residue analysis in food from animal origin. *Appl. High Resolut. Mass Spectrom. Food Saf. Pestic. Residue Anal.* 2017, 203–232.

38. Nácher-Mestre, J.; Ibáñez, M.; Serrano, R.; Pérez-Sánchez, J.; Hernández, F. Qualitative screening of undesirable compounds from feeds to fish by liquid chromatography coupled to mass spectrometry. *J. Agric. Food Chem.* 2013, *61*(9), 2077–2087.

39. Al Naggar, Y.; Codling, G.; Vogt, A.; Naiem, E.; Mona, M.; Seif, A.; Giesy, J. P. Organophosphorus insecticides in honey, pollen and bees (Apis mellifera L.) and their potential hazard to bee colonies in Egypt. *Ecotoxicol. Environ. Saf.* 2015, *114*, 1–8.

40. Lichtmannegger, K.; Fischer, R.; Steemann, F. X.; Unterluggauer, H.; Masseller, S. Alternative QuEChERS-based modular approach for pesticide residue analysis in food of animal origin. *Anal. Bioanal. Chem.* 2015, *407*(13), 3727–3742.

41. Tian, H. Determination of chloramphenicol, enrofloxacin and 29 pesticides residues in bovine milk by liquid chromatography–tandem mass spectrometry. *Chemosphere* 2011, *83*(3), 349–355.

42. Blasco, C.; Vazquez-Roig, P.; Onghena, M.; Masia, A.; Picó, Y. Analysis of insecticides in honey by liquid chromatography–ion trap-mass spectrometry: Comparison of different extraction procedures. *J. Chromatogr. A* 2011, *1218*(30), 4892–4901.

43. Filigenzi, M. S.; Ehrke, N.; Aston, L. S.; Poppenga, R. H. Evaluation of a rapid screening method for chemical contaminants of concern in four food-related matrices using QuEChERS extraction, UHPLC and high resolution mass spectrometry. *Food Addit. Contam. Part A. Chem. Anal. Control Expo. Risk Assess.* 2011, *28*(10), 1324–1339.

44. Hayward, D. G.; Pisano, T. S.; Wong, J. W.; Scudder, R. J. Multiresidue method for pesticides and persistent organic pollutants (POPs) in milk and cream using comprehensive two-dimensional capillary gas chromatography–time-of-flight mass spectrometry. *J. Agric. Food Chem.* 2010, *58*(9), 5248–5256.

45. Farré, M.; Picó, Y.; Barceló, D. Application of ultra-high pressure liquid chromatography linear ion-trap orbitrap to qualitative and quantitative assessment of pesticide residues. *J. Chromatogr. A* 2014, *1328*, 66–79.

46. Gómez-Pérez, M. L.; Romero-González, R.; Vidal, J. L. M.; Frenich, A. G. Identification of transformation products of pesticides and veterinary drugs in food and related matrices: Use of retrospective analysis. *J. Chromatogr. A* 2015, *1389*, 133–138.

47. Dallegrave, A.; Pizzolato, T. M.; Barreto, F.; Eljarrat, E.; Barceló, D. Methodology for trace analysis of 17 pyrethroids and chlorpyrifos in foodstuff by gas chromatography–tandem mass spectrometry. *Anal. Bioanal. Chem.* 2016, *408*(27), 7689–7697.

48. Panseri, S.; Catalano, A.; Giorgi, A.; Arioli, F.; Procopio, A.; Britti, D.; Chiesa, L. M. Occurrence of pesticide residues in Italian honey from different areas in relation to its potential contamination sources. *Food Control* 2014, *38*(1), 150–156.

49. Rodríguez López, D.; Ahumada, D. A.; Díaz, A. C.; Guerrero, J. A. Evaluation of pesticide residues in honey from different geographic regions of Colombia. *Food Control* 2014, *37*(1), 33–40.

50. Tomasini, D.; Sampaio, M. R. F.; Caldas, S. S.; Buffon, J. G.; Duarte, F. A.; Primel, E. G. Simultaneous determination of pesticides and 5-hydroxymethylfurfural in honey by the modified QuEChERS method and liquid chromatography coupled to tandem mass spectrometry. *Talanta* 2012, *99*, 380–386.

51. Tanner, G.; Czerwenka, C. LC-MS/MS analysis of neonicotinoid insecticides in honey: Methodology and residue findings in Austrian honeys. *J. Agric. Food Chem.* 2011, *59*(23), 12271–12277.

52. Laaniste, A.; Leito, I.; Rebane, R.; Lõhmus, R.; Lõhmus, A.; Punga, F.; Kruve, A. Determination of neonicotinoids in Estonian honey by liquid chromatography–electrospray mass spectrometry. *J. Environ. Sci. Heal. - Part B Pestic. Food Contam. Agric. Wastes* 2016, *51*(7), 455–464.

53. Cotton, J.; Leroux, F.; Broudin, S.; Marie, M.; Corman, B.; Tabet, J. C.; Ducruix, C.; Junot, C. High-resolution mass spectrometry associated with data mining tools for the detection of pollutants and chemical characterization of honey samples. *J. Agric. Food Chem.* 2014, *62*(46), 11335–11345.

54. Jia, W.; Chu, X.; Ling, Y.; Huang, J.; Chang, J. High-throughput screening of pesticide and veterinary drug residues in baby food by liquid chromatography coupled to quadrupole Orbitrap mass spectrometry. *J. Chromatogr. A* 2014, *1347*, 122–128.

55. Xie, J.; Peng, T.; Zhu, A. et al. Multi-residue analysis of veterinary drugs, pesticides and mycotoxins in dairy products by liquid chromatography–tandem mass spectrometry using low-temperature cleanup and solid phase extraction. *J. Chromatogr. B Anal. Technol. Biomed. Life Sci.* 2015, *1002*, 19–29.

56. Beyer, A.; Biziuk, M. Comparison of efficiency of different sorbents used during clean-up of extracts for determination of polychlorinated biphenyls and pesticide residues in low-fat food. *Food Res. Int.* 2010, *43*(3), 831–837.

57. Jeong, I. S.; Kwak, B. M.; Ahn, J. H.; Jeong, S. H. Determination of pesticide residues in milk using a QuEChERS-based method developed by response surface methodology. *Food Chem.* 2012, *133*(2), 473–481.

58. Malhat, F. M.; Haggag, M. N.; Loutfy, N. M.; Osman, M. A. M.; Ahmed, M. T. Residues of organochlorine and synthetic pyrethroid pesticides in honey, an indicator of ambient environment, a pilot study. *Chemosphere* 2015, *120*, 457–461.

59. Orso, D.; Martins, M. L.; Donato, F. F.; Rizzetti, T. M.; Kemmerich, M.; Adaime, M. B.; Zanella, R. Multiresidue determination of pesticide residues in honey by modified QuEChERS method and gas chromatography with electron capture detection. *J. Braz. Chem. Soc.* 2014, *25*(8), 1355–1364.

60. Paulino de Pinho G.; Neves, A. A.; de Queiroz, M. E. L. R.; Silvério, F. O. Optimization of the liquid–liquid extraction method and low temperature purification (LLE-LTP) for pesticide residue analysis in honey samples by gas chromatography. *Food Control* 2010, *21*(10), 1307–1311.

61. Yavuz, H.; Guler, G. O.; Aktumsek, A.; Cakmak, Y. S.; Ozparlak, H. Determination of some organochlorine pesticide residues in honeys from Konya, Turkey. *Environ. Monit. Assess.* 2010, *168*(1–4), 277–283.

62. Yu, H.; Tao, Y.; Le, T.; Chen, D.; Ishsan, A.; Liu, Y.; Wang, Y.; Yuan, Z. Simultaneous determination of amitraz and its metabolite residue in food animal tissues by gas chromatography–electron capture detector and gas chromatography–mass spectrometry with accelerated solvent extraction. *J. Chromatogr. B Anal. Technol. Biomed. Life Sci.* 2010, *878*(21), 1746–1752.

63. Eissa, F.; El-sawi, S.; Zidan, N. E. Determining pesticide residues in honey and their potential risk to consumers. *Polish J. Environ. Stud.* 2014, *23*(5), 1573–1580.

64. Farajzadeh, M. A.; Mogaddam, M. R. A.; Ghorbanpour, H. Development of a new microextraction method based on elevated temperature dispersive liquid–liquid micro-extraction for determination of triazole pesticides residues in honey by gas chromatog-raphy–nitrogen phosphorus detection. *J. Chromatogr. A* 2014, *1347*, 8–16.

65. Fan, Y. B.; Yin, Y. M.; Jiang, W. B.; Chen, Y. P.; Yang, J. W.; Wu, J.; Xie, M. X. Simultaneous determination of ten steroid hormones in animal origin food by matrix solid-phase dispersion and liquid chromatography–electrospray tandem mass spec-trometry. *Food Chem.* 2014, *142*, 170–177.

66. Moniruzzaman, M.; Chowdhury, M. A. Z.; Rahman, M. A.; Sulaiman, S. A.; Gan, S. H. Determination of mineral, trace element, and pesticide levels in honey samples origi-nating from different regions of Malaysia compared to Manuka honey. *Biomed Res. Int.* 2014, *2014*, 4–7.

67. Vichapong, J.; Burakham, R.; Srijaranai, S. In-coupled syringe assisted octanol–water partition microextraction coupled with high-performance liquid chromatography for simultaneous determination of neonicotinoid insecticide residues in honey. *Talanta* 2015, *139*, 21–26.

68. Vichapong, J.; Burakham, R.; Santaladchaiyakit, Y.; Srijaranai, S. A preconcentration method for analysis of neonicotinoids in honey samples by ionic liquid-based cold-induced aggregation microextraction. *Talanta* 2016, *155*, 216–221.

69. Jovanov, P.; Guzsvány, V.; Lazić, S.; Franko, M.; Sakač, M.; Šarić, L.; Kos, J. Development of HPLC–DAD method for determination of neonicotinoids in honey. *J. Food Compos. Anal.* 2015, *40*, 106–113.

70. Sampaio, M. R. F.; Tomasini, D.; Cardoso, L. V.; Caldas, S. S.; Primel, E. G. Determination of pesticide residues in sugarcane honey by QuEChERS and liquid chromatography. *J. Braz. Chem. Soc.* 2012, *23*(2), 197–205.

71. Zhang, J.; Gao, H.; Peng, B.; Li, S.; Zhou, Z. Comparison of the performance of conventional, temperature-controlled, and ultrasound-assisted ionic liquid dispersive liquid–liquid microextraction combined with high-performance liquid chromatography in analyzing pyrethroid pesticides in honey samples. *J. Chromatogr. A* 2011, *1218*(38), 6621–6629.

72. Li, M.; Zhang, J.; Li, Y.; Peng, B.; Zhou, W.; Gao, H. Ionic liquid-linked dual magnetic microextraction: A novel and facile procedure for the determination of pyrethroids in honey samples. *Talanta* 2013, *107*, 81–87.

73. Commission Directive of 12 August 2002 implementing Council Directive 96/23/EC concerning the performance of analytical methods and the interpretation of results. Official Journal of the European Communities.

74. Elbashir, A. A.; Aboul-Enein, H. Y. Application of gas and liquid chromatography coupled to time-of-flight mass spectrometry in pesticides: Multiresidue analysis. *Biomed. Chromatogr.* 2017, 32(2), 1–7.

75. Zacharis, C. K.; Rotsias, I.; Zachariadis, P. G.; Zotos, A. Dispersive liquid–liquid microextraction for the determination of organochlorine pesticides residues in honey by gas chromatography-electron capture and ion trap mass spectrometric detection. *Food Chem.* 2012, *134*(3), 1665–1672.

76. Wiest, L.; Buleté, A.; Giroud, B.; Fratta, C.; Amic, S.; Lambert, O.; Pouliquen, H.; Arnaudguilhem, C. Multi-residue analysis of 80 environmental contaminants in honeys, honeybees and pollens by one extraction procedure followed by liquid and gas chromatography coupled with mass spectrometric detection. *J. Chromatogr. A* 2011, *1218*(34), 5743–5756.

77. Chung, S. W. C.; Chen, B. L. S. Determination of organochlorine pesticide residues in fatty foods: A critical review on the analytical methods and their testing capabilities. *J. Chromatogr. A* 2011, *1218*(33), 5555–5567.

78. Masiá, A.; Blasco, C.; Picó, Y. Last trends in pesticide residue determination by liquid chromatography–mass spectrometry. *Trends Environ. Anal. Chem.* 2014, 2(2455), 11–24.

79. Stachniuk, A.; Fornal, E. Liquid chromatography–mass spectrometry in the analysis of pesticide residues in food. *Food Anal. Methods* 2016, 9(6), 1654–1665.

80. Guidance document on analytical quality control and method validation procedures for pesticide residues and analysis in food and feed. SANTE/11813/2017. European Commission Directorate General For Health And Food Safety.

81. Bargańska, Z.; Ślebioda, M.; Namieśnik, J. Pesticide residues levels in honey from apiaries located of Northern Poland. *Food Control* 2013, *31*(1), 196–201.

82. Choi, S.; Kim, S.; Shin, J. Y.; Kim, M.; Kim, J. H. Development and verification for analysis of pesticides in eggs and egg products using QuEChERS and LC–MS/MS. *Food Chem.* 2015, *173*, 1236–1242.

83. Danezis, G. P.; Anagnostopoulos, C. J.; Liapis, K.; Koupparis, M. A. Multi-residue analysis of pesticides, plant hormones, veterinary drugs and mycotoxins using HILIC chromatography–MS/MS in various food matrices. *Anal. Chim. Acta* 2016, *942*, 121–138.

84. Gan, J.; Lv, L.; Peng, J.; Li, J.; Xiong, Z.; Chen, D.; He, L. Multi-residue method for the determination of organofluorine pesticides in fish tissue by liquid chromatography triple quadrupole tandem mass spectrometry. *Food Chem.* 2016, *207*, 195–204.

85. Gbylik-Sikorska, M.; Sniegocki, T.; Posyniak, A. Determination of neonicotinoid insecticides and their metabolites in honeybee and honey by liquid chromatography tandem mass spectrometry. *J. Chromatogr. B Anal. Technol. Biomed. Life Sci.* 2015, *990*, 132–140.

86. Kamel, A. Refined methodology for the determination of neonicotinoid pesticides and their metabolites in honeybees and bee products by liquid chromatography–tandem mass spectrometry (LC–MS/MS). *J. Agric. Food Chem.* 2010, *58*(10), 5926–5931.

87. Kasiotis, K. M.; Anagnostopoulos, C.; Anastasiadou, P.; Machera, K. Pesticide residues in honeybees, honey and bee pollen by LC–MS/MS screening: Reported death incidents in honeybees. *Sci. Total Environ.* 2014, *485–486*(1), 633–642.

88. Kujawski, M. W.; Bargańska, Z.; Marciniak, K.; Miedzianowska, E.; Kujawski, J. K.; Ślebioda, M.; Namieśnik, J. Determining pesticide contamination in honey by LC–ESI-MS/MS - Comparison of pesticide recoveries of two liquid–liquid extraction based approaches. *LWT - Food Sci. Technol.* 2014, *56*(2), 517–523.

89. Nakajima, T.; Tsuruoka, Y.; Kanda, M.; Hayashi, H.; Hashimoto, T.; Matsushima, Y.; Yoshikawa, S.; Nagano, C.; Okutomi, Y.; Takano, I. Determination and surveillance of nine acaricides and one metabolite in honey by liquid chromatography–tandem mass spectrometry. *Food Addit. Contam. - Part A Chem. Anal. Control Expo. Risk Assess.* 2015, *32*(7), 1099–1104.

90. Proietto Galeano, M.; Scordino, M.; Sabatino, L.; Pantò, V.; Morabito, G.; Chiappara, E.; Traulo, P.; Gagliano, G. UHPLC/MS-MS analysis of six neonicotinoids in honey by modified QuEChERS: Method development, validation, and uncertainty measurement. *Int. J. Food Sci.* 2013 .

91. Wei, H.; Tao, Y.; Chen, D.; Xie, S.; Pan, Y.; Liu, Z.; Huang, L.; Yuan, Z. Development and validation of a multi-residue screening method for veterinary drugs, their metabolites and pesticide in meat using liquid chromatography–tandem mass spectrometry. *Food Addit. Contam. Part A. Chem. Anal. Control Expo. Risk Assess.* 2015, *32*(5):686–701.

92. Yáñez, K. P.; Martín, M. T.; Bernal, J. L.; Nozal, M. J.; Bernal, J. Determination of spinosad at trace levels in bee pollen and beeswax with solid-liquid extraction and LC–ESI-MS. *J. Sep. Sci.* 2014, *37*(3), 204–210.

93. Gilbert-López, B.; García-Reyes, J. F.; Molina-Díaz, A. Determination of fungicide residues in baby food by liquid chromatography–ion trap tandem mass spectrometry. *Food Chem.* 2012, *135*(2), 780–786.

94. Kujawski, M. W.; Namieśnik, J. Levels of 13 multi-class pesticide residues in Polish honeys determined by LC–ESI-MS/MS. *Food Control* 2011, *22*(6), 914–919.

95. Lazartigues, A.; Wiest, L.; Baudot, R.; Thomas, M.; Feidt, C.; Cren-Olivé, C. Multiresidue method to quantify pesticides in fish muscle by QuEChERS-based extraction and LC–MS/MS. *Anal. Bioanal. Chem.* 2011, *400*(7), 2185–2193.

96. Danezis, G. P.; Anagnostopoulos, C. J.; Liapis, K.; Koupparis, M. A. Multi-residue analysis of pesticides, plant hormones, veterinary drugs and mycotoxins using HILIC chromatography–MS/MS in various food matrices. *Anal. Chim. Acta* 2016, *942*, 121–138.

97. De Dominicis, E.; Commissati, I.; Suman, M. Targeted screening of pesticides, veterinary drugs and mycotoxins in bakery ingredients and food commodities by liquid chromatography-high-resolution single-stage Orbitrap mass spectrometry. *J. Mass Spectrom.* 2012, *47*(9), 1232–1241.

98. *ISO 90000:2015 Quality management-Fundamentals and vocabulary.* International Organization for Standardization, Geneva, Switzerland

99. Magnusson, B.; Örnemark, U. Eurachem Guide: The Fitness for Purpose of Analytical Methods- A Laboratory Guide to Method Validation and Related Topics. 2014. Available from: https://www.eurachem.org/images/stories/Guides/pdf/MV_guide_2nd _ed_EN.pdf.

100. Martínez Vidal, J. L.; Garrido Frenich, A.; Egea González, F. J. Internal quality control criteria for environmental monitoring of organic micro-contamination in water. *Trends Anal. Chem.* 2003, *22*, 34.

101. Shamsipur, M.; Yazdanfar, N.; Ghambarian, M. Combination of solid-phase extraction with dispersive liquid–liquid microextraction followed by GC–MS for determination of pesticide residues from water, milk, honey and fruit juice. *Food Chem.* 2016, *204*, 289–297.

102. Li, J.; Li, H.; Zhang, W.-J.; Wang, Y.-B.; Su, Q.; Wu, L. Hollow fiber–stir bar sorptive extraction and gas chromatography–mass spectrometry for determination of organochlorine pesticide residues in environmental and food matrices. *Food Anal. Methods* 2018, *11*(3), 883–891.

103. Sánchez-Rojas, F.; Bosch-Ojeda, C.; Cano-Pavón, J. M. A review of stir bar sorptive extraction. *Chromatographia* 2009, *69*(S1), 79–94.

104. Sapozhnikova, Y.; Lehotay, S. J. Multi-class, multi-residue analysis of pesticides, polychlorinated biphenyls, polycyclic aromatic hydrocarbons, polybrominated diphenyl ethers and novel flame retardants in fish using fast, low-pressure gas chromatography–tandem mass spectrometry. *Anal. Chim. Acta* 2013, *758*, 80–92.

105. Santaladchaiyakit, Y.; Srijaranai, S.; Burakham, R. Methodological aspects of sample preparation for the determination of carbamate residues: A review. *J. Sep. Sci.* 2012, *35*(18), 2373–2389.

106. Hamed, A. M.; Moreno-González, D.; Gámiz-Gracia, L.; García-Campaña, A. M. Evaluation of a new modified QuEChERS method for the monitoring of carbamate residues in high-fat cheeses by using UHPLC–MS/MS. *J. Sep. Sci.* 2016, *40* (2): 488–496.

107. Liu, S.-Y.; Jin, Q.; Huang, X.-H.; Zhu, G.-N. Determination of residues of seven carbamate pesticides in milk using ultra-performance liquid chromatography/tandem mass spectrometry and QuEChERS methods. *J. AOAC Int.* 2013, *96*(3), 657–662.

108. Meneghini, L. Z.; Rübensam, G.; Bica, V. C.; Ceccon, A.; Barreto, F.; Ferrão, M. F.; Bergold, A. M. Multivariate optimization for extraction of pyrethroids in milk and validation for GC–ECD and CG–MS/MS analysis. *Int. J. Environ. Res. Public Health* 2014, *11*(11), 11421–11437.

109. Wilkowska, A.; Biziuk, M. Determination of pesticide residues in food matrices using the QuEChERS methodology. *Food Chem.* 2011, *125*(3), 803–812.

110. Liu, S.; Zheng, Z.; Wei, F.; Ren, Y.; Gui, W.; Wu, H.; Zhu, G. Simultaneous determination of seven neonicotinoid pesticide residues in food by ultraperformance liquid chromatography tandem mass spectrometry. *J. Agric. Food Chem.* 2010, *58*(6), 3271–3278.

111. EFSA. EFSA identifies risks to bees from neonicotinoids. 2013. Available from: https://www.efsa.europa.eu/en/press/news/130116.

112. European Commission. Commission implementing regulation (EU) No 485/2013. *Off. J. Eur. Union* 2013.

113. EFSA. Conclusion on the peer review of the pesticide risk assessment for bees for the active substance thiamethoxam. *EFSA J.* 2013, *11*(1), 3067–3068.

114. Seccia, S.; Fidente, P.; Montesano, D.; Morrica, P. Determination of neonicotinoid insecticides residues in bovine milk samples by solid-phase extraction clean-up and liquid chromatography with diode-array detection. *J. Chromatogr. A* 2008, *1214*(1–2), 115–120.

115. Tranchida, P. Q.; Purcaro, G.; Maimone, M.; Mondello, L. Impact of comprehensive two-dimensional gas chromatography with mass spectrometry on food analysis. *J. Sep. Sci.* 2016, *39*(1), 149–161.

116. Sapozhnikova, Y.; Lehotay, S. J. Review of recent developments and applications in low-pressure (vacuum outlet) gas chromatography. *Anal. Chim. Acta* 2015, *899*, 13–22.

117. Zoccali, M.; Purcaro, G.; Schepis, A.; Tranchida, P. Q.; Mondello, L. Miniaturization of the QuEChERS method in the fast gas chromatography-tandem mass spectrometry analysis of pesticide residues in vegetables. *Food Anal. Methods* 2017, *10*(8), 2636–2645.

118. dos Souza, M. R.; Moreira, C. O.; De Lima, T. G.; Aquino, A.; Dórea, H. S. Validation of a matrix solid phase dispersion (MSPD) technique for determination of pesticides in lyophilized eggs of the chicken Gallus gallus domesticus. *Microchem. J.* 2013, *110*, 395–401.

119. Hamed, A. M.; Moreno-González, D.; Gámiz-Gracia, L.; García-Campaña, A. M. Evaluation of a new modified QuEChERS method for the monitoring of carbamate residues in high-fat cheeses by using UHPLC–MS/MS. *J. Sep. Sci.* 2016, *40*(2): 488–496.

8 Determination of Pesticides in Soil

Beatriz Albero, Rosa Ana Pérez, and José L. Tadeo

CONTENTS

8.1 Introduction 246
8.2 Sample Preparation...247
    8.2.1 Sampling and Preparation of Soil Samples 247
    8.2.2 Extraction..247
        8.2.2.1 Herbicides 249
        8.2.2.2 Insecticides and Fungicides 250
        8.2.2.3 Multiresidue 251
    8.2.3 Cleanup...252
        8.2.3.1 Herbicides 252
        8.2.3.2 Insecticides and Fungicides 254
        8.2.3.3 Multiresidue 254
    8.2.4 Derivatization 254
        8.2.4.1 Glyphosate 255
        8.2.4.2 Phenoxy Acid Herbicides..255
        8.2.4.3 Phenylureas 255
        8.2.4.4 Sulfonylureas 255
        8.2.4.5 Carbamates 256
8.3 Determination of Pesticide Residues..256
8.4 Application to Real Samples...260
    8.4.1 Glyphosate 260
    8.4.2 Sulfonylureas 260
    8.4.3 Triazines 260
    8.4.4 Organophosphorus 261
    8.4.5 Pyrethroids..261
    8.4.6 Neonicotinoids..261
    8.4.7 Pyrimethanil and Kresoxim-Methyl Fungicides 262
    8.4.8 Multiresidue..262
8.5 Future Trends..262
References...263

8.1 INTRODUCTION

Pesticides may reach the soil compartment in different ways. Direct soil application is normally employed for the control of weeds, insects, or microorganisms, the use of herbicides being a typical example. Pesticides may also reach the soil indirectly, when they are applied to the aerial part of plants (to control weeds or crop pests and diseases), and a pesticide fraction drops to the soil during application, or afterward, when it lixiviates from crops or falls absorbed on vegetal material. Other ways the pesticides reach the soil are by transportation from a different compartment, e.g., with the irrigation water, or by atmospheric deposition.

Once in the soil, pesticides may undergo a series of transformation and distribution processes. These transformation processes may have a biotic or abiotic origin and cause the degradation of pesticides through several mechanisms, such as oxidation, reduction, or hydrolysis. The distribution of pesticides can be originated by various processes, such as volatilization, leaching, runoff, and absorption by plants. In these processes, the physical–chemical properties of pesticides and the adsorption–desorption equilibrium in soil are the main factors involved. Figure 8.1 shows the most important pathways of pesticide distribution and transformation in soil.

The fate of pesticides and their degradation products in soil will depend on different factors, such as the agricultural practices, the climate, and the type of soil. Pesticides and their degradation or transformation products may cause toxic effects to man and the environment, making it necessary to evaluate if their application may cause an unacceptable risk. Consequently, many developed countries have regulated pesticide use in agriculture [1, 2].

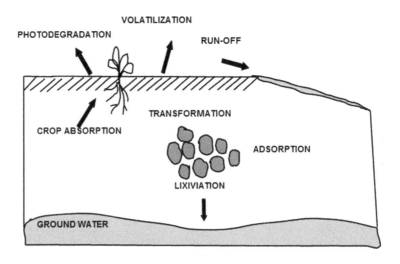

FIGURE 8.1   Distribution and transformation pathways of pesticides in soil.

8.2 SAMPLE PREPARATION

8.2.1 Sampling and Preparation of Soil Samples

The plough layer of soil (0–20 cm) is generally sampled for the determination of pesticides in this compartment. Nevertheless, other layers may be sampled at different depths to study the distribution of these compounds in soil and, in addition, soil solution may be sometimes sampled to find the bioavailability of pesticides. After field sampling, soil is usually air dried and sieved through a 2 mm mesh in the laboratory. Then, soil samples are placed in closed glass flasks and stored frozen until the analysis of pesticides. The addition of known amounts of pesticides to blank soil samples is a normal practice to study the recovery of these compounds. However, the recovery of pesticides from soil may be different in freshly spiked than in aged soil samples. Pesticides in soil may undergo transformation processes that lead to the formation of bound residues, which cannot be extracted even after exhaustive extraction with organic solvents. The use of reference soil samples with certified concentrations of the studied pesticides is recommended for the validation of the analytical methods, but these reference materials are difficult to prepare and maintain and are available only for a few pesticides.

8.2.2 Extraction

The liquid–solid extraction of pesticides from soil is generally carried out by organic solvents. Two classical analytical techniques have been widely used, the shaking and filter method and the Soxhlet extraction method. These techniques have the advantage of being simple and low-cost methods, but they are time consuming, laborious, difficult to automate, and nonselective methods. In addition, they suffer from various disadvantages, such as the use of large volume of organic solvents and the need for cleanup steps.

Several modern analytical techniques have been developed to overcome these problems. Accelerated solvent extraction (ASE), also named pressurized liquid extraction (PLE), is a fast technique that uses low volumes of solvents and can be automated, although the high temperatures used to accelerate the process may degrade some pesticides. Supercritical fluid extraction (SFE) uses fluids above their critical temperature and pressure. In these conditions, supercritical fluids behave similar to liquids, CO_2 being widely employed because of its reduced cost and low critical temperature (31°C) and pressure (73 atm). Microwave-assisted extraction (MAE) is also a fast technique that is able to extract multiple samples at the same time, but the extraction vessels are expensive and must be cooled at room temperature before opening. Matrix solid-phase dispersion (MSPD) is a simultaneous extraction and cleanup technique developed for solid or semisolid samples in which the sample is disintegrated in the presence of a solid support employing a pestle and a mortar, creating a new phase that is transferred to a column for the elution of the analytes with a suitable solvent. Ultrasonic assisted extraction (UAE) with various organic solvents has also been employed to extract pesticides from soil. The ultrasound radiation promotes the penetration of solvent into soil improving efficiency of the extraction step.

Solid-phase microextraction (SPME) uses a fiber coated with a stationary phase that sorbs the analytes from the matrix followed by desorption of the retained compounds before its analytical determination. This procedure can be carried out exposing the sample to the headspace or with the fiber immersed in the solution.

The QuEChERS method, an acronym for "quick, easy, cheap, effective, rugged, and safe", is a sample preparation technique that was developed in 2003 for the analysis of pesticides in fresh fruits and vegetables and consists of a partitioning with acetonitrile in the presence of salts, followed by dispersive solid-phase extraction. Its first application in the analysis of soil was reported in 2008 and it has become very popular taking into account the number of works in the scientific literature. Tables 8.1 through 8.3

TABLE 8.1
Extraction Methods of Herbicides from Soil

Technique	Class	Solvent	Ref.
Shaking	Phenoxyacids, Glyphosate	Water, basic pH	[3–5]
	Sulfonylureas	Acetonitrile	[6]
		Acetonitrile–0.15M NaHCO$_3$ (2:8, v/v)	[7]
	Triazines	Water–methanol (1:1, v/v)	[8]
	Imidazolinones	Ammonium acetate	[9]
	Multiclass	Acetonitrile, acetonitrile–phosphate buffer pH 7.5 (9:1,v/v)	[10]
Soxhlet	Triazines, Benzonitriles	Methanol	[11–13]
UAE	Pyrimidines	Water, basic pH	[14]
	Triazines	Hexane–acetone (2:1, v/v)	[15]
		Acetone–water (3:1, v/v)	[16]
		Acetonitrile	[17]
	Sulfonylureas	Methanol	[18]
		Acetonitrile	[19]
	Multiclass	Cyclohexane–acetone (3:1, v/v)	[20]
PLE	Phenoxyacids	Water	[21]
	Triazines	Acetone	[22]
	Multiclass	Acetone	[23]
MAE	Phenoxyacids	Water–methanol, pH 7	[24]
	Triazines	Water–methanol (1:1, v/v) pH 7	[25, 26]
		Dichloromethane	[27]
	Sulfonylureas	Dichloromethane–acetonitrile (2:1, v/v)	[28]
	Quaternary ammonium compounds	HNO$_3$–HCl–HF (4:1:1, v/v/v)	[29]
	Multiclass	Acetonitrile	[30]
QuEChERS	Phenoxyacids	Water–acetonitrile with 1% acetic acid (1:1, v/v)	[31]

Abbreviations: MAE: microwave assisted extraction; PLE: pressurized liquid extraction; UAE: ultrasound-assisted extraction

TABLE 8.2
Extraction Methods of Insecticides and Fungicides from Soil

Technique	Class	Solvent	Ref.
Shaking	Strobilurins	Acetone	[32]
	Benzimidazoles	Ethyl acetate	[33]
	Pyrethroids	Acetonitrile	[34]
	Carbamates	Monochloroacetic acid buffer	[35]
	Multiclass-insecticides	Acetonitrile–water (1:1, v/v)	[36]
	Multiclass-fungicides	Acetone	[37]
Soxhlet	Multiclass-insecticides	Dichloromethane	[38]
UAE	Organochlorines	Petroleum ether–acetone (1:1, v/v)	[39]
	Organophosphorus	Acetonitrile	[40]
		Methanol	[41, 42]
	Benzoylureas	Acetone	[43]
	Pyrethroids	Isooctane–dichloromethane (15:85, v/v)	[44]
	Carbamates	Methanol	[45]
	Multiclass-fungicides	Water, acetone	[46]
	Multiclass-insecticides	Ethyl acetate	[47]
	Multiclass-fungicides	Ethyl acetate	[48]
SFE	Carbamates, Pyrethroids	CO_2 –3% methanol	[49, 50]
PLE	Organochlorines	Acetone–hexane (1:1, v/v)	[51, 52]
	Pyrethroids	Acetone–dichloromethane (1:1, v/v)	[53]
MAE	Carbamates	Methanol	[49]
	Organochlorines	Acetone–hexane (1:1, v/v)	[54]
	Pyrethroids	Toluene	[55]
SPME	Organochlorines		[56, 57]
	Multiclass-fungicides		[58]
MSPD	Organochlorines	Dichloromethane	[59]
QuEChERS	Neonicotinoids	Acetonitrile–dichloromethane (1:2, v/v)	[60]
	Organochlorines	Acetonitrile–water (7:3, v/v)	[61]
		Hexane–acetone (9:1, v/v) –water (2:1, v/v)	[62]

MAE: microwave assisted extraction; MSPD: matrix solid-phase dispersion; PLE: pressurized liquid extraction; SFE: supercritical fluid extraction; SPME: solid-phase microextraction; UAE: ultrasound-assisted extraction

summarize representative published papers on the analysis of pesticides in soil using those extraction techniques.

8.2.2.1 Herbicides

Analyses of herbicide residues in soil have been frequently performed because of the wide application of these compounds. Initially, polar herbicides, such as benzonitriles and phenoxy acids, were extracted from soil with organic solvents of low–medium polarity at acidic pH, using manual or mechanical shaking or sonication.

TABLE 8.3
Multiresidue Methods of Pesticide Extraction from Soil

Technique	Number of Analytes	Solvent	Ref.
Shaking	5	Methanol–acetone (1:1, v/v)	[63]
UAE	54	Acetonitrile–water (1:1, v/v)	[64]
	32	Methanol–water (4:1, v/v)	[65]
	44, 52	Ethyl acetate	[66, 67]
SFE	20	CO_2–3% methanol	[68]
PLE	17	Water	[69]
	30	Acetone–dichloromethane (1:1, v/v)	[70]
	126	Ethyl acetate–methanol (3:1, v/v)	[71]
SPME	4		[72]
QuEChERS	36, 193	Water–acetonitrile (3:10, v/v)	[73, 74]
	216	Water–acetonitrile containing 1% formic acid (1:1, v/v)	[75]

PLE: pressurized liquid extraction; SFE: supercritical fluid extraction; SPME: solid-phase microextraction; UAE: ultrasound-assisted extraction

For less polar herbicides, such as triazines, chloroacetamides, and dinitroanilines, organic solvents such as acetone, ethyl acetate, methanol, and acetonitrile, alone or in mixtures with water, were commonly used.

More recently, a considerable reduction in solvent consumption has been achieved by miniaturizing the scale of sample extraction. MAE has been successfully applied to the extraction of various herbicides from soil. MAE is a technique with a reduced consumption of solvent, which is normally acetonitrile or methanol, alone or in mixtures with water.

In multiclass herbicide analysis, soil samples were generally extracted with a polar or medium polarity solvent, such as acetone or acetonitrile. PLE is a modern technique used successfully for the extraction of herbicides, such as triazines, chloroacetanilides, phenylureas, and phenoxy acids, using water and acetone as solvents.

8.2.2.2 Insecticides and Fungicides

Conventional methods have been widely used in the extraction of organochlorine (OC) insecticides from soil, although the use of new extraction techniques has increased during the last number of years. In PLE, the soil sample is placed in a cartridge and extracted with mixtures of acetone and hexane. The use of MAE has also increased because of the good recoveries obtained. Moreover, headspace SPME has been successfully used to determine OC insecticides in soil with limits of detection (LOD) similar to other extraction techniques.

Organophosphorus (OP) pesticides are compounds highly polar and soluble in water that have been extracted from soil applying ultrasound radiation with organic solvents such as methanol or acetonitrile.

Carbamates were initially extracted from soil by conventional methods using mechanical shaking with different solvents. SFE and MAE were afterward successfully applied to soil as a practical alternative to traditional methods. In recent years, analyses by means of UAE have obtained good results.

Pyrethroid insecticides are a class of natural and synthetic compounds that are retained in soils because of their high lipophility and low water solubility and extracted from soil samples by sonication with organic solvents, alone or in binary mixtures. Investigations with fortified samples showed that good and similar recoveries of these compounds were obtained with MAE and SFE.

Neonicotinoids are a class of insecticides that have gained an increasing interest in the last 15 years. These compounds are used as foliar application or as seed coating and once they reach the plant, they are easily translocated through the plant tissues. QuEChERS has been successfully applied in the determination of neonicotinoids and metabolites in soil with low LODs. The analysis of multiclass mixtures of insecticides was initially carried out by Soxhlet or shaking methods with low or medium polarity solvents. UAE with ethyl acetate was used more recently in the analysis multiclass mixtures.

The analysis of fungicides in soil was initially accomplished by classical extraction methods, such as the shaking and filter method using acetone. UAE and SPME have been other techniques used more recently for the determination of fungicides in soil samples.

8.2.2.3 Multiresidue

Reliable multiresidue analytical methods are needed for monitoring programs of pesticide residues in soil. The classical procedure for pesticide extraction from soil was to shake soil samples with an organic solvent, usually ethyl acetate or acetonitrile, alone or in mixtures with water. Thus, SFE with carbon dioxide (CO_2) containing 3% methanol, as a modifier used to improve recoveries of polar pesticides, has been employed for the multiresidue extraction of pesticides having a wide range of polarities and molecular weights. SFE using CO_2 is essentially a solvent-free extraction wherein the CO_2 is easily removed at atmospheric pressure. This method has been used for the simultaneous determination of 20 pesticides using CO_2 and 3% methanol [68]. Moreover, UAE has been used for the simultaneous determination of more than 30 compounds belonging to different classes of pesticides. The good reproducibility and detection limits achieved with this method allow its application to the monitoring of pesticide residues in soil [67].

SPME is an extraction technique that has been mainly used for the extraction of pesticides from aqueous samples; however, headspace SPME has been used for the determination of pesticides volatilized from soil [72]. The application of MAE for the extraction of pesticide residues is increasing in the last number of years and together with other modern techniques, such as UAE and PLE, are the most widely used methods at present. Finally, although QuEChERS is a methodology developed for the analysis of pesticides in fruits and vegetables, nowadays it is used to analyze pesticides in other types of samples, including soil. Thus, in a recent study this technique allowed the determination of 193 pesticides in soil samples [73].

8.2.3 CLEANUP

Soil sample extracts, obtained with any of the methods described earlier, generally contain a considerable amount of other components that may interfere in the subsequent analysis. Therefore, the determination of pesticides at residue level frequently requires a further cleanup of soil extracts. Liquid–liquid partition (LLP) between an aqueous and an organic phase, at modulated pH in some cases, has been the most common first step in the cleanup of extracts. However, this technique has been replaced by solid phase extraction (SPE) in which pesticides are separated from interferences in the soil extract using an adequate sorbent packed in a cartridge or column followed by elution with a solvent of adequate polarity. An alternative to SPE is dispersive solid-phase extraction (dSPE) which has been widely used since its invention in 2000, particularly since it is the cleanup step of the QuEChERS method. The main difference between it and SPE is that the sorbent is added directly to the extract to remove the matrix interferences and then is separated by centrifugation. Tables 8.4 through 8.6 summarize the cleanup procedures employed in the determination of pesticides in soil.

8.2.3.1 Herbicides

Phenoxy acid herbicides are normally formulated as amine salts or esters, which are rapidly hydrolyzed in soil to the acidic form. Cleanup techniques for the purification of soil extracts include liquid–liquid partitioning, at acidic pH, and dSPE using acidic alumina.

TABLE 8.4
Cleanup Techniques Used in the Analysis of Herbicides

Class	Technique	Solvent	Ref.
Phenoxyacids	LLP, pH 2	MTBE	[21]
	dSPE: Acidic alumina	Acetonitrile	[31]
Imidazolines	dSPE: PSA	0.5M Ammonium acetate	[9]
Sulfonylureas	dSPE: MIP nanoparticles	Methanol–acetic acid (99:1, v/v)	[19]
	SPME: multiple monolithic fiber	Acetonitrile–formic acid (99:1, v/v)	[6]
	SPE: Silica	Dichloromethane–acetone (3:1, v/v)	[28]
	UASEME	1-octanol	[18]
Triazines	SPE: MIP	Methanol	[16, 22]
	SPE: MWCNT	Ethyl acetate	[8]
	mSPE: MIP	Acetonitrile	[17]
	SPME: MIP	Benzene	[27]
Ammonium quaternary compounds	SPE: Silica	Methanol–6.5M HCl (7:3, v/v)	[29]
Glyphosate	SBSE: MIP	10mM NaH$_2$PO$_4$	[76]

dSPE: dispersive solid-phase extraction; LLP: liquid-liquid partition; MIP: molecularly imprinted polymer; mSPE: magnetic solid-phase extraction; MTBE: methyl tert-butyl ether; MWCNT: multiwall carbon nanotubes PSA: primary secondary amines; SBSE: stir bar sorptive extraction; SPE: solid-phase extraction; SPME: solid-phase microextraction; UASEME: ultrasound-assisted surfactant-enhanced emulsification microextraction

TABLE 8.5
Cleanup Techniques Used in the Analysis of Insecticides and Fungicides

Class	Technique	Solvent	Ref.
Insecticides			
Organochlorines	SPE: alumina	Hexane–ethyl acetate (7:3, v/v)	[39]
	SPE: GCB	Hexane–ethyl acetate (80:20, v/v)	[51]
	SPE: Florisil	Heptane–ethyl acetate (1:1, v/v)	[52]
	dSPE: PSA+C18	Acetonitrile	[61]
	DLLME	Ethylene tetrachloride	[59]
Organophosphorus	USAEME	Toluene	[41]
	DLLME	Toluene	[42]
Benzoylureas	DLLME	1-undecanol	[43]
Pyrethroids	SPE: GCB	Dichloromethane–n-hexane (3:7, v/v)	[53]
	SPE: Florisil	Hexane–ethyl acetate (2:1, v/v)	[55]
	SPME	Water	[34]
Neonicotinoids	dSPE: C18	Acetonitrinile	[60]
Carbamates	DLLME	Tetrachloroethane	[77]
Multiclass	DLLME	Carbon tetrachloride	[36]
	LLP	Dichloromethane	[37]
Fungicides			
Strobilurins	SPE: Florisil	Toluene–ethyl acetate (20:1, v/v)	[32]

DLLME: dispersive liquid-liquid microextraction; dSPE: dispersive solid-phase extraction; GCB: graphitized carbon black; LLP: liquid-liquid partition; PSA: primary secondary amines; SPE: solid-phase extraction; SPME: solid-phase microextraction; USAEME: ultrasound assisted emulsification microextraction

TABLE 8.6
Cleanup Techniques Used in the Multiresidue Analysis of Pesticides

Technique	Solvent	Ref.
SPME	Water	[63]
SPE: divinybenzene-vinyl pyrrolidinone copolymer	Dichloromethane–methanol (1:1, v/v)	[65]
dSPE: $MgSO_4$+C18 +PSA	Acetonitrile	[73, 74]
HF-LPME	1-octanol	[78]

dSPE: dispersive solid-phase extraction; HF-LPME: hollow fiber liquid-phase microextraction; PSA: primary secondary amines; SPE: solid-phase extraction; SPME: solid-phase microextraction;

The cleanup of triazine herbicides in soil extracts has been carried out by SPE with alumina or Florisil but recently new sorbents have been used including nanomaterials such as multi-walled carbon nanotubes (MWCNT) or molecularly imprinted polymers (MIPs). MIPs are selective sorbents with molecular recognition sites that are synthetized for a particular analyte or group of analytes. This type of material has been employed not only as sorbent for SPE but as coating on SPME fibers or as coating in magnetic nanoparticles to retain triazines before performing magnetic solid-phase extraction. MIP nanoparticles are also used as dSPE sorbent in the cleanup of soil extracts to determine sulfonylureas. The cleanup is usually carried out with solid sorbents and elution with a suitable solvent, but with the miniaturization of liquid–liquid partitioning, a number of liquid-phase microextraction techniques have been developed. The principle of these miniaturized techniques is the extraction of the analytes from an aqueous solution using a very low volume of an organic solvent immiscible with water in the presence of a dispersing solvent, such as methanol, acetone, or acetonitrile. An alternative to the use of a disperser solvent is the use of a surfactant and the application of ultrasound radiation to facilitate the dispersion of the extraction solvent.

Glyphosate is the most widely used herbicide in the world. Its determination can be considered challenging due to its high polarity and insolubility in most organic solvents. Due to the ionic character of glyphosate, the cleanup of soil extracts has been performed with SPE containing anion-exchange resins. Recently, a magnetic stir bar was coated with an MIP designed specifically for glyphosate that allows the selective extraction of this compound in the presence of structural analogues [76].

8.2.3.2 Insecticides and Fungicides

In general, extracts from soil samples have been cleaned up by means of SPE with alumina or Florisil as adsorbents and pesticides have been eluted with nonpolar or low polarity solvents (such as hexane or ethyl acetate). In some cases, more hydrophobic sorbents, such as graphitized carbon, have been used for low polarity insecticides. In addition, LLP of soil extracts with immiscible solvents is a method sometimes used. A modern alternative to LLP is dispersive liquid–liquid microextraction (DLLME) which employs a very low volume (only a few microliters) of a water-immiscible solvent, such as toluene, ethylene tetrachloride, and carbon tetrachloride. This procedure has been applied in the determination of different classes of insecticides.

8.2.3.3 Multiresidue

Analysis of complex mixtures of pesticides in soil is a difficult problem to solve because of the presence of a wide variety of compounds with different physical–chemical properties.

In modern analytical techniques, the classical methodology for the cleanup of extracts, based on LLP, has been replaced by miniaturized techniques for multiresidue analysis that are less solvent consuming. SPE is a technique widely used in the multiresidue determination of pesticides in soil after their extraction with water or aqueous mixtures with organic solvents. Octyl and octadecyl-bonded silica sorbents have been frequently used in the analysis of nonpolar and medium polarity pesticides in soil extracts. SPE is being replaced by dSPE with the gaining popularity of the QuEChERS procedure to determine multiresidue pesticides in soil.

8.2.4 Derivatization

The thermal instability and low volatility of some pesticides make analysis by gas chromatography (GC) difficult. Consequently, methods of analysis based on GC require, in some cases, the derivatization of pesticides to increase their volatility or thermal stability.

In addition, pesticide derivatives are sometimes prepared to enhance the response of a pesticide to a specific detector in GC or liquid chromatography (LC) analyses.

8.2.4.1 Glyphosate

This compound is very polar and has a high solubility in water so direct determination by GC is difficult. In addition, the lack of chromophores or fluorophores in its chemical structure makes impossible its analysis by LC using ultraviolet (UV) or fluorescence detection. In the GC analysis of glyphosate and its main metabolites, isopropyl chloroformate or the mixture of trifluoroacetic anhydride and trifluoro-ethanol are used as derivatizing agents [79]. Derivatives for LC determination are prepared using pre- or postcolumn reactions to improve the pesticide response. In postcolumn derivatization, the reaction is produced with o-phthalaldehyde (OPA) and mercaptoethanol, but 9-fluorenylmethyl chloroformate (FMOC-Cl) is used in precolumn derivatization, to form fluorescent derivatives with an improvement in the chromatographic determination [5].

8.2.4.2 Phenoxy Acid Herbicides

Because of their highly polar nature and low volatility, phenoxy acid herbicides cannot be directly determined by GC and have to be derivatized to their corresponding esters. Several derivatization procedures have been applied to make phenoxy acid herbicides amenable to GC analysis.

Thus, the carboxylic group is converted to the corresponding methyl ester by reacting with diazomethane or by alternative, less toxic methods, such as esterification with methanol using an acid catalyst (for example boron trifluoride) or with trimethylphenylammonium hydroxide. Alkylsilyl reagents are also very commonly used for the derivatization of polar compounds containing labile hydrogen atoms.

8.2.4.3 Phenylureas

The analysis by direct GC of phenylureas is difficult because of their thermal instability caused by the NH group. These compounds decompose in the sample inlet port and produce several peaks in the chromatogram (phenyl isocyanates).

Several analytical methods have been developed based on their ability to obtain stable derivatives for GC determination, such as alkyl, acyl, and silyl derivatives. Another derivatization mode for phenylureas is ethylation using ethyl iodide and hydrolysis to N-ethyl derivatives.

8.2.4.4 Sulfonylureas

Due to the low volatility and thermal instability of sulfonylureas, they are not detected by GC without a derivatization step. Thus, the formation of the methyl derivative suitable for GC analysis can be carried out using diazomethane or boron trifluoride diethyl ether, the latter having the advantages of better catalytic effect and stability than the former [80]. In addition, these compounds can be detected by LC, although the analysis of derivatized sulfonylureas by GC is more sensitive.

8.2.4.5 Carbamates

Carbamates are thermally decomposed into the corresponding phenols and methyl isocyanate; for this reason, LC methods for carbamates are preferred over GC determination. These methods are based on postcolumn basic hydrolysis to release methylamine, which subsequently reacts with the OPA reagent to form isoindol derivatives, which are determined by fluorescence (FL) detection [45].

8.3 DETERMINATION OF PESTICIDE RESIDUES

GC and LC are the most widely used analytical techniques for the determination of pesticide residues in soil. Thermal stability and volatility are the main characteristics that a pesticide must possess in order to be suitable for gas chromatographic (GC) analysis but after the adequate derivatization procedure this drawback may be overcome. Table 8.7 summarizes the GC methods used to determine pesticide residues in soil. Electron-capture detection (ECD) is adequate for halogenated compounds or those that contain electronegative atoms such as oxygen or sulfur, pyrethroids and OC pesticides being typical examples. A chromatogram of a mixture of fungicides analyzed by GC–ECD is depicted in Figure 8.2. On the other hand, the determination of pesticides that contain nitrogen or phosphorus atoms, such as triazines and OP pesticides, has been carried out with nitrogen–phosphorus detection (NPD). Although these selective detectors allow quantitating residues at trace levels, they have been replaced by mass spectrometry (MS) detection because it offers the possibility of the simultaneous determination and confirmation of the identity of the target compounds.

TABLE 8.7
GC Methods Used for the Determination of Pesticide Residues in Soil

Detector	Compound	LOD (µg/kg)	Ref
ECD	Organochlorines	0.01–1.6	[57, 59]
	Pyrethroids	0.5–190	[55]
	Multiresidue	2–10	[48]
NPD	Multiresidue	0.1–10	[81]
MS	Pyrethroids	0.1–1.3	[55]
	Triazines	2–4	[26]
	Imidazolinones	0.03–0.46 µg/L	[82]
	Organophosphorus	0.002–0.125	[42]
	Multiresidue	0.02–3.2	[63, 67, 83]
QTOF	Organochlorines	0.3–3 µg/L	[62]
MS/MS	Organochlorines	0.02–3.6	[54]
	Pyrethroids	0.08–0.9	[53, 55]
	Multiresidue	0.2–10	[71, 74, 75]

ECD: electron-capture detector; NPD: nitrogen-phosphorus detector; MS: mass spectrometry; MS/MS: tandem mass spectrometry; QTOF: quadrupole-time of flight; LOD: limit of detection.

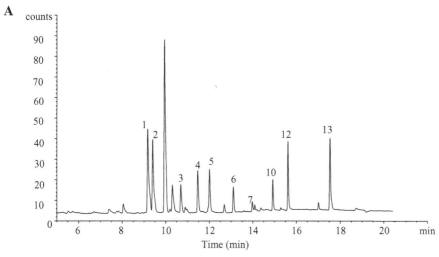

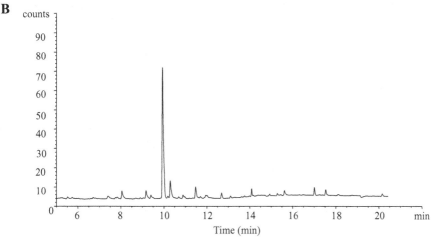

FIGURE 8.2 GC–ECD chromatograms. (a) A soil sample fortified at 0.05 µg/g and (b) a blank soil sample. Peak identification: 1=quintozene; 2=chlorothalonil; 3=tolclofos-methyl; 4=dichlofluanid; 5=triadimefon; 6=procymidone; 7=myclobutanil; 10=ofurace; 12=nuarimol; and 13=fenarimol. (Reprinted from Sánchez-Brunete, C. et al., J. Chromatogr. A, 976, 319, 2002, with permission from Elsevier.)

The ionization technique most commonly used in GC–MS analysis is electron impact (EI), which produces characteristic ion fragments of compounds that are collected in spectral libraries. Full scan and selected ion monitoring (SIM) are the two working modes for EI–MS; SIM mode is more sensitive and selective than full scan. Most of the multiresidue methods developed in the last few years use MS as detection system as it offers the possibility of the simultaneous determination and identity confirmation of a large number of pesticides from different chemical classes in a single injection. Chemical ionization (CI) is a soft ionization technique that can work with two different polarities, positive (PCI) and negative (NCI), with minimal fragmentation which is useful to obtain molecular ions that are not observed in EI mass spectra. Atmospheric pressure ionization, commonly used in liquid chromatography (LC), is also a soft ionization in comparison to EI, resulting in less fragmentation, that is starting to be used coupled to GC. The generation of higher m/z ions as abundant ions makes the selection of precursor ions for tandem mass spectrometry (MS/MS) analysis easier and compound specific. GC–MS/MS allows the determination of pesticides in soil with good selectivity and high sensibility, reducing the sample treatment steps. Time of flight mass spectrometry (TOF-MS) is the result of the significant advances undergone by the analytical instrumentation that is being applied in the determination of pesticides since full mass-range spectrum and exact mass determination can be obtained for each pesticide without compromising sensitivity. Quadrupole time of flight (QTOF) is a hybrid analyzer that combines the benefits of obtaining accurate mass determination with performing a full spectral acquisition of the entire product ion-profile.

LC is an analytical tool adequate for the determination of pesticides that are not thermally stable or not volatile. The LC methods developed for the determination of pesticides in soil are summarized in Table 8.8. Ultraviolet (UV) detection has been the most frequently used technique in LC, although other selective detectors such as fluorescence (FL) present higher selectivity and sensitivity. The drawback of FL detection is that it is limited to compounds that fluoresce. Otherwise, derivatization is required to obtain a fluorescent compound.

TABLE 8.8
LC Methods Used for the Determination of Pesticide Residues in Soil

Detector	Compound	LOD (µg/kg)	Ref
UV	Carbamates	0.01–0.4	[84]
	Sulfonylureas	0.1–1.2	[7, 19]
	Benzoylureas	0.08–0.6	[43]
	Triazines	0.1–3	[17]
	Quaternary ammonium compounds	5–100	[29]
	Multiresidue	0.1–29	[85]
FL	Carbamates	0.2–3.7	[35, 45]
	Glyphosate	250	[76]
	Multiresidue	0.001–6.9	[78]
MS	Multiresidue	0.5–2.5	[65]
MS/MS	Phenoxyacids	2–5	[31]
	Imidazolinones	1.5	[9]
	Neonicotinoids	0.08–6.1	[60]
	Glyphosate	5	[5]
	Multiresidue	0.03–33	[10, 36, 71, 86, 87]
QTOF	Multiresidue		[88]

UV: ultraviolet detector; FL: fluorescence detector; MS: mass spectrometry; MS/MS: tandem mass spectrometry; QTOF: quadrupole-time of flight

The use of liquid chromatography coupled with mass spectrometry (LC–MS) in pesticide residue analysis has increased exponentially in the last decade, particularly liquid chromatography coupled to tandem mass spectrometry (LC–MS/MS), which has become the most popular technique for the identification and confirmation of pesticide residues. The implementation of robust ionization interfaces, such as electrospray ionization (ESI) and atmospheric pressure chemical ionization (APCI), is considered one of the main instrumental improvements. The selection of the ionization interface depends on the nature of the analyzed pesticide; APCI is adequate for moderately nonpolar pesticides such as triazines and phenylureas, whereas ESI is suitable for polar and ionic pesticides.

LC–QTOF-MS has become a popular technique because it allows the analysis of pesticide residues together with unknown transformation products or metabolites. Figure 8.3 shows the chromatogram and mass spectra obtained for linuron in a soil sample obtained by ultra-high performance liquid chromatography (UHPLC) –QTOF-MS. The spectrum acquired at low collision energy provides information on the parent compounds, and the spectrum obtained at high collision energy is the result of the fragmentation of the analyte, which is very useful in identifying it.

The analysis of pesticides has also been carried out with non-chromatographic methods. Capillary electrophoresis (CE) is an alternative analytical tool that has been

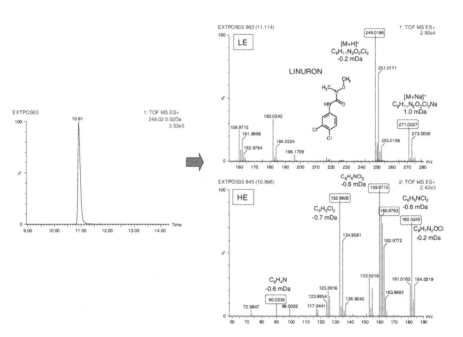

FIGURE 8.3 Detection and identification of linuron in a soil sample using UHPLC–QTOF under MS^E acquisition mode. Chromatogram (left) and LE (low collision energy) and HE (high collision energy) spectra (right) of the sample. Mass errors in mDa (ppm, in brackets): m/z 271.0027: 1.0 (3.7); 249.0196: −0.2 (0.8); 182.0245: −0.2 (1.1); 160.9793: −0.6 (−3.7); 159.9715: −0.6 (−3.7); 132.9605: −0.7 (5.3); 90.0338: −0.6 (6.7). (Reprinted from Hernández, F. et al., *Sci. Total Environ.* 439, 249, 2012, with permission from Elsevier.)

applied in the determination of residues in soil samples [14, 89, 90]. CE presents different working modes, and micellar electrokinetic chromatography (MECK), capillary zone electrophoresis (CZE), and capillary electrochromatography (CEC) are the most frequently used. The application of sensors and biosensors in the determination of pesticides in environmental samples is also rapidly increasing. These portable analytical devices offer the possibility of *in situ* analysis [91]. Immunoassays, such as enzyme-linked immunoabsorbent assay (ELISA), have also been used to determine pesticides [92, 93]. This technique, as well as the biosensors, is usually applied as a screening test rather than to quantitate residue levels, and the chromatographic methods are a more suitable alternative for the latter purpose.

8.4 APPLICATION TO REAL SAMPLES

In this section, principles of the main methods used in the determination of representative pesticide classes in soils are given.

8.4.1 GLYPHOSATE

Glyphosate is a highly polar herbicide, very soluble in water and insoluble in most organic solvents. GC analysis is normally carried out after obtaining acetyl derivatives and LC analysis after derivatization with FMOC.

Soil samples (5 g) are extracted by shaking with 10 mL of 0.6 M KOH. The extract is neutralized by adding an adequate volume of HCl until the extract reaches pH 7, and derivatized with 120 µL of FMOC-Cl reagent. The derivative is acidified to pH 1.5 and analyzed by LC–ESI-MS/MS with a limit of detection of 5 µg/kg. The method is rapid and selective for the determination of glyphosate at very low levels [5].

8.4.2 SULFONYLUREAS

GC of sulfonylurea herbicides is very difficult because of their thermally labile properties and strongly polar nature.

Soil (10 g) is transferred to Teflon vessels for its microwave-assisted extraction with 10 mL of dichloromethane–acetonitrile (2:1, v/v) acidified with acetic acid (0.8%) and 0.3 g of urea. Two extraction cycles of 10 min each are carried out with fresh solvent. The sample is centrifuged and the extract is evaporated to dryness. The dry residue is re-dissolved in dichloromethane and analyzed by LC–UV without further cleanup. Limits of detection of 0.9 µg/kg were achieved and the results of this method were in agreement with those obtained using an LC–MS/MS method [28].

8.4.3 TRIAZINES

Triazines are herbicides that have been heavily applied to control weeds in agricultural, industrial, and urban areas. These herbicides are persistent in the environment and due to their water solubility, they may contaminate groundwater.

Triazines are extracted from soil (15 g) by applying pressurized liquid extraction with acetone, performing three cycles at 60°C. The extract is evaporated to dryness and reconstituted in 10 mL dichloromethane–methanol (99:1, v/v) for its purification using 170 mg of terbutylazine MIP as SPE sorbent. The analytes are eluted from the cartridge with 3 mL of methanol and then concentrated to 0.5 mL before LC–MS/MS. This selective method allowed limits of quantification at sub-ppb levels [22].

8.4.4 Organophosphorus

These insecticides have high polarity and solubility in water and are frequently analyzed by GC and LC. Soil (20 g) is extracted for 10 min by ultrasonic agitation with acetonitrile (20 mL). The acetonitrile is evaporated to dryness and the residue reconstituted in 0.4 mL of mobile phase (acetonitrile–water, 65:35, v/v). The determination of diazinon and fenitrothion is performed by LC and UV detection at 245 and 267 nm, respectively. The quantification limits are 1 and 2 ng/g for fenitrothion and diazinon, respectively, with a good level of reproducibility and accuracy [40].

8.4.5 Pyrethroids

These compounds are retained in soil due to their low solubility in water. Chromatographic methods, GC as well as LC, are used for the determination of pyrethroids in soil.

Soil (10 g) is mixed with diatomaceous earth (2 g) for pressurized liquid extraction with dichloromethane–acetone (1:1, v/v) performing three cycles at 100°C. The extract is concentrated to near dryness under a nitrogen stream and re-dissolved in n-hexane (1 mL). Then the hexane extract is passed through a 500 mg graphitized carbon black (GCB) column and pyrethroids are eluted with 10 mL dichloromethane–n-hexane (3:7, v/v). The determination of pyrethroid residues is carried out by GC–MS/MS. This method provides a high sensitivity and selectivity with limits of detection from 0.26 to 0.87 ng/g [53].

8.4.6 Neonicotinoids

Neonicotinoids, water-soluble and polar compounds, are determined by LC methods.

Soil samples (10 g) are shaken for 2 min in a Teflon tube with 25 mL acetonitrile–dichloromethane (8.3:16.7, v/v). Sodium chloride (5 g) is added and the sample vortexed for 1 min before a 10 min centrifugation. The upper layer of the supernatant is transferred to a tube containing 0.4 g C18 that is then vortexed for 1 min and centrifuged for 10 min. The purified extract was evaporated to dryness in a water bath at 40°C and reconstituted in the mobile phase (water–acetonitrile, 95:5, v/v both containing 0.1% formic acid) for LC–MS/MS analysis. This method allowed the determination of eight neonicotinoid insecticides and two metabolites [60].

8.4.7 Pyrimethanil and Kresoxim-Methyl Fungicides

Pyrimethanil (anilino-pyrimidine) and kresoxim-methyl (strobilurin) are two novel fungicides with broad-spectrum activity.

Soil (2 g) is placed in a vial with phosphate buffer solution (pH 7) and NaCl, and immersed in a temperature-controlled oil bath at 100°C. The sample is agitated with a magnetic stir bar during the headspace SPME extraction. The polyacrilate (PA) fiber is exposed to the headspace for 25 min, and then inserted in the injector of a GC where the fungicides are desorbed for 5 min. A low polarity capillary column of 30 m is used for the determination of fungicides with a temperature program from 100°C to 300°C and carrier gas at a flow rate of 2 mL/min. The detection limits are 1 and 4 µg/kg for pyrimethanil and kresoxim-methyl, respectively [58].

8.4.8 Multiresidue

Due to the large number of pesticides used, multiresidue analytical methods require techniques able to determine the greatest possible number of these compounds in a single analysis.

Soil (5 g) is shaken for 1 min with 10 mL of acetonitrile and 3 mL of water, followed by the addition of 6 g anhydrous $MgSO_4$, 1.5 g NaCl, 1.5 g trisodium citrate dehydrate, and 0.75 g disodium hydrogen citrate sesquihydrate to carry out the extraction step of the QuEChERS method. The samples are sonicated for 5 min in an ultrasonic bath before centrifugation. An aliquot of the upper layer (1.5 mL) is transferred to a mini-centrifuge tube containing 150 mg primary secondary amine (PSA) sorbent, 50 mg C18, and 150 mg anhydrous $MgSO_4$ for the dSPE cleanup. LC–MS/MS and GC–MS/MS were used to determine 193 targeted pesticides whereas two-dimensional gas chromatography time-of-flight mass spectrometry was applied to screen over 600 pesticides in a targeted and untargeted approach [73].

8.5 FUTURE TRENDS

Determination of pesticides in soils usually involves conventional extraction methods that demand large volumes of hazardous organic solvents. Therefore, substantial efforts have been made to develop sample preparation techniques that could alleviate the drawbacks associated with the conventional methods. Various modern extraction techniques have yielded good results, although they still require optimization for multiresidue analysis of pesticides in soil because of the disparity of chemical compounds involved. Automation of sample preparation and coupling with instrumental analysis are also important goals to reach.

Analytical methodologies employed must be capable of residue measurement at very low levels and must also provide unambiguous evidence to confirm the identity of any residue detected. GC–MS/MS is a powerful tool to identify thermally stable pesticides in soils with high sensitivity and selectivity. However, the number of compounds that cannot be determined by GC because of their poor volatility and thermal instability has grown dramatically in the last few years. Thus, LC–MS has become one important technique for the determination of pesticide residues, and LC–MS/MS is capable of discriminating more efficiently than LC–MS. Another analyzer employed is TOF-MS. This results in an improved mass spectrometric resolution, which is important in the detection of unknown compounds. Further optimization of sensitivity and quality is accomplished when mass spectrometers with very fast MS–MS/MS switching and scanning capabilities are used. Most of the methods based on LC–MS/MS achieve satisfactory results even without making use of any cleanup step.

REFERENCES

1. Food Quality Protection Act (FQPA), Summary of the Food Quality Protection Act. Available from https://www.epa.gov/laws-regulations/summary-food-quality-protection-act, 1996.
2. European Parliament, Regulation (EC) 396/2005 of the European Parliament and of the Council of 23 February 2005 on maximum residue levels of pesticides in or on food and feed of plant and animal origin and amending Council Directive 91/414/EEC. *Official Journal of the European Union*. Available from http://eur-lex.europa.eu/.
3. S. Moret, M. Hidalgo, J.M. Sánchez, Development of an ion-pairing liquid chromatography method for the determination of phenoxyacetic herbicides and their main metabolites: Application to the analysis of soil samples, *Chromatographia*, 63 (2006) 109–115.
4. S. Moret, J.M. Sánchez, V. Salvado, M. Hidalgo, The evaluation of different sorbents for the preconcentration of phenoxyacetic acid herbicides and their metabolites from soils, *J. Chromatogr. A*, 1099 (2005) 55–63.
5. M. Ibáñez, O.J. Pozo, J.V. Sancho, F.J. López, F. Hernández, Residue determination of glyphosate, glufosinate and aminomethylphosphonic acid in water and soil samples by liquid chromatography coupled to electrospray tandem mass spectrometry, *J. Chromatogr. A*, 1081 (2005) 145–155.
6. M. Pei, X. Zhu, X. Huang, Mixed functional monomers-based monolithic adsorbent for the effective extraction of sulfonylurea herbicides in water and soil samples, *J. Chromatogr. A*, 1531 (2018) 13–21.
7. Q. Wu, C. Wang, Z. Liu, C. Wu, X. Zeng, J. Wen, Z. Wang, Dispersive solid-phase extraction followed by dispersive liquid–liquid microextraction for the determination of some sulfonylurea herbicides in soil by high-performance liquid chromatography, *J. Chromatogr. A*, 1216 (2009) 5504–5510.
8. G. Min, S. Wang, H. Zhu, G. Fang, Y. Zhang, Multi-walled carbon nanotubes as solid-phase extraction adsorbents for determination of atrazine and its principal metabolites in water and soil samples by gas chromatography–mass spectrometry, *Sci. Total Environ.*, 396 (2008) 79–85.
9. M. Kemmerich, G. Bernardi, M.B. Adaime, R. Zanella, O.D. Prestes, A simple and efficient method for imidazolinone herbicides determination in soil by ultra-high performance liquid chromatography–tandem mass spectrometry, *J. Chromatogr. A*, 1412 (2015) 82–89.
10. X. Dong, S. Liang, Z. Shi, H. Sun, Multi-residue analysis of herbicides in soil with an UPLC–ESI-MS method, *Soil Sediment Contam.*, 24 (2015) 573–587.
11. J. Abián, G. Durand, D. Barceló, Analysis of chlorotriazines and their degradation products in environmental-samples by selecting various operating modes in thermospray HPLC/MS/MS, *J. Agric. Food Chem.*, 41 (1993) 1264–1273.

12. G. Durand, P. Gille, D. Fraisse, D. Barceló, Comparison of gas-chromatographic mass-spectrometric methods for screening of chlorotriazine pesticides in soil, *J.Chromatogr.*, 603 (1992) 175–184.

13. R.V. Crouch, E.M. Pullin, Analytical method for residues of bromoxynil octanoate and bromoxynil in soil, *Pestic. Sci.*, 5 (1974) 281–285.

14. J. Hernández-Borges, F.J. García-Montelongo, A. Cifuentes, M.A. Rodríguez-Delgado, Analysis of triazolopyrimidine herbicides in soils using field-enhanced sample injection-coelectroosmotic capillary electrophoresis combined with solid-phase extraction, *J. Chromatogr. A*, 1100 (2005) 236–242.

15. S. Stipičević, S. Fingler, L. Zupančič-Kralj, V. Drevenkar, Comparison of gas and high performance liquid chromatography with selective detection for determination of triazine herbicides and their degradation products extracted ultrasonically from soil, *J. Sep. Sci.*, 26 (2003) 1237–1246.

16. H.R. Geng, S.S. Miao, S.F. Jin, H. Yang, A newly developed molecularly imprinted polymer on the surface of TiO_2 for selective extraction of triazine herbicides residues in maize, water, and soil, *Anal. Bioanal. Chem.*, 407 (2015) 8803–8812.

17. M. José Patino-Ropero, M. Díaz-Álvarez, A. Martín-Esteban, Molecularly imprinted core-shell magnetic nanoparticles for selective extraction of triazines in soils, *J. Mol. Recognit.*, 30 (2017).

18. M. Ghobadi, Y. Yamini, B. Ebrahimpour, Extraction and determination of sulfonylurea herbicides in water and soil samples by using ultrasound-assisted surfactant-enhanced emulsification microextraction and analysis by high-performance liquid chromatography, *Ecotoxicol. Environ. Saf.*, 112 (2015) 68–73.

19. Y. Peng, Y. Xie, J. Luo, L. Nie, Y. Chen, L. Chen, S. Du, Z. Zhang, Molecularly imprinted polymer layer-coated silica nanoparticles toward dispersive solid-phase extraction of trace sulfonylurea herbicides from soil and crop samples, *Anal. Chim. Acta*, 674 (2010) 190–200.

20. M. Guardia Rubio, V. Banegas Font, A. Molina Díaz, M.J. Ayora Cañada, Determination of triazine herbicides and diuron in mud from olive washing devices and soils using gas chromatography with selective detectors, *Anal. Lett.*, 39 (2006) 835–850.

21. E. Kremer, M. Rompa, B. Zygmunt, Extraction of acidic herbicides from soil by means of accelerated solvent extraction, *Chromatographia*, 60 (2004) S169–S174.

22. L. Amalric, C. Mouvet, V. Pichon, S. Bristeau, Molecularly imprinted polymer applied to the determination of the residual mass of atrazine and metabolites within an agricultural catchment (Brevilles, France), *J. Chromatogr. A*, 1206 (2008) 95–104.

23. T. Dagnac, S. Bristeau, R. Jeannot, C. Mouvet, N. Baran, Determination of chloroacetanilides, triazines and phenylureas and some of their metabolites in soils by pressurised liquid extraction, GC–MS/MS, LC–MS and LC–MS/MS, *J. Chromatogr. A*, 1067 (2005) 225–233.

24. J. Patsias, E.N. Papadakis, E. Papadopoulou-Mourkidou, Analysis of phenoxyalkanoic acid herbicides and their phenolic conversion products in soil by microwave assisted solvent extraction and subsequent analysis of extracts by on-line solid-phase extraction-liquid chromatography, *J. Chromatogr. A*, 959 (2002) 153–161.

25. E.N. Papadakis, E. Papadopoulou-Mourkidou, LC–UV determination of atrazine and its principal conversion products in soil after combined microwave-assisted and solid-phase extraction, *Intern. J. Environ. Anal. Chem.*, 86 (2006) 573–582.

26. G. Shen, H.K. Lee, Determination of triazines in soil by microwave-assisted extraction followed by solid-phase microextraction and gas chromatography–mass spectrometry, *J. Chromatogr. A*, 985 (2003) 167–174.

27. X. Hu, Y. Hu, G. Li, Development of novel molecularly imprinted solid-phase microextraction fiber and its application for the determination of triazines in complicated samples coupled with high-performance liquid chromatography, *J. Chromatogr. A*, 1147 (2007) 1–9.

28. N.L. Grahovac, Z.S. Stojanovic, S.Ž. Kravić, D.Z. Orčić, Z.J. Suturović, A.D. Kondić-Špika, J.R. Vasin, D.B. Šunjka, S.P. Jakšić, M.M. Rajković, N.M. Grahovac, Determination of residues of sulfonylurea herbicides in soil by using microwave-assisted extraction and high performance liquid chromatographic method, *Hem. Ind.*, 71 (2017) 289–298.

29. M. Pateiro-Moure, E. Martínez-Carballo, M. Arias-Estévez, J. Simal-Gándara, Determination of quaternary ammonium herbicides in soils – Comparison of digestion, shaking and microwave-assisted extractions, *J. Chromatogr. A*, 1196 (2008) 110–116.

30. Z. Vryzas, E. Papadopoulou-Mourkidou, Determination of triazine and chloroacetanilide herbicides in soils by microwave-assisted extraction (MAE) coupled to gas chromatographic analysis with either GC–NPD or GC–MS, *J. Agric. Food Chem.*, 50 (2002) 5026–5033.

31. P. Kaczyński, B. Łozowicka, M. Jankowska, I. Hrynko, Rapid determination of acid herbicides in soil by liquid chromatography with tandem mass spectrometric detection based on dispersive solid phase extraction, *Talanta*, 152 (2016) 127–136.

32. J.Z. Li, X. Wu, J.Y. Hu, Determination of fungicide kresoxim-methyl residues in cucumber and soil by capillary gas chromatography with nitrogen-phosphorus detection, *J. Environ. Sci. Health B*, 41 (2006) 427–436.

33. S.A. Sassman, L.S. Lee, M. Bischoff, R.F. Turco, Assessing N,N-dibutylurea(DBU) formation in soils after application of n-butylisocyanate and benlate fungicides, *J. Agric. Food Chem.*, 52 (2004) 747–754.

34. C. Jia, X. Zhu, E. Zhao, P. Yu, M. He, L. Chen, Application of SPME based on a stainless steel wire for the determination of pyrethroid insecticide residues in water and soil, *Chromatographia*, 72 (2010) 1219–1223.

35. H. Lu, Y. Lin, P.C. Wilson, Organic-solvent-free extraction method for determination of carbamate and carbamoyloxime pesticides in soil and sediment samples, *Bull. Environ. Contam. Toxicol.*, 83 (2009) 621–625.

36. M. Pastor-Belda, I. Garrido, N. Campillo, P. Viñas, P. Hellín, P. Flores, J. Fenoll, Dispersive liquid–liquid microextraction for the determination of new generation pesticides in soils by liquid chromatography and tandem mass spectrometry, *J. Chromatogr. A*, 1394 (2015) 1–8.

37. K. Vig, D.K. Singh, H.C. Agarwal, A.K. Dhawan, P. Dureja, Insecticide residues in cotton crop soil, *J. Environ. Sci. Health B*, 36 (2001) 421–434.

38. J. Błądek, A. Rostkowski, M. Miszczak, Application of instrumental thin-layer chromatography and solid-phase extraction to the analyses of pesticide residues in grossly contaminated samples of soil, *J. Chromatogr. A*, 754 (1996) 273–278.

39. A. Tor, M.E. Aydin, S. Ozcan, Ultrasonic solvent extraction of organochlorine pesticides from soil, *Anal. Chim. Acta*, 559 (2006) 173–180.

40. M.E. Sánchez, R. Méndez, X. Gómez, J. Martín-Villacorta, Determination of diazinon and fenitrothion in environmental water and soil samples by HPLC, *J. Liq. Chromatogr. Relat. Technol.*, 26 (2003) 483–497.

41. Y. Abdollahzadeh, Y. Yamini, A. Jabbari, A. Esrafili, M. Rezaee, Application of ultrasound-assisted emulsification microextraction followed by gas chromatography for determination of organophosphorus pesticides in water and soil samples, *Anal. Methods*, 4 (2012) 830–837.

42. M. Mohammadi, H. Tavakoli, Y. Abdollahzadeh, A. Khosravi, R. Torkaman, A. Mashayekhi, Ultra-preconcentration and determination of organophosphorus pesticides in soil samples by a combination of ultrasound assisted leaching-solid phase extraction and low-density solvent based dispersive liquid-liquid microextraction, *RSC Adv.*, 5 (2015) 75174–75181.

43. G. Peng, Q. He, D. Mmereki, Y. Lu, Z. Zhong, H. Liu, W. Pan, G. Zhou, J. Chen, Dispersive solid-phase extraction followed by vortex-assisted dispersive liquid–liquid microextraction based on the solidification of a floating organic droplet for the determination of benzoylurea insecticides in soil and sewage sludge, *J. Sep. Sci.*, 39 (2016) 1258–1265.

44. M.A. Ali, P.J. Baugh, Pyrethroid soil extraction, properties of mixed solvents and time profiles using GC/MS-NICI analysis, *Intern. J. Environ. Anal. Chem.*, 83 (2003) 909–922.

45. C. Sánchez-Brunete, A. Rodriguez, J.L. Tadeo, Multiresidue analysis of carbamate pesticides in soil by sonication-assisted extraction in small columns and liquid chromatography, *J. Chromatogr. A*, 1007 (2003) 85–91.

46. R. Rial-Otero, R.M. González-Rodríguez, B. Cancho-Grande, J. Simal-Gándara, Parameters affecting extraction of selected fungicides from vineyard soils, *J. Agric. Food Chem.*, 52 (2004) 7227–7234.

47. J. Castro, C. Sánchez-Brunete, J.L. Tadeo, Multiresidue analysis of insecticides in soil by gas chromatography with electron-capture detection and confirmation by gas chromatography–mass spectrometry, *J. Chromatogr. A*, 918 (2001) 371–380.

48. C. Sánchez-Brunete, E. Miguel, J.L. Tadeo, Multiresidue analysis of fungicides in soil by sonication-assisted extraction in small columns and gas chromatography, *J. Chromatogr. A*, 976 (2002) 319–327.

49. L. Sun, H.K. Lee, Optimization of microwave-assisted extraction and supercritical fluid extraction of carbamate pesticides in soil by experimental design methodology, *J. Chromatogr. A*, 1014 (2003) 165–177.

50. T. O'Mahony, S. Moore, B. Brosnan, J.D. Glennon, Monitoring the supercritical fluid extraction of pyrethroid pesticides using capillary electrochromatography, *Intern. J. Environ. Anal. Chem.*, 83 (2003) 681–691.

51. E. Concha-Grana, M.I. Turnes-Carou, S. Muniategui-Lorenzo, P. López-Mahía, E. Fernández-Fernández, D. Prada-Rodríguez, Development of pressurized liquid extraction and cleanup procedures for determination of organochlorine pesticides in soils, *J. Chromatogr. A*, 1047 (2004) 147–155.

52. A. Hussen, R. Westbom, N. Megersa, L. Mathiasson, E. Bjorklund, Development of a pressurized liquid extraction and clean-up procedure for the determination of alpha-endosulfan, beta-endosulfan and endosulfan sulfate in aged contaminated Ethiopian soils, *J. Chromatogr. A*, 1103 (2006) 202–210.

53. L. Luo, B. Shao, J. Zhang, Pressurized liquid extraction and cleanup procedure for the determination of pyrethroids in soils using gas chromatography/tandem mass spectrometry, *Anal. Sci.*, 26 (2010) 461–465.

54. P. Herbert, S. Morais, P. Paíga, A. Alves, L. Santos, Development and validation of a novel method for the analysis of chlorinated pesticides in soils using microwave-assisted extraction-headspace solid phase microextraction and gas chromatography–tandem mass spectrometry, *Anal. Bioanal. Chem.*, 384 (2006) 810–816.

55. F.A. Esteve-Turrillas, A. Pastor, M. de la Guardia, Comparison of different mass spectrometric detection techniques in the gas chromatographic analysis of pyrethroid insecticide residues in soil after microwave-assisted extraction, *Anal. Bioanal. Chem.*, 384 (2006) 801–809.

56. R.S. Zhao, X. Wang, S. Fu, J.P. Yuan, T. Jiang, S.B. Xu, A novel headspace solid-phase microextraction method for the exact determination of organochlorine pesticides in environmental soil samples, *Anal. Bioanal. Chem.*, 384 (2006) 1584–1589.

57. X. Chai, J. Jia, T. Sun, Y. Wang, L. Liao, Application of a novel cold activated carbon fiber-solid phase microextraction for analysis of organochlorine pesticides in soil, *J. Environ. Sci. Health B*, 42 (2007) 629–634.

58. A. Navalón, A. Prieto, L. Araujo, L.L. Vílchez, Determination of pyrimethanil and kresoxim-methyl in soils by headspace solid-phase microextraction and gas chromatography–mass spectrometry, *Anal. Bioanal. Chem.*, 379 (2004) 1100–1105.

59. A. Ghadiri, A. Salemi, Matrix solid-phase dispersion based on carbon nanotube coupled with dispersive liquid–liquid microextraction for determination of organochlorine pesticides in soil, *J. Chromatogr. Sci.*, 55 (2017) 578–585.

60. M.F. Abdel-Ghany, L.A. Hussein, N.F. El Azab, A.H. El-Khatib, M.W. Linscheid, Simultaneous determination of eight neonicotinoid insecticide residues and two primary metabolites in cucumbers and soil by liquid chromatography–tandem mass spectrometry coupled with QuEChERS, *J. Chromatogr. B*, 1031 (2016) 15–28.

61. L. Correia-Sá, V.C. Fernandes, M. Carvalho, C. Calhau, V.F. Domingues, C. Delerue-Matos, Optimization of QuEChERS method for the analysis of organochlorine pesticides in soils with diverse organic matter, *J. Sep. Sci.*, 35 (2012) 1521–1530.

62. Z. Cheng, F. Dong, J. Xu, X. Liu, X. Wu, Z. Chen, X. Pan, Y. Zheng, Atmospheric pressure gas chromatography quadrupole-time-of-flight mass spectrometry for simultaneous determination of fifteen organochlorine pesticides in soil and water, *J. Chromatogr. A*, 1435 (2016) 115–124.

63. R.D. Đurović, T.M. Đorđević, L.R. Šantrić, S.M. Gašić, L.M. Ignjatović, Headspace solid phase microextraction method for determination of triazine and organophosphorus pesticides in soil, *J Environ Sci Health BJ. Environ. Sci. Health B*, 45 (2010) 626–632.

64. J. Fenoll, P. Hellín, C.M. Martínez, P. Flores, Multiresidue analysis of pesticides in soil by high-performance liquid chromatography with tandem mass spectrometry, *J. AOAC Int.*, 92 (2009) 1566–1575.

65. A.B. Vega, A.G. Frenich, J.L.M. Vidal, Monitoring of pesticides in agricultural water and soil samples from Andalusia by liquid chromatography coupled to mass spectrometry, *Anal. Chim. Acta*, 538 (2005) 117–127.

66. C. Gonçalves, M.F. Alpendurada, Assessment of pesticide contamination in soil samples from an intensive horticulture area, using ultrasonic extraction and gas chromatography–mass spectrometry, *Talanta*, 65 (2005) 1179–1189.

67. C. Sánchez-Brunete, B. Albero, J.L. Tadeo, Multiresidue determination of pesticides in soil by gas chromatography–mass spectrometry detection, *J. Agric. Food Chem.*, 52 (2004) 1445–1451.

68. C. Gonçalves, J.J. Carvalho, M.A. Azenha, M.F. Alpendurada, Optimization of supercritical fluid extraction of pesticide residues in soil by means of central composite design and analysis by gas chromatography–tandem mass spectrometry, *J. Chromatogr. A*, 1110 (2006) 6–14.

69. P. Richter, B. Sepúlveda, R. Oliva, K. Calderón, R. Seguel, Screening and determination of pesticides in soil using continuous subcritical water extraction and gas chromatography–mass spectrometry, *J. Chromatogr. A*, 994 (2003) 169–177.

70. A. Hildebrandt, S. Lacorte, D. Barceló, Assessment of priority pesticides, degradation products, and pesticide adjuvants in groundwaters and top soils from agricultural areas of the Ebro river basin, *Anal. Bioanal. Chem.*, 387 (2007) 1459–1468.

71. J.L. Martínez Vidal, J.A. Padilla Sánchez, P. Plaza-Bolaños, A. Garrido Frenich, R. Romero-González, Use of pressurized liquid extraction for the simultaneous analysis of 28 polar and 94 non-polar pesticides in agricultural soils by GC/QqQ-MS/MS and UPLC/QqQ-MS/MS, *J. AOAC Int.*, 93 (2010) 1715–1731.

72. J. Castro, R.A. Pérez, C. Sánchez-Brunete, J.L. Tadeo, Analysis of pesticides volatilised from plants and soil by headspace solid-phase microextraction and gas chromatography, *Chromatographia*, 53 (2001) S361–S365.

73. V.C. Fernandes, S.J. Lehotay, L. Geis-Asteggiante, H. Kwon, H.G.J. Mol, H. van der Kamp, N. Mateus, V.F. Domingues, C. Delerue-Matos, Analysis of pesticide residues in strawberries and soils by GC–MS/MS, LC–MS/MS and two dimensional GC–time-of-flight MS comparing organic and integrated pest management farming, *Food Addit. Contam. A*, 31 (2014) 262–270.

74. V.C. Fernandes, V.F. Domingues, N. Mateus, C. Delerue-Matos, Multiresidue pesticides analysis in soils using modified QuEChERS with disposable pipette extraction and dispersive solid-phase extraction, *J. Sep. Sci.*, 36 (2013) 376–382.

75. B. Łozowicka, E. Rutkowska, M. Jankowska, Influence of QuEChERS modifications on recovery and matrix effect during the multi-residue pesticide analysis in soil by GC/MS/MS and GC/ECD/NPD, *Environ. Sci. Pollut. R.*, 24 (2017) 7124–7138.

76. A. Gómez-Caballero, G. Diaz-Diaz, O. Bengoetxea, A. Quintela, N. Unceta, M.A. Goicolea, R.J. Barrio, Water compatible stir-bar devices imprinted with underivatised glyphosate for selective sample clean-up, *J. Chromatogr. A*, 1451 (2016) 23–32.

77. L. Fu, X. Liu, J. Hu, X. Zhao, H. Wang, C. Huang, X. Wang, Determination of two pesticides in soils by dispersive liquid–liquid microextraction combined with LC–fluorescence detection, *Chromatographia*, 70 (2009) 1697–1701.

78. M. Asensio-Ramos, J. Hernández-Borges, G. González-Hernández, M. Angel Rodríguez-Delgado, Hollow-fiber liquid-phase microextraction for the determination of pesticides and metabolites in soils and water samples using HPLC and fluorescence detection, *Electrophoresis*, 33 (2012) 2184–2191.

79. J.P.K. Gill, N. Sethi, A. Mohan, Analysis of the glyphosate herbicide in water, soil and food using derivatising agents, *Environ. Chem. Lett.*, 15 (2017) 85–100.

80. Y. Zhang, J. Wang, G. Wang, C. Gao, Y. Yan, B. Wen, Optimization of derivatization procedure and gas chromatography–mass spectrometry method for determination of bensulfuron-methyl herbicide residues in water, *J. Chromatogr. B*, 995 (2015) 31–37.

81. J. Fenoll, P. Hellín, C. Marín, C.M. Martínez, P. Flores, Multiresidue analysis of pesticides in soil by gas chromatography with nitrogen-phosphorus detection and gas chromatography mass spectrometry, *J. Agric. Food Chem.*, 53 (2005) 7661–7666.

82. C. Ozcan, U.K. Cebi, M.A. Gurbuz, S. Ozer, Residue analysis and determination of IMI herbicides in sunflower and soil by GC–MS, *Chromatographia*, 80 (2017) 941–950.

83. N. Mantzos, A. Karakitsou, I. Zioris, E. Leneti, I. Konstantinou, QuEChERS and solid phase extraction methods for the determination of energy crop pesticides in soil, plant and runoff water matrices, *Intern. J. Environ. Anal. Chem.*, 93 (2013) 1566–1584.

84. C. Basheer, A.A. Alnedhary, B.S.M. Rao, H.K. Lee, Determination of carbamate pesticides using micro-solid-phase extraction combined with high-performance liquid chromatography, *J. Chromatogr. A*, 1216 (2009) 211–216.

85. M. Hutta, M. Chalányová, R. Halko, R. Góra, S. Dokupilová, I. Rybár, Reversed phase liquid chromatography trace analysis of pesticides in soil by on-column sample pumping large volume injection and UV detection, *J. Sep. Sci.*, 32 (2009) 2034–2042.

86. J.L. de Oliveira Arias, C. Rombaldi, S.S. Caldas, E.G. Primel, Alternative sorbents for the dispersive solid-phase extraction step in quick, easy, cheap, effective, rugged and safe method for extraction of pesticides from rice paddy soils with determination by liquid chromatography tandem mass spectrometry, *J. Chromatogr. A*, 1360 (2014) 66–75.

87. X. Feng, Z. He, L. Wang, Y. Peng, M. Luo, X. Liu, Multiresidue analysis of 36 pesticides in soil using a modified quick, easy, cheap, effective, rugged, and safe method by liquid chromatography with tandem quadruple linear ion trap mass spectrometry, *J. Sep. Sci.*, 38 (2015) 3047–3054.

88. F. Hernández, T. Portolés, M. Ibáñez, M.C. Bustos-López, R. Díaz, A.M. Botero-Coy, C.L. Fuentes, G. Peñuela, Use of time-of-flight mass spectrometry for large screening of organic pollutants in surface waters and soils from a rice production area in Colombia, *Sci. Total Environ.*, 439 (2012) 249–259.

89. E. Orejuela, M. Silva, Rapid and sensitive determination of phosphorus-containing amino acid herbicides in soil samples by capillary zone electrophoresis with diode laser-induced fluorescence detection, *Electrophoresis*, 26 (2005) 4478–4485.

90. M.E. Sánchez, B. Rabanal, M. Otero, J. Martín-Villacorta, Solid-phase extraction for the determination of dimethoate in environmental water and soil samples by micellar electrokinetic capillary chromatography (MEKC), *J. Liq. Chromatogr.Relat. Technol.*, 26 (2003) 545–557.

91. C.I.L. Justino, A.C. Duarte, T.A.P. Rocha-Santos, Recent progress in biosensors for environmental monitoring: A review, *Sensors*, 17 (2017).

92. E. Watanabe, S. Miyake, Y. Yogo, Review of enzyme-linked immunosorbent assays (ELISAs) for analyses of neonicotinoid insecticides in agro-environments, *J. Agric. Food Chem.*, 61 (2013) 12459–12472.

93. S. Conde, K. Suyama, K. Itoh, H. Yamamoto, Application of commercially available fenitrothion-ELISA kit for soil residue analysis, *J. Pestic. Sci.*, 33 (2008) 51–57.

9 Determination of Pesticides in Water

Rosa Ana Pérez, Beatriz Albero, and José L. Tadeo

CONTENTS

List of Abbreviations...271
9.1 Introduction 273
9.2 Sample Preparation Techniques in Pesticide Residue Analysis from
  Water Samples 274
  9.2.1 Liquid–Liquid Extraction 274
  9.2.2 Liquid–Liquid Microextraction...275
    9.2.2.1 Dispersive Liquid–Liquid Microextraction (DLLME)......275
    9.2.2.2 Single-Drop Microextraction (SDME) 279
    9.2.2.3 Hollow Fiber Membrane Liquid Phase Microextraction
      (HF-LPME)..280
  9.2.3 Solid-Phase Extraction...281
    9.2.3.1 Standard SPE 281
    9.2.3.2 Solid-Phase Microextraction (SPME).............................283
    9.2.3.3 Stir Bar Sorptive Extraction (SBSE)...............................286
    9.2.3.4 Magnetic Solid-Phase Extraction (mSPE) 287
    9.2.3.5 Molecularly Imprinted Solid-Phase Extraction (MISPE).....289
  9.2.4 Other Techniques...290
9.3 Analytical Techniques for Pesticide Determination in Water Samples........292
References..293

LIST OF ABBREVIATIONS

ACN	acetonitrile
C18	octadecyl modified silica
DI	direct immersion
DLLME	dispersive liquid–liquid microextraction
DLLME-SFO	dispersive liquid–liquid microextraction based on the solidification of a floating organic drop
DSD	directly suspended droplet
DVB	divinylbenzene
ECD	electron capture detector
FL	fluorescence
FPD	flame photometric detector
GC	gas chromatography

GC×GC	comprehensive two-dimensional gas chromatography
HF-LPME	hollow fiber membrane liquid phase microextraction
HLLME	homogeneous liquid–liquid microextraction
HS	headspace
ILs	ionic liquids
IT	in-tube
LC	liquid chromatography
LD	low density solvent
LLE	liquid–liquid extraction
LLL	liquid–liquid–liquid
LLME	liquid–liquid microextraction
LODs	limits of detection
LOQs	limits of quantification
MECK	micellar electrokinetic chromatography
MEPS	microextraction in packed syringe
MIP	molecularly imprinted polymers
MISPE	molecularly imprinted solid-phase extraction
MNPs	magnetic nanoparticles
MS	mass spectrometry
MS/MS	tandem mass spectrometry
MSPD	Matrix solid-phase dispersion
mSPE	magnetic solid phase microextraction
MWCNTs	multi-walled carbon nanotubes
Na[N(CN)$_2$]	sodium dicyanamide
NPD	nitrogen–phosphorus detector
OCs	organochlorine pesticides
OPPs	organophosphorus pesticides
PA	polyacrylate
[P$_{4448}$][Br]	tributyloctylphosphonium bromide
PCBs	polychlorinated biphenyls
PDMS	polydimethylsiloxane
PLRP-s	rigid macroporous styrene/divinylbenzene
QTOF	time-of-flight mass spectrometry
SBSE	stir bar sorptive extraction
SD-DLLME	solvent based de-emulsification DLLME
SDME	single-drop microextraction
SPDE	solid-phase disk extraction
SPE	solid-phase extraction
SPME	solid-phase microextraction
TFME	thin film microextraction
TOF-MS	time-of-flight mass spectrometry
UHPLC	ultrahigh performance liquid chromatography
USAEME	ultrasound-assisted emulsification-microextraction
UV	ultraviolet
VALLME	vortex-assisted liquid–liquid microextraction

9.1 INTRODUCTION

The high global use of pesticides is closely related to agricultural needs in response to global food demand. Although the agricultural use of pesticides is undoubtedly their main use, pesticides are also used with non-agricultural purpose (as for example as wood preservative, in household uses, or in urban pest control or park maintenance). However, pesticides can be toxic to non-target organisms and a non-adequate use of these compounds could result in environmental pollution. Pesticides may reach surface and groundwater in three main ways: by runoff, by leaching, and by effluents from sewage treatment plants. Runoff is the transport of pesticides over the ground surface of agricultural fields, landfill sites, hazardous waste disposal sites, or animal-breeding farms; leaching is a process whereby pesticides are washed out from soil or wastes by the action of percolating water from rain or irrigation water. Surface runoff from agricultural areas used to be the main source of pesticide pollution; but in urban areas, urban wastewaters are important sources of pesticide pollution in the aquatic environment due to often poor pesticides removal in the wastewater treatment plants [1].

Due to the potential toxicity of pesticides in humans, as well as their persistence and possible bioaccumulation, pesticides have received special attention and numerous regulations on the permissible levels of pesticide residues have entered into force [2]. From the point of view of the multiresidue analysis of pesticides, the efforts to carry out their analysis in water samples at low levels have been numerous. These efforts have resulted in the development of new analytical techniques and the improvement of the existing ones. These improvements have been mainly focused on miniaturization, automation, simplification, and the use of low amounts of organic solvent (or solvent-free techniques) during the sample preparation step.

Sample preparation is a key step in the analysis of pesticides. The need to detect trace levels of pesticides in water samples usually requires a selective preconcentration to produce samples suitable for analysis. The extraction techniques are based on the enrichment in liquid (liquid–liquid extraction, LLE), or in solid phases (solid-phase extraction, SPE) and sometimes, this step can be coupled to liquid or gas chromatography (LC or GC, respectively). In the last decade, liquid–liquid microextraction (LLME) methods have emerged as powerful procedures for the extraction of pesticides from liquid samples. Figure 9.1 shows the relative percentages of published articles on the different techniques used for the extraction of pesticides from water found in the available scientific literature from 2007 to 2018, containing in the title the words pesticides, water, and determination or analysis. When different extraction techniques were combined in the sample preparation (usually SPE together with a microextraction technique), they are included in Figure 9.1 as "other techniques."

It is important to remark that in the last ten years, the number of publications on the analysis of pesticides in water continues to be enormous (>5200 articles). For this reason, the most commonly used methods for the analysis of pesticides in water are presented in this chapter in order to provide the reader with an overview on the state of the art including some representative works.

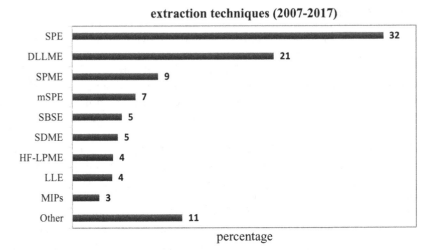

extraction techniques (2007-2017)

percentage

FIGURE 9.1 Publications on extraction techniques for the analysis of pesticides in water from 2007 to 2018. Total number of articles: 217.

9.2 SAMPLE PREPARATION TECHNIQUES IN PESTICIDE RESIDUE ANALYSIS FROM WATER SAMPLES

9.2.1 LIQUID–LIQUID EXTRACTION

Conventional LLE is a solvent extraction technique that is the traditional method used to extract and pre-concentrate pesticides from water samples using organic solvents. The LLE process can be accomplished by shaking both phases (aqueous and organic) in a separatory funnel. This technique has been long used in standard analytical methods for routine analysis of pesticide due to its simplicity, good recoveries, and precision. When LLE was used, the main objective of the work was, in general, to develop a multiresidue analytical method to monitor a large number of pesticides, which was simple, fast, and cost-effective [3,4], or to evaluate the presence of a class of pesticides to know their distribution in a specific place [5,6]. In traditional LLE, analyses are carried out using high volumes of water sample and organic solvent (from 500 to 1000 mL and from 120 to 180 mL, respectively) [3,4,7]. Nevertheless, lower sample and organic solvent volumes are used when the extraction process is assisted by sonication (about 10–20 mL and 10 mL, respectively) [5,8]. The extracted organic phase is dried adding anhydrous sodium sulfate or by passing through a glass funnel containing the desiccant. In some cases, the extracts are subjected to a SPE cleanup procedure, usually using Florisil [6,7]. Finally, the organic phase must be concentrated using a vacuum evaporator due the high volume of organic solvent used. This sample preparation technique has some disadvantages such as the formation of emulsions, multiple extraction steps required to obtain optimum output, the consumption of large quantities of organic solvents, which makes the process not environment-friendly, and the difficulty of automation. For these reasons, the application of this traditional technique has decreased in the last years. The need for extraction procedures that provide a fast extraction of the analytes, with a low consumption of toxic organic solvents during the extraction process, has led to adaptations of existing methods and the development of new procedures. Consequently, miniaturization of the extraction process has become a key factor and new sample preparation methods have been developed. Thus, miniaturized procedures for the extraction of liquids, such as dispersive liquid–liquid microextraction (DLLME), DLLME based on the solidification of a floating organic drop (DLLME-SFO), single drop microextraction (SDME), or hollow fiber membrane liquid-phase microextraction (HF-LPME) have been introduced. These micro-LLE methods are easy, fast, and use low amounts of organic solvent (microliters), and they have been widely used in the last number of years for the analysis of pesticides in water samples (see Figure 9.1).

9.2.2 LIQUID–LIQUID MICROEXTRACTION

9.2.2.1 Dispersive Liquid–Liquid Microextraction (DLLME)

The DLLME technique was developed in 2006 by Rezaee et al. [9]. In summary, the technique is based on the formation of tiny droplets of the extraction solvent in the aqueous solution to analyze, after a rapid injection of an immiscible organic solvent (extractant). The extraction solvent is usually injected together with another organic solvent miscible with water that acts as disperser solvent. Then, the mixture of extracting and dispersing solvents is rapidly injected into the aqueous sample, the dispersion is formed, and a cloudy solution is obtained. Finally, a centrifugation step allows the separation of the phases and the extraction solvent containing analytes is withdrawn with a microsyringe for chromatographic analysis. The basic principle of the method is that the dispersion forming tiny droplets generates a very high contact area between the aqueous phase and the extraction solvent, and therefore, the extraction equilibrium can be rapidly reached, reducing the extraction times and increasing the enrichment factors. Figure 9.2 shows the schematic diagram of the DLLME procedure. Since its

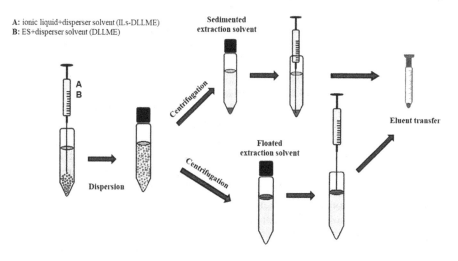

FIGURE 9.2   Schematic diagram of DLLME procedure.

introduction, it has fast become one of the most popular sample preparation techniques for the analysis of pollutants in aqueous samples (together with SPE discussed below). The advantages of DLLME method are the rapidity, low cost, high enrichment factor, and the use of low quantities of organic solvents. Thus, many methods have been developed recently for the analysis of pesticides in water samples by different DLLME techniques or by combination of DLLME with other extraction techniques.

According with the initial DLLME method, the extraction solvent should be immiscible and denser than water to enable the separation of the phase by centrifugation and to collect it from the bottom of a conical vial. As a result, only a few solvents (such as chlorobenzene, chloroform, tetrachloroethylene, tetrachloroethane, or 1-dodecanol) have been used as extractant in this modality, due to their high density. Acetone, methanol, and acetonitrile are usually selected as disperser solvents. The DLLME methods use low sample volumes (usually from 5 to 20 mL), extractant organic solvents (10–220 µL), and disperser solvent (0.2–2 mL), when a dispersant is needed. The use of an auxiliary solvent must be optimized because it affects the extraction efficiency, since the type and volume of dispersing solvent can significantly affect the volume of the sedimented phase.

Thus, as with other extraction techniques, the extraction efficiency of DLLME is influenced by many factors, such as the selection of the extraction and disperser solvents, their respective volumes, the extraction time, and the ionic strength. The biggest challenge of the DLLME is its application to the multiresidue analysis of a broad number of contaminants in a single analytical run. In this way, Martins et al. [10] developed a method for determination of 22 pesticides and related compounds in water by DLLME and GC–MS/MS. The proposed DLLME method uses 5 mL of water sample aliquots, 2 mL of acetone as a dispersive solvent, and 50 µL of carbon tetrachloride as extraction solvent, at low temperature. Briefly, the water sample was placed into a 10 mL glass tube with conical bottom and the extraction solvent mixture (acetone and carbon tetrachloride, cooled between 0 and 4°C) was rapidly injected into the sample. Then an emulsion containing fine droplets of the extraction solvent dispersed in the aqueous phase was obtained and the cloudy solution was separated after 10 min and kept at a cold temperature, without agitation or centrifugation steps. Finally, the sedimented phase (around 25 µL) was collected and transferred to an insert vial and the pesticide concentrations were determined by GC–MS/MS. This method was successfully applied for the determination of the target compounds in water, the limits of quantification (LOQs) were between 0.03 and 0.10 µg/L and mean recoveries between 70 and 120% for most of the compounds, with good precision despite the different chemical nature of the compounds analyzed. Recently, Tankiewicz and Biziuk [2] developed a multiresidue DLLME procedure for the determination by GC–MS of 34 commonly used pesticides belonging to various chemical groups, some of them not previously determined and considered difficult to extract. In the proposed method, 10 mL of water were added into a 12 mL glass tube with conical bottom and the mixture of the disperser solvent (1 mL methanol) and the extraction solvent (40 µL tetrachloroethylene) was rapidly injected into the sample solution with a syringe, and the mixture was shaken for 30 s. A cloudy emulsion was formed. Then, the mixture was centrifuged and the extraction phase was collected with a microsyringe, transferred to the vial, and injected into the GC–MS system. Under the optimum conditions, the LOQs ranged from 0.0096 to 0.052 µg/L, and recoveries ranged between 84 and 115%.

To improve the efficiency in DLLME, the application of ultrasound radiation is very useful, and this procedure is named ultrasound-assisted emulsification-microextraction (USAEME). Ultrasonic radiation is an efficient tool to facilitate the emulsification and increase the speed of the mass transfer process between the two immiscible phases implied because smaller droplets are generated when sonication is applied, which enlarges the contact surface between both phases. After a short extraction time, the emulsion is separated into two layers by centrifugation. When USAEME is used, a disperser solvent is usually not required. Sample volumes of 5 to 10 mL of sample and between 20 to 200 µL of extraction solvent are injected without a dispersant [11,12] or with a surfactant (such as Tween 20 or Triton X-100) as emulsifier [13,14].

Other approaches have been taken in DLLME to use other organic solvents as extractants and to avoid the centrifugation step. So, procedures using DLLME with low density solvent (LD) have increased in the last years. LD-DLLME is the general name of the extraction technique when low density solvent is used. However, more specific names, such as dispersive liquid–liquid microextraction based on the solidification of a floating organic drop (DLLME-SFO) or solvent based de-emulsification DLLME (SD-DLLME) are used when the separation of the organic phase is carried out after cooling or when a demulsification solvent is injected to break up the emulsion after DLLME, respectively. In order to minimize the use of organic solvents in the procedure, the extraction could be done using ionic liquids (ILs) and the methods are named ILs-DLLME.

In the LD-DLLME procedures, the separation of low-density solvents (1-octanol or toluene) could be difficult. After extraction, the organic solvent could be separated removing the lower aqueous phase with a 5 mL syringe leaving a volume equivalent to the organic phase at the bottom of the conical tube [1]; removing with a pipette the immiscible extraction solvent droplet formed in the middle of the vortex stirred with a magnetic stirrer [15] or using special devices [16].

In DLLME-SFO, the solvents used (1-dodecanol, 1-undecanol, n-hexadecane, or bromohexadecane) have lower density than water and a melting point that allows the solidification at low temperature of the floated droplet that is easily removed from the surface of the aqueous sample. Ultrasound was employed to shorten the extraction time and improve the extraction efficiency [17]. In those cases, there is no need for conical-bottom glass tubes and stirring during extraction is not needed. Nevertheless, centrifugation to separate the extraction droplet from the aqueous sample and an additional cooling step are required. The selection of appropriate extraction and disperser solvents are key to an efficient DLLME-SFO process. For the analysis of five pesticides of different classes [17] and five OPPs [18] the optimum extraction solvent was 1-dodecanol together with methanol or acetonitrile as disperser solvent, respectively. In both methods, the addition of salt to the sample solution increased the extraction recovery of some pesticides. Therefore, 1% or 5% NaCl was added to the sample prior to the extraction, respectively [17,18].

In order to remove the centrifugation step, the use of a solvent as demulsifier is considered in the DLLME procedure. In SD-DLLME, after the formation of the emulsion, another aliquot of disperser solvent (which serves as demulsifier) is introduced leading to the breakdown of the emulsion so the solution quickly clears into two phases. However, an appropriate extraction solvent must be selected in order to break the emulsion and minimize potential adverse effects on the coefficient of partition of the analytes, because the demulsifiers commonly used are water-miscible organic solvents [2]. This technique has been successfully applied by Caldas et al. [19] in a multiresidue method for the determination of 32 pesticides and 26 pharmaceutical and personal care products in water samples using LC–MS/MS. In the optimization of the SD-DLLME method, the type and volume of the extraction and disperser solvents, pH, salt addition, amount of salt, and type of demulsification solvent were evaluated. In summary, the extraction was carried out by dividing the sample into two subsamples: one of them with the pH adjusted to 2 and the other one with the pH adjusted to 8. Then 10 mL of the sample (containing 1% $MgSO_4$, w/v) was placed in a 10 mL volumetric flask, and the mixture of 120 µL of octanol and 750 µL of acetone (disperser solvent) was quickly injected into the sample with a syringe. Then, the emulsion formed was broken up after the addition of 750 µL of water on the upper surface. Afterward, the upper layer was collected with a microsyringe for its analysis by LC–MS/MS. The LOQs and recoveries were in the range from 0.0125 to 1.25 µg/L and from 56 to 130%, respectively.

The use of solvent-free techniques at the sample preparation stage, in order to make them consistent with "Green Chemistry" principles, is another of the aims of the new extraction procedures. ILs are a group of non-molecular solvents that can be defined as organic salts that remain in a liquid state at room temperature. They have high density and low volatility, which facilitates phase separation, and they have stable droplets [16]. ILs are known as "green solvents," which have a high thermal stability and, based on different combinations of cations and anions, can be hydrophobic or hydrophilic, as well as miscible or immiscible with the disperser solvent in DLLME. 1-alkyl-3-methyl-imidazolium hexafluorophosphate ([CnMIM] PF6) is the IL most commonly employed in ILs-DLLME, with two to eight carbons in the alkyl chain. In general, ILs-DLLME methods were developed for the analysis of four or five pesticides of the same group (such as OPPs, phenylureas, or benzoylureas). Besides the standard ILs-DLLME using an organic solvent as dispersing agent [20], ultrasound-assisted-ILs-DLLME (UA-ILs-DLLME) has been applied in order to obtain a dispersion of the extraction solvent avoiding the use of a disperser. In general, good recoveries (from 89 to 118%) and low detection limits (0.05 to 0.45 µg/L) were achieved with UA-ILs-DLLME methods developed for the analysis of pesticides [21–23]. Nevertheless, Tankiewicz and Biziuk [2] pointed out that the use of ILs as extractants has some disadvantages such as that only water-immiscible ILs can be used in DLLME and that background peaks can interfere in the chromatographic analysis of pesticides. ILs are relatively expensive, and often need purification before application; their choice is based mainly on empirical experimentation due to the lack of precise physicochemical properties and most ILs are incompatible with GC analysis.

In DLLME methods, various techniques can be used to enhance the dispersion: manual shaking, air-assisted, ultrasound-assisted, surfactant-assisted, vortex-assisted, or microwave-assisted emulsifications [24]. Recently, Liu et al. [25] applied effervescence to assist DLLME for the analysis of triazine herbicides and triazole fungicides in water. Effervescence consists of the generation of bubbles of CO_2 by reaction of an acid with carbonate/bicarbonate. In this method, a mixed solution of methanol, citric acid, and 1-undecanol was quickly added into the bottom of the glass tube that contained the aqueous sample with sodium carbonate, producing bubbles that eased the dispersion of the extraction solvent. Compared with methods assisted with ultrasound radiation or vortex, effervescence is a simple and effective system that does not need energy, but the mixture has to be centrifuged in order to have a large drop of 1-undecanol floating on the surface.

In some studies, mainly when high preconcentration factors are needed, DLLME is used in combination with other extraction methods, such as magnetic solid-phase extraction (mSPE) [26]. This extraction technique is described below in Section 9.2.3.

9.2.2.2 Single-Drop Microextraction (SDME)

SDME is an extraction technique based on the exposure of a microdrop of an organic solvent into the matrix containing the analytes, which was developed as an alternative to SPME (discussed in Section 9.2.3.3). SDME avoids some problems of the SPME method such as sample carryover and fiber degradation. It is also fast, inexpensive, and it uses very simple equipment, although, the technique has some disadvantages. SDME includes several modes, such as direct immersion (DI), headspace (HS), continuous flow, drop-to-drop, directly suspended droplet (DSD), liquid–liquid–liquid (LLL), or a combination of LLL and DSD. However, DI-SDME is the mode most frequently found in the literature for the extraction of pesticides from water.

For DI-SDME, a microdrop (usually less than 3.5 μL) of an immiscible organic solvent is generally suspended at the tip of a microsyringe needle, which is immersed in an aqueous sample under stirring. The analytes are transferred from the aqueous solution to the organic phase, and after an adequate extraction time, the microdrop is retracted back into the microsyringe and transferred to a vial to be chromatographically analyzed. Usually, the extraction time, organic solvent, stirring speed, ionic strength, pH, temperature of the aqueous phase, and the drop volume of the extraction solvent are the parameters to be optimized. As with other extraction techniques, DI-SDME has advantages and drawbacks. Thus, DI-SDME uses a low amount of organic solvent (1–3.5 μL), it is fast, the cost is low in comparison to other techniques, such as SPE or SPME, and the pre-concentration and extraction are carried out in the same step. However, some disadvantages are the drop instability and solvent loss during the extraction that reduce the sensibility and repeatability of the method. In order to avoid some of the drawbacks, SDME can be carried out using a modified microsyringe or holder. In this way, a novel improved SDME method was proposed by Wang et al. [27] focusing on holding more solvent and overcoming the problem of the instability of the microdrop at high stirring rates for the analysis of carbamates and OPPs in water. In this method, an oval-shaped polychloroprene rubber tube was used as the extraction device and 15 μL of extraction solvent (chlorobenzene) could be loaded. In the same way, Tian et al. [28] developed a SDME procedure with a modified microsyringe, by attaching a 2 mm cone onto the needle tip end, and 3.5 μL of toluene for the analysis of six OPPs in water. In both methods, recoveries from 71 to 112% were obtained for the target compounds.

9.2.2.3 Hollow Fiber Membrane Liquid Phase Microextraction (HF-LPME)

SDME minimized the solvent used in the extraction procedure but, as it has been reported above, hanging small organic solvent droplets can lead to solvent loss during the extraction that reduces the sensibility and repeatability of the method. HF-LPME has a microporous structure and can overcome this difficulty, because the organic solvent is fixed in the hollow fiber membrane. Therefore, the sample solution may be stirred or vibrated vigorously without loss of the extraction phase since it is not in direct contact with the sample, and the small pore size of the fiber prevents large molecules from entering into the acceptor phase; thus, extraction and cleanup of samples can be done simultaneously.

HF-LPME has been used for the determination of different pollutants in water samples due to the low consumption of organic solvents in the extraction and its low cost, which enables the disposal of the material after use, eliminating problems of cross-contamination and low reproducibility. Moreover, the process is simple, a cleanup step is not necessary, and high enrichment factors can be obtained. In this technique, the surface of the capillary porous hydrophilic fiber is impregnated with an organic solvent and its lumen is filled with an acceptor phase that does not come into direct contact with the sample, which allows the application of agitation during the extraction. Various organic solvents such as toluene, 1-octanol, ethyl octanoate, or di-n-hexyl ether were used as acceptor phase in HF-LPME for the analysis of pesticides in water samples, toluene being the most popular. However, other extraction solvents or mixture of solvents have been reported. After extraction, the acceptor is removed with a microsyringe for its chromatographic determination.

There are two alternatives for HF-LPME: two- or three-phase modes, the former being the most used for the analysis of pesticides in water. In the 2-phase mode, the organic solvent is present both in the porous wall and inside the lumen of the hollow fiber; whereas, in a 3-phase mode, the acceptor solution may be an aqueous or an organic solvent that is immiscible with the supported liquid membrane solvent. HF-LPME is performed by direct immersion; however, extraction has been occasionally carried out by headspace extraction [29]. In the latter, a drop of extraction solvent is suspended at the tip of a microsyringe needle, inserted in a hollow fiber, and exposed to the headspace of the sample solution. Thus, this extraction mode is suitable for volatile compounds; nevertheless, in direct immersion the solvent must be water-immiscible. Most of the HF-LPME methods use polypropylene fibers with fiber lengths from 1.2 to 6 cm, sample volumes from 4 to 15 mL, and a volume of extraction solvent from 3 to 30 mL. Usually, the optimum extraction time is from 30 to 60 min under stirring. Menezes et al. [30] described a two-phase HF-LPME for the analysis of ten pesticides in water. In this work, factors such as extraction mode, time, solvents, agitation, and salt addition were investigated and the optimal conditions were: 6 cm of polypropylene hollow fiber, ethyl octanoate as acceptor phase, and extraction of 30 min under stirring at 200 rpm. The optimized method showed LOQs ranging from 0.14 to 1.69 µg/L and recoveries between 85 and 115%. Nevertheless, a long fiber length (32.8 cm), sample volume up to 250 mL, an extraction time of 4 h, and a binary mixture of tri-n-octylphosphine oxide (10%) and tri-n-butyl phosphate (10%) in di-n-hexyl ether were reported by Trtić-Petrović [31] as optimal conditions for the multiresidue analysis of 16 pesticides of different chemical classes in natural water samples. Recoveries ranged from 30 to 100% and LOQs from 0.09 to 0.79 μg/L.

9.2.3   Solid-Phase Extraction

9.2.3.1   Standard SPE

SPE was developed in 1970 as an alternative to LLE and, now, SPE is one of the most widely used sample preparation methods for the extraction and purification of target analytes from a sample matrix or from an extract with the analytes. Thus, SPE is the most used extraction method for the determination of pesticides in water samples. This is due to its good extraction efficiencies and selectivity, ease of use, and its potential for automation.

SPE is based on the different affinity of the analytes for the liquid phase and a solid phase. A typical SPE extraction involves five main steps: selection of a suitable SPE sorbent, column conditioning, sample loading, washing of the sorbent in order to remove impurities bound less strongly than the compounds of interest, and, finally, elution of the target analytes. Samples are usually filtered prior to the SPE step to remove suspended solids or particulate matter but this may result in an undesirable loss of target compounds.

Sorbent selection requires consideration of sample volume, the nature of the analyte, analyte concentration, and the inherent properties of the sorbent itself. Cartridges pre-packed with known quantities of adsorbent are on the market, and they are ready to use after simple conditioning. The SPE products are available in a wide variety of chemistries, adsorbents, and sizes. Selecting the most suitable product for each application and sample is important. The sorbent features provided by the cartridge supplier allow an initial consideration for appropriate sorbent selection. The most popular sorbents for the SPE analysis of pesticides in water samples are a copolymer of divinyl-benzene and N-vinyl-pyrrolidone (Oasis cartridges), and octadecyl modified silica (C18 cartridges). Although less widely used, other popular sorbents are the multi-walled carbon nanotubes (MWCNTs) followed by styrene-divinylbenzene copolymer (Strata X).

The conditioning of the sorbent is necessary to ensure a reproducible interaction with the analytes, activate the sorbent, and remove any residual process materials. In summary, conditioning results in a wetting of the sorbent to make it suitable for the adsorption of the analytes. Nonpolar adsorbents are usually conditioned with 2–3 column volumes of a solvent miscible with water (e.g., methanol) to ensure consistent interaction, followed by the addition of a solution similar to the sample matrix (e.g., water) to maximize analyte retention. Polar sorbents are conditioned with nonpolar solvents. The cartridge is never allowed to dry during the conditioning process. Then, the sample is loaded and its volume can vary from few milliliters to liters (5–1500 mL) depending on the maximum sample capacity of the cartridge. Thus, the extraction efficiency decreases when an excessive sample volume is added. The sample is usually pumped into the cartridge at a flow rate ≤ 10 mL/min. After the sample is loaded, the cartridge may be washed to remove interferences while still retaining the analytes. In this step, it is important to properly understand the relationship between the analyte and the stationary phase in order to optimize the strength of the washing solvent. If the polarity difference between the washing solution and the eluant is very high, or if both are not miscible, drying of the adsorbent after washing is recommended. Finally, analytes are eluted with a small volume of an appropriate solvent and elution speed should not be too fast.

One reason for the widespread use of SPE for the analysis of pesticide in water is the high enrichment factor that is achieved due to the low volume of the elution solvent used in comparison to the sample volume. Additionally, SPE offers the great advantage of easier transportation between the sampling point and the laboratory or between laboratories. Thus, for example, water samples can be processed at a remote site, and after a drying step, the cartridges containing the analytes can be transported to the laboratory for analysis. However, SPE methods can be tedious and time consuming when performed manually so the efforts made to automatize SPE have allowed the development of faster methods by reducing the sample preparation time and increasing the sample throughput. Nowadays, the different steps of the SPE process can be performed automatically and some systems permit the extraction of one sample while another one is being analyzed. Automation decreases the risk of sample or extract contamination and analyte losses during sample pre-concentration, while improving precision, accuracy, and sensibility. Furthermore, with the on-line configuration the totality of the extracted analytes is transferred into the analytical system whereas only an aliquot is injected with the off-line approach, which leads to lower limits of detection [32]. However, although the automation or semi-automation of the SPE process is possible, off-line SPE is still the most used.

SPE is usually chosen when a high number of pesticides are evaluated simultaneously in a multiresidue analysis. Thus, the methods developed by Ferrer and Thurman [33] and Charalampous et al. [34] are two examples of the application of off-line SPE for the analysis of more than 100 pesticides in water. On the other hand, the method described by Cotton et al. [35] for the analysis of over 500 pesticides and drugs in water is an example of an on-line SPE method. In the first method, a Sep-Pak C18 column (500 mg) was used for the analysis of 101 pesticides and their degradation products in water samples [33], using 100 mL of sample, 3 mL of ethyl acetate for the elution of the analytes from the cartridge, and liquid chromatography/time-of-flight mass spectrometry (LC-TOF-MS) for the analysis. In the other work, Charalampous et al. [34] described a multiresidue method for the determination of more than 200 pesticides and degradation products and 13 polychlorinated biphenyls (PCBs) in water using LC–MS/MS and GC–MS/MS. In this study, 500 mL of sample were loaded to a previously conditioned and equilibrated Oasis HLB cartridge (200 mg), at a flow rate of 15 mL/min. Then, the cartridges were dried for 45 min and the adsorbed compounds were eluted with 5 mL of methanol and 5 mL of ethyl acetate. For the target pesticides, recoveries from 60 to 140% were obtained and LOQs were set at 0.01 µg/L (except for 29 compounds with an LOQ of 0.1 µg/L). The on-line SPE method developed by Cotton et al. [35] was based on on-line SPE and ultra-high performance liquid chromatography–mass spectrometry (UHPLC–MS). The selection of the analytical column and the cartridge was carried out by testing four analytical columns and one on-line preconcentration cartridge with different mobile phase composition. Taking into account that the on-line SPE systems, in general, are incompatible with the high back pressures generated by the high flow rates required by UHPLC systems, the column Oasis HLB Direct Connect was the only SPE column available that could handle those kinds of backpressures. The developed method carried out the analysis in 36 min with only 5 mL sample. It was applied to the analysis of 20 tap water samples, and 34 compounds were detected at concentrations below 0.1 µg/L, the EU limit for drinking water. Pesticides such as atrazine and its metabolites, hexazinone, oxadixyl, propazine, and simazine were detected in the samples.

New SPE materials are being developed and, nowadays, the efforts are aimed mostly at finding a universal sorbent suitable for every purpose. However, due to its great flexibility, simplicity, and the variety of commercial cartridges, the off-line procedures are still a valuable tool for the multiresidue analysis of pesticides in water. Recently, García et al. [36] reported a novel phase sorbent, named MCM-21, for the pre-concentration of eight pesticides in water using 100 mL of sample at flow rate of 5 mL/min, eluting with 7 mL of ACN, and detection by micro-LC–MS/MS. Another example of these SPE trends is the research carried out by Mann et al. [37] of an automated online-SPE-LC–MS/MS multiresidue method for the determination of three pesticides and six thiamethoxam metabolites in water using a polystyrene divinylbenzene copolymer (PLRP-s) sorbent.

Solid-phase disk extraction (SPDE) is another form of SPE where the sorbent is bonded to a solid support that is configured as a disk. SPE disks are recommended for large sample volumes, for samples containing high amounts of particulates, or when a high flow rate is required during sampling. US-EPA and the EU have incorporated the use of SPDE in some methods for the determination of pesticides in water. Thus, EPA method 527 (2005) describes the determination of selected pesticides and flame retardants in drinking water and the European Document EN 16693 (2015) specifies a method for the determination of selected organochlorine pesticides (OCs) in water samples; both methods use SPE disks followed by GC–MS.

A previous filtration of the sample could eliminate some analytes bound to the particles and result in the quantification of a false negative or a lower concentration of the pollutant in the sample. This problem can be avoided using SPE disks, because the particles are kept in the disk. SPE disks can handle a higher percentage of solid material without clogging or taking an excessively long time to flow through the disk. Disks bonded with C18 or C8 solid phase are the most commonly use.

9.2.3.2   Solid-Phase Microextraction (SPME)

Solid-phase microextraction (SPME) was introduced in 1990 as an alternative extraction procedure to SPE for the analysis of volatile and semivolatile compounds in different matrices. This technique does not require the use of solvents for the extraction of the analytes. This solvent-free extraction technique is based on the exposure of a thin fiber coated with a sorbent to the headspace (HS-SPME) or into the aqueous sample (DI-SPME). The analysis by GC is the most usual. Thus, after equilibration, the compounds are thermally desorbed in the injection port of the chromatograph and transferred onto the head of a capillary column for subsequent separation and identification. SPME is an equilibrium process that involves the partitioning of analytes between the sample and the extraction phase. SPME uses a fiber coated with a liquid (polymer), a solid (sorbent), or a combination of both. To date, the commercialized SPME coatings most used for the analysis of pesticides in water samples are: polydimethylsiloxane (PDMS) fibers with films of different thicknesses (30 and 100 μm); polyacrylate (PA) of 85 μm; mixed phases of PDMS-divinylbenzene (DVB) of 65 and 60 μm; and 50/30 μm DVB/Carboxen on PDMS. The selection of the appropriate fiber coating is one of the most critical steps of SPME method development. After the fiber selection, sampling conditions must be optimized to increase the partitioning of analytes in the coated fiber. Thus, for high accuracy and precision from SPME, homogenization time before sampling, sampling time, and desorption parameters (such as injection port temperature, depth of fiber insertion in the injection port, and desorption time) are the main parameters to optimize. It is also important to keep constant the vial size, the sample volume, and the depth to which the fiber is immersed or exposed to the HS. To carry out the samplings at a same depth, the fiber holder includes an adjustable depth guide that positions the fiber for correct placement in the sample and in the heated zone of the GC injection port. It is important to minimize the headspace in the sample vial when DI-SPME is applied. It was described that a combination of salt and pH modification often enhances the extraction of analytes from the headspace. With the optimized parameters, consistent, quantifiable results can be obtained from SPME.

SPME methods using other coatings, such as polythiophene for the HS-SPME of five endocrine-disruptor pesticides [38] or polypyrrole/sol-gel composite and sol-gel/nanoclay composite for DI-SPME analysis of OPPs [39,40] have been reported, but these were laboratory-made fibers. When non-commercialized fibers are used only a small number of pesticides are analyzed (from three to five), and usually of the same family (e.g., OPPs). In general, many of the SPME methods reported using commercial SPME fibers were developed for the simultaneous determination of different types of pesticides in water by GC. Thus, Beceiro-González et al. [41] reported a DI-SPME method for the simultaneous determination of 46 pesticides with a wide range of polarities and chemical structures in water samples. In this work, the extraction was carried out using a PDMS/DVB fiber (60 μm) followed by GC–MS analysis. The optimal conditions were: 45 min of extraction time, sample agitation, and temperature control at 60°C, neither pH adjustment nor ionic strength correction, and a sample volume of 18 mL. The method reported good detection limits, linearity, and repeatability for the pesticides studied. Chaturvedi et al. [42] described a HS-SPME method for the extraction of 24 pesticides (15 OCs, 6 OPPs, and 3 pyrethroids) in water with PDMS fiber (100 μm) followed by GC–ECD analysis. The method was carried out with 2.0 mL sample and a stir bar in a vial capped with a septum and placed in a stirring hot plate at 90°C. The PDMS fiber was exposed to the headspace for 30 min, and then it was inserted into the GC injection port (at 280°C) for 10 min. The method detection limits and recovery percentages were found in the ranges of 0.05–0.20 μg/L and 87–95% respectively, and the method was used in the analysis of more than 147 water samples.

Recently, Rodríguez-Lafuente et al. [43] reported an inter-laboratory validation of an automated SPME-GC–MS method for the determination of 25 semi-volatile pesticides. The automated SPME was performed in an SPME Multi Fiber Exchange Autosampler, and for SPME analysis, aqueous samples (15.5 ml) were placed into amber glass vials with magnetic metal screw tops and silicone septa for automated SPME. Sample pH was adjusted to 3.0 and NaCl (4 g) was added to each sample. The vial was capped and vortexed for 30 seconds and the sample was pre-incubated for 2 min at 30°C, followed by direct immersion of a PDMS/DVB (65 μm) fiber for 30 min at the same temperature. After extraction with agitation, the fiber was desorbed in the injector port of the GC–MS instrument, where it was kept for 10 minutes at 270°C. The developed SPME method was compared with a US EPA method based on LLE and both methods showed to be very accurate, though, the SPME proved to be more sensitive and required a lower sample volume than the LLE method.

Although SPME is solvent-free, rapid, and easy to use, it also suffers from some drawbacks such as: sample carryover, fiber fragility, and limited lifetime, the partial loss of stationary phase that may affect precision, low sample throughput, and the fact that experimental conditions must be precisely controlled to achieve good reproducibility. Nevertheless, with automation of SPME methods, the commercially available autosamplers can be programmed to perform various sample preparation steps (such as dilution, agitation, and extraction), providing the reduction of times for routine analysis and development of analytic methods, faster sample throughput, and greater reproducibility.

Perhaps the most well-known and most applied SPME technique is the fiber-SPME discussed above, though there are other types of SPME techniques, such as in-tube SPME (IT-SPME), which was developed for the miniaturization and on-line coupling with the analytical instrument. This extraction technique uses a capillary column as an extraction device, extracting the analytes directly onto the inner capillary coatings. In order to improve the extraction efficiency and specificity, the capillary tube can be packed with some polymer. IT-SPME employed in combination with either HPLC or GC allows on-line analysis by combining sample treatment and analytical determination in a single step. Alternative microextraction techniques have been developed using a coated magnetic stir bar, a thin film polymer, microsyringes, or pipette tips.

Recently, De Toffoli et al. [44] reported an on-line IT-SPME method coupled to HPLC–MS/MS for the determination of three triazines in water. The method employed a packed column containing graphene oxide supported on aminopropyl silica that showed a high potential for triazine extraction due to its physical-chemical properties including ultrahigh specific surface area, good mechanical and thermal stability, and high fracture strength. Injection volume and loading time were optimized and the validated method showed satisfactory results, good sensitivity, and low detection limits (1.1–2.9 ng/L).

In the past years, substantial research efforts have been made in the pursuit of the optimum SPME coating/fiber/device. In this sense, the developments of new extraction phases using sol/gel technology, employing ILs, or developing molecularly imprinted polymer (MIP) coatings allowed the implementation of higher temperatures for desorption, enhanced the robustness of the coating toward organic solvent and acidic/alkali solutions, or enhanced method sensitivity and selectivity [45].

In most papers, after SPME the analyses were carried out by GC–MS; although ECD and other detection systems, such as nitrogen–phosphorus (NPD) and flame photometric detectors (FPD), have also been used.

9.2.3.3 Stir Bar Sorptive Extraction (SBSE)

SBSE was developed in 1999 and commercialized under the name "Twister." This technique usually employs a glass-enveloped magnet (10 mm or 20 mm length) coated with PDMS (film thickness, 0.5 mm and 1 mm). Thus, although the extraction mechanism and advantages of SBSE are similar to those of SPME, its enrichment factor is up to 100 times higher. In summary, in SBSE, analytes are adsorbed on the coated stir bar during stirring for a set time. Then, the stir bar is removed and can be placed in a glass tube, which is transferred to a thermal desorption system to be desorbed and analyzed on-line by GC, or can be desorbed with an organic solvent for subsequent injection into a GC or LC system. Although SBSE is based on the same principles as SPME, the higher volume of sorbent phase involved in SBSE promotes a lower phase ratio between this phase and the sample, this technique is superior to SPME in terms of sensitivity and accuracy for trace level quantities in difficult matrices. Nevertheless, SPME can be fully automated, while automatization is not complete for SBSE.

The SBSE technique has two major steps: the extraction of the analytes from the sample toward the coated sorbent, and desorption from the polymeric phase. The most important variables that influence the extraction process must be optimized (such as sample and PDMS volumes, equilibrium time, agitation speed, pH adjustment, and ionic strength of the matrix). Equilibrium time for SBSE is extremely long and, in order to minimize the extraction time, the sensitivity and precision can be sacrificed by working under non-equilibrium conditions with sufficient performance. After the extraction step, the stir bar is carefully removed, rinsed with distilled water, and dried with a paper tissue to remove any interference. In thermal desorption, on-line mode, several instrumental parameters should be optimized, particularly desorption temperature, purge flow, and the GC injector temperature, a programmed temperature vaporization inlet being essential. No organic solvent is used in the thermal desorption mode but, although automation promotes a high sensitivity, the desorption unit is expensive and only allows combination with GC systems. In the second desorption mode, a low quantity of organic solvent is needed (from 50 µL to 1.5 mL) and mechanical or sonication treatments are usually required to improve desorption efficiency. In this mode, variables such as the solvent (methanol, acetone, ACN, or mixtures) or soak time have to be evaluated and a concentration of the extract is usually needed. Liquid desorption allows extracts to be analyzed more than once and not only by GC, so it is more versatile than the thermal desorption. Although other phases have been assayed for SBSE [46], only PDMS, PDMS/ethylene glycol (EG), and PA stir bars are available in the market. PA and PDMS/EG stir bars can extract polar compounds more efficiently than the PDMS stir bar due to their polar nature. In addition, the PDMS/EG stir bar can also extract non-polar compounds efficiently. However, there is a lack of applications for both stir bars and the reported methods usually use the PDMS ones.

There is no general method for the analysis of pesticides in water by SBSE using PDMS-coated stir bars. Thus, the multiresidue analysis of pesticides (13 to 37 compounds) can be carried out using 10 to 500 mL of sample, with long extraction times (one to three hours), with the addition of salt and stirring (600 to 1500 rpm).

Recently, a method based on SBSE technology with GC–MS was developed to analyze 20 OCs pesticides, together with 18 PCBs in aqueous samples [47]. In this work, the problems of long adsorption time and small sample volume, found in conventional SBSE methods, were solved using several stir bars simultaneously for enrichment, with thermal desorption and cryofocusing within a cooled programmed temperature vaporization. Thus, four stir bars were attached to a large Teflon magnetic stir bar to stir simultaneously 500 mL of sample. An extraction time of 1 h was selected, although an increase of the peak areas was observed even until 6 h. Afterward, the stir bars were removed, rinsed with water, dried with a tissue, and placed in two tubes to be desorbed in a thermal desorption device. A cryofocusing temperature of 20°C was selected that did not increase until the complete desorption in the latter tube. The desorption time from the stir bars was established in 5 min. Recoveries from 64 to 111% and LODs from 0.12 to 2.1 ng/g were obtained for the pesticides evaluated. The use of multiple stir bars in the SBSE process was previously reported by Sampedro et al. [48] for the extraction of 37 pesticides from 10 mL of water. In this work, the two stir bars were used sequentially involving a first extraction of salt-free sample and a second extraction of the sample after a 20% salt addition.

The work developed by Margoum et al. [49] used SBSE coupled to LC–MS/MS for the multiresidue determination of 15 pesticides or selected metabolites from different families in water samples. Optimization of key parameters was done by experimental design. The optimal condition for the extraction of 20 mL of sample were: 3 h of extraction time at 800 rpm, addition of 10% NaCl, no addition of methanol as organic modifier, and a desorption time of 15 min under sonication with ACN–MeOH (50:50, v/v). Recoveries from 93 to 101% and LOQs ranging from 0.02 to 1 µg/L were obtained for the target analytes.

Although stir bars coated with PDMS are the most frequently used in SBSE, new sorption coatings were developed and used for the determination of pesticides. In these works, the extraction of a few pesticides of the same group (usually OPPs) was carried out using stir bars coated with PDMS/polythiophene [50], MWCNTs [51], or PDMS/metal-organic frameworks (MIL-101-Cr-NH$_2$) [52] prepared by the sol-gel technique.

9.2.3.4 Magnetic Solid-Phase Extraction (mSPE)

Magnetic solid-phase extraction (mSPE) methods using nanoparticles (NPs) of magnetite (Fe$_3$O$_4$) have been described as an interesting technique for the analysis of pollutants in liquid samples. The use of magnetic nanoparticles (MNPs) in the extraction process is a very interesting tool because the sorbent is added directly to the solution and, after the adsorption of the analytes onto the magnetic sorbent, the MNPs (with the captured analytes) are isolated by placing a magnet on the wall of the flask and discarding the solution. Finally, the target compounds can be eluted from the sorbent with a low quantity of an adequate organic solvent to be analyzed. Figure 9.3 shows the schematic diagram of the mSPE procedure.

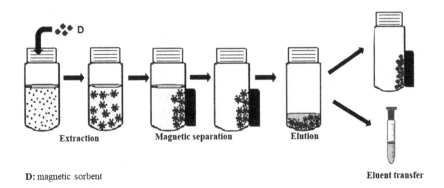

Extraction   Magnetic separation   Elution

D: magnetic sorbent   Eluent transfer

FIGURE 9.3   Schematic diagram of mSPE procedure.

However, the drawbacks of the use of MNPs are their aggregation and the lack of target selectivity. Therefore, in order to solve these limitations, the surface of MNPs can be functionalized. Thus, for example, titanium dioxide MNPs [53,54], metal organic frameworks [55], graphene-based tetraethoxysilane–methyltrimethoxysilane sol-gel hybrid magnetic nanocomposite [56], n-octylated NPs [57], or poly(p-phenylenediamine-co-tiophene) coated Fe_3O_4 NPs [58] have been assayed for the determination of some pesticides (mainly OPPs or pyrethroids) in water samples. In these works, sample volumes of 10 to 30 mL were used, although with titanium dioxide MNPs and graphene-based tetraethoxysilane–methyltrimethoxysilane sol-gel hybrid MNPs the sample volume was higher (500 and 800 mL, respectively). In all these works, the extraction is quick (from 8 to 50 min), a low amount of MNPs is necessary (10–80 mg), and desorption is usually done with a low amount of organic solvent. Recently, zero valent Fe-reduced graphene oxide quantum dots were described as a new and effective sorbent for the extraction of five OPPs with recoveries in the range of 82.9–113.2% [59].

Two-step microextraction techniques, such as DLLME combined with mSPE, have been described as an alternative to the conventional DLLME, in which the retrieval of the extraction solvent is based on the adsorption by MNPs. Thus, *in situ* ILs-DLLME combined with Fe_3O_4 MNPs was reported by Fan et al. [60] for the analysis of four pyrethroids in water. The *in situ* forming hydrophobic IL obtained by reaction of the ionic liquid [P_{4448}][Br] with the anion-exchange reagent Na [N(CN)$_2$] was the extraction agent, and Fe_3O_4 (30 mg) was added to water samples. Then, the MNPs adsorbing ILs were easily isolated using a magnet and the pesticides were eluted with 50 µL ACN. The proposed method has high recoveries (80–117%), short pretreatment time (< 3 min), and low extraction solvent consumption.

Although the functionalized magnetic materials demonstrated good extraction performance, they are not commercialized, their synthetic procedures are relatively tedious, and the nanocomposite needs to be well characterized. Recently, Peng et al. [61] described the detection of seven phenoxy acid herbicides in water samples based on mSPE, using magnetic amino-functionalized MWCNTs, followed by LC–MS/MS. In this work, the MWCNTs were simply mixed with bare Fe_3O_4 MNPs to fabricate a novel kind of magnetic NH_2–MWCNTs composites, which avoids the tedious modification steps of Fe_3O_4. Under the optimized conditions, recoveries from 92 to 103% and LOQs ranging from 0.03 to 0.06 ng/mL were obtained for the target pesticides.

The lack of a scalable and standard process for the synthesis of MNPs restricts the access of this technology to possible practical and commercial applications. The continuous advances in the manufacture of MNPs have produced a growing range of products, each with their own unique properties. Thus, the reproducibility of different batches in NP synthesis is a significant challenge, as it is difficult to produce particles with predictable and reproducible properties, and another challenge is standardization in vocabulary, methods, and measurement techniques [62].

9.2.3.5 Molecularly Imprinted Solid-Phase Extraction (MISPE)

Molecularly imprinted polymers (MIPs) are synthetic materials prepared by co-polymerizing functional and cross-linking monomers in the presence of a template molecule. After polymerization, the template molecule is extracted, leaving cavities that are complementary in size, shape, and chemical functionality to the template. Thus, MIPs are selective sorbent materials that can be used in the enrichment and cleanup of target analytes from complex samples. Use of SPE sorbents based on MIPs introduces "tailor-made selectivity" into the sample preparation procedure. For this reason, during the last number of years MIPs have been combined with sample preparation techniques such as SPME, SBSE, mSPE, and matrix solid-phase dispersion. However, MISPE is the most developed technical application of MIPs. The conventional off-line MISPE is similar to the traditional SPE procedure, where the sorbent is a MIP packed into the cartridges that are conditioned, loaded, and washed, and finally the analytes are eluted for subsequent chromatographic or spectrophotometric analysis. An interesting advantage of MISPE methods is the high selectivity of the extraction process that improves sample cleanup. Thus, where selectivity is required MIP phases have distinct advantages over standard SPE phases.

Bulk polymerization was the first strategy used for the synthesis of MIPs and nowadays is still used to prepare them [63,64]. However, more recently there is a trend to develop MIPs in NPs sizes, in order to increase the total surface area, and provide high accessibility to the imprinted sites and fast adsorption equilibrium. For that reason, the coating of NPs with MIPs to use in sample preparation has increased. Nano-sized MIPs can be used together with other extraction techniques, such as the ultrasound-assisted dispersive solid phase microextraction. Thus, Bazrafshan et al. [65] reported the application of nano-sized MIPs for the selective ultrasound-assisted dispersive solid-phase extraction of carbaryl from water samples, using only 9.4 mg of sorbent and 400 μL of MeOH in the elution. Another example is the use of MIP-NPs prepared by surface grafting polymerization on nanosilica for the extraction of four phosphorothionate pesticides in water by molecularly imprinted matrix solid-phase dispersion coupled with GC–NPD [66].

The trend of using nano-size MIPs has been also applied in the development of electrochemical sensors for selective and sensitive determination of pesticides [67]. Electrodes modified with MIP-NPs show better response than electrodes modified with MIP-microparticles, mostly due to an increase in the number of recognition sites. Additionally, compared with conventional MIPs, magnetic MIPs have attracted considerable attention due to the fact that they not only exhibit outstanding magnetism but also have specific selective binding for template molecules. The use of core-shell structured amphiphilic magnetic MIPs for the selective extraction of five chloroacetamide herbicides in water was proposed by Ji et al. [68]. In this work, magnetic MIPs were prepared by using Fe_3O_4 microspheres as the magnetic core, and 4-vinyl pyridine and alkenyl glycosides glucose as functional co-monomers via the distillation–precipitation polymerization, and low LODs (0.03–0.06 µg/L), and good recoveries (82 to 103%) were obtained.

The use of MIPs provides a promising technology to the selective extraction of pesticides in environmental samples. Nevertheless, the extraction methods reported by now only have been applied to a low number of pesticides (from one to six) usually belonging to the same chemical group.

9.2.4 OTHER TECHNIQUES

The concept of combining several extraction techniques to overcome the deficiencies of each technique, or to obtain the advantages of both, has been applied in some methods to extract pesticides from water samples. Thus, methods combining SPE and SPME [69], LLE and SPE [6], SPE and DLLME [70], or LLE and dispersive SPE [71] have been described. The combination of LLE and DLLME allows the handling of large sample volumes (an advantage of LLE) along with obtaining of high enrichment factors (an advantage of DLLME), and something similar occurs with the combination of SPE and DLLME. In addition, there are some miniaturized techniques developed as an alternative approach to LLE or SPE that have not been described in the previous sections (these are either modifications of a microextraction technique or of the extraction device). Some of these alternative microextraction techniques are vortex-assisted liquid–liquid microextraction (VALLME), homogeneous liquid–liquid microextraction (HLLME), thin film microextraction (TFME), or microextraction in packed syringe (MEPS).

The term VALLME was introduced in 2010 by Yiantzi et al. [72] as a new and fast microextraction method whereby dispersion of the extractant phase into the aqueous phase is achieved using vortex mixing. They used the VALLME for the determination of octylphenol, nonylphenol, and bisphenol-A in water and wastewater samples. This extraction method, followed by GC–MS analysis, was successfully applied by Ozcan [73] for the analysis of 20 OCs in water and wastewater samples. In this work, the main parameters involved in the extraction process (extraction solvent, rotational speed of the vortex, vortex extraction time, solvent volume, and ionic strength of the sample) were optimized, and the optimized extraction conditions for 5 mL water sample were: extraction solvent, bromoform; solvent volume, 50 mL; vortex extraction time, 2 min at 3000 rpm with no ionic strength adjustment and 5 min of centrifugation at 4000 rpm. LOQs were in the range of 0.037–0.173 µg/L and the mean recoveries of OCs from fortified water samples were 96%.

HLLME extracts the analytes from a homogeneous solution into a very small phase formed from the solution by the phase separation phenomenon. The procedure is simple and requires only addition of a reagent. In HLLME, there are no interfaces between the water, the pesticides, and the extracting solvent. For this reason, the interface surface area is infinitely large and no vigorous mechanical shaking is necessary, although the extraction process can be assisted by ultrasonic or microwaves radiation. This technique was successfully applied for the determination in water of 16 OCs [74], 15 OPPs [75], and 5 triazoles [76]. In the last one, the homogenous mixture solution was obtained with ultrasound assistance for 30 sec, previous to the salting out of the homogenous solution to obtain a phase separation [76].

Membrane techniques exist in various forms and may be geometrically classified into film/flat sheet or hollow fiber configurations. The application of hollow fiber membranes to the extraction of pesticides in water was discussed in Section 9.2.2.3. The TFME technique was developed as an alternative approach to LLE. The use of films in the extraction step allows lower detection limits by increasing the relation surface area-to-volume ratio. The initial development of TFME employed a thin sheet of PDMS to be used as a membrane-based SPME device. Subsequently, PDMS/DVB and Carboxen/PDMS sheets were also developed in order to further enhance the extraction efficiency of pure PDMS membranes. Recently, Piri-Moghadam et al. [77] reported the use of PDMS/DVB and PDMS/DVB-carbon mesh supported membranes for the extraction of 23 pesticides, using a thermal desorption unit to transfer the analytes to the GC–MS. After optimization of the most critical parameters, both membranes were able to achieve LOQs between 0.025 and 0.50 µg/L and demonstrated excellent robustness, withstanding up to 100 extractions/desorption cycles. Additionally, 18 surface water samples were analyzed using TFME and the results were compared with those obtained by LLE. The comparison of TFME and LLE from several analytical aspects demonstrated that the TFME method gave similar accuracy to LLE, while providing additional advantages including higher sensitivity, lower sample volume, reduction of waste production, and faster analytical throughput.

MEPS is a micro-version of SPE. In this extraction method, the packing is integrated directly into the syringe and not in a separate column as in SPE. This form of miniaturized SPE uses a procedure similar to in-tube SPME and SPDE. The small amount of the sorbent (1–6 mg) enables the reduction of sample volume and organic solvent for elution of the compounds. Moreover, if the cleaning conditions are properly optimized, the same packed syringe may be reused many times. Many materials such as silica-based (C2, C8, C18) sorbents, strong cation exchangers, or MWCNTs can be used as sorbents in MEPS [78]. The direct injection of the extracts into the liquid or gas chromatograph can be done, and this technique can be coupled on-line with GC or LC. Recently, Szarka et al. [79] developed a MEPS method for the simultaneous extraction of seven pesticides (including endocrine disrupting chemicals) from environmental water samples, in combination with GC–MS. The developed methodology requires a very low sample volume (60 µL) and also low consumption of organic solvent (21 µL). Recoveries ranged between 77 and 119% and LOQs from 1.3 to 50 ng/L.

Some extraction techniques, initially developed for other matrices, have been used for pesticide determination in water samples. Thus, Ramos et al. [80] described the first method for the determination of 23 pesticides in water using a passive sampler named VERAM for monitoring atmospheric pollutants. The VERAM devices consist of a low-density polyethylene lay-flat tube filled with a solid phase, such as active carbon and Florisil, which is able to adsorb a wide range of compounds with different physicochemical properties.

9.3 ANALYTICAL TECHNIQUES FOR PESTICIDE DETERMINATION IN WATER SAMPLES

The analysis of pesticide residues in water is usually done by GC and LC. To be suitable for GC, the pesticide must possess thermal stability and volatility. Though due to the high polarity, non-volatility, and thermal lability of some pesticides, a derivatization step before GC analysis is necessary to improve their volatility and chromatographic behavior. Classic GC detectors, such as NPD, FPD, and ECD often provide LODs low enough for some pesticide groups, as a consequence of their high selectivity. Thus, the determination of pesticides that contain nitrogen or phosphorus atoms, such as OPPs and triazines, has been satisfactorily carried out with NPD [81], whereas FPD is commonly used for the selective detection of organosulfur and OPPs [52,82]. ECD is another classic detector that provides a very sensitive response in the determination of halogenated compounds or those that contain electronegative atoms, such as pyrethroids and OCs [42,74], but this detector also responds to other organic molecules, making necessary a good cleanup of the extract before the analysis. On the other hand, LC is an analytical tool adequate for the determination of pesticides that are not thermally stable or not volatile. The classic and most frequently used detection methods in liquid chromatography are ultraviolet (UV) together with other selective detectors such as fluorescence (FL). FL presents high selectivity and sensitivity, but either it is limited to compounds that fluoresce or a derivatization step to obtain a fluorescent compound is required. However, with these detection systems it not possible to have the simultaneous determination and confirmation of the analytes. For this reason, GC–MS, GC–MS/MS, and LC–MS/MS have gained popularity in the multiresidue analysis of pesticides, and these are nowadays the most widely used techniques in the analysis of pesticides in environmental samples.

Time of flight mass spectrometry (TOF-MS) is the result of the significant advances undergone in analytical instrumentation. Thus, GC, LC, or comprehensive two-dimensional gas chromatography (GC×GC) coupled to TOF-MS have been used in the multiresidue analysis of a broad number of pesticides in water. Hernandez et al. [83] reported that TOF-MS hyphenated to LC and GC allows large screening of many organic pollutants in surface water and soil samples. Whereas, Ferrer and Thurman [33] reported a multiresidue method for the chromatographic separation and accurate mass identification of 101 pesticides and their degradation products using LC–TOF-MS and Matamoros et al. [84] reported an analytical procedure based on GC×GC coupled with TOF-MS for the simultaneous determination of 97 organic contaminants, many of them pesticides, at trace concentration in river water.

With the improving sensitivity of MS/MS in the multiple-reaction monitoring mode, direct injection of water samples for the quantitation of pesticides at the EU drinking water standard of 0.1 µg/L is possible. Direct injection is especially promising for multiresidue analyses without preparative procedures, becoming a key tool for the routine quality control of priority pesticides in water. Thus, Diaz et al. [85] developed a LC–MS/MS method for the detection of 31 pesticides in tap water and treated wastewater by direct injection in 11 min of analysis, making possible to analyze up to 80 water samples per day, by the direct injection of 100 µL of sample with LODs < 10 ng/L for most of the compounds. Moreover, based on the information available on 293 pesticides and 210 pesticide metabolites, Reemtsma et al. [86] developed an analytical method for the analysis of 150 pesticide metabolites from groundwater and surface water by direct injection-electrospray ionization-tandem mass spectrometry with multiple-reaction monitoring. A broad overview of the analysis of pesticides by chromatographic techniques coupled with mass spectrometry is included in Chapter 3.

Pesticides have also been analyzed with non-chromatographic methods. Thus, the determination of the pesticide can be done by electro and electroluminescence methods [87,88], UV-spectrophotometry [89], fluorescence [90], luminescence, and chemiluminescence [91,92] methods. Recently, Duarte et al. [93] reported a method for the determination of three pesticides (diuron, 2,4-dichlorophenoxyacetic acid, and tebuthiuron) by differential pulse voltammetry associated with SPE using a cathodically pretreated boron-doped diamond electrode. The authors reported that the association of SPE with electroanalytical techniques, although rarely exploited, shows to be a promising tool to obtain simple, precise, accurate, and interference-free electroanalytical methods for the determination of herbicides. The results obtained were in agreement with those obtained by HPLC at a 95% confidence level.

Capillary electrophoresis (CE) is an alternative analytical technique that is complementary to GC and LC and has been applied in the determination of pesticide residues in water samples. The most frequently used working mode by CE is micellar electrokinetic chromatography (MECK), although capillary zone electrophoresis and capillary electrochromatography are also applied. Some examples of the application of MECK for the determination of pesticides in water samples are the methods developed by Santalad et al. [94] and Amelin et al. [95] using this technique in combination with SPE and/or DLLME for the determination of 6 carbamates pesticides and 38 polar pesticides in water samples, respectively. The application of sensors and biosensors in the determination of pesticides in environmental samples is also rapidly increasing. The main characteristics of these technologies and their application to pesticide analysis are included in Chapter 4. A recent example of the application of sensors is the method reported by Figueiredo-Filho et al. [87] in which a bismuth film on a disposable minisensor was developed for the determination of two dipyridyl pesticides (diquat and paraquat) using voltammetric techniques.

The application of non-chromatographic methods, except MEKC, has been generally focused on the detection of a low number of pesticides. Thus, the multiresidue analysis of pesticides is clearly a future trend for the application of non-chromatographic analytical methods.

REFERENCES

1. T. Bedassa, A. Gure, N. Megersa, Low density solvent based dispersive liquid-liquid microextraction and preconcentration of multiresidue pesticides in environmental waters for liquid chromatographic analysis, *J. Anal. Chem.*, 70 (2015) 1199–1206.
2. M. Tankiewicz, M. Biziuk, Fast, sensitive and reliable multi-residue method for routine determination of 34 pesticides from various chemical groups in water samples by using dispersive liquid-liquid microextraction coupled with gas chromatography-mass spectrometry, *Anal. Bioanal. Chem.*, 410 (2018) 1533–1550.
3. J.H. Park, M.I.R. Mamun, A.M. Abd El-Aty, T.W. Na, J.H. Choi, M.W. Ghafar, W.J. Choi, K.S. Kim, S.D. Kim, J.H. Shim, Simultaneous multiresidue determination of 48

pesticides in Yeongsan and Sumjin River water using GC–NPD and confirmation via GC–MS, *Biomed. Chromatogr.*, 25 (2011) 155–163.

4. M.I.R. Mamun, J.H. Park, J.H. Choi, H.K. Kim, W.J. Choi, S.S. Han, K. Hwang, N.I. Jang, M.E. Assayed, M.A. El-Dib, H.C. Shin, A.M. Abd El-Aty, J.H. Shim, Development and validation of a multiresidue method for determination of 82 pesticides in water using GC, *J. Sep. Sci.*, 32 (2009) 559–574.

5. R. Aznar, C. Sánchez-Brunete, B. Albero, H. Moreno-Ramón, J.L. Tadeo, Pyrethroids levels in paddy field water under Mediterranean conditions: measurements and distribution modelling, *Paddy Water Environ.*, 15 (2017) 307–316.

6. H. Kuranchie-Mensah, S.M. Atiemo, L. Palm, S. Blankson-Arthur, A.O. Tutu, P. Fosu, Determination of organochlorine pesticide residue in sediment and water from the Densu river basin, Ghana, *Chemosphere*, 86 (2012) 286–292.

7. J. Wu, J.A. Lu, C. Wilson, Y.J. Lin, H. Lu, Effective liquid–liquid extraction method for analysis of pyrethroid and phenylpyrazole pesticides in emulsion-prone surface water samples, *J. Chromatogr. A*, 1217 (2010) 6327–6333.

8. J. Fenoll, P. Hellín, C.M. Martínez, P. Flores, S. Navarro, Determination of 48 pesticides and their main metabolites in water samples by employing sonication and liquid chromatography–tandem mass spectrometry, *Talanta*, 85 (2011) 975–982.

9. M. Rezaee, Y. Assadi, M.-R.M. Hosseinia, E. Aghaee, F. Ahmadi, S. Berijani, Determination of organic compounds in water using dispersive liquid–liquid microextraction, *J. Chromatogr. A*, 1116 (2006) 1–9.

10. M.L. Martins, O.D. Prestes, M.B. Adaime, R. Zanella, Determination of pesticides and related compounds in water by dispersive liquid-liquid microextraction and gas chromatography–triple quadrupole mass spectrometry, *Anal. Methods*, 6 (2014) 5020–5027.

11. Y.S. Su, J.F. Jen, Determination of organophosphorous pesticides in water using in-syringe ultrasound-assisted emulsification and gas chromatography with electron-capture detection, *J. Chromatogr. A*, 1217 (2010) 5043–5049.

12. S. Ozcan, A. Tor, M.E. Aydin, Application of ultrasound-assisted emulsification-microextraction for the analysis of organochlorine pesticides in waters, *Water Res.*, 43 (2009) 4269–4277.

13. C.X. Wu, N. Liu, Q.H. Wu, C. Wang, Z. Wang, Application of ultrasound-assisted surfactant-enhanced emulsification microextraction for the determination of some organophosphorus pesticides in water samples, *Anal. Chim. Acta*, 679 (2010) 56–62.

14. Q.H. Wu, Q.Y. Chang, C.X. Wu, H. Rao, X. Zeng, C. Wang, Z. Wang, Ultrasound-assisted surfactant-enhanced emulsification microextraction for the determination of carbamate pesticides in water samples by high performance liquid chromatography, *J. Chromatogr. A*, 1217 (2010) 1773–1778.

15. W.T. Zhao, J.D. Li, T. Wu, P. Wang, Z.Q. Zhou, Determination of organochlorine pesticides in snow water samples by low density solvent based dispersive liquid–liquid microextraction, *J. Sep. Sci.*, 37 (2014) 2599–2604.

16. H. Yan, H. Wang, Recent development and applications of dispersive liquid–liquid microextraction, *J. Chromatogr. A*, 1295 (2013) 1–15.

17. W.P. Wang, H.L. Zhu, S.M. Cui, J.G. Miao, J.R. Chen, Ultrasound-assisted dispersive liquid–liquid microextraction based on solidification of floating organic droplets coupled with gas chromatography for the determination of pesticide residues in water samples, *Anal. Methods*, 6 (2014) 3388–3394.

18. C.X. Wu, H.M. Liu, W.H. Liu, Q.H. Wu, C. Wang, Z. Wang, Determination of organophosphorus pesticides in environmental water samples by dispersive liquid–liquid microextraction with solidification of floating organic droplet followed by high-performance liquid chromatography, *Anal. Bioanal. Chem.*, 397 (2010) 2543–2549.

19. S.S. Caldas, C. Rombaldi, J.L.D. Arias, L.C. Marube, E.G. Primel, Multi-residue method for determination of 58 pesticides, pharmaceuticals and personal care products

in water using solvent demulsification dispersive liquid-liquid microextraction combined with liquid chromatography–tandem mass spectrometry, *Talanta*, 146 (2016) 676–688.

20. L.J. He, X.L. Luo, H.X. Xie, C.J. Wang, X.M. Jiang, K. Lu, Ionic liquid-based dispersive liquid–liquid microextraction followed high-performance liquid chromatography for the determination of organophosphorus pesticides in water sample, *Anal. Chim. Acta*, 655 (2009) 52–59.

21. H.M. Albishri, N.A.M. Aldawsari, D. Abd El-Hady, Ultrasound-assisted temperature-controlled ionic liquid dispersive liquid-phase microextraction combined with reversed-phase liquid chromatography for determination of organophosphorus pesticides in water samples, *Electrophoresis*, 37 (2016) 2462–2469.

22. J.H. Zhang, Z. Liang, S.Q. Li, Y.B. Li, B. Peng, W.F. Zhou, H.X. Gao, In-situ metathesis reaction combined with ultrasound-assisted ionic liquid dispersive liquid–liquid microextraction method for the determination of phenylurea pesticides in water samples, *Talanta*, 98 (2012) 145–151.

23. Q.X. Zhou, X.G. Zhang, Combination of ultrasound-assisted ionic liquid dispersive liquid-phase microextraction and high performance liquid chromatography for the sensitive determination of benzoylureas pesticides in environmental water samples, *J. Sep. Sci.*, 33 (2010) 3734–3740.

24. M.I. Leong, M.R. Fuh, S.D. Huang, Beyond dispersive liquid–liquid microextraction, *J. Chromatogr. A*, 1335 (2014) 2–14.

25. X. Liu, C. Liu, P. Wang, G. Yao, D. Liu, Z. Zhou, Effervescence assisted dispersive liquid–liquid microextraction based on cohesive floating organic drop for the determination of herbicides and fungicides in water and grape juice, *Food Chem.*, 245 (2018) 653–658.

26. J. Yang, C. Fan, D. Kong, G. Tang, W. Zhang, H. Dong, Y. Liang, D. Wang, Y. Cao, Synthesis and application of imidazolium-based ionic liquids as extraction solvent for pretreatment of triazole fungicides in water samples, *Anal. Bioanal. Chem.*, 410 (2018) 1647–1656.

27. X.H. Wang, J. Cheng, X.F. Wang, M. Wu, M. Cheng, Development of an improved single-drop microextraction method and its application for the analysis of carbamate and organophosphorus pesticides in water samples, *Analyst*, 137 (2012) 5339–5345.

28. F. Tian, W.J. Liu, H.S. Fang, M. An, S.S. Duan, Determination of six organophosphorus pesticides in water by single-drop microextraction coupled with GC–NPD, *Chromatographia*, 77 (2014) 487–492.

29. S.P. Huang, S.D. Huang, Determination of organochlorine pesticides in water using solvent cooling assisted dynamic hollow-fiber-supported headspace liquid-phase microextraction, *J. Chromatogr. A*, 1176 (2007) 19–25.

30. H.C. Menezes, B.P. Paulo, M.J.N. Paiva, Z.L. Cardeal, A simple and quick method for the determination of pesticides in environmental water by HF-LPME-GC/MS, *J. Anal. Methods Chem.*, (2016).

31. T. Trtić-Petrović, J. Đorđević, N. Dujaković, K. Kumrić, T. Vasiljević, M. Laušević, Determination of selected pesticides in environmental water by employing liquid-phase microextraction and liquid chromatography–tandem mass spectrometry, *Anal. Bioanal. Chem.*, 397 (2010) 2233–2243.

32. S. Rodríguez-Mozaz, M.J. López de Alda, D. Barceló, Advantages and limitations of on-line solid phase extraction coupled to liquid chromatography–mass spectrometry technologies versus biosensors for monitoring of emerging contaminants in water, *J. Chromatogr. A*, 1152 (2007) 97–115.

33. I. Ferrer, E.M. Thurman, Multi-residue method for the analysis of 101 pesticides and their degradates in food and water samples by liquid chromatography/time-of-flight mass spectrometry, *J. Chromatogr. A*, 1175 (2007) 24–37.

34. A.C. Charalampous, G.E. Miliadis, M.A. Koupparis, A new multiresidue method for the determination of multiclass pesticides, degradation products and PCBs in water using LC–MS/MS and GC–MS(n) systems, *Int. J. Environ. Anal. Chem.*, 95 (2015) 1283–1298.

35. J. Cotton, F. Leroux, S. Broudin, M. Poirel, B. Corman, C. Junot, C. Ducruix, Development and validation of a multiresidue method for the analysis of more than 500 pesticides and drugs in water based on on-line and liquid chromatography coupled to high resolution mass spectrometry, *Water Res.*, 104 (2016) 20–27.

36. M.D. Gil García, S. Dahane, F.M. Arrabal-Campos, M.M. SocíasViciana, M.A. García, I. Fernández, M.M. Galera, MCM-41 as novel solid phase sorbent for the pre-concentration of pesticides in environmental waters and determination by microflow liquid chromatography–quadrupole linear ion trap mass spectrometry, *Microchem. J.*, 134 (2017) 181–190.

37. O. Mann, E. Pock, K. Wruss, W. Wruss, R. Krska, Development and validation of a fully automated online-SPE-ESI-LC–MS/MS multi-residue method for the determination of different classes of pesticides in drinking, ground and surface water, *Int. J. Environ. Anal. Chem.*, 96 (2016) 353–372.

38. L. Pelit, T.N. Dizdas, Preparation and application of a polythiophene solid-phase microextraction fiber for the determination of endocrine-disruptor pesticides in well waters, *J. Sep. Sci.*, 36 (2013) 3234–3241.

39. M. Saraji, M.T. Jafari, H. Sherafatmand, Sol-gel/nanoclay composite as a solid-phase microextraction fiber coating for the determination of organophosphorus pesticides in water samples, *Anal. Bioanal. Chem.*, 407 (2015) 1241–1252.

40. M. Saraji, B. Rezaei, M.K. Boroujeni, A.A.H. Bidgoli, Polypyrrole/sol-gel composite as a solid-phase microextraction fiber coating for the determination of organophosphorus pesticides in water and vegetable samples, *J. Chromatogr. A*, 1279 (2013) 20–26.

41. E. Beceiro-González, E. Concha-Graña, A. Guimaraes, C. Gonçalves, S. Muniategui-Lorenzo, M.F. Alpendurada, Optimisation and validation of a solid-phase microextraction method for simultaneous determination of different types of pesticides in water by gas chromatography–mass spectrometry, *J. Chromatogr. A*, 1141 (2007) 165–173.

42. P. Chaturvedi, R. Kumari, R.C. Murthy, D.K. Patel, Analysis of pesticide residues in drinking water samples using solid-phase micro-extraction (SPME) coupled to a gas chromatography–electron-capture detector (GC-ECD), *Water Sci. Tech.-W. Sup.*, 11 (2011) 754–764.

43. A. Rodríguez-Lafuente, H. Piri-Moghadam, H.L. Lord, T. Obal, J. Pawliszyn, Inter-laboratory validation of automated SPME-GC/MS for determination of pesticides in surface and ground water samples: sensitive and green alternative to liquid–liquid extraction, *Water Qual. Res. J. Can.*, 51 (2016) 331–343.

44. A.L. De Toffoli, B.H. Fumes, F.M. Lanças, Packed in-tube solid phase microextraction with graphene oxide supported on aminopropyl silica: determination of target triazines in water samples, *J. Environ. Sci. Health Part B*, 53 (2018) 1–7.

45. A. Souza-Silva, J. Pawliszyn, Chapter sixteen - recent advances in solid-phase micro-extraction for contaminant analysis in food matrices, in: E. Ibáñez, A. Cifuentes (Eds.), *Comprehensive Analytical Chemistry*, Elsevier: Oxford, UK, 2017, pp. 483–517.

46. J.M.F. Nogueira, Stir-bar sorptive extraction: 15 years making sample preparation more environment-friendly, *Trends Anal. Chem.*, 71 (2015) 214–223.

47. L. Feng, S.J. Zhang, G.H. Zhu, M.F. Li, J.S. Liu, Determination of trace polychlorinated biphenyls and organochlorine pesticides in water samples through large-volume stir bar sorptive extraction method with thermal desorption gas chromatography, *J. Sep. Sci.*, 40 (2017) 4583–4590.

48. M.C. Sampedro, M.A. Goicolea, N. Unceta, A. Sánchez-Ortega, R.J. Barrio, Sequential stir bar extraction, thermal desorption and retention time locked GC–MS for determination of pesticides in water, *J. Sep. Sci.*, 32 (2009) 3449–3456.

49. C. Margoum, C. Guillemain, X. Yang, M. Coquery, Stir bar sorptive extraction coupled to liquid chromatography–tandem mass spectrometry for the determination of pesticides in water samples: method validation and measurement uncertainty, *Talanta*, 116 (2013) 1–7.

50. C. Hu, M. He, B.B. Chen, B. Hu, A sol-gel polydimethylsilsoxane/polythiophene coated stir bar sorptive extraction combined with gas chromatography–flame photometric detection for the determination of organophosphorus pesticides in environmental water samples, *J. Chromatogr. A*, 1275 (2013) 25–31.

51. R. Ahmadkhaniha, N. Rastkari, Development of a carbon nanotube-coated stir bar for determination of organophosphorus pesticides in water, *Asia-Pac. J. Chem. Eng.*, 11 (2016) 893–900.

52. Z.W. Xiao, M. He, B.B. Chen, B. Hu, Polydimethylsiloxane/metal-organic frameworks coated stir bar sorptive extraction coupled to gas chromatography–flame photometric detection for the determination of organophosphorus pesticides in environmental water samples, *Talanta*, 156 (2016) 126–133.

53. C.Y. Li, L.G. Chen, Determination of pyrethroid pesticides in environmental waters based on magnetic titanium dioxide nanoparticles extraction followed by HPLC analysis, *Chromatographia*, 76 (2013) 409–417.

54. C.Y. Li, L.G. Chen, W. Li, Magnetic titanium oxide nanoparticles for hemimicelle extraction and HPLC determination of organophosphorus pesticides in environmental water, *Microchim. Acta*, 180 (2013) 1109–1116.

55. J.P. Ma, Z.D. Yao, L.W. Hou, W.H. Lu, Q.P. Yang, J.H. Li, L.X. Chen, Metal organic frameworks (MOFs) for magnetic solid-phase extraction of pyrazole/pyrrole pesticides in environmental water samples followed by HPLC–DAD determination, *Talanta*, 161 (2016) 686–692.

56. R. Mohammad-Rezaei, H. Razmi, V. Abdollahi, A.A. Matin, Preparation and characterization of Fe$_3$O$_4$/graphene quantum dots nanocomposite as an efficient adsorbent in magnetic solid phase extraction: application to determination of bisphenol A in water samples, *Anal. Methods*, 6 (2014) 8413–8419.

57. Y.X. Soon, K.S. Tay, n-Octylated magnetic nanoparticle-based microextraction for the determination of organophosphorus pesticides in water, *Anal. Lett.*, 48 (2015) 1604–1618.

58. A. Targhoo, A. Amiri, M. Baghayeri, Magnetic nanoparticles coated with poly(p-phenylenediamine-co-thiophene) as a sorbent for preconcentration of organophosphorus pesticides, *Microchim. Acta*, 185 (2018) 15.

59. S. Akbarzade, M. Chamsaz, G.H. Rounaghi, M. Ghorbani, Zero valent Fe-reduced graphene oxide quantum dots as a novel magnetic dispersive solid phase microextraction sorbent for extraction of organophosphorus pesticides in real water and fruit juice samples prior to analysis by gas chromatography–mass spectrometry, *Anal. Bioanal. Chem.*, 410 (2018) 429–439.

60. C. Fan, Y. Liang, H.Q. Dong, G.L. Ding, W.B. Zhang, G. Tang, J.L. Yang, D.D. Kong, D. Wang, Y.S. Cao, In-situ ionic liquid dispersive liquid–liquid microextraction using a new anion-exchange reagent combined Fe$_3$O$_4$ magnetic nanoparticles for determination of pyrethroid pesticides in water samples, *Anal. Chim. Acta*, 975 (2017) 20–29.

61. M.M. Peng, Y.Q. Han, H. Xia, X.Z. Hu, Y.X. Zhou, L.J. Peng, X.T. Peng, Rapid and sensitive detection of the phenoxy acid herbicides in environmental water samples by magnetic solid-phase extraction combined with liquid chromatography–tandem mass spectrometry, *J. Sep. Sci.*, 41 (2018) 2221–2228.

62. J. Wells, O. Kazakova, O. Posth, U. Steinhoff, S. Petronis, L.K. Bogart, P. Southern, Q. Pankhurst, C. Johansson, Standardisation of magnetic nanoparticles in liquid suspension, *J. Phys. D Appl. Phys.*, 50 (2017).

63. P.P. Qi, X.Y. Wang, X.Q. Wang, H. Zhang, H. Xu, K.Z. Jiang, Q. Wang, Computer-assisted design and synthesis of molecularly imprinted polymers for the simultaneous

determination of six carbamate pesticides from environmental water, *J. Sep. Sci.*, 37 (2014) 2955–2965.

64. L. Zhao, Z.H. Ma, L.G. Pan, J.H. Wang, MISPE combined with GCMS for analysis of organophosphorus pesticides from environmental water sample, in: Z. Cao, X.Q. Cao, L. Sun, Y.H. He (Eds.), *Advanced Materials*, Trans Tech Publications: Zurich, Switzerland, Pts 1–4, 2011, pp. 3216–3220.

65. A.A. Bazrafshan, M. Ghaedi, Z. Rafiee, S. Hajati, A. Ostovan, Nano-sized molecularly imprinted polymer for selective ultrasound assisted microextraction of pesticide Carbaryl from water samples: spectrophotometric determination, *J. Colloid Interf. Sci.*, 498 (2017) 313–322.

66. M.C. Zhou, F. Hu, H. He, S.H. Shu, M. Wang, Determination of phosphorothioate pesticides in environmental water by molecularly imprinted matrix solid-phase dispersion coupled with gas chromatography and a nitrogen phosphorus detector, *Instrum. Sci. Technol.*, 43 (2015) 669–680.

67. A. Motaharian, F. Motaharian, K. Abnous, M.R.M. Hosseini, M. Hassanzadeh-Khayyat, Molecularly imprinted polymer nanoparticles-based electrochemical sensor for determination of diazinon pesticide in well water and apple fruit samples, *Anal. Bioanal. Chem.*, 408 (2016) 6769–6779.

68. W.H. Ji, R.H. Sun, W.J. Duan, X. Wang, T. Wang, Y. Mu, L.P. Guo, Selective solid phase extraction of chloroacetamide herbicides from environmental water samples by amphiphilic magnetic molecularly imprinted polymers, *Talanta*, 170 (2017) 111–118.

69. C.R. Qiu, M.G. Cai, Ultra trace analysis of 17 organochlorine pesticides in water samples from the Arctic based on the combination of solid-phase extraction and headspace solid-phase microextraction-gas chromatography–electron-capture detector, *J. Chromatogr. A*, 1217 (2010) 1191–1202.

70. M. Shamsipur, N. Yazdanfar, M. Ghambarian, Combination of solid-phase extraction with dispersive liquid–liquid microextraction followed by GC–MS for determination of pesticide residues from water, milk, honey and fruit juice, *Food Chem.*, 204 (2016) 289–297.

71. S.B.A. Ghani, A.H. Hanafi, QuEChERS method combined with GC–MS for pesticide residues determination in water, *J. Anal. Chem.*, 71 (2016) 508–512.

72. E. Yiantzi, E. Psillakis, K. Tyrovola, N. Kalogerakis, Vortex-assisted liquid–liquid microextraction of octylphenol, nonylphenol and bisphenol-A, *Talanta*, 80 (2010) 2057–2062.

73. S. Ozcan, Viable and rapid determination of organochlorine pesticides in water, *CLEAN Soil Air Water*, 38 (2010) 457–465.

74. N. Yazdanfar, Y. Yamini, M. Ghambarian, Homogeneous liquid–liquid microextraction for determination of organochlorine pesticides in water and fruit samples, *Chromatographia*, 77 (2014) 329–336.

75. S. Berijani, M. Sadigh, E. Pournamdari, Homogeneous liquid–liquid microextraction for determination of organophosphorus pesticides in environmental water samples prior to gas chromatography-flame photometric detection, *J. Chromatogr. Sci.*, 54 (2016) 1061–1067.

76. X.Y. Xu, J.Q. Ye, J. Nie, Z.G. Li, M.R. Lee, A new liquid–liquid microextraction method by ultrasound assisted salting-out for determination of triazole pesticides in water samples coupled by gas chromatography–mass spectrometry, *Anal. Methods*, 7 (2015) 1194–1199.

77. H. Piri-Moghadam, E. Gionfriddo, A. Rodríguez-Lafuente, J.J. Grandy, H.L. Lord, T. Obal, J. Pawliszyn, Inter-laboratory validation of a thin film microextraction technique for determination of pesticides in surface water samples, *Anal. Chim. Acta*, 964 (2017) 74–84.

78. A. Taghani, N. Goudarzi, G. Bagherian, Application of multiwalled carbon nanotubes for the preconcentration and determination of organochlorine pesticides in water samples by gas chromatography with mass spectrometry, *J. Sep. Sci.*, 39 (2016) 4219–4226.

79. A. Szarka, S. Hrouzkova, T.M. Deszatova, S. Zichova, Microextraction in packed syringe coupled with GC–MS for the determination of pesticides in environmental water samples, *Fresen. Environ. Bull.*, 26 (2017) 2664–2671.

80. T.D. Ramos, R.J. Cassella, M. de la Guardia, A. Pastor, F.A. Esteve-Turrillas, Use of a versatile, easy, and rapid atmospheric monitor (VERAM) passive samplers for pesticide determination in continental waters, *Anal. Bioanal. Chem.*, 408 (2016) 8495–8503.

81. M.A. González-Curbelo, A.V. Herrera-Herrera, J. Hernández-Borges, M.A. Rodríguez-Delgado, Analysis of pesticides residues in environmental water samples using multiwalled carbon nanotubes dispersive solid-phase extraction, *J. Sep. Sci.*, 36 (2013) 556–563.

82. J. Xiong, B. Hu, Comparison of hollow fiber liquid phase microextraction and dispersive liquid-liquid microextraction for the determination of organosulfur pesticides in environmental and beverage samples by gas chromatography with flame photometric detection, *J. Chromatogr. A*, 1193 (2008) 7–18.

83. F. Hernández, T. Portolés, M. Ibáñez, M.C. Bustos-López, R. Díaz, A.M. Botero-Coy, C.L. Fuentes, G. Penuela, Use of time-of-flight mass spectrometry for large screening of organic pollutants in surface waters and soils from a rice production area in Colombia, *Sci. Total Environ.*, 439 (2012) 249–259.

84. V. Matamoros, E. Jover, J.M. Bayona, Part-per-trillion determination of pharmaceuticals, pesticides, and related organic contaminants in river water by solid-phase extraction followed by comprehensive two-dimensional gas chromatography time-of-flight mass spectrometry, *Anal. Chem.*, 82 (2010) 699–706.

85. L. Díaz, J. Llorca-Pórcel, I. Valor, Ultra trace determination of 31 pesticides in water samples by direct injection-rapid resolution liquid chromatography–electrospray tandem mass spectrometry, *Anal. Chim. Acta*, 624 (2008) 90–96.

86. T. Reemtsma, L. Alder, U. Banasiak, A multimethod for the determination of 150 pesticide metabolites in surface water and groundwater using direct injection liquid chromatography–mass spectrometry, *J. Chromatogr. A*, 1271 (2013) 95–104.

87. L.C.S. de Figueiredo, M. Baccarin, B.C. Janegitz, O. Fatibello, A disposable and inexpensive bismuth film minisensor for a voltammetric determination of diquat and paraquat pesticides in natural water samples, *Sens. Actuators B Chem.*, 240 (2017) 749–756.

88. V.A. Pedrosa, J. Caetano, S.A.S. Machado, M. Bertotti, Determination of parathion and carbaryl pesticides in water and food samples using a self assembled monolayer/acetylcholinesterase electrochemical biosensor, *Sensors*, 8 (2008) 4600–4610.

89. Y.S. Al-Degs, M.A. Al-Ghouti, A.H. El-Sheikh, Simultaneous determination of pesticides at trace levels in water using multiwalled carbon nanotubes as solid-phase extractant and multivariate calibration, *J. Hazard. Mater.*, 169 (2009) 128–135.

90. H.A. Azab, Z.M. Anwar, M.A. Rizk, G.M. Khairy, M.H. El-Asfoury, Determination of organophosphorus pesticides in water samples by using a new sensitive luminescent probe of Eu (III) complex, *J. Lumin.*, 157 (2015) 371–382.

91. M. Catalá-Icardo, L. Lahuerta-Zamora, S. Torres-Cartas, S. Meseguer-Lloret, Determination of organothiophosphorus pesticides in water by liquid chromatography and post-column chemiluminescence with cerium (IV), *J. Chromatogr. A*, 1341 (2014) 31–40.

92. J.F. Huertas-Pérez, A.M. García-Campaña, Determination of N-methylcarbamate pesticides in water and vegetable samples by HPLC with post-column chemiluminescence detection using the luminol reaction, *Anal. Chim. Acta*, 630 (2008) 194–204.

93. E.H. Duarte, J. Casarin, E.R. Sartori, C.R. Teixeira Tarley, Highly improved simultaneous herbicides determination in water samples by differential pulse voltammetry using boron-doped diamond electrode and solid phase extraction on cross-linked poly(vinylimidazole), *Sens. Actuators B Chem.*, 255 (2018) 166–175.

94. A. Santalad, L. Zhou, F. Shang, D. Fitzpatrick, R. Burakham, S. Srijaranai, J.D. Glennon, J.H.T. Luong, Micellar electrokinetic chromatography with amperometric detection and off-line solid-phase extraction for analysis of carbamate insecticides, *J. Chromatogr. A*, 1217 (2010) 5288–5297.

95. V.G. Amelin, D.S. Bol'shakov, A.V. Tret'yakov, Dispersive liquid–liquid microextraction and solid-phase extraction of polar pesticides from natural water and their determination by micellar electrokinetic chromatography, *J. Anal. Chem.*, 68 (2013) 386–397.

10 Sampling and Analysis of Pesticides in the Atmosphere

Maurice Millet

CONTENTS

10.1 Introduction 301
10.2 Monitoring of Pesticides in the Atmosphere 304
    10.2.1 Sampling and Extraction of Pesticides in Ambient Air.................. 304
        10.2.1.1 Sampling of Pesticides in Ambient Air 304
        10.2.1.2 Extraction of Pesticides in Ambient Air 307
        10.2.1.3 Cleaning of Traps for the Sampling of Pesticides in
                Ambient Air... 308
    10.2.2 Sampling and Extraction of Pesticides in Rainwater Samples 308
        10.2.2.1 Sampling of Rainwater 308
        10.2.2.2 Extraction of Pesticides from Rainwater 309
    10.2.3 Evaluation of Soil–Air Transfer of Pesticides (Spray Drift and
        Volatilization) 314
        10.2.3.1 Method Performances.. 318
10.3 Analysis of Pesticides in the Atmosphere.................................... 321
    10.3.1 Analysis by Gas Chromatography 321
        10.3.1.1 Analysis by GC–ECD and GC–NPD 321
        10.3.1.2 Analysis by GC–MS 322
    10.3.2 Derivatization 322
    10.3.3 Analysis by High Performance Liquid Chromatography 323
        10.3.3.1 Analysis by LC–UV or LC–DAD 323
        10.3.3.2 Analysis by LC–MS... 323
References... 324

10.1 INTRODUCTION

The intensive use of pesticide has led to a contamination of all compartments of the environment. The atmosphere is known to be a good pathway for the worldwide dissemination of pesticides. Pesticides can enter into the atmosphere by "spray drift" during application, by post-application volatilization from soils and leaves, and by wind erosion when pesticides are sorbed to soil particles and entrained into the atmosphere on wind-blown particles.[1] There are few data on the significance of this pathway, and on the quantitative effects of soil and environmental factors that influence this process.[2] This process is most important for herbicides as they are applied either at pre-emergence or post-emergence at an early growth stage of the crops (e.g., summer cereals, maize) when there is low soil coverage.[3]

When in the atmosphere, pesticides can be distributed between the gas and particle phases depending on their physical and chemical properties (vapor pressure, Henry's law constant, etc.) and on environmental and climatic conditions (concentration of particles, temperature, air humidity, etc.). The knowledge of the gas/particle partitioning of pesticides is important since this process affects the potential removal of pesticides by wet and dry deposition and by photolysis. It can also, together with photolysis, play a role in the atmospheric transport of pesticides over short or long distances.

The FOCUS-Air group[4] has defined that substances that are applied to plants and have a vapor pressure less than 10^{-5} Pa (at $20°C$), or are applied to soil and have a vapor pressure less than 10^{-4} Pa (at $20°C$), need not be considered in the short-range risk assessment scheme. Substances that exceed these limits require evaluation at the second tier, which is done by modeling.

Compounds adsorbed to particulate matter are mostly found in wet deposition.[5] Compounds mostly in the vapor phase are likely to be more evenly divided between wet and dry deposition. Pesticides in the gas phase generally have longer atmospheric residence time. In this case, the rate of removal is strongly influenced by the Henry's law constant (H). Compounds with a low H value will be more selectively washed out by rain.

On the other hand, the gaseous organic compounds with high H values will demonstrate long atmospheric residence time since they will not be removed either by precipitation or by particle deposition.[6]

The capacity for pesticides to be transported over long distances is also a function of their atmospheric lifetime, which is the result of emission and removal processes. In fact, long-range transport of pesticides will occur when compounds have a significant lifetime.[7] Photooxidative processes (indirect photolysis) and light-induced reactions (direct photolysis) are the main transformation pathways for pesticides in the atmosphere. According to Finlayson-Pitts and Pitts[8] four processes can be considered (the first three being photooxidative processes and the fourth being direct photolysis): reactions with OH-radicals which are considered to be the major sink for most air pollutants, including pesticides,[9,10] due to the reaction with double bonds, the H abstractive power of hydroxyl and its high electrophilicity,[11–13] reactions with O_3 (ozone), which are only efficient with molecules with multiple bonds,[9] reactions with NO_3-radicals, which are potentially important for compounds containing double bonds,[7] and direct photolysis which acts only with molecules absorbing at $\lambda > ca$ 290 nm which corresponds to the cut-off region of sunlight UV radiation.

"Deposition" is defined as the entry path for transport of airborne substances from the air as an environmental compartment to the earth's surface, i.e., to an aquatic or terrestrial compartment. It is also a loss pathway for substances from the air. Dry and wet deposition should be considered separately because they are subject to different atmospheric physical processes. In essence, wet deposition is the removal of pesticides in precipitation, while dry deposition of particulates is due to a settling out effect (often referred to as the deposition velocity). Indeed, the removal rate of pesticides from the atmosphere by dry and wet deposition depends partly on the Henry's law coefficient, and to some extent on their diffusivity in air, on meteorological conditions (wind speed, atmospheric stability, precipitation), and on the conditions of the surface (for dry deposition only).

The presence of modern pesticides, like 2,4-D, in rainwater was published for the first time in the mid-1960s by Cohen and Pinkerton[14] but until the late 1980s, no special attention was given to this problem. Van Dijk and Guicherit[15] and Dubus et al.[16] published in the beginning of the 2000s reviews on monitoring data of current-use pesticides in rainwater for European countries. Some other measurements were also performed in the US[17,18] and in Japan,[19] in France,[20] Germany,[21,22] Poland,[23] Belgium,[24] and Denmark.[25]

Pesticides are generally present in precipitation from a few ng.L^{-1} to several μg. L^{-1} [18] and the highest concentrations were detected during application of pesticides to crops.

Generally, a local contamination of rainwater by pesticides was observed but some data shows a contamination of rainwater by pesticides in regions where they are not in use.[18] These data confirm the potentiality of transport and consequently the potentiality of the contamination of ecosystems far from their application.

The actual concentration of a pesticide in rainwater or wet deposition of a pesticide does not only depend on its properties and the meteorological conditions at the observational site, but also on the geographical distribution of the amount of pesticide applied, the type of surface onto which it is applied, and the meteorological conditions in the area from which the emissions contribute to the concentration at the measuring site.

From studies performed on the monitoring of the contamination of the atmosphere by pesticides, it appears that atmospheric concentrations were a function of applied quantities, physical-chemical properties of pesticides, climatic and soils conditions, and site localization.

Many data on the atmospheric contamination have been obtained in different countries like Japan,[26] USA,[27–29] France,[30–34] Spain,[35] etc. In France, since 2000, air quality networks surveys have performed monitoring of pesticides in their in different contexts like background air, rural areas, and areas close to pesticides applications (large crops, vineyards, etc.), and recently the ANSES have published a report on recommendations for a national continuous survey of pesticides in air in France.

In general, all of the year, residues of pesticides in the atmosphere were very low in comparison to volatile organic compounds (VOCs) or polycyclic aromatic hydrocarbons (PAHs) in atmospheric concentrations. Some very punctual peaks of pollution have been observed with levels sometimes higher than other pollutants during application periods. However, this strong contamination remains very short in term of duration. These assumptions are in accordance with the EPCA report[36] which concludes that: extremely low levels of Crop Protection Products can be detected in rain and fog, re-deposition rates are about 1000 times lower than normal application rates less than 1 g per year, levels detected in precipitation and air pose no risk to man, and any environmental impact, particularly to aquatic organisms, is extremely unlikely.

10.2 MONITORING OF PESTICIDES IN THE ATMOSPHERE

Pesticides are present in the atmosphere at very low concentrations, except when measurements are performed near the field where treatments are performed. Due to these low concentrations, high volumes of air, rain, or fog are needed to assess the atmospheric levels together with concentration and purification steps before analysis.

10.2.1 SAMPLING AND EXTRACTION OF PESTICIDES IN AMBIENT AIR

Methods used for the sampling and extraction of pesticides in the atmosphere are not diverse and generally the pumping of air onto traps is used. Extraction of pesticides on traps are generally performed by solid–liquid extraction.

10.2.1.1 Sampling of Pesticides in Ambient Air

Pesticides in ambient air are sampled by using conventional high-volume samplers on glass fiber or quartz filters followed by solid adsorbents, mainly polyurethane foam (PUF) or polymeric resin (XAD-2 or XAD-4) for the collection of particle and gas phases respectively.

Depending on the high-volume sampler used, the length or diameter of filters varies generally between 200 × 250 mm (Andersen sampler), 102 mm diameter (PS-1 Tisch Environmental, Inc., Village of Cleves, OH), to 300 mm (LPCA collector, homemade) diameter (Figure 10.1). Generally, 10 to 20 g of XAD-2 resin, a styrene–divinylbenzene sorbent that retains all but the most volatile organic compounds, is employed to trap the gaseous phase and is used alone or sandwiched between PUF plugs (75 mm × 37 mm). White et al. (2006)[37] used 100 g of XAD-2 resin between two polyurethane foam (PUF) plugs.

XAD has been previously used to collect a variety of pesticides including diazinon, chlorpyrifos, diazinon, disulfoton, fonofos, mevinphos, phorate, terbufos, cyanazine, alachlor, metolachlor, simazine, atrazine, deethyl atrazine, deisopropyl atrazine, molinate, hexachlorobenzene, trifluralin, methyl parathion, dichlorvos, and isofenphos.[31–35,38]

In a recent study, the efficiency of trapping gaseous currently used pesticides on different traps including PUF, XAD-2 resin, XAD-4 resin, PUF/XAD-2/PUF, and PUF/XAD-4/PUF sandwich was determined.[34] From this study it appears that XAD-2 and PUF/XAD-2/PUF are the better adsorbents for currently used pesticides (27 pesticides tested) and the sandwich form is slightly more efficient than XAD-2 alone, while PUF plugs are the least efficient.

The duration of sampling depends mainly on the purpose of the sampling and on the detection limits of the analytical method used. Generally, sampling varies between 24 h to one week, and the total air pumped varies between 250 m^3,[31–39] 525 to 1081 m^3,[37] and 2500 m^3 of air.[40] A sampling time of about 24 h is generally sufficient to reach the detection limit of pesticides in middle latitude atmosphere and avoid clogging up the filters.[26,41,42]

These devices are very effective although expensive and heavy to use when the objective is, for example, tracking the spatial and temporal variations of concentration

FIGURE 10.1 High-Volume sampler developed in the LPCA. (Scheyer, 2004)[89]

levels. Indeed, this objective requires several collectors spread over the studied area and their maintenance and implementation mobilize significant human resources. In addition, the transfer of traps after sampling and their treatment in the laboratory must not be overlooked.

Due to heavy investments, both financial and human, induced by high-volume samplers, one must consider alternative methods for original and innovative collection and analysis of SVOCs in the atmosphere, for example, in a study with the objective of evaluation of the exposure of populations.

Passive sampling constitutes a simple, easy-to-use, and inexpensive way for the study of variations of COSVs concentrations levels in the atmosphere. This technique is already used routinely for certain VOCs such as BTEX (Benzene, Toluene, Ethylbenzene, and Xylenes), formaldehyde, and other primary pollutants (NOx, ammonia NH_3, etc.) by air quality survey networks (AASQA).

Recent developments have been driven to use this technology to track spatio-temporal variations of Persistent Organic Pollutants such as PCBs, organochlorine pesticides (OCs), and PAHs in the UK, in Scandinavia, and in North America.[43–45] Registered pesticides were little studied.[46–48] In France, to our knowledge no study using passive samplers has been implemented for semi-volatile organic pollutants such as OCs, PAHs, and pesticides. However, recent studies with conventional

sampling systems have shown that the atmospheres of urban and rural areas are contaminated by low levels of pesticides and there is a high variability of contamination
levels.[31,32,49]

If the concept and theory of passive samplers are now well understood, there is
much conceptual and experimental work to be done, particularly on the choice of the
adsorbent, calibration, and interpretation of data. In addition, the process of extraction and purification of passive samplers after sampling of polar organic compounds
such as pesticides require analytical development.

Any sampling technique can be described as "passive" when a non-forced flow
of analytes is established between the sample and the sample holder. This flow is
induced by a potential difference between the two environments. These techniques
are commonly referred to as "diffusive sampler", "passive sensors", or "integrative
sampling techniques" according to the researchers. This flow continues until reaching an equilibrium between the two environments or until sampling is interrupted
by the user.[50]

Passive samplers were used for the first time in 1927 for the semi-quantitative
analysis of carbon monoxide.[51] In 1973, sampling systems for passive quantitative
analysis of SO_2 and NO_2 in the air have been developed.[50] Since these systems are
commonly used by air quality monitoring networks for SO_2, NO_2, or NH_3.[52] In
1982, a theory was adapted to describe the accumulation of VOCs by such a sampler.[53] They are now used routinely for many families of molecules (VOCs, nitrogen
oxides, etc.).

The development of their use for semi-volatile compounds starts in 1993 when
Petty et al.[54] deployed a semi-permeable membrane (SPMDs) for sampling certain
compounds in a laboratory. Ockenden et al.[43,55] then continued with the same SPMDs
to explore the concept of global distillation of PCBs and their atmospheric transport
over long distances. New samplers (stationary phase + configuration) were developed rapidly thereafter. In 2002, Shoeib and Harner[56] proposed the use of polyurethane disks (PUFS) for passive sampling of PCBs. These were then used to study
spatio-temporal atmospheric contamination, e.g., PCBs and organochlorine pesticides along a north–south axis in Chile or along a rural–urban axis in Toronto,
Canada.[57,58] Moreover, Jaward et al. (2004)[45] used them to provide a European map
of atmospheric contamination by PAHs, polychloronapthalenes (NCPs), and PCBs.
The results of the first tests on registered pesticides were also published.[46,48] Another
passive sampler, resin XAD-2 (Figure 10.2) was developed by Wania et al. (2003).[59]
Deployed throughout the American continent for a year, they provided a "picture"
of the average contamination of atmospheric hexachlorocyclohexanes (HCHS).[44]
Correlated with other information (use of pesticides, α/γ-HCH ratio, etc.), the sources
and transport of these molecules can be traced. XAD-2 passive samples have been
used for the survey of and the evaluation of spatial and temporal variations of pesticides in Luxembourg and France.[47,60] Other studies of atmospheric contamination
on a smaller scale were also performed: vertical gradient of pollution in Toronto[61]
using a polymer film, or detection of "hot spots" of contamination in Germany using
a polydimethylsiloxane coated stir bar.[62] More recently, polyurethane foams associated with the XAD-2 were used to assess the contamination of air by currently used
pesticides.[63]

FIGURE 10.2 XAD-2 passive sampler.

10.2.1.2 Extraction of Pesticides in Ambient Air

After sampling, traps are separately extracted by using Soxhlet extraction with different solvents used alone, like acetone,[64] or as a mixture, like: 36% ethyl acetate in n-hexane,[65] (85:15) n-hexane–CH_2Cl_2,[65,66] 25% CH_2Cl_2 in n-hexane,[67] (50:50) n-hexane–acetone,[38] or (50:50) n-hexane–methylene chloride[27,68] for 12 h to 24 h. In some studies, the ASTM D4861-91 method was followed.[37]

After Soxhlet extraction, extracts were dried with sodium sulfate and reduced to 0.5 mL using a Kuderna-Danish concentrator followed by nitrogen gas evaporation[29] or were simply concentrated to about 1 mL by using a conventional rotary evaporator.[31–42,68]

Accelerated Solvent Extraction (ASE) is now used for the extraction of pesticides from filters and adsorbents.[32,33,47,60] This technique permits the strong reduction of the duration of extraction and the amount of solvent used.

Depending on the authors and on the analytical method used, a cleanup procedure can be performed after concentration. Foreman et al.[29] passed extracts through a Pasteur pipet column containing 0.75 g of fully activated Florisil overlaid with 1 cm of powdered sodium sulfate. Pesticides were eluted using 4 mL of ethyl acetate into a test tube containing 0.1 mL of a perdeuterated polycyclic aromatic hydrocarbon used as internal standard. The extract was evaporated to approximately 150 mL using nitrogen gas, transferred to autosampler vial inserts using a 100 mL toluene rinse. Sauret et al.[42] and Scheyer et al.[31,68] used GC–MS/MS for the analysis of airborne pesticides and they did not perform a cleanup procedure.

Badawy,[67] who used GC–ECD for the analysis of pesticides in particulate samples, concentrated Soxhlet extracts to 5 mL and firstly removed elemental sulfur by reaction with mercury. After that, extracts were quantitatively transferred to a column chromatography for separation into two fractions using 3 gr of 5% deactivated alumina. Fraction one (FI), which contains chlorobiphenyls, chlorobenzenes, and hexachlorocyclohexane, was eluted with 16 mL of n-hexane. Second fraction (FII), includes permethrin, cypermethrin, deltamethrin, and chloropyrophos (rosfin) and was eluted with 6 mL of 20% ether in hexane.

In the 1990s, a method using fractionation by HPLC on a silica column was used for the cleanup of atmospheric extracts.[69,70] After extraction, samples were fractionated on a silica column using an n-hexane/MTBE gradient to isolate non-polar, medium-polar, and polar pesticides which were analyzed by specific methods including GC–ECD and HPLC–UV. In the method developed by Millet et al.,[70] three fractions were obtained; the first one contained pp'DDT, pp'DDD, pp'DDE, aldrin, dieldrin, HCB, fenpropathrin, and mecoprop, the second one contained methyl-parathion, and the third one contains aldicarb, atrazine, and isoproturon. This step was necessary since the fraction two and three were analyzed by HPLC–UV, a non-specific method.

Solid phase micro-extraction (SPME) was also used after ASE extraction in order to decrease the detection limits.[34,71] Liaud et al.[34] have used also coupled derivatization with MtBSTFA (Methyl *tert*-butyl Sylyl trifluoroacetamide) and SPME for the quantification of pesticides in air samples.

10.2.1.3 Cleaning of Traps for the Sampling of Pesticides in Ambient Air

Traps (XAD and PUF foam) were pre-cleaned before use by Soxhlet successive cleaning steps or by one cleaning step, depending on authors. Scheyer et al.[31,68] pre-cleaned the filters and the XAD-2 resin by 24-h Soxhlet (50:50) with n-hexane–CH_2Cl_2 and stored them in clean bags before use, while Peck and Hornbuckle[38] pre-cleaned the XAD-2 resin with successive 24-h Soxhlet extractions with methanol, acetone, dichloromethane, hexane, and 50:50 hexane–acetone prior to sampling. ASE was also used for the cleaning of traps.[32,34]

Some authors (e.g., Coupe et al.[18]) used a heater to clean filters (backing at 450°C, for example). In all cases, a blank analysis is required to check the efficiency of the cleaning and storage before use.

The ultrasonic bath is rarely used for the extraction of filters and resins after sampling. Haraguchi et al.[26] used this technique for their study of pesticides in the atmosphere in Japan.

10.2.2 Sampling and Extraction of Pesticides in Rainwater Samples

10.2.2.1 Sampling of Rainwater

Rainwater samples are collected using different systems depending on studies and authors. Asman et al.[25] and Epple et al.[21] used, for their study on pesticides in rainwater in Denmark and Germany, respectively, a cooled wet-only collector of the type NSA 181/KE made by G.K. Walter Eigenbrodt Environmental Measurements Systems (Konigsmoor, Germany). It consists of a glass 2(Duran) funnel of diameter of about 500 cm connected to a glass bottle that is kept in a dark refrigerator below the funnel at a constant temperature of 4–8°C. A conductivity sensor is activated when it starts to rain and then the lid on top of the funnel is removed. At the end of the rain period the lid is again moved back onto the funnel. With this system, no dry deposit to the funnel during dry periods is collected. Millet et al.[72] and Scheyer et al.[31,73] also used a wet-only rainwater sampler built by Précis Mécanique (France). This collector is approved by the French Meteorological Society (Figure 10.3). It consists of a PVC funnel of 250 mm diameter connected to a glass bottle kept in the dark. No freezing of the bottle was installed and the stability of the sample was

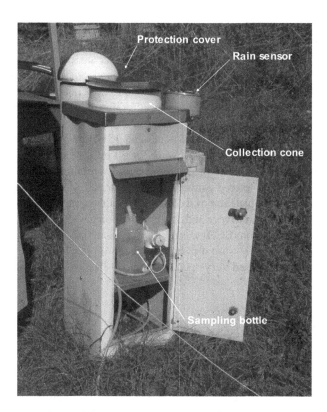

FIGURE 10.3 Wet-only rainwater collector. (Scheyer, 2004)[89]

checked to be stable for one week in warm months. This collector is equipped with a moisture sensor which promotes the opening of the lid when rain occurs.

Quaghebeur et al.[27] used for their study in Belgium, a bulk collector made of stainless steel by the FEA (Flemish Environmental Agency, Ghent, Belgium). The sampler consists of a funnel (D~0.5 m) the sides of which meet at an angle of 120°. The outlet of the funnel is equipped with a perforated plate (D~0.05 m). The holes have a diameter of 0.002 m. The funnel is connected with a collecting flask.

Haraguchi et al.[23] and Grynkiewicz et al.[26] used a very simple bulk sampler which consists of a 40 cm or 0.5 m² diameter, respectively, stainless steel funnel inserted in a glass bottle for their study of pesticides in rainwater in Japan and Poland respectively.

10.2.2.2 Extraction of Pesticides from Rainwater

Extraction of pesticides was made using the conventional method used for water: liquid–liquid extraction (LLE), solid phase extraction (SPE), and solid phase micro-extraction (SPME).

10.2.2.2.1 Liquid–Liquid Extraction

This method was used by many authors. Chevreuil et al.[30] extracted pesticides from rainwater by LLE three times with a mixture of 85% n-hexane–15% methylene chloride. Recoveries obtained were higher than 95% except for atrazine degradation metabolites

(> 75%). Depending on the chemical nature of the pesticide, Quaghebeur et al.[24] used different LLE extraction methods. Organochlorine pesticides, polychlorinated biphenyls, and trifluralin were extracted from the rainwater sample with petroleum ether (extraction yield > 80%) while organophosphorus and organonitrogen compounds (e.g., atrazine) were extracted with dichloromethane (extraction yield > 80%).

Kumari et al.[74] for their study of pesticides in rainwater in India used the following procedure to extract pesticides from rainwater. A representative (500 mL) sample of water was taken in 1 L separatory funnel and 15–20 gr of sodium chloride was added. Liquid–liquid extraction (LLE) with 3 × 50 mL of 15% dichloromethane in hexane was performed. The combined organic phases were filtered through anhydrous sodium sulfate and this filtered extract was concentrated to near dryness on rotary vacuum evaporator. Complete removal of dichloromethane traces was ensured by adding 5 mL fractions of hexane twice and concentrating on gas manifold evaporator since Electron-Capture Detection (ECD) was used for the analysis of some pesticides.

All these authors did not use a cleanup procedure after LLE of rainwater samples mainly since they used very specific methods like GC–ECD, GC–NPD, and GC–MS.

10.2.2.2.2 Solid Phase Extraction

Solid Phase Extraction (SPE) was used by Haraguchi et al.,[26] Millet et al.,[72] Coupe et al.,[18] Grynkiewicz et al.,[23] Bossi et al.,[75] and Asman et al.[25]

These authors used XAD-2 resin or C_{18} cartridges and they followed the classical procedure of SPE extraction consisting of conditioning of the cartridge, loading of the sample, and elution of pesticides by different solvents. Haraguchi et al.[26] used dichloromethane for the elution of pesticides trapped on XAD-2 cartridge while Asman et al.[76] used 5 mL of ethylacetate–hexane mixture (99:1 v/v) for the elution of pesticides from Oasis HLB 1000 mg cartridges (Waters) before GC–MS analysis. A 200 μL volume of isooctane was added to the extract as a keeper in order to avoid losses of the more volatile compounds during evaporation. For LC–MS/MS analysis, these authors used Oasis HLB 200 mg cartridges (Waters) and pesticides were eluted with 8 mL methanol. The extracts were evaporated to dryness and then redissolved in 1 mL of a Millipore water–methanol mixture (90:10 v/v) before LC–MS/MS in ESI mode analysis.

Grynkiewicz et al.[26] used Lichrolut EN 200 mg cartridges (Merck) for the extraction of pesticides in rainwater. Pesticides were eluted with 6 mL of a mixture of methanol and acetonitrile (1:1). After it, a gentle evaporation to dryness under nitrogen was performed before analysis by GC–ECD (organochlorine pesticides) and GC–NPD (organophosphorus and organonitrogen).

Epple et al.[24] have compared two kind of SPE cartridges for the extraction of pesticides in rainwater samples and their analysis by GC–NPD; Bakerbond C_{18} solid-phase extraction cartridges (Baker, Phillipsburg, NJ, USA) and Chromabond HR-P SDB (styrene–divinyl–benzene copolymer) cartridges 200 mg (Macherey-Nagel, Duren, Germany). The latter one is more efficient for polar compounds such as the triazine metabolites. Prior to SPE extraction, rainwater samples were filtered by a glass fiber prefilter followed by a nylon membrane filter 0.45 nm. After that, filtered rainwater was filled with 5% of tetrahydrofuran (THF).

Elution was carried out with 5 mL of THF, the solvent evaporated, and the residue dried with a gentle stream of nitrogen, and then dissolved in 750 µL of ethyl acetate. The sample was then cleaned by small silica-gel columns in order to remove polar components from precipitation samples. For this purpose, 3 mL silica-gel columns (5 × 0.9 cm boro silicate glass) with Teflon frits were used. The silica-gel type (60, 70–230 mesh, Merck) was dried overnight at 130°C, mixed with 5% by weight of water and transferred into glass tubes as a mixture with ethyl acetate, so that each column contained 0.8 g of silica-gel. The sample (750 µL) was transferred to the column and eluted with 4 mL of ethyl acetate before GC–NPD analysis.

Recoveries of the method for all the pesticides studied are summarized in Table 10.1.

Millet et al.[72] also used SPE extraction on Sep-Pak C_{18} cartridges (Waters) and elution with methanol for the analysis of pesticides in rainwater. Before analysis, they performed a HPLC fractionation as described earlier.[70]

10.2.2.2.3 Solid Phase Micro-Extraction

Among studies on pesticides in precipitation, extraction of pesticides was performed using classical developed methods for surface water. No special development was specifically done for atmospheric water. More recently, Scheyer et al.[39,73] used solid phase micro-extraction (SPME) for the analysis of pesticides in rainwater by GC–MS/MS. They used direct extraction for stable pesticides and a derivatization step coupled to SPME extraction for highly polar pesticides or thermo labile pesticides. These developments were derived from studies in water. SPME is a very interesting method for a fast and inexpensive determination of organic pollutants in water including rainwater. The main advantage of the SPME technique is that it integrates sampling, extraction, and concentration in one step. This method is actually rarely used for the extraction of organic pollutants in atmospheric water probably because of the low levels commonly found in precipitation.

For the evaluation of the spatial and temporal variations of pesticides concentrations in rainwater between urban (Strasbourg, East of France) and rural (Erstein, East of France) areas, Scheyer et al.[39] have developed a method using solid-phase microextraction and ion trap GC–MS/MS for the analysis of 20 pesticides (alachlor, atrazine, azinphos-ethyl, azinphos-methyl, captan, chlorfenvinphos, dichlorvos, diflufenican, α and β-endosulfan, iprodione, lindane, metolachlor, mevinphos, parathion-methyl, phosalone, phosmet, tebuconazole, triadimefon, and trifluralin) easily analyzable by gas chromatography. For some seven other pesticides (bromoxynil, chlorotoluron, diuron, isoproturon, 2,4-MCPA, MCPP, and 2,4-D), Scheyer et al.[73] used SPME and GC–MS/MS but they added, prior to GC analysis, a derivatization step. SPME was chosen because it permits with accuracy a rapid extraction and analysis of a great number of samples and MS/MS enables the analysis of pesticides at trace level in the presence of interfering compounds without losing identification capability due to a drastic reduction of the background noise.

The first step in developing a method for SPME is the choice of the type of fiber. In order to do that, all other parameters are fixed (temperature, pH, ionic strength, etc.). The fiber depth in the injector was set at 3.4 cm and the time of the thermal desorption in the split-splitless injector was 5 min at 250°C, as recommended by

TABLE 10.1
Relative Standard Deviations RSD, Recoveries Rec. and Determination Limits DL, (n=10), (P=95%) for Determination of Pesticides in Wet-Deposition Samples (from Epple et al., 2002)[21]

Pesticide	Bakerbond C$_{18}$			Chromabond HR-P SDB		
	RSD (%)	Rec (%)	DL (ng.L^{-1})	RSD (%)	Rec. (%)	DL (ng.L^{-1})
Desethyl atrazine 2	1.64	31	15	1.39	102	13
Desethyl terbuthylazine 2	1.91	95	19	1.20	102	12
Simazine 2	1.09	98	10	1.15	98	11
Atrazine 2	1.10	99	10	1.19	99	11
Propazine 2	1.38	101	13	1.56	98	15
Terbuthylazine 2	1.84	97	18	1.48	97	15
Diazinon 1	2.58	89	5	3.62	87	6
Triallate 3	2.64	106	110	3.36	85	130
Sebuthylazine 2	1.23	96	11	–	–	–
Metribuzin 3	4.21	81	95	3.35	79	75
Parathion-methyl 1	3.14	105	6	2.21	83	4.5
Metalaxyl 4	1.04	99	75	1.82	93	120
Prosulfocarb 3	2.48	100	60	3.81	94	90
Metolachlor 4	1.35	104	105	1.03	98	80
Parathion 1	3.11	103	6	3.16	82	6
Metazachlor 3	1.36	98	30	1.03	103	25
Pendimethalin 3	2.17	90	50	52.8	41	1300
Triadimenol 3	1.45	100	60	1.82	87	75
Triadimenol 3	1.42	101	60	2.44	88	100
Napropamide 3	1.13	101	35	–	–	–
Flusilazol 3	1.70	97	40	2.52	81	60
Propiconazol 3	1.90	94	75	2.74	77	110
Propiconazol 3	1.25	98	50	1.52	92	60
Tebuconazole 3	2.26	93	50	3.46	74	75
Bifenox 4	0.95	83	210	–	–	–
Pyrazophos 1	6.10	103	25	3.31	95	15
Prochloraz 4	3.32	86	300	–	–	–

Concentration ranges (1) 5–50 ng.L^{-1}; (2) 20–200 ng.L^{-1}; (3) 100–1000 ng.L^{-1}; (4) 250–2500 ng.L^{-1}.
Enantiomeric pairs numbered in the order of their elution times.

Supelco and confirmed by Scheyer et al.[73] Deeper fiber in the injector gave rise to carryover effects and less deep fiber caused loss of response. The liner purge was closed during the desorption of the analytes from the SPME fiber in the split-splitless injector (2 min delay time). A blank must be carried out with the same fiber in order to confirm that all the compounds were desorbed with 5 min of thermal desorption.

In the method of Scheyer et al.,[73] extractions were performed by immersion of the fiber in 3 mL of sample, with permanent stirring and temperature control at 40°C,

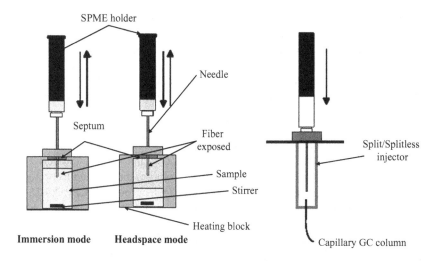

FIGURE 10.4 Principle of SPME extraction. (Scheyer, 2004)[89]

for 30 min. Indeed, a headspace coating of the fiber is possible but, in the case of pesticides, this method cannot be used with efficiency due to the general low volatility of pesticides from water (Figure 10.4). However, for some volatile pesticides like some organophosphorus pesticides, headspace coating of the fiber can be developed.

Since the SPME technique depends on an equilibrium process that involves the adsorption of analytes from a liquid sample into the polymeric phase according to their partition coefficient, the determination of the time (duration of extraction) required to reach this equilibrium for each compound is required.

The equilibration rate is limited by the mass transfer rate of the analyte through a thin static aqueous layer at the fiber-solution interface, the distribution constant of the analyte, and the thickness and the kind of fiber coating.[77] Moreover analytes with high molecular masses are expected to need longer equilibrium times, due to their lower diffusion coefficient, since the equilibrium time is inversely proportional to the diffusion coefficient.[78]

The temperature and the duration of extraction are associated since when increasing the temperature, it is possible to reach the equilibrium faster. Temperature can also modify the partition coefficient of the fiber and consequently decrease the amount of extracted compound.[77] A compromise has to be determined between the temperature and the duration of the extraction in order to obtain a sensitive method for the analysis of pesticides in rainwater.

In order to increase the extraction efficiencies, it is possible to add salts which have the effect of modifying the ionic strength and decreasing the solubility of the molecules in the water.

10.2.2.2.3.1 SPME of Pesticides in Rainwater with a Derivatization Step
The SPME technique, firstly developed for GC analysis, integrates sampling, extraction, and concentration in one step followed by analysis by gas chromatography (GC); even the use of HPLC is possible.

However, many pesticides like phenyl ureas (PUHs), phenoxy acids, or carbamates cannot be analyzed directly by GC due to their low volatility or thermal instability. GC analysis of these molecules requires a derivatization step to stabilize or increase their volatilities.

The use of SPME with derivatization is not commonly used for pesticides, especially in the simultaneous determination of many classes of pesticides like phenyl ureas, phenoxy acids, phenolic herbicides, etc.

Derivatization (sylilation, alkylation, acylation) is employed for molecules where properties cannot permit their direct analysis by gas chromatography.[79,80]

Alkylation with PFBBr is a very common reaction and permits the derivatization of molecules containing -NH- groups (chlorotoluron, diuron, and isoproturon), -OH groups on aromatic ring (bromoxynil), and -COOH groups (MCPP, 2,4-D, 2,4-MCPA). The mechanism of reaction on a molecule containing a hydrogen acid is a bimolecular nucleophile substitution (SN_2).[31]

After extraction, samples present in organic solvents are derivatized by addition of a small amount of derivatizing agent. In the case of SPME, no solvent is present and some approaches have been tested for combining derivatization and SPME:[77]

- Derivatization directly in the aqueous phase followed by SPME extraction (direct technique).
- Derivatization on the fiber. This method consists of headspace coating of PFBBr for 10' of the fiber followed by SPME extraction. In this case, extraction and derivatization are carried out simultaneously.
- Extraction of the analytes present in water followed by derivatization on the fiber or onto the GC injector.

For the direct technique, it is necessary to adjust the pH of the water below the pKa of the molecules to be derivatized (i.e., < 2.73 which is the lowest pKa value for 2,4-D) since in this case they are protonated and consequently derivatization becomes possible.

Scheyer et al.[31] have clearly shown that the exposure of the fiber to the derivatization reagent followed by extraction gave the better results and this method was used for the analysis of the seven pesticides, which required derivatization before analysis by GC, in rainwater.

10.2.3   EVALUATION OF SOIL–AIR TRANSFER OF PESTICIDES (SPRAY DRIFT AND VOLATILIZATION)

As shown in the preceding paragraph, pesticides in ambient air are commonly sampled by high-volume samplers on filters and adsorbents (PUFs, XAD-2). After sampling, compounds trapped on the adsorbent must be released before determination. For this purpose, the use of a solvent for desorption with Soxhlet or ultrasonic extraction, followed by a concentration step, is common. It is generally time consuming and the different steps (extraction, clean-up, concentration, etc.) induce many losses and subsequently increase detection limits.

Even if the association of high-volume sampling and solvent extraction is accurate for the measurement of ambient level of trace contaminants, this method cannot be applied to assess spray drift and volatilization processes. Indeed, this kind of study requires a short sampling periodicity to be close to the variation of atmospheric dissipation processes.

As quoted by Majewski,[81] estimation of volatilization rate in the field is classically carried out using the aerodynamic profile. It gives an estimate of this mass transfer under actual field conditions and its variation with time. This method, based on the measurements of vertical profiles of pesticide concentrations in the atmosphere, needs a good precision for the estimation of these concentrations. Also, the determination of concentration gradients requires the measurements of concentrations at four heights at least and consequently greatly increases the number of samples to analyze.

Thermal desorption (Figure 10.5) can present a novel approach since: it substantially simplifies analyses (no concentration step is needed), it increases sensitivity (a large part of the preconcentrated material may be recovered for determination), and detection limits and background noise are lower due to the disappearance of solvent components. Moreover, this technique is easily automatable. Due to these aspects, it seems to be an interesting alternative to solvent extraction to assess atmospheric transfer of pesticides during and after application. Thermal desorption has often been used for the analysis of volatile organic compounds (VOCs) in indoor and outdoor atmospheres. Thermal desorption for the analysis of pesticides has already been described for volatile and stable pesticides trifluralin and triallate[82,83] in the case of field measurements, and atrazine in the case of laboratory volatilization experiments.[84]

Thermal desorption was extended to six pesticides in order to evaluate atmospheric transfer of pesticides following application (spray drift and volatilization).[85]

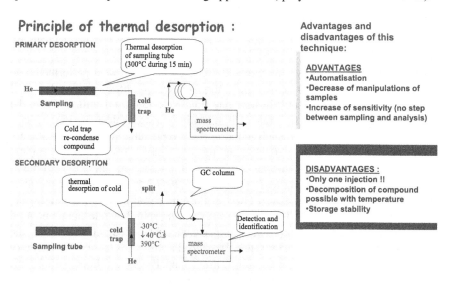

FIGURE 10.5   Principle of thermal desorption.

To the best of our knowledge, this was the first time that a thermal desorption unit–gas chromatography was interfaced with a mass selective detector to provide both pesticide quantification and confirmation. From the first results obtained in this study, it appears that thermal desorption followed by gas chromatography–mass spectroscopy analysis is accurate and sensitive but presents some limitations, especially as a result of the physicochemical properties of pesticides such as thermal stability and/or low volatility.

The principle of thermal desorption is detailed in Figure 10.5 and consists of two steps; one primary desorption which consists of desorption of pesticides adsorbed on the resin of the sampling tube and accumulation on a trap maintained at -30°C by Peltier effect, and one secondary step which consists of the rapid heating of the trap before introduction on the GC column maintained at 50°C.

Application of thermal desorption for pesticides presents some difficulties mainly due to the very low volatility of some of them. Briand et al.[84] have extended the method developed by Clément et al.[85] to deethylatrazine (DEA), deisopropylatrazine, carbofuran (DIA), cyprodinil, epoxyconazole, iprodione, 3,5-dichloroaniline, lindane, α-HCH, metolachlor, terbuconazole, and trifluralin.

The main problem in dealing with the extraction of pesticides is the memory effects on the thermal desorption system. This problem was located in the cold trap containing glass wool.

To visualize an eventual memory effect for the ten pesticides and metabolites under study, first experiments were performed as follows: a 400 ng amount of each compound was deposited at the end of a tube which was placed in the thermal desorption unit followed by four empty tubes (tubes without adsorbent) which were analyzed following the spiked tube.

To check the influence of the amount of pesticides accumulated on the cold trap, two parameters can be modified in the ATD system, the inlet-split flow rate (initially at 0 mL min^{-1}), situated between the tube and the trap and the outlet-split flow rate (initially at 20 mL min^{-1}), located after the cold trap just before injection into the analytical column. This last flow rate imposes the gas velocity in the cold trap.

To evaluate the memory effect, two kinds of experiments were performed: one modifying inlet-split flow rate of 10 mL min^{-1} (outlet-split flow rate 20 mL min^{-1}) and second modifying outlet-split flow rate (inlet-split rate 0 mL min^{-1}). From the first experiment (inlet-split flow rate of 10 mL min^{-1}), a strong decrease of the memory effect in all empty tubes, analyzed after the sample tube, was observed since it remained only for cyprodinyl (0.93%) and tebuconazole (1.70%). Thus, the amount of pesticides reaching the cold trap seems to be the reason for the observed memory effect. However, a strong loss of sensitivity (20–60%) especially for the most volatile compounds (DIA, DEA, α-HCH, trifluralin, carbofuran, lindane, atrazine, and alachlor) was observed. Thus, increasing the inlet-split flow rate cannot be used to resolve the memory effect problem. Experiments conducted with increasing the outlet-split flow rate (30 and 35 mL min^{-1}) induced a strong decrease of the memory effect: 0.90% for cyprodynil with 30 mL min^{-1}, and 1% for iprodione with 35 mL min^{-1} in the first empty tube. Percentages obtained in the second tube were not significant and can be neglected.

From these experiments, it appears that increasing the outlet-split flow rate from 20 to 30 mL min⁻¹ limits the memory effect. These outlet splits correspond to 5 and 3.3% respectively of the total amount of spiked compound in the tube actually injected into the GC column. Increasing the outlet-split flow rate to 35 mL min⁻¹ would not be accurate, since a too great loss of sensitivity was observed.

Loss of sensitivity when increasing outlet-split can be compared to the principle of GC split/splitless injector, in which more volatile compounds (especially solvent) are preferentially removed before entering the column.

Experiments were conducted without glass wool in the cold trap to remove the memory effect completely. These experiments showed that the memory effect was very low (maximum 0.10% for epoxyconazole) and disappeared completely in the second empty tube. The resolution and sensitivity of each pesticide and metabolite under study were not affected by this removal since no significant decrease of areas was observed.

An experiment was performed by changing desorption rate of the cold trap from > 40°C s⁻¹ to 5°C s⁻¹. This change greatly improved the peak resolution. From the different tests performed, it appeared that the memory effect was located in the cold trap, and that it could be partially removed by using an empty trap. Following these observations, complementary tests were performed with decreased outlet-split flow rates (to 25 and 20 mL min⁻¹) to increase the method sensitivity. Tests performed with a spiked tube at 400 ng showed recurrence of the memory effect with an outlet-split flow under 25 mL min⁻¹; therefore, no more tests were conducted. Decreasing the outlet-split flow rate could be envisaged in the case of very low amounts of pesticides to improve method performances.

ATD optimal conditions for the quantitative desorption of the ten pesticides and metabolites under study are presented in Table 10.2.

This study used an ATD 400 from Perkin-Elmer Corp. (Norwalk, CT, USA) where some temperature ranges are limited (transfer line, valve). With new Turbomatrix systems, the temperature can be increased and it is possible to improve the efficiency of thermal desorption for pesticides analysis.

TABLE 10.2
ATD Conditions (Briand et al., 2002)[84]

Parameter	Initial Conditions	Optimal Conditions
Oven temperature for tube	350°C	350°C
Desorb flow and time for tube	60 mL.min⁻¹; 15 min	60 mL.min⁻¹
Inlet-split	0 mL.min⁻¹	0 mL.min⁻¹
Temperature of cold trap	−30°C	−30°C
Temperature of desorption for the trap	390°C	390°C
Desorb time for the trap	15 min	15 min
Trap fast (−30 to 390°C)	Yes (>40°C.s⁻¹)	No
Outlet-split	20 mL.min⁻¹	25 mL.min⁻¹
Temperature of the transfer valve	250°C	250°C
Temperature of the transfer line	225°C	225°C

10.2.3.1 Method Performances

10.2.3.1.1 ATD-GC–MS Repeatability and Calibration Range

For repeatability experiments, five assays were conducted successively with conditions defined in Table 10.2. From this experiment, it appears that repeatability (determined by five replications) was good for each compound, with a relative standard deviation of 9–12% (deviation due to the manual tube spiking step is included in this result).

A calibration range was performed between 1 and 100 ng deposited on tubes. Linear range was observed at:

- 1–100 ng for carbofuran and epoxyconazole
- 2–100 ng for alachlor and cyprodinyl and -HCH and trifluralin
- 4–100 ng for atrazine, iprodione, metolachlor, and tebuconazole
- 10–100 ng for desethylatrazine, disopropylatrazine, and 3,5-dichloroaniline

Detection limits were determined as two times lower than values of the quantification limit. No memory effect was observed in these ranges of concentrations.

10.2.3.1.2 Pesticides Recoveries from Tenax

The optimal temperature for sampling tube desorption was 350°C. No trace of compounds had been observed during the second desorption of the tube. Recovery efficiencies obtained from Equation (10.1) equal 100%.

$$R.E_i \left(\% \right) = \frac{A_{i,1} - A_{Bi}}{\left(A_{i,1} + A_{i,2} \right) - A_{Bi}} \times 100 \ \% \tag{10.1}$$

where:

$R.E_i$ is the recovery efficiency for the analyte i

$A_{i,1}$ is the peak area of analyte i for the first desorption of the spiked tube

$A_{i,2}$ is the peak area of analyte i for the second desorption

A_{Bi} is the count of analyte i from the adsorbent blank (if any)

No additional peak was observed in GC–MS, which seems to indicate that no thermal degradation occurs during tube desorption. Recovery efficiencies obtained at the other temperatures were lower than those at 350°C and were directly correlated to desorption temperature. Recovery efficiencies ranged from 17, 22, and 35% at 225°C for low volatile pesticides (respectively iprodione, epoxyconazole, and tebuconazole), to more than 90% at 300°C. The other compounds gave recovery efficiencies of 60–95% at 225–300°C.

10.2.3.1.3 Resin Efficiency

Tenax TA performances in retaining pesticides under study were tested by an experiment with three tubes in series and a GC oven. This technique offers some advantages such as simplicity and low cost, and the possibility of investigating two parameters at the same time, to evaluate adsorbent performances or reliabilities, retention efficiency, and breakthrough percentage.

For this purpose, three tubes were connected in series. A heating system was combined with a stream of gas to sweep volatile pesticides from a solid (125 mg of Tenax® enclosed in tube 1) into the vapor phase. Pesticides were then adsorbed on sample tubes (tubes 2 and 3), also packed with 125 mg of Tenax®.

Tube 1, located in a GC oven, spiked with a known amount of pesticides, was connected with Teflon tubes to a pump at one extremity and to two preconditioned Tenax® tubes (kept at room temperature) in the other. Tube 1 was then heated in the GC oven at the same temperature as the first step of the ATD desorption (350°C). After 15 minutes, the GC oven was brought down to 60°C for 2h 45 min. During the whole experiment, a stream of clean air was continuously passed through the first tube to carry volatiles to subsequent tubes 2 and 3. In total, 300 L of air was passed through the tubes in three hours to simulate field conditions. Tubes 2 and 3 were maintained at ambient temperature (20–25°C) with a stream of compressed air on their surface.

At the end of the experiment, the tubes were separated and analyzed by ATD-GC–MS. For each compound, peak areas were then compared to a reference value (achieved by direct injection on the top of Tenax® tube just before analysis).

With this experiment, it was possible to calculate the actual quantity of pesticides which was volatilized (Equation 10.2), and Tenax® retention efficiency (Equation 10.3), and to collect non-retained pesticides with the third tube in order to estimate breakthrough percentage (Equation 10.4).

$$V.E_i\,(\%) = \frac{A_{i,ref} - A_{i,T1}}{A_{i,ref}} \times 100\% \tag{10.2}$$

where:

$V.E_i$ is the volatilization efficiency for the analyte i

$A_{i,ref}$ is the peak area of analyte $_i$ for the reference desorption (20 ng injected); and

$A_{i,T1}$ is the peak area of analyte i for the tube 1 analyzed.

$$T_R.E_i\,(\%) = \frac{A_{i,T2}}{A_{i,ref} \times V.E_{\cdot i}} \times 100\% \tag{10.3}$$

where:

$T_R.E_{\cdot i}$ is the Tenax® retention efficiency for the analyte i

$A_{i,T2}$ is the peak area of analyte i for the tube 2

$A_{i,ref} \times V.E_{\cdot i}$ represents the actual volatilized quantity

$$B.P_i\,(\%) = \frac{A_{i,T3}}{A_{i,ref} \times V.E_{\cdot i}} \times 100\% \tag{10.4}$$

where:

$B.P_i$ is the breakthrough percentage for the analyte i

$A_{i,T3}$ is the peak area of analyte i for the tube 3.

10.2.3.1.4 Tenax® TA Retention Efficiency

Testing the capacity of an adsorbent to quantitatively retain all molecules present in the air during the sampling duration is fundamental in terms of the accuracy and precision of the method.

Determining the maximum quantity of air passed through the adsorbent with 100% retention of molecules or breakthrough volume is required when air sampling is performed. Generally, breakthrough is determined by using two sampling tubes in series. When molecules go to the second indicates the limit of the sampling method.

In order to test the Tenax® TA retention efficiency, the same device as the one used for the resin efficiency was used. This experiment refers to the physical interaction between a molecule of gas, coming from tube 1, and a solid surface, the porous polymer of the adsorbent. The sorption capacity was determined by passing a known amount M_i of analyte i through the sorbent bed and then analyzing the tube and measuring the amount of retained pesticides.

Vapor pesticide mixture came from tube 1, where compounds were first, in adsorbed form, volatilized by heating action, and transferred to tube 2.

In case of a lack of suitable standard gaseous mixtures, this test is an alternative to a direct liquid injection on the cartridge, and must be more representative of field experiments where pesticides are in vapor phase or coming from an aerosol. Values of efficiency obtained ranged between 68.4 and 99.1%. Two phenomena could explain this variability: a competitive adsorption whereby the molecules with the highest affinity for Tenax® displace those of lowest affinity previously adsorbed and produce a migration in the sorbent bed, or kinetics of capture are different for each compound.

The presence of pesticides in the third tube indicated that some of them had penetrated through the front section. Thus, in the first tube, the adsorption capacity was exceeded so that some layers of the sorbent bed were partially or completely saturated and breakthrough occurred.

However, the breakthrough percentage gives an indicative value of non-retained pesticides for a known volume of gas passed through the tube, but cannot replace breakthrough volume or breakthrough time measurements using stable standard atmosphere and a continuous effluent monitoring with an appropriate detector. These conditions are rarely obtained for pesticide studies.

The breakthrough percentage was never more than 0.75%, whatever the compound, for about 300 L passed through the tubes. This appeared to be very low and to have a direct application for field experiments, since this volume greatly exceeds all field-sampling volumes.

Nevertheless, an increase to 10% of the breakthrough was observed in relation to increasing the ambient temperature from 20 to 60°C.

10.2.3.1.5 Recoveries and Method Detection Limits

From the previous results described (resin retention efficiency and recoveries from Tenax®), no corrections of the atmospheric concentrations were needed.

According to the type of studies, determination of spray-drift, characterization of post-application transfers, or determination of volatilization fluxes, sampling periods can be very different and conduct to variable detection limits of the method.

In the case of spray-drift, sampling periods are short, about a few minutes. Detection limits ranged from 50 to 500 ng.m^{-3} (carbofuran, epoxyconazole, and metabolites respectively) based on a 20 L air volume sampled.

In the case of post-application, on account of night–day cycles, sampling periods are longer, generally a few hours. For this study, they were fixed at 4 h, so that detection limits ranged from 2 to 20 $ng.m^{-3}$, based on a 500 L air volume sampled. These results illustrate the effectiveness of this present method to assess atmospheric pesticide concentrations. Performances could be compared to the conventional method (liquid extraction). For example, Demel et al.[65] have obtained detection limits between 1 and 9 $\mu g.m^{-3}$ based on 1 m^3 air volume sampled (trapping on Tenax® of propiconazole, deltamethrine, etc.). These differences confirm the interest in thermodesorption to analyze atmospheric pesticides in exposed area.

10.3 ANALYSIS OF PESTICIDES IN THE ATMOSPHERE

Pesticides are analyzed after extraction by conventional gas chromatography (GC) or high performance liquid chromatography (HPLC). The detectors used in GC are electron capture detectors (ECD) for the analysis of pesticides containing halogens (organochlorines, pyrethrenoids, alachlor, etc.), nitrogen–phosphorous detectors (organophosphates, triazines, etc.), and mass detection in the single ion monitoring mode (SIM). For HPLC, detectors are diode array detectors, fluorescence detectors for carbamates after post-column derivatization, and mass spectrometry.

10.3.1 ANALYSIS BY GAS CHROMATOGRAPHY

ECD and NPD are very sensible and selective detectors, but few recent studies have used these detectors. They have been used, for example, by Millet et al.,[70] Sanusi et al.,[41] Epple et al.,[21] Quaghebeur et al.,[24] and Kumari et al.[74], for the analysis of currently used pesticides in air and rainwater. The detection and analysis by ECD were commonly used for the analysis of organochlorine pesticides, and GC–NIMS (negative ionization mass spectroscopy) has tended to replace this detector, especially because of the uncertainty in identification with ECD.

The use of GC–MS is more developed since it provides sensitivity, specificity, and selectivity. Indeed, with mass spectroscopic detection, the identification of the compound can be done together with the identification of co-eluted compounds.

Columns used are generally non-polar or semi polar columns (30 m×0.25 mm, 0.25 μm film thickness) and helium is used as carrier gas. A 5% phenyl–95 % polydimethylsiloxane (type DB-5, HP-5, Optima-5, etc. depending on manufacturers) was used in many cases.[23,39,68,70,73,84,86]

10.3.1.1 Analysis by GC–ECD and GC–NPD

Epple et al.,[21] for the separation and analysis of pesticides (see Table 10.1 for the list) by GC–NPD, used an SE-54 column (30 m×0.25 mm, 0.25 μm film thickness; J&W Scientific, Folsom, CA, USA) and helium as the carrier gas. The injection (1 μL) was made in the splitless mode and the temperature of the injector and detector was maintained at 250°C. Due to the fluctuating sensitivity of the detector, quantification of pesticides extracted by C_{18} cartridges was carried out by the internal standard method by using 2,3-diethyl-5-methylpyrazine and quinazoline. Detection limits and uncertainty of the whole method (extraction and GC–NPD analysis) are presented in Table 10.1.

Authors, due to the uncertainty of the identification by GC–NPD, have performed, for most of the GC–NPD analysis, a verification by GC–MS by using a GC HP 5890 II Plus, a MS 5989 B Engine, and a column HP 5 MS (30 m×0.25 mm, 0.25 μm film thickness) crosslinked (Hewlett-Packard, Palo Alto, CA, USA). Identification was performed by comparing the retention time and mass–peak relations with the standard substance.

Millet et al.[70] and Sanusi et al.[41] used GC–ECD for the analysis of organochlorine pesticides in atmospheric samples (air, fog, and rainwater) after fractionation of the samples by HPLC.

Detection limits obtained by Millet et al.[70] varied between 0.01 and 0.8 mg.L^{-1} corresponding to 33 and 333 pg.m^{-3} for a 24 h sampling at 12.5 m^3.h^{-1}.

10.3.1.2 Analysis by GC–MS

GC–MS is employed for the analysis of pesticides in atmospheric samples for its capacity to deliver results with high sensitivity and guarantee on the identification. In many cases, quadrupole GC–MS in single ion monitoring (SIM) is employed and quantification is performed by the internal standard method by using various deuterated compounds including pesticides.[26,37,81,87]

Ion trap was used in the MS/MS mode by Sauret et al.,[42] Scheyer et al.,[68] and Schummer et al.[47] for air samples, and by Liaud et al.[60] and Scheyer et al.[39,73] for the analysis of pesticides in rainwater after SPME extraction. Triple quadrupole is also commonly used.[33] The use of MS/MS permits better sensitivity, higher specificity, and more important structural information on molecules in comparison to single MS, and is also better for quantification. To improve the specificity of the detection, in MS/MS only the daughter ions characteristic of the studied pesticides are used for quantification. The parent ion is systematically excluded from the quantitative analysis, since this parent ion could be obtained from several molecules and consequently have a low specificity. Indeed, the presence of the parent ion on the MS/MS spectrum means that a fraction of this ion had not been fragmented by the Collision Induced Dissociation (CID) phenomenon, necessary to produce daughter ions.

10.3.2 DERIVATIZATION

Some pesticides cannot be analyzed directly by gas chromatography. This is the case for some phenoxy acids (2,4 D, MCPA, and MCPP) and ureas herbicides (chlorotoluron, diuron, and isoproturon). Prior to their analysis by GC, a derivatization step is required. Scheyer et al.[73] have used pentafluorobenzyl bromide (PFBBr) for the derivatization of seven herbicides before their analysis by GC–MS/MS while Liaud et al.[60] used sylilation with MtBSTFA.

Phenyl ureas (PUHs), phenoxy acids, and bromoxynil show very different physicochemical properties and molecular structures. It was necessary to find a derivatization agent which can react simultaneously and easily with all the pesticides studied. An alkylation reaction with pentafluorobenzyl bromide (PFBBr) seems to be a good compromise. The mechanism allows a nucleophilic substitution with a bimolecular mechanism, without formation of carbocations (Figure 10.6).

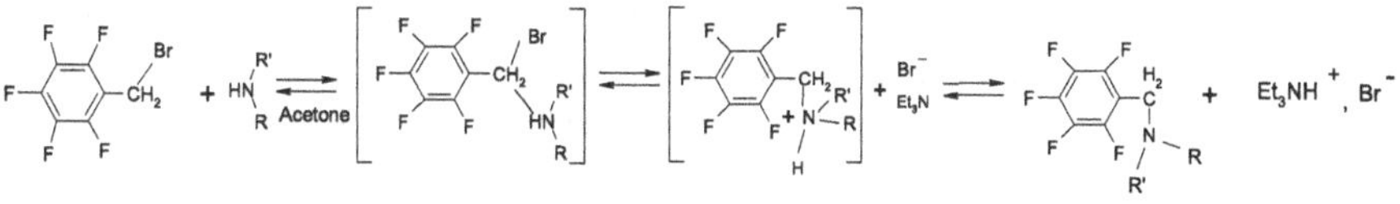

FIGURE 10.6 Mechanism of reaction with pentafluorobenzyl bromide (PFBBr) for amine group as an example. (Scheyer et al., 2005)[6]

The mechanism for the reaction of PFBBr with a molecule that has an acidic hydrogen atom is a bimolecular nucleophilic substitution SN_2. The functional groups present on pesticides which can react with PFBBr are **–NH** in the α position with respect to a carbonyl group (chlorotoluron, diuron, and isoproturon), **–OH** on an aromatic ring (i.e., bromoxynil), and **–COOH** (2,4 D, MCPA, and MCPP). The better solvent for this reaction must be aprotic and polar. This is why acetone was used. The reaction can also be performed in the presence of a base, such as triethylamine, which plays the role of proton acceptor.

PFBBr seems to be a good derivatization agent for the phenoxy acid herbicides, but the method is less efficient for PUHs.

This method was employed for the analysis of these herbicides in rainwater after SPME as mentioned in the section on extraction by SPME.[73]

MtBSTFA present the same mechanism than PFBBr but have the advantage to deliver, when mass spectrometry is used in electron impact mode, a specific m/z M-57 at 100% intensity (M is the molecular mass of the pesticides plus 114 corresponding to the fixation of the *tert*-butyl dimethyl sylyl group from MtBSTFA).[88] This permits good sensitivity and specificity even in single quadrupole.

10.3.3 Analysis by High Performance Liquid Chromatography

Liquid chromatography is used especially for the analysis of polar or acidic compounds. Detectors used are UV-Visible or Diode Array detectors or mass spectrometry. Columns used are mainly C_{18} phases.

10.3.3.1 Analysis by LC–UV or LC–DAD

HPLC–UV or HPLC–DAD is used especially for triazines, ureas herbicides, and carbamates. Generally, a fractionation step is performed since the detection used is not specific.

HPLC–UV was used by Millet et al.[70] and Sanusi et al.[66] for the analysis of triazines herbicides (i.e., atrazine), urea herbicides (i.e., isoproturon), and carbamates (aldicarb). The HPLC quantification was done by the internal standard method after fractionation by normal phase HPLC.

Quaghebeur et al.[24] used HPLC–DAD for the analysis of phenylureas herbicides and their aniline degradation products in rainwater samples after solid phase extraction. Quantification of the results is obtained using an internal standard.

10.3.3.2 Analysis by LC–MS

Bossi et al.[75] have developed and validated a LC–MS/MS method for the analysis of 53 pesticides, degradation products, and selected nitrophenols in rainwater.

After extraction of rainwater by solid-phase extraction on Oasis HLB columns, extracts were analyzed by LC–MS/MS with electrospray ionization. All samples were analyzed in negative and in positive ionization mode, respectively, for acidic and neutral compounds.

Indeed, most of the modern pesticides and their degradation products are characterized by medium polarity and thermal lability. For these reasons liquid chromatography (LC) is the most appropriate analytical method. In order to quantify and identify the target analytes at trace levels, mass spectrometry (MS) in the selected ion monitoring (SIM) mode has to be employed.

REFERENCES

1. Glotfelty, D.E. et al. Volatilisation and wind erosion of soil surface applied atrazine, simazine, alachlor, and toxaphene. *J. Agric. Food Chem.*, 37, 546, 1984.
2. Van den Berg, F. et al. Emission of pesticides in the air. In: Fate of Pesticides in the Atmosphere, Implications for Environmental Risk Assessment. *Water Air Soil Pollut.*, 115, 195, 1999.
3. Fritz, R. Pflanzenschutzmittel in der Atmosphäre. *Pflanzenschutznachr. Bayer*, 46, 229, 1993.
4. Kubiak, R. et al. Pesticides in Air: Considerations for Exposure Assessment, FOCUS Air Group, SANCO/xxx/xxxx, 2007; see also Miller, P. The measurement of spray drift. *Pestic. Outlook*, October, 205, 2003.
5. Unsworth, J.B. et al. Significance of long range transport of pesticides in the atmosphere. *Pure Appl. Chem.*, 71, 1359, 1999.
6. Sanusi, A. et al. Gas-particles partitioning of pesticides in atmospheric samples. *Atmos. Environ.*, 33, 4941, 1999.
7. Atkinson, R. et al. Transformation of pesticides in the atmosphere: A state of the art. *Water Air Soil Pollut.*, 115, 219, 1999.
8. Finlayson-Pitts, B.J. and Pitts, J.N. Jr. *Atmospheric Chemistry*. Wiley Ed., New York, 1986.
9. Klöpffer, W. et al. Testing of the abiotic degradation of chemicals in the atmosphere: The smog chamber approach. *Ecotox. Environ. Safety*, 15, 298, 1988.
10. Klöpffer, W., Kaufmann, G. and Frank, R. Phototransformation of air pollutants: Rapid test for the determination of k_{OH}. *Z. Naturforsch.*, 40a, 686, 1985.
11. Atkinson, R. Kinetics and mechanisms of the gas phase reactions of the hydroxyl radical with organic compounds under atmospheric conditions. *Chem. Rev.*, 86, 69, 1986.
12. Becker, K.H. et al. Methods for ecotoxicological evaluation of chemicals. Photochemical degradation in the gas phase, Vol. 6, Report 1980–1983, Kernforschunganlage Jülich GmbH, Projektträgerschaft Umveltchemikalien, Jül-Spez-279, 1984.
13. Atkinson, R. et al. Kinetics and mechanisms of the reaction of the hydroxyl radical with organic compounds in the gas phase. In: *Advances in Photochemistry*, Vol. 11, 375–488, Wiley Ed., New York, 1979.
14. Cohen, J.M. and Pinkerton, C. *150th Meeting Am. Chem. Soc. Div. Water, Air, and Waste Chem.*, Atlantic City, New J. 1965, 5(20).
15. Van Dijk, H.F.G. and Guicherit, R. Atmospheric dispersion of current-use pesticides - a review of the evidence from monitoring studies. *Water Air Soil Pollut.*, 115, 21, 1999.
16. Dubus, I.G., Hollis, J.M. and Brown, C.D. Pesticides in rainfall in Europe. *Environ. Pollut.*, 110, 331, 2000.
17. McConnell, L.L. et al. Chlorpyrifos in the air and surface water of Chesapeake Bay: Prediction of atmospheric deposition fluxes. *Environ. Sci. Technol.*, 31, 1390, 1997.

18. Coupe, R.H. et al. Occurence of pesticides in rain and air in urban and agricultural areas of Mississippi. *Sci. total Environ.*, 248, 227, 1998.
19. Haragushi, K. et al. Simultaneous determination of trace pesticides in urban precipitation. *Atmos. Environ.*, 29, 247, 1995.
20. Briand, O. et al. Influence de la pluviométrie sur la contamination de l'atmosphère et des eaux de pluie par les pesticides. *Revue des Sciences de l'Eau*, 15, 767, 2002.
21. Epple, J. et al. Input of pesticides by atmospheric deposition. *Geoderma*, 105, 327, 2002.
22. de Rossi, C., Bierl, R. and Riefstahl, J. Organic pollutants in precipitation: Monitoring of pesticides and polycyclic aromatic hydrocarbons in the region of Trier (Germany). *Phys. Chem. Earth*, 28, 307, 2003.
23. Grybkiewicz, M. et al. Pesticides in precipitation from an urban region in Poland (Gdańsk-Sopot-Gdynia tricity) between 1998 and 2000. *Water Air Soil Pollut.*, 149, 3, 2003.
24. Quaghebeur, D. et al. Pesticides in rainwater in Flanders, Belgium: Results from the monitoring program 1997–2001. *J. Environ. Monit.*, 6, 182, 2004.
25. Asman, W. et al. Wet deposition of pesticides and nitrophenols at two sites in Denmark: Measurements and contributions from regional sources. *Chemosphere*, 59, 1023, 2005.
26. Haragushi, K. et al. Simultaneous determination of trace pesticides in urban air. *Atmos. Environ.*, 28, 1319, 1994.
27. Aston, L.S. and Seiber, J.N. Fate of summertime airborne organophosphate pesticide residues in the Sierra Nevada mountains. *J. Environ. Qual.*, 26, 1483, 1997.
28. Bidleman, T.F. and Leonard, R. Aerial transport of pesticides over the northern Indian ocean and adjacent seas. *Atmos. Environ.*, 16, 1099, 1982.
29. Foreman, W.T. et al. Pesticides in the atmosphere of the Mississippi River Valley part II- air. *Sci. Total Environ.*, 248, 213, 2000.
30. Chevreuil, M. et al. Occurrence of organochlorines (PCBs, pesticides) and herbicides (Triazines, phenylureas) in the atmosphere and in the fallout from urban and rural stations of the Paris area. *Sci. Total Environ.*, 182, 25, 1996.
31. Scheyer, A. et al. Variability of atmospheric pesticide concentrations between urban and rural areas during intensive pesticide application. *Atmos. Environ.*, 41, 3604, 2007.
32. Schummer, Cl. et al. Temporal variations of concentrations of currently used pesticides in the atmosphere of an urban area (Strasbourg, France). *Environ. Pollut.*, 158, 576, 2010.
33. Coscollà, C. et al. Occurrence of currently used pesticides in ambient air of Centre Region (France). *Atmos. Environ.*, 44, 3915, 2010.
34. Liaud, C. et al. Application of long duration high-volume sampling coupled to SPME-GC–MS/MS for the assessment of airborne pesticides variability in an urban area (Strasbourg, France) during agricultural application. *J. Environ. Sci. Health Part B*, 51, 703, 2016.
35. Coscollà, C. et al. LC–MS characterization of contemporary pesticides in PM_{10} of Valencia Region, Spain. *Atmos. Environ.*, 77, 394, 2013.
36. EPCA. Residues of crop protection products in precipitation and air. 1998, D/98/GRG/3255.
37. White, L.M. et al. Ambient air concentrations of pesticides used in potato cultivation in Prince Edward Island, Canada. *Pest Manag. Sci.*, 62, 126, 2006.
38. Peck, M. and Hornbuckle, K.C. Gas-phase concentrations of current-use pesticides in Iowa. *Environ. Sci. Technol.*, 39, 2952, 2005.
39. Scheyer, A. et al. Analysis of trace levels of pesticides in rainwater by SPME and GC–tandem mass spectrometry after derivatisation with PFFBr. *Anal. Bioanal. Chem.*, 387, 359, 2007.
40. Yao, Y. et al. Spatial and temporal distribution of pesticide air concentrations in Canadian agricultural regions. *Atmos. Environ.*, 40, 4339, 2006.

41. Sanusi, A. et al. A multiresidue method for determination of trace levels of pesticides in atmosphere. *Analusis*, 25, 302, 1997.
42. Sauret, N. et al. Analytical method using gas chromatography and ion trap tandem mass spectrometry for the determination of S-triazines and their metabolites in the atmosphere. *Environ. Pollut.*, 110, 243, 2000.
43. Ockenden, W.A. et al. Toward an understanding of the global atmospheric distribution of persistent organic pollutants: The use of semipermeable membrane devices as time-integrated passive samplers. *Environ. Sci. Technol.*, 32, 2795, 1998.
44. Shen, L. et al. Atmospheric distribution and long-range transport behavior of organochlorine pesticides in North America. *Environ. Sci. Technol.*, 38, 965, 2004.
45. Jaward, F. et al. Passive air sampling of PCBs, PBDEs, and organochlorine pesticides across Europe. *Environ. Toxicol. Chem.*, 23, 1355, 2004.
46. Hayward, S.J. et al. Comparison of four active and passive sampling techniques for pesticides in air. *Environ. Sci. Technol.*, 44, 3410, 2010.
47. Schummer, Cl. et al. Application of XAD-2 resin-based passive samplers and SPME-GC–MS/MS analysis for the monitoring of spatial and temporal variations of atmospheric pesticides in Luxembourg. *Environ. Pollut.*, 170, 88, 2012.
48. Waite, D.T. et al. Comparison of active versus passive atmospheric samplers for some current-use pesticides. *Bull. Environ. Contam. Toxicol.*, 74, 1011, 2005.
49. Rousseau, P. et al. XXXIVᵉ Congrès du Groupe Français des pesticides. Produits phytosanitaires: Concilier efficacité et gestion durable. Dijon, France, 2004.
50. Gorecki, T. and Namiesnik, J. Passive sampling. *Trends Anal. Chem.*, 21, 276, 2002.
51. Gordon, C.S. and Lowe, J.T. US Patent 1,644,014, 1927.
52. Thoni, L. et al. A passive sampling method to determine ammonia in ambient air. *J. Environ. Monit.*, 5, 96, 2003.
53. Fowler, W.K. Fundamentals of passive vapor sampling. *Am. Lab.*, 14, 80, 1982.
54. Petty, J.D. et al. Application of semipermeable membrane devices (SPMDs) as passive air samplers. *Chemosphere*, 27, 1609, 1993.
55. Ockenden, W.A. et al. Further developments in the use of semipermeable membrane devices as passive air samplers: Application to PCBs. *Environ. Sci. Technol.*, 35, 4536, 2001.
56. Shoeib, M. and Harner, T. Characterization and comparison of three passive air samplers for persistent organic pollutants. *Environ. Sci. Technol.*, 36, 4142, 2002.
57. Pozo, K. et al. Passive-sampler derived air concentrations of persistent organic pollutants on a north–south transect in Chile . *Environ. Sci. Technol.*, 38, 6529–6537, 2004.
58. Motelay-Masséi, A. et al. Passive air sampling of PCBs, PBDEs, and organochlorine pesticides across Europe. *Environ. Sci. Technol.*, 39, 5763, 2005.
59. Wania, F. et al. Development and calibration of a resin-based passive sampling system for monitoring persistent organic pollutants in the atmosphere. *Environ. Sci. Technol.*, 37, 1352, 2003.
60. Liaud, C. et al. Comparison of atmospheric concentrations of current-used pesticides and lindane between urban and rural areas during intensive application period in Alsace (France) by using XAD-2® based passive samplers. *J. Environ. Sci. Health Part B*, 52, 458, 2017.
61. Farrar, N.J. et al. Field deployment of thin film passive air samplers for persistent organic pollutants: A study in the urban atmospheric boundary layer. *Environ. Sci. Technol.*, 39, 42, 2005.
62. Wennrich, L. et al. Novel integrative passive samplers for the long-term monitoring of semivolatile organic air pollutant. *J. Environ. Monit.*, 4, 371, 2002.
63. Koblizkova, M. et al. Sorbent impregnated polyurethane foam disk passive air samplers for investigating current-use pesticides at the global scale. *Atmos. Pollut. Res.*, 3, 456, 2012.

64. Yao, Y. et al. Spatial and temporal distribution of pesticide air concentrations in Canadian agricultural regions. *Atmos. Environ.*, 40, 4339, 2006.

65. Albanis, T.A., Pomonis, P.J. and Sdoukos, A.Th. Seasonal fluctuations of organochlorine and triazines pesticides in the aquatic system of Ioannina basin. *Sci. Total Environ.*, 58, 243, 1986.

66. Sanusi, A. et al. A multiresidue method for determination of trace levels of pesticides in atmosphere. *Analusis*, 25, 302, 1997.

67. Badawy, M.I. Organic insecticides in airborne suspended particulates. *Bull. Environ. Contam. Toxicol.*, 60, 693, 1998.

68. Scheyer, A. et al. A multiresidue method using ion trap GC–MS/MS by direct injection or after derivatization with PFBBr for the analysis of pesticides in the atmosphere. *Anal. Bioanal. Chem.*, 381, 1226, 2005.

69. Seiber, J.N. et al. A multiresidue method by high performance liquid chromatography-based fractionation and gas chromatographic determination of trace levels pesticides in the air and water. *Arch. Environ. Contam. Toxicol.*, 19, 583, 1990.

70. Millet, M. et al. A multiresidue method for determination of trace levels of pesticides in air and water. *Arch. Environ. Contam.Toxicol.*, 31, 543, 1996.

71. Raeppel, C. et al. Coupling ASE, silylation and SPME-GC/MS for the analysis of current-used pesticides in atmosphere. *Talanta*, 121, 24, 2014.

72. Millet, M. et al. Atmospheric contamination by pesticides: Determination in the liquid, gaseous and particulate phases. *Environ. Sci. Pollut. Res.*, 4, 172, 1997.

73. Scheyer, A. et al. Analysis of trace levels of pesticides in rainwater using SPME and GC–tandem mass spectrometry. *Anal. Bioanal. Chem.*, 384, 475, 2006.

74. Kumari, B., Madan, V.K. and Kathpal, T.S. Pesticide residues in rain water from Hisar, India. *Environ. Monit. Assess.*, 133, 467, 2007.

75. Bossi, R. et al. Analysis of polar pesticides in rainwater in Denmark by liquid chromatography–tandem mass spectrometry. *Chemosphere*, 59, 1023, 2002.

76. Asman, W.A.H. et al. Wet deposition of pesticides and nitrophenols at two sites in Denmark: Measurements and contributions from regional sources. *Chemosphere*, 59, 1023, 2005.

77. Lord, H. and Pawliszyn, J. Evolution of solid-phase microextraction technology. *J. Chromatogr. A*, 885, 153, 2000.

78. Louch, D., Motlagu, S. and Pawliszyn, J. Dynamics of organic compound extraction from water using liquid-coated fused silica fibers. *Anal. Chem.*, 64, 1187, 1992.

79. Cserhati, T. and Forgacs, E. Phenoxyacetic acids: Separation and quantitative determination. *J. Chromatogr. B*, 717, 157, 1998.

80. Boucharat, C. et al. Experimental design for the study of two derivatization procedures for simultaneous GC analysis of acidic herbicides and water chlorination by-product. *Talanta*, 47, 311, 1998.

81. Majewski, M.S. Micrometeorological methods for measuring the post-application volatilisation of pesticides. *Water Air Soil Pollut.*, 115, 83, 1999.

82. Pattey, E. et al. Herbicides volatilization measured by the relaxed eddy-accumulation technique using two trapping media. *Agric. For. Meteorol.*, 76, 201, 1995.

83. Foster, W., Ferrari, C. and Turloni, S. Environmental behaviour of herbicides. Atrazine volatilisation study. *Fresenius Environ. Bull.*, 4, 256, 1995.

84. Briand, O. et al. Assessing transfer of pesticide to the atmosphere during and after application. Development of a multiresidue method using adsorption on Tenax/Thermal Desorption-GC/MS. *Anal. Bioanal. Chem.*, 374, 848, 2002.

85. Clément, M. et al. Adsorption/thermal desorption-GC/MS for the analysis of pesticides in the atmosphere. *Chemosphere*, 40, 49, 2000.

86. Demel, J., Buchberger, W. and Malissa, H. Jr. Multiclass/multiresidue method for monitoring widely applied plant protecting agents in air during field dispersion work. *J. Chromatogr. A*, 931, 107, 2001.

87. Cessna, A.J. and Kerr, L.A. Use of an automated thermal desorption system for gas chromatographic analysis of the herbicides trifluralin and triallate in air samples. *J. Chromatogr. A*, 642, 417, 1993.
88. Schummer, Cl. et al. Comparison of MTBSTFA and BSTFA in derivatization reactions of polar compounds prior to GC/MS analysis. *Talanta*, 77, 1473, 2009.
89. Scheyer, A. Développement d'une méthode d'analyse par CPG/MS/MS de 27 pesticides identifiés dans les phases gazeuse, particulaire et liquide de l'atmosphère. Application à l'étude des variations spatio-temporelles des concentrations dans l'air et dans les eaux de pluie. Ph.D thesis, University of Strasbourg I (2004).

11 Levels of Pesticides in Food and Food Safety Aspects

Kit Granby, Annette Petersen, Susan Strange Herrmann, and Mette Erecius Poulsen

CONTENTS

11.1 Introduction 329
11.2 Monitoring Programs; Residue Levels in Food 330
    11.2.1 Legislation 330
    11.2.2 Monitoring Programs; General Aspects 331
    11.2.3 Results from Monitoring Programs 332
        11.2.3.1 Fruit and Vegetables 332
        11.2.3.2 Processed Fruit and Vegetables Including Processing
                 Studies 335
        11.2.3.3 Cereals 337
        11.2.3.4 Food of Animal Origin 342
        11.2.3.5 Infant and Baby Food 346
11.3 Consumer Exposure and Risk Assessment 350
    11.3.1 Dietary Intake Estimation 350
        11.3.1.1 Deterministic Approach (Chronic and Acute Intake) 351
        11.3.1.2 Probabilistic Approach 351
        11.3.1.3 Cumulative Exposure 352
    11.3.2 Intake Calculations of Pesticide Residues 353
        11.3.2.1 Deterministic Approach 353
        11.3.2.2 Total Diet and Duplicate Diet Studies 354
        11.3.2.3 Cumulative Exposure 355
References 358

11.1 INTRODUCTION

Monitoring programs for pesticide residues in food are performed in many countries around the world to ensure that consumers are not exposed to unacceptable levels of pesticides and that only pesticides approved by the authorities are used and for the right applications with respect to crop, application dose, time, and intervals. The food products are permitted as long as they comply with the maximum residue levels (MRLs) set by the authorities. Another purpose of pesticide residue monitoring in food may be to assess the food safety risk due to the dietary exposure of the population to pesticides.

The present chapter deals with monitoring programs for pesticide residues in food in general. It also covers monitoring results in fruit, vegetables, cereals, food of animal origin, and processed food like, e.g., drinks, and infant and baby food. In addition, risk assessments of consumer exposure based on dietary intake estimates are described and examples of exposure assessment studies are shown.

11.2 MONITORING PROGRAMS; RESIDUE LEVELS IN FOOD

11.2.1 LEGISLATION

In many countries, there is national legislation regulation on which pesticides are authorized. Many countries also have national legislation on the maximum amounts of pesticide residues permitted in different food commodities. Such upper limits are also referred to as MRLs or tolerances (in the United States). In countries with no national legislation, the MRLs set by the Codex system are often used. MRLs are normally set for raw agricultural commodities (RAC), for example, banana with peel, lettuce, and apples.

The Codex Alimentarius Commission (CAC) is an international body that aims to protect the health of consumers, ensure fair trade practices in the food trade, and promote co-ordination of all food standards work undertaken by international governmental and non governmental organizations. CAC also sets MRLs, which are indicative and not statutory. The Codex MRLs are to be used as guidance on acceptable levels when there is no other legislation in place, for example, in countries without their own national MRLs, or they can be used if national MRLs have not been set for a particular pesticide/food combination.

MRLs set by Codex are evaluated and negotiated through a stepwise procedure. Initially, the Joint FAO/WHO Meeting on Pesticide Residues (JMPR)[1] considers recognized use patterns or good agricultural practice (GAP) and evaluates the fate of residues, animal and plant metabolism data, and analytical methodology as well as residue data from supervised trials conducted according to GAP. Based on these data, MRLs are proposed for individual pesticides. Toxicologists evaluate the toxicological data related to the pesticide and propose an acceptable daily intake (ADI) and an acute reference dose (ARfD). The toxicological data originate from animal studies and include both studies on the short-term and long-term effects. The ADI is a measure of the amount of a specific substance (in this case a pesticide) in food and drink that can be consumed over a lifetime without any appreciable health risk. ADIs are expressed as milligram/kilogram body weight/day. The ARfD of a substance (here pesticide) is an estimate of the amount a substance in food or drink, normally expressed on a body weight basis, that can be ingested in a period of 24 h or less without appreciable health risks to the consumer on the basis of all known facts at the time of the evaluation. ARfD apply only to pesticides that cause acute effects, for example phosphorus pesticides that are acetylcholinesterase inhibitors.

The Codex Committee on Pesticide Residues (CCPR) considers at their annual meetings the MRLs proposed by the JMPR. CCPR is an intergovernmental meeting with the prime objective to reach agreement on proposed MRLs. The MRLs are discussed in an eight-step procedure and after the final step the CCPR recommends MRLs to CAC for adoption as Codex MRLs. To protect the health of the consumers, the intake calculated using the proposed MRLs is compared to the ADI or the ARfD and if the calculated intake exceeds one of these two values the MRL cannot be accepted.

Often when national MRLs are set, an evaluation is performed on a national level, that in many ways is similar to the evaluation performed by JMPR. Some countries also set their own ADIs or ARfDs. As part of the evaluation of pesticides within the European Union (EU) ADIs and ARfDs are set on the EU level which then apply in all Member States. These values can differ from the values set by Codex.

The Member States within the EU set harmonized EU MRLs for pesticides. All harmonized legislation can be found on the Web site of the EU Commision.[2] According to Regulation 396/2005[2] only harmonized EU MRLs can be set and hence previous national legislations have been changed into EU legislation. Specific legislation is laid down in the EU for food intended for infants and young children (Directive 2006/125, Regulation 609/2013, Regulation 2016/127).[3]

Some countries publish their MRLs on the Internet, for example, the United States,[4] Canada,[5] Australia,[6] New Zealand,[7] India,[8] Japan[9] South Africa,[10] and Brazil.[11] In Australia[6] and New Zealand[12] authorities have compiled information about legislation and MRLs worldwide. Other countries do not have their own legislation or MRLs published on websites but the information can be gathered by contacting the relevant authorities. For countries that have published MRLs on websites, be aware that addresses change and the most recent legislation is often not yet published.

11.2.2 Monitoring Programs; General Aspects

There is a growing interest in pesticide residues in food from all aspects of the food chain from 'farm to fork'. It is the national governments that are responsible for regular monitoring of pesticide residues in food. Besides the national governments, monitoring activities or surveillance are also performed by nongovernmental organizations and by scientists studying the occurrence and fate of pesticides in relation to the environment, agriculture, food, and human health. Food companies may also monitor pesticide residues in their products to ensure and demonstrate good food safety quality of their products and/or prevent economic losses.

The monitoring sampling may be surveillance sampling where there is no prior knowledge or evidence that a specific food shipment contains samples exceeding the MRLs. Surveillance sampling may also include more frequent sampling of food groups with samples frequently exceeding the MRLs. Compliance sampling is defined as a direct follow-up enforcement sampling, where the samples are taken in case of suspicion for previously found violations. The follow-up enforcement may be directed to a specific grower/producer or to a specific consignment. To cover both the control aspect and the food safety aspect regarding exposure assessments, the design of a monitoring program may be a mixture of a program where the different food types are weighted relative to the consumption or sale and one where the food groups with samples exceeding the MRLs are weighted higher. In order to have more samples of the same type for representativeness, all sample types may not be monitored annually as the selection of some (minor) sample types may change from year to year.

The monitoring programs do often include imported as well as domestically produced foods. Domestic samples may be collected as close to the point of production as possible; for food crops, the sampling may be at the farm or at wholesaler or retailer. Imported samples may be collected by the customs authorities or at the import companies or retailers. The samples are often raw food, for example fruits, vegetables, cereals, or food of animal origin. In addition, different kinds of processed foods are monitored, e.g., dried, extracted, fermented, heated, milled, peeled, pressed, washed, or otherwise prepared products. The different kinds of processing, in most cases, lead to a decrease in levels of pesticides compared to the contents in the raw food.

An important parameter for a monitoring program is the choice and the number of pesticides investigated. To cover as many pesticides as possible both multi-methods and single residue methods may have to be included in the monitoring program. In 2015, the US Food and Drug Administration (USFDA) was able to monitor the majority of the approximately 400 pesticides for which the US Environmental Protection Agency (USEPA) had set tolerances, and many others that have no tolerances.[13] The same year all states participating in the EU monitoring program together analyzed for 774 different pesticides. However, on average 220 pesticides were analyzed for per sample.[14] In 2015, the number of EU approved pesticides for conventional production and with an established ADI was 318.[15] In addition to the selection and number of pesticides analyzed for, the detection limits of the pesticides in the different foods determine how frequent findings of pesticide residues are.

On a worldwide scale the two major monitoring programs including many states are the EU monitoring program and the USFDA program, both programs publishing annual results at their Web sites. As an example, the '2015 European Union Report on Pesticide Residues in Food' included a total of 84,341 samples of fresh fruit, vegetables, and cereals as well as processed foods and organically labeled foods.[14] The USFDA program included 5987 samples of fruit, vegetables, cereals, and food of animal origin, etc.[13]

11.2.3　Results from Monitoring Programs

The results of pesticide residues in different foods were found in internationally published surveys and monitoring programs on pesticide residues. The results attempted to reflect the pesticide residue results in food worldwide. However, many countries either do not have monitoring results of pesticide residues, or they do not publish, or their results are not internationally available.

11.2.3.1　Fruit and Vegetables

In general, fresh fruit and vegetables account for the largest proportion of samples analyzed within pesticide monitoring programs. In 2015, the USFDA monitoring program included 835 domestic samples and 4737 imported samples, where the majority of the import shipment from 111 countries came from Mexico, China, Canada,

India, and Chile.[13] The samples of vegetables included 266 domestic samples and 224 imported samples. Pesticide residues ≤MRL were detected in 58% of the domestic and 38% of the imported vegetable samples while non-compliant residue contents were detected in 3.8% of the domestic and 9.7% of the imported vegetable samples. The frequency of fruit samples with detected pesticide residues ≤MRL is somewhat higher, i.e., 80% of 224 domestic samples and 42% of 1443 imported samples. Non-compliant fruit samples comprised 2.2% of the domestic and 9.4% of the imported samples. Most of the non-compliant samples revealed no-tolerance pesticides, i.e., pesticides not approved to be used in the commodity.

The '2015 European Union Report on Pesticide Residues in Food' included 58,448 samples (69%) originating from EU and 21,747 samples (26%) imported from third countries outside EU.[14] Among these were 33,744 conventional vegetable samples for surveillance monitoring of which 43% contained residues ≤MRL, and in 3.45% of the samples, the residue concentrations exceeded the MRLs. For the category, 'fruit and nuts', 26,214 conventional samples were monitored, 67% of which contained residues ≤MRL, and 7.5% contained residues exceeding the MRLs. Included in the monitoring program 2015 were ~5% enforcement samples. Compared to the surveillance samples with 2.3% MRL exceedances, the rate was 11.8% for the enforcement samples.

The trend in the presence of pesticide residues shows that the percentage of samples with residues ≤MRL increased from 32% in 1999, to 42% in 2004,[16] to 44% by 2015.[14] The percentage of samples with residues above the MRL changed from 3% in 1996, to over 5% in 2004, to 3% by 2015. Various factors may have influenced the development in these results. The average number of pesticides detected per sample increased from approximately 126, over 169 to 220 (1999–2015), which may result in more findings. Legislation has also changed and will continue to change toward more MRLs set at the limit of detection (LOD).

The EU monitoring program has also compared pesticide residues in organic versus conventionally produced food samples.[15] For the period 2013–2015, the number of samples with residues was 45% for the 28,912 conventional samples compared to 6.4% for the 1940 organic samples. The MRL exceedance rate was 1.2% for conventional compared to 0.2% for organic products.

Within the EU monitoring program, the Commission has designed a coordinated control program (EUCP) where alternating commodities and products are analyzed for a certain number of pesticides. The 2015 EUCP covered 165 pesticides in 10,884 samples of 11 different food products (aubergines, bananas, broccoli, virgin olive oil, orange juice, peas without pods, sweet peppers, table grapes, wheat, butter, and eggs). Of these 11 food products, 61% did not contain detectable pesticide residues and 0.8% exceeded the MRLs.[14] The same food categories were also measured in the 2012 EUCP with comparable level of exceedances (0.9%)[17]; the frequencies of findings in the individual food products of the programs appear on Table 11.1. Among the pesticides that had to be analyzed in all plant products, the most frequently found were imazalil (7.1%), thiabendazole (6.8%), azoxystrobin (6.4%), boscalid (6.3%), and chlorpyrifos (4.1%). Dithiocarbamates (a group of sulfur containing fungicides measured as CS_2) were detected in 61% of the broccoli samples. However, some of the findings may be false positives, as broccoli belongs to the brassica vegetables

TABLE 11.1

Examples of Frequencies of Pesticide Residues and Residues Exceeding the MRLs Detected in Foods of the EU Coordinated Monitoring Program

Commodity	2012 %≤ MRL	2012 %> MRL	2015 %≤ MRL	2015 %> MRL
Aubergine	30.6	1.0	29.8	0.4
Broccoli	31.8	2.8	42.0	3.4
Peas (without pods)	21.4	0.1	24.6	0.6
Peppers (sweet)	46.0	1.4	47.0	0.8
Bananas	77.1	0.7	72.8	0.3
Table grapes	75.1	1.8	75.6	1.7
Olive oil	21.9	0.1	15.4	0.1
Orange juice	31.2	0.0	15.1	0.1
Wheat	39.0	0.7	37.5	0.6
Butter	16.6	0.0	12.8	0.0
Chicken egg	5.4	0.0	4.0	0.2

Source: European Food Safety Authority 2017.

that naturally contain CS_2 precursor compounds that may interfere with the results. Table grapes had the highest frequency of pesticide residues of 77.3%, of which 58.3% contained multiple residues; up to 19 different pesticides were reported in an individual table grapes sample from Turkey. Compared to the banana which also has a high detection frequency of pesticide residues, the table grapes will be eaten without prior peeling, hence the actual dietary exposure will be higher. The five pesticides most frequently found in table grapes (all ~20% with residues) were boscalid, ethephone, dimethomorph, dithiocarbamates, and fenhexamid; the exceedances of MRLs mostly being due to the growth regulator ethephone.

As regards pesticide surveys from other continents, in South Africa they studied pesticide residues in fresh products for 2009–2014 export markets; 37,838 fruit (99%) and vegetable (0.73%) samples were analyzed for the presence of 73 pesticides, 56% of these samples were with residues and 1.05% were non-compliant. From this number, the majority (0.73%) was non-compliant due to unregistered pesticides use. The most frequently detected pesticide group was imidazoles (for example imazalil) (48%), mostly in mangoes (97%), avocados (97%), and oranges (84%).[18]

In Brazil 2001–2010, 48% of 13,556 samples of 22 fruit and vegetable, crops, rice, and beans contained pesticide residues. Dithiocarbamates and organophosphorus compounds were found in 42% and 31% of the samples, respectively. Carbendazim (27%) and chlorpyrifos (16%) were the pesticides most frequently found.[19]

Pesticide residue monitoring has been performed for many Asian countries. In Taiwan, 1997–2003, pesticide residues were detected in 14% of 9955 samples (analyzed for 79 pesticides) and 1.2% of the samples were violating the MRLs.[20]

In 2008–2009, 120 fruit samples (peeled if common practice), collected in Karachi Pakistan, for example, apple, apricot, persimmon, chiku, citrus, grapes, guava, mango, papaya, peach, plum, and pomegranate were analyzed for pesticides.[21] On average 62.5% of the samples contained residues of pesticides while 22% of the samples exceeded the Codex MRLs. The pesticides found were the organochlorine endosulfan, pyrethroids; carbamates; and three additional fungicides and phosphorous insecticides in the order methamidophos > ethion > profenofos > chlorpyrifos > dimethoate > fenitrothion > methyl parathion.

In 2011, pesticide results of 721 samples of 63 different commodities imported to EU from Southeast Asia (ten countries; however, 80% of samples from Thailand) revealed residues in 40% of the samples of which 12% exceeded the EU MRLs. Of the 111 different pesticides found in the samples, the insecticides cypermethrin, chlorpyrifos, imidacloprid, and the fungicides carbendazin/benomyl, and metalaxyl were the most frequently detected.[22]

The growth regulator chlormequat is an example of a pesticide that has been regulated during the period 2001–2006. Due to the systemic effect the residues remained in the pear trees from one year to another. This caused residues in the pears even in harvest seasons without application of chlormequat. Chlormequat was also studied in UK foods.[23] In 2001 the MRL of 3 mg/kg for pears was changed to a temporary MRL of 0.5 mg/kg, which was reduced in two steps to 0.1 mg/kg from 2006. Surveys in 1997 and 1998 showed chlormequat contents of 0.05–16 mg/kg (n=54) and 0.05–11 mg/kg (n=48), respectively. In 1999 half the 1997 pear samples contained chlormequat and 10% exceeded the MRL of 3 mg/kg. In 2000 79% of 136 samples contained chlormequat, but none of the samples exceeded the MRL of 3 mg/kg. By 2018 chlormequat still occurs in pears due to former use. Therefore, the EU has set a temporary MRL of 0.07 mg/kg.[24]

11.2.3.2 Processed Fruit and Vegetables Including Processing Studies

The MRL is established for residues in the whole commodity. Hence for control purposes in the monitoring program the pesticide residues are mostly determined in raw commodities. However, many foods are eaten after various kinds of processing. The processing factor (Pf) of the food is defined as any operation performed on a food or food product from the point of harvest of the raw agricultural commodity (RAC):

$$Pf = \frac{Residue\,of\,processed\,fraction\,(mg/kg)}{Residue\,of\,RAC\,(mg/kg)}$$

The processing may take place when preparing the food at home or may be commercial food processing. Typical home processing includes washing, peeling, heating, and/or juicing while the commercial food processing additionally may include drying, canning, fermenting, oil extraction, refining, preserving, jamming, and/or mixing with other ingredients, etc.

The processing may affect the pesticide residue levels in the food products mainly by reducing the levels. The extent to which a pesticide is removed during processing depends on a variety of factors such as chemical properties of the pesticide, the nature of the food commodity, the processing step, and time of processing. The reductions may be predicted by the solubility, sensitivity toward hydrolysis, octanol–water proportioning, and the volatility, e.g., lipophilic pesticides tend to concentrate in tissues rich in lipids, etc. Examples to the contrary are increased pesticide levels after drying or refining. The effects of processing on pesticide residues in food are compiled in a database by the German Institute for Risk assessment (BfR), including more than 2600 studies on the effect of processing on the pesticide levels.[25,26] Information on processing may also be found in the annual pesticide evaluations reported by Joint FAO/WHO Meeting on Pesticide Residues (JMPR)[27] and in the EFSA conclusions performed as part of the overall evaluation in the EU. During these joint meetings on pesticide residues (JMPR), selected pesticides are reviewed including the effects of processing but the company data presented here may be in a compiled form without detailed information.

A majority of the pesticides applied directly to crops are mainly found on the surface of the crops[28] as the crops cuticular wax serves as a transport barrier for pesticides. Hence the majorities of the pesticide residues may be found in the peel and when the peel is not an edible part this will reduce the pesticide levels taken in through the diet. This is often the case for citrus fruits, where an investigation showed that more than 90% of the pesticide residues were found in the peel.[29]

In a study on apple processing, juicing and peeling significantly reduced the levels of the fourteen pesticides investigated compared to the unprocessed apple.[30] However none of the pesticide residues were significantly reduced when the apples were subject to simple washing or coring. The effect of processing was compared for two different apple varieties: Discovery and Jonagold. The pesticides selected for field application were the most commonly used in the Danish apple orchards or those most often detected in the national monitoring program. The concentrations of chlorpyriphos in unprocessed and processed apples (Figure 11.1), show, for example,

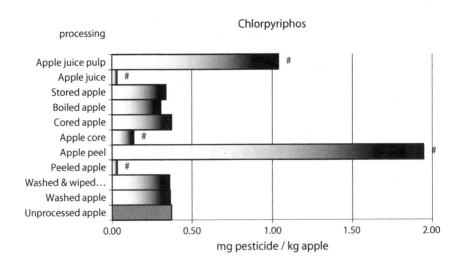

FIGURE 11.1 Concentrations (mg/kg) of chlorpyriphos in apples of the variety Discovery before and after different kind of processing (after Rasmussen et al. 2003).[30] #: significant changes at the 95% confidence level (n=5)

that peeling reduced the chlorpyriphos concentration by 93% and juicing reduced the chlorpyriphos concentrations by 95% compared to the unprocessed samples.

In a study on commercial processing, samples of tomatoes, peppers, asparagus, spinach, and peaches were exposed to three insecticides and four dithiocarbamates.[31] In most cases canning operations gradually decreased the residue levels in the finished product, particularly through washing, blanching, peeling, and cooking processes. Washing and blanching led to more than 50% loss in pesticide residues except for peaches. The total amount of pesticide reduced by all the combined canning operations ranged from 90 to 100% in most products.

The reduction of pesticides is not necessarily beneficial. The pesticide may be degraded to a metabolite more hazardous like the ethylenethiourea (ETU) formed during degradation of dithiocarbamates. The formation of ETU was studied for different food processing steps, e.g., 80% of ethylenebisdithiocarbamates was metabolized to ETU in drinkable beer.[32] The persistence of the ETU varies in different matrices and it can be stable for up to 200 days in canned tomato puree.[33]

Both the reduction due to processing of wine and the pesticide residues in wine were compiled in an Italian study.[34] The different pesticides behaved differently according to their physicochemical properties. Some pesticides disappeared totally or partly during the wine making, due to either degradation in the acidic environment, degradation during the fermentation process, or adsorption by the lees and the cake. Only a few pesticides passed from the grape to the wine without showing appreciable reduction. Among these pesticides were dimethoate, omethoate, metalaxyl, and pyrimethanil. In 1998–1999, 449 wine samples were analyzed for ~120 pesticides. Only very few pesticides were found in wine and at low levels.

Pesticide residues in processed food are monitored, e.g., within the EU. In the 2015-coordinated program, 1045 samples of olive oil and 756 samples of orange juice were analyzed.[14] The percentages of monitoring samples with residues were significantly lower in processed food than in fresh products. In 2015 residues ≤MRL were, e.g., found in 15% of the samples, and residues exceeding the MRL were found in 0.1% of the samples, i.e., results were the same for olive oil and orange juice. In comparison, in 2004 residues ≤MRL were detected in 23% of 704 orange juice samples and in 2.3% of the juice samples the pesticide residues exceeded the MRLs.[16]

11.2.3.3   Cereals

Cereals cover a range of crops like wheat, rye, barley, rice, maize, and millet. Cereals are treated with growth regulators, insecticides, fungicides, and herbicides throughout the entire growing period. To protect against insects, stored cereals are often postharvest treated with insecticides. That is the reason for frequently findings of insecticides like malathion, pirimiphos-methyl, chlorpyrifos-methyl, deltamethrin, and dichlorvos.[35] However, despite the high use of pesticides in cereal production, residues can be found less frequently than, for example, in fruits. The reason may be that the laboratories do not analyze for the whole range of pesticides used in the production, especially not those pesticides that have to be analyzed by single residues methods, like glyphosate and chlormequat. Another reason could be that the samples are typically collected at the mills, and might then be mixtures from different producers with different usage of pesticides; so the individual pesticide residues are diluted to below the analytical limit of determination. Previously, the frequently found pesticides were therefore the insecticides used postharvest like malathion, pirimiphos-methyl, chlorpyrifos-methyl, deltamethrin, and dichlorvos.[35]

Published data on pesticide residues in cereals are relatively scattered. The major part of the data found and presented below, covered from 2000 onwards, is from the United States and Europe. No data was found either from South America, Africa, or Australia. Data from Asia is mainly from the two biggest nations, India and China, and cover therefore the majority of the population of this region.[36–39] The majority of the results consist of data on DDTs and HCHs. This reflects not only the usage pattern of these compounds, which are effective and cheap although in some cases not allowed anymore, but also the instrumental capacity of the laboratories to analyze for newer types of pesticides. These types of pesticides are likely used but still not included in the monitoring programs. However, monitoring of seven triazolopyrimidine sulfonamide herbicides (flumetsulam, pyroxsulam, florasulam, metosulam, diclosulam, cloransulam-methyl, and penoxsulam) in 40 wheat samples from eight Chinese provinces only showed residues of florasulam in three samples and penoxsulam in two samples (from 0.005 to 0.007 mg/kg).[40] Likewise, in 2007 samples from 12 Chinese provinces, analyzed for 19 triazine pesticides, showed 11 triazines (cyromazine, metamitron, atrazine-2-hydroxy, simeton, hexazinone, simazine, atrazine, methoprotryne, ametryn, prometryn, and dipropetryn) detected in 6 samples in the range of 0.013–0.987 mg/kg.[41] Pesticide residue monitoring in Korean agricultural products from 2003 to 2005 was reported to have a detection rate of 0.3% for grains (28 samples). However, no specific information has been given about which pesticide residues were detected in the different cereals.[42] From Table 11.2 it is observed that DDTs and HCHs were frequently found in rice and wheat.

Since 1991 the U.S. Department of Agriculture (USDA) has been responsible for the pesticide residues testing program in cereals produced in the United States and imported cereals. The data for 2012–2015 for import of the four major cereal types, barley, corn, rice, and wheat and product hereof are shown in Figure 11.2.[48–51] On average 366 (ranging from 278 to 470) samples were analyzed each year; most of them being rice. Residues above the MRLs were found in 28% of the rice samples of 2012–2015. For wheat and barley, the frequencies were 7.6 and 5.0%, respectively. No exceedances of the MRLs were found for corn. Figure 11.2 shows that the sample types with the highest frequency of residues were wheat and rice, while barley and corn had the lowest. No decrease has been detected from 2012 to 2015. No information is given on the specific pesticides found in the cereals as only a common list for all the commodities is given. Nevertheless, the growth regulator chlormequat and the herbicide glyphosate were not included in the list and were likely frequently used. Therefore, the number of samples with residues may have been higher than reported.

Reynold et al.[23] investigated the levels of chlormequat in UK produced cereals from 1997 to 2002 and found residues in 50% of 59 wheat samples at 0.05–0.7 mg/kg and in 41% of 45 barley samples at 0.06–1.1 mg/kg; none of the samples exceeded the UK MRLs for grain. A similar study from Denmark showed that chlormequat was found in wheat, rye, and oat in 71, 60 and 100% of samples respectively.[52] Danish monitoring results from 2004 to 2011 showed residues of chlormequat in 25% of imported cereal samples and 13% of the Danish produced samples.[53] For samples

TABLE 11.2

Pesticide Residue Results in Cereals from Asia. Numbers in Brackets Are Percentage of Samples with Residues

	DDT mg/kg	HCH mg/kg	Other Pesticides mg/kg	Number of Samples	Year of Sampling	Reference
India – rice	0.023	0.066		30		36
India – wheat	0.22	2.99	7.97 – sum of heptachlor 0.17 – aldrin	150		37
India – rice	0.01 (57.7)	0.013 (64.4)		2000		38
India (Hyderabad) – rice	0.0001	0.0001	N.D. – chlorpyifos and cypermethrin	?	2010	43
India (Hyderabad) – wheat	0.00004	0.0044		?	2010	43
India – rice			0.5 – endosulfan, beta-0.008 – endosulfan, beta-	10	2008–2009	44
India – maize			0.02 – endosulfan, alpha, 0.02 – endosulfan, alpha-	10	2008–2009	
China – cereals	0.0045 (5.0)	0.0011 (53.0)		60	2002	39
China – cereals	0.0252	0.0053		–	1999	39
China – cereals	0.0019	0.0048		–	1992	39
China – cereals	N.D.	N.D.	0.21 – HCB0.05 – sum of chlordanes	12	2011–12	45
Bangladesh – rice	0.21	N.D.		2		46
Thailand – rice	N.D.	0.075	0.032 – aldrin	5	2009–10	47

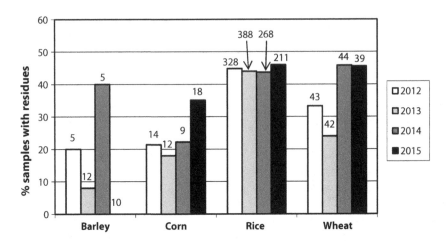

FIGURE 11.2 Frequencies of samples with residues from 2012–15 in barley, corn, rice, and wheat imported to the US. The values above the bars are the numbers of samples analyzed. (Data from USFDA[48–51].)

collected in 2016, residues of chlormequat were found in 40% of 25 imported wheat samples, while only 6% of the Danish produced wheat contained residues. Glyphosate was found in more than half of the monitored cereal samples produced in the Denmark from 1998 to 1999.[54] However, this frequency has decreased dramatically to 1.5% from 2004 to 2011.[53]

The European Commission has compiled data of 5117 conventionally produced cereal samples analyzed by the Member States in 2015.[14] The result showed that the percentages of samples with residues ≤MRL and exceeding the MRL were 43.2% and 3.4%, respectively. The data do not include information on the analyzed cereal type or the most frequently found pesticides explicitly for cereals. However, 1079 of the samples were analyzed for glyphosate and it was detected in 10.1% of the samples.

Apart from the national monitoring program, the Commission conducts a coordinated control program, where cereals are included regularly. Wheat flour and rice were included in 2011 and 2014, wheat kernels in 2012 and 2015, and rye/oats in 2013. The data show that residues were found in 41–52% of the wheat samples (Table 11.3) and the most frequently found pesticides included chlormequat, pirimiphos-methyl, glyphosate chlorpyriphos-methyl, tebuconazole, and deltamethrin. For rice, the percentages of samples with residues were 27–29% and the three most frequent pesticides were bromide ion, pirimiphos-methyl, and deltamethrin. In rye pesticide residues were found in 41% of the samples and most frequently found were chlormequat, bromide ion, mepiquat, and glyphosate. Finally, pesticides residues were found in 46% of the oat samples and the most frequently found pesticides were chlormequat, glyphosate, dithiocarbamates, and pyraclostrobin.

Storage may have only very little effect on the degradation of pesticide residues. However, temperature and humidity influence the degradation during storage. For example, malathion residues can be decreased by 30–40% over the course of 32 weeks at 30°C.[28] For wheat Uygun et al.[58] have reported 50% degradation over 127 days of malathion and 30% of fenitrothion over 55 days. Residues of pesticides are

TABLE 11.3
Results from the EU Coordinated Monitoring Program in Relation to Cereals

Cereal Type	Year of Sample Collection	Number of Samples	% Samples with Residues below MRL	% Samples with Multiple Residues	% Samples above MRL	Number of Pesticides Included in the Analytical Program	Reference
Oat	2013	232	46	1.0	8	180	55
Rice	2011	1060	29.4	2.0		164	56
Rice	2014	763	27.4	2.1	9.2	180	57
Rye	2013	424	41	0	16	180	55
Wheat flour	2011	605	52.1	0.3		164	56
Wheat flour	2014	702	41.2	0.4	11.8	180	57
Wheat Kernels	2012	862	39.7	0.7	17.2	177	17
Wheat kernels	2015	851	38	0.6	14	143	14

greatly reduced by milling. Most residues are present in the outer part of the grain, and consequently the reduction, for example, from wheat to sifted flour can be as high as 90%, whereas the concentration in bran increases compared to the whole grain. Cooking further reduced malathion and its degradation compounds if the grains were boiled in water.[59]

11.2.3.4　Food of Animal Origin

Pesticide residues occur in animals as a result of both previous and present uses of pesticides for agricultural purposes. The residue levels in products of animal origin are however generally low or non-detectable (<0.01 mg/kg). The residues ingested by, for example, livestock via the feed are metabolized by the animals and for most pesticides, in particular in the case of the more polar pesticides (log P_{ow}<3), the major part of the pesticides/metabolites is excreted. The highest levels of pesticide residues (including metabolites) are most often observed in organs involved in the metabolism and excretion of the pesticides, for example, the liver and kidney. For lipophilic compounds (log P_{ow}>3) however, the highest residues may be found in adipose tissues.

In reference to products of animal origin, the focus has mainly been on persistent pesticides, authorized and used in large amounts and for a wide range of purposes in the 1950s to 1970s. The very efficient pesticides such as OCPs were later found to be very stable in the environment, to bioaccumulate through the food chain, and to pose a risk of causing adverse human health effects.

In most parts of the world, the use of the environmentally persistent pesticides has been reduced dramatically during the last number of decades. The use of, for example, DDT was restricted in the United States, Canada, and most European countries in the early 1970s. In several countries with the need for malaria control, DDT was used until the end of the 1990s. Other OCPs such as dicofol were still in use in 2003 in, for example, China, and DDT is still authorized in different parts of the world for, e.g., malaria control.[60,61] Increasing resistance in malaria vectors against pyrethroid and carbamate insecticides has increased the dependence on DDT for malaria control though the global production of DDT has declined by 32% from 2001 to 2014.[62]

OCPs are detectable in most matrices of animal origin especially matrices with high fat content like butter[61,14] (Table 11.4), cheese,[62,63] milk,[63–66] and meat.[67–69] The levels are dependent on the age of the animals at the time of slaughter[64] and the fat content of the product,[65] that is, the older the animal and the higher the fat content, the higher, in general, the residue level of OCPs. In areas where the organochlorine compounds were recently or are still in use, legally or illegally, the residue levels are in some cases at or above the MRLs (Table 11.4).

Results from the Danish monitoring program including analysis of samples of Danish origin as well as imported products (1995–1996,[52] 1998–2003,[70] and 2004–2011[71]) have shown that OCPs are detectable, but below the MRL, in more than half of the animal product samples analyzed (n=1829). The animal products include meat, fish, butter, mixed products of butter and vegetable oils, cheese, animal fat, and eggs. In animal fat, ΣDDT was detected in the majority of the samples (about 65%) but at low levels. The mean levels were ≤ 19µg/kg fat, except for garfish and cod liver. The highest level of DDT, 700 µg/kg fat was found in cod liver from the

TABLE 11.4

Residue Levels (μg/kg Lipid Unless Otherwise Stated) of the Sum of DDT and its Degradation Products DDE and DDD (ΣDDT) and Sum of HCH Isomers (ΣHCH) in Bovine Butter. Mean Values Are Presented and Minimum and Maximum Values Are Presented in Brackets

Country (Year of Sampling)	ΣDDT (μg/kg Lipid)	ΣHCH (μg/kg Lipid)	No. of Samples/no. of Samples with Detectable Residues	Reference
Europe (2015)	<2.5 (mean)	<LOQ	DDT: 537/58 HCH: 537/0	14
Denmark (2004–11)	2.3 (mean)	α-HCH: 0.2 (mean) β-HCH: <LOQ	DDT: 421/295 α-HCH: 421/10 β-HCH:421/0	71
Spain (≤2000)	p,p'-DDE: 7.3 (0.02–52.5) μg/kg ww	γ-HCH: 10.8 (0.0039–19.59) β-HCH: 3.2 (0.01–9.1) μg/kg ww	γ-HCH: 36/36 β-HCH: 36/34 p,p'-DDE: 36/35 HCB: 36/32	73
Turkey (~2000)	p,p'-DDT, p,p'-DDE and p,p'-DDD all <0.001 mg/kg	γ-HCH <0.001 mg/kg	100/0	74
Canada (≤2000)	5.77 (0.38–16.92)	1.21 (0.13–2.10)	6	70
USA (≤2000)	23.61 (0.41–141.26)	1.33 (0–2.17)	18	70
Australia (≤2000)	5.96 (1.44–13.78)	0.31 (0–0.86)	5	70
India (~2004)	120	0.132	46	65

Baltic Sea. α-HCH was detected in less than 1% of the samples of animal origin (excluding seafood) and at mean levels ≤ 0.5 μg/kg fat. Dieldrin was detected in < 10% of the samples of animal origin at mean levels ≤ 6 μg/kg fat.

In 2004 the EU coordinated official control program was amended to also include products of animal origin. In 2014 and 2015 a total of 16974 samples of animal origin were analyzed as part of the coordinated program. No pesticide residues were found in 85% of the samples. The most frequently found pesticides were the persistent environmental pollutants or compounds not originating from pesticides.[14,57]

The frequencies of DDT, HCH, and dieldrin found in Japanese samples from 2000 to 2004[75] are higher compared to the European data. ΣDDT was detected in 64%, 90%, and 90% of beef (n=25), pork (n=30), and poultry (20) samples. ΣHCH was detected in 24%, 23%, and 20% of the Japanese samples, respectively. Dieldrin was detected in 24%, 23%, and 45% of the samples.

In several studies, butter has been analyzed as a representative of animal products with high fat contents and the levels found can be used as an indicator of the general OCP levels in animal products. Weiss et al. 2005[72] have performed a worldwide survey of, among other compounds, DDT and HCB. One sample of butter was sampled from 39 European countries and from 25 non-European countries. It was found that the average level of ΣDDT in butter from all the participating countries was 10.8 µg/kg fat. The average level of HCB in butter from all the participating countries was 3.5 µg/kg fat. Overall the study shows that the levels found in butter originating from countries like India and Mexico are higher compared to butter originating from countries such as Denmark and Germany.[73] The results are in good agreement with the fact that the persistent OCPs were banned earlier in the latter countries than they were in the former countries.

Butter is also included in the EU coordinated monitoring program though not every year. In 2015 a total of 616 samples of butter were analyzed and 87% of these contained no quantifiable levels of pesticide residues while 13% of the samples contained one or several quantifiable pesticide residues. The most frequently found pesticides were DDT (10.7%), hexachlorobenzene (7.2%), and dieldrin (1.1%). The samples were analyzed for a total of 22 pesticides, but residues of pesticides other than those mentioned above occurred very rarely. Compared with the results obtained for butter in the EU monitoring program in 2012, there was a slight decrease in the frequency of pesticides found.[14,57]

11.2.3.4.1   Pesticides in Fish

The residue levels of organochlorine pesticides (OCPs) in fish vary greatly depending on the origin. In general, higher levels are observed in seafood caught in waters close to pollution sources, e.g., some coastal waters. Furthermore, the levels may be positively correlated with age and lipid content of the organism.

OCPs can be found in large fractions of seafood even from waters of countries were the compounds have been banned for several decades. In Table 11.5 are presented some reported levels of DDT, HCH, and dieldrin in different seafood samples caught or produced in aquaculture in different parts of the world. A large study has been performed on the levels of OCPs in seafood from Taiwan (2001–2003)[78] showing that OCPs were detectable in 24% of the fish samples and organophosphorus compounds in 11% of the fish samples (n = 607). The detection rate was lower in shellfish, i.e., OCPs in 6% and organophosphorus compounds in none (n = 62). The mean residue level of ΣDDT in all of the sampled seafoods with detected residues was 32.5 µg/kg wet weight (ww). A later study also from Taiwan (2013–14)[79] found ΣDDT in two of 31 seafood samples (6%): 33.4 µg/kg ww and 134 µg/kg ww in barracuda and tuna, respectively.

In a study of farmed pangasius from India and Vietnam, in addition to OCPs, other pesticides such as quinalphos, malathion, and parathion-methyl, were detected in 38% of the samples.[80] In comparison with Indian pangasius, fewer contaminants at low residue level were detected in pangasius fillets imported from Vietnam.[80] Some of the detected pesticides in the study (e.g., malathion, fipronil) have been listed as

TABLE 11.5

Examples of Reported Residue Levels in µg/kg Wet Weight (/ww) or Lipid Weight(/lw) of ΣDDT, ΣHCH and Dieldrin in Fish and Seafood. Mean Values or Ranges Are Presented (Maximum Values in Brackets)

Place and Year of Sampling	Sample Type	ΣDDT µg/kg	ΣHCH µg/kg	Dieldrin	No. Samples	Reference
Greenland, 1994	Polar Cod liver	12–83 /ww	6–16 /ww		77	
	Cod liver	60–98 /ww	7–9 /ww		25	
Denmark, 2004–11	Cod liver	281(max 1771)/lw	4.7(max 24) /lw	19 (max 72)/lw	156	71
	Herring	12 (max 143)/lw	0.8 (max 4.4)/lw	2.5 (max 3.8)/lw	236	
	Trout, aquaculture	3.7 (max 17) /lw	0.2 (max 3.2) /lw	1.0 (max 4.8) /lw	484	
	Trout, marine aquaculture	9.3 (max 31) /lw	1.4 (max 7.8) /lw	2.0 (max 8.1) /lw	116	
Norway, 2012	Salmon	7.98 (20.7) /lw	<0.08–0.90 /lw	2.12 (max 4.2) /lw	76–111	77
Norway, 2015–16	Salmon, aquaculture	5.70 (14.0) /lw	<0.08–0.29 /lw	1.22 (max 3.5) /lw	77–293	77
Taiwan, 2001–3	Fish, shellfish, bivalve, crustacean, cephalopod.	32.5 (max 169) /ww		0.9 (max 31) /ww	920	78
Taiwan, 2013–14	Marine fish (e.g., barracuda, tuna)	6.2 (max 134) /ww			31/2	79
India, Vietnam 2014–15	Pangasius aquaculture	29.2 /ww			148	80
Kenya, 2011	Tilapie, aquaculture	2.57 /ww	0.072 ± 0.011 /ww	0.071 ± 0.075 /ww	125	81

endocrine disrupters which emphasizes the importance of pesticide residue monitoring in fish.

In salmon farmed in Norway, OCPs were detected, though in lower concentrations compared to the wild salmon (Table 11.5).[77] The presence in the farmed salmon is due to transfer and accumulation from the part of the composite fish feed consisting of marine fish constituents. In addition to ΣDDT, ΣHCH, and dieldrin, endosulfan and cypermethrin also occurred in farmed salmon. Diflubenzuron, an insecticide, was found in farmed salmon, however only in few of the analyzed samples and in concentrations far below the maximum levels.

The residue levels of OCPs in seafood have been declining for the last decades and may decline further in the future, for example as shown in cod liver from the Arctic.[82]

11.2.3.5 Infant and Baby Food

Infants and children consume more food per kilo of body weight per day than adults do. Furthermore, the detoxification systems of infants are not fully developed. These are some of the factors that make infants and young children a sensitive group of consumers. The primary food intake for infants (0–6 months of age) is accounted for by either human breast milk or formulae. As the child gets older, an increasing proportion of the daily food intake is accounted for by vegetables, fruits, and cereals and to some extent also food of animal origin, either prepared at home from raw products or as pre-processed products.

Because infants and young children are sensitive consumers, special attention has been directed toward pesticide residues in not only breast milk but also infant and weaning foods marketed as such. In order to ensure low levels of pesticides and increase the protection level for this consumer group, specific rules were established in EU in the late 1990s for the occurrence of pesticides residues in baby food. Accordingly, Directive 2006/125/EC[3] sets lower MRLs for specific pesticides of 0.01 mg/kg, which in many cases correspond to the analytical quantification limit. In addition, low MRLs of 0.004–0.008 mg/kg are put into force for five pesticides of relatively high toxicity (cadusafos, demeton-s-methyl, ethoprophos, fipronil including fipronil-desulfinyl, and propineb/propylenethiourea) and nine other pesticides (e.g., hexachlorobenzene, nitrofen, and omethoate) are prohibited for treatment of crops intended to be used for the production of baby foods (Directive 2006/125/EC). The directive only applies to infant and baby food products on the European market.

11.2.3.5.1 Residues in Human Breast Milk

Human breast milk has a high fat content and for that reason a major concern in relation to pesticide residues in human breast milk worldwide is the environmentally stable pesticides, for example, OCPs. During breast-feeding, OCPs from the mother are excreted via the milk to the baby.

The levels of OCPs vary and depend on the age of the mother,[83,84] whether the mother has breast-fed before,[84] her eating habits (e.g., the amount of fatty fish[84,85]), and place of living, that is whether there are OCPs in the local environment including the food.[85] The OCPs are of concern since they are under suspicion for having the potential to affect, for example, the birth weight of infants, the risk of cancer, and the neurodevelopment of infants.

The levels of OCPs have been shown to be higher in human breast milk from the populations of Asian countries as China, India, Cambodia, and Indonesia compared to European/North American countries such as the United Kingdom, Germany, Sweden, Spain, and Canada.[83] Wong et al. (2005)[83] has reported that the levels of DDT, DDE, and β-HCH in human breast milk are 2–15-fold higher in samples from China compared to samples from several European countries.[83] Examples of residue levels of ΣDDT and ΣHCH in human breast milk from different parts of the world are shown in Table 11.6.

The levels of OCPs in breast milk are in general decreasing as a result of prohibition of the compounds and/or restrictions in use.[85-88] Recently it was concluded, in a review of the global contamination levels of DDT and other persistent organic pollutants, that the global distribution for ΣDDTs is almost exclusively associated with countries where malaria is still a significant health problem.[60] The ΣDDT levels in breast milk from women in Taiwan sampled in 2001 have, for example, been found on average to contain 333 µg/kg milk fat (36 samples), whereas the levels of ΣDDT in breast milk sampled in Taiwan in the previous two decades on average amounted

---

TABLE 11.6

Examples of Reported Residue Levels (µg/kg fat) of the Sum of DDT and Its Degradation Products DDE and DDD (ΣDDT) and Sum of HCH Isomers (ΣHCH) in Human Breast Milk. Mean Values Are Presented and Minimum and Maximum Values Are Presented in Brackets

Origin of Samples (Year of Sampling)	Level of ΣDDT µg/kg Fat Unless Otherwise Stated	Level of ΣHCH	Number of Samples Analyzed	Reference
Finland and Denmark (1997–2001)	129 (31–443)	β-HCH 13 (2,7–66) µg/kg fat	130	86
Germany (1995–7)	DDT: 240 (27–1540)	β-HCH: 40 (4–50)	246	87
UK (1997–8)	DDT: 40 DDE: 430		168	83
USA (2004)	ΣDDT: 65 (< 0.6–1910)	ΣHCH:19 (< 1.6–74)	38	87
China (2011–12)	ΣDDT: 316	ΣHCH: 42	142	91
China (1999–2000)	DDT: 545 DDE: 2665	β-HCH: 1030	169	88
Vietnam (2000–1)	DDT: 218 (34–6900) DDE: 1950 (340–16000) ΣDDT: 2200 (440–17000)	β-HCH: 36 (4–160)	86	83
Tanzania (2012)	ΣDDT: 205 (26–2490)	ΣHCH: 1 (< LOD–24.5)	150	92
Zimbabwe (1999)	*p,p'*-DDE: 4863 *p,p'*-DDT: 1149 ΣDDT: 6314	β-HCH: 216 γ-HCH: 99 ΣHCH: 383	116	93

to 3595 µg/kg milk fat.[89] In milk from women in Indonesia (sampled 2001–2003),[85] great differences in the levels of ΣDDT and ΣHCH have been observed. The higher levels were observed in suburban and rural areas and the lower levels in the urban areas. In breast milk from German women,[87] the level of DDT was approximately 81% lower in 1995–1997 (240 µg/kg milk fat) than it was ten years earlier.

A review of data on the occurrence of POPs including DDT/DDE and HCH published between 1995 and 2011 by Fång et al. 2015,[94] reveals that the lowest average levels of DDT/DDE were found in breast milk from Western European mothers and that the average levels are higher in samples from the Asian and Pacific region, the Americas, and highest in Africa. The data also reveal that the HCH levels in milk from European mothers generally are lower than in milk from Asia, Australia, and the Pacific regions but higher than the levels reported for Africa and the Americas. Data from Sweden (covering 1972–2010) and Japan (covering 1972–1998) allowed for the study of a possible time trend, and these show significantly decreasing trends in the levels of DDT/DDE, HCH, dieldrin, and heptachlorepoxide.[94]

Even though OCPs, as well as other persistent organic pollutants, occur in human breast milk and therefore are consumed by infants, a literature search and Web search do not reveal any authorities or researchers that recommend avoiding breast-feeding. Recently a risk–benefit evaluation of breast-feeding based on the three most recent surveys of WHO/UNEP was performed by van der Berg et al. 2017. The conclusion was that the advantages of breast-feeding by far out weight the potential adverse effects. The primary concerns related to breast-feeding were the occurrence of polychlorinated compounds (PCBs, PCDDs, and PCDFs) whereas the levels of DDT were found to be below or around the levels that are considered safe.[60]

11.2.3.5.2   Residues in Formulae and Weaning Products

Formulae and weaning foods are highly processed foods and processing most often reduces the levels of the pesticides. Especially, thermolabile pesticides are not expected to be detectable in infant formulae or weaning foods, since these products have been heat-treated during processing and for preservation. Furthermore, raw products for the production of weaning foods are washed and perhaps also peeled. The formulae available on the market are based on cow's milk or soya or a combination of the two. Weaning food are, for example, fruit and vegetable purée, fruit juices, cereal-based meals, biscuits, and complete meals composed of, for example, vegetables, pasta, meat, and biscuits.

Formulae has a relatively high fat content (~25 g/100 g). OCPs are therefore of relevance in reference to pesticide residues also in these products. Pesticides such as organophosphorus, carbamates, and pyrethroids have not been found to accumulate in the fat and milk of livestock to any significant degree and no residues of these pesticides have been detected in 1008 samples of US manufactured milk-based infant formulae samples.[95]

Lackmann et al.[96] have shown that in Germany the intake of OCPs, for example, DDT and DDE, are significantly higher for breast-fed infants than for bottle-fed infants. The serum concentration of DDE in breast-fed infants was about six times higher after six weeks of feeding compared to the serum concentration in bottle-fed infants. Whether this relatively large difference is maintained after a longer period of breast-feeding has not yet been reported.

The occurrence of OCPs in formulae and weaning products follows the same trends as seen for the levels in breast milk, that is, higher levels of OCPs are in general found in products produced in previously developing countries and lower levels in products from developed countries. In infant formulae collected from the Indian market in 1989 residues of ΣDDT and ΣHCH were found in 94% and 70% of the samples, respectively. A total of 186 samples of 20 different brands were analyzed. The mean level of ΣDDT was found to be 300 µg/kg fat and the mean level of ΣHCH was found to be 490 µg/kg fat.[97]

In weaning foods, it is also likely to detect pesticide residues due to use of pesticides during cultivation of the raw products. Baby foods including infant formulae, follow-on formulae, processed cereal-based foods for infants and young children, as well as other baby foods are included in the EU coordinated control program. In 2014 and 2015 a total of 3358 of such samples available on the European market were analyzed for pesticide residues. 90% of the samples contained no detectable residues of pesticides.[14,57] The most frequently detected pesticides were copper (89 findings), fosetyl-Al (57 findings), and the biocidal products DDAC (four findings) and BAC (four findings). Thus, few different pesticides were found and copper accounted for the majority of the findings. However, the finding of copper is often related to the use of copper as added nutrient and not the use for plant protection.[14,57]

Of a total of 181 samples of baby food collected within the Danish Food Monitoring Program 1998–2003 residues were only detectable in one sample.[98] The sample was a cereal powder with a residue of chlormequat (0.025 mg/kg) and mepiquat (0.019 mg/kg).

In the US Total Diet Studies (2000), a total of 78 items of different baby foods were analyzed. The most frequently found pesticide residues were the insecticides carbaryl (18%), endosulfan (17%), malathion (12%), and chlorpyriphos-methyl (10%), and the fungicide iprodione (12%). The highest level of 0.096 mg/kg was found for iprodione.[99]

Cressey and Vannoort (2003)[100] analyzed 25 infant formulae and 30 weaning foods commercially available in New Zealand in 1996. Soy-based formulae and weaning products were screened for about 140 pesticides and the milk-based formulae for OCPs (*p,p'*-DDE, *p,p'*-DDT and dieldrin). *p,p'*-DDE was found in seven of 20 milk-based infant formulae and residues of *p,p'*-DDT were found in one milk-based infant formula. Dieldrin was detected in four of five soy-based formulae. Dithiocarbamates (LOD of 100 µg/kg) were not found in any of the soy-based formulae or any of the weaning foods. Cressey and Vannoort did not analyze for ETU, the degradation product of dithiocarbamates. Two organophosphorus pesticides, azinphos-methyl and pirimiphos-methyl, were detected in one soy-based formula and in two out of nine cereal-based weaning foods, respectively.

In the Australian 19th Total Diet Survey, residues were found in cereal-based infant foods but neither in formulae, infant desserts, or dinners. The pesticides detected in the cereal-based products were chlorpyrifos-methyl (4 µg/kg), fenitrothion (2 µg/kg), iprodione (4 µg/kg), and piperonyl butoxide (8 µg/kg). The residue levels (nine samples) were low and the mean level ranged from 2 to 8 µg/kg. Thus, no residues of DDT, DDE, or other OCPs were detected in formulae or infant foods but some organophosphorus pesticides were found in cereal-based products.[101]

Thus, generally the frequency of pesticides found in formulae and baby foods is low as are the levels of residues found. Setting low MRLs for pesticide residues in formulae and baby food, for example, within EU with the implementation of Directive 2006/125/EC of 5 December 2006, provides a higher level of safety. However, these low MRLs do not increase the safety of baby foods prepared in the home from conventional raw produce.

11.3 CONSUMER EXPOSURE AND RISK ASSESSMENT

11.3.1 DIETARY INTAKE ESTIMATION

To perform a dietary intake estimation is basically easy. The consumption is multiplied with the content:

$$Intake = Consumption \times Content$$

The question is, however, what consumption and which content should be used? Should it be mean values? Should it be high values, a kind of worst-case situation? And how is the consumption respective to the content estimated? There is no one way of performing dietary intake estimations and in the literature different ways of performing the estimations have been used and in addition the data collection has been very diverse. Therefore, it is also often very difficult to directly compare the dietary estimations.

Dietary surveys can be performed in many ways. In some surveys, participants are asked to fill out a diary about what they have been eating and the amounts; in others people are interviewed about what they have been eating as, for example, yesterday. In some surveys the food bought in the household is used for estimation of the consumption. Here the total amount of, for example, potatoes is divided by the numbers of people in the household and number of meals where the potatoes are eaten. Both number of participants, number of days, and the details concerning the food eaten differ between dietary surveys. In many circumstances the food as eaten is calculated back by using recipes to ingredients or raw agricultural commodities (RAC); e.g., an apple pie is divided into flour (grain), apple, and other ingredients. Both Codex and the EU have published consumption data.[102,103]

For total diet studies and duplicate diet studies, however, the content is directly determined by the food eaten. In total diet studies a certain number of raw and prepared foods are chosen to represent the total diet of the population. The foods are then bought and prepared according to recipes and the content of the pesticides or other substances are directly determined in the foods. In the duplicate diet studies, the participants collect exactly the same amount of food as they eat and the pesticides are then determined in the collected foods.

Data concerning the content of the pesticides often comes from monitoring or surveillance. These studies differ widely with regard to which pesticides that are included, the number of pesticides included, and the number of commodities included.

In the calculation of the dietary exposure, other factors such as correction for undetectable residues or processing also influence the result. Although a pesticide is not detected in a commodity this does not necessarily mean that it is not present, just that the level could be lower than the analytical Limit of Detection (LOD) or Limit of Quantification (LOQ). In some calculations the undetectable residues are set at, e.g., ½ LOD or ½ LOQ. Using '0' for non-detects are sometimes reported as the Lower Bound (LB) estimate while using a value is denoted as the Upper Bound (UB) estimate. It is known that for examples peeling or boiling can reduce the amount of pesticides, while drying (e.g., grapes to raisins) can increase the content of the pesticides. To perform the most reliable estimation of the dietary exposure, processing factors should be included if available.

Dietary exposure calculations can be performed with different approaches, deterministic or probabilistic, and for both chronic and acute intake. The chronic intake or the long-term intake is the possible intake over a long time, e.g., a whole life, and in the risk assessment this intake is compared with the ADI. The acute intake or the short-term intake, on the other hand, is the intake within 24 hours or less, e.g., a meal. The acute intake is compared with the ARfD in the risk assessment.

11.3.1.1 Deterministic Approach (Chronic and Acute Intake)

A calculation of the chronic intake by the deterministic approach only yields a single value for the intake and is also called a point estimate. In this approach a single value of the consumption of a commodity is multiplied with a single value of the concentrations of residues. Often consumption and concentrations are average values, but they can also be high percentiles if a worst-case calculation is performed. If the chronic intake for a certain pesticide from all commodities is calculated the single intakes for each possible commodity are summed.

The acute intake in the deterministic approach is always calculated for a single commodity. Depending on the commodity, different equations defined by JMPR[104] are used for calculation of the acute intake. In the two most often used equations the so-called variability factor is included. This factor is based on the variation of the residues in a composite sample. In monitoring usually analyzed samples are composite samples while all the content of a pesticide found can be from just one sample. In an estimation of the acute intake the intake from this one sample is of interest and the variability factor is an expression used to estimate the content in a single sample from the content in a composite sample.

The deterministic approach is the absolute most often used method for the calculation of pesticide intakes. The advantages of the approach are that the approach is easy and simple to perform and the results are easy to interpret. The drawback of the approach is that the exposure is expressed as single values because single values are used for both consumption and content regardless of the variability in both variables. Thus, intakes determined by the deterministic approach are generally highly overestimated.

11.3.1.2 Probabilistic Approach

Probabilistic modeling is called so because this approach yields the probability for an intake. In this approach the whole distributions of consumption data and

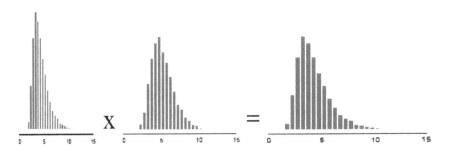

FIGURE 11.3   How a distribution of intake is performed from a distribution of consumption and of contents. (from Pieters et al.[105])

concentrations are used in the calculations, resulting in a new distribution for the intake; a consumption of a commodity is chosen and a residue in this commodity is chosen and the two values are multiplied to yield an intake. Then a new consumption value and a new residue value are chosen. This is done several times, e.g., 100,000 times, resulting in a distribution of intakes[105] (Figure 11.3). In this way percentiles of the intake can be determined. The probabilistic modeling determines the acute or short-term intake, if it is the consumption for a meal or a day that is used in the calculation. Algorithms to calculate the chronic intake have become a part of some programs and the chronic intake can be compared with the ADI. The advantage of this approach is that all data of both consumptions and concentrations are used, the whole distribution of the intake is shown, and the uncertainties in the calculation can be estimated.

In connection with authorization of pesticides, the EPA in the United States use probabilistic modeling as part of their evaluations of pesticides and have published guidelines for the work.[104] Both in Codex and in the EU the uses of probabilistic modeling are discussed. EFSA (European Food Safety Authority) has published a guidance document on the use of the probabilistic methodology for dietary exposure to pesticides[106] and research projects concerning the subject have been initiated, for example, in the EU project Acropolis.[107]

11.3.1.3   Cumulative Exposure

The term 'cumulative exposure' can be used in different ways. Some use it as the total intake of a single pesticide from all commodities. The most often applied definition of the term, and the definition used in this book, is that the cumulative exposure is the total intake of a group of pesticides from all commodities with a combined action. Several approaches can be used[108] but at the moment there is no common worldwide agreement on which approach to use to calculate the cumulative exposure for pesticides in our food.

Examples using three different approaches are summarized here, namely the so-called Hazard Index approach for the Danish population, the TEF/RPF approach for the Brazilian population, and the Margin of Exposure for the US population. In the last two examples the cumulative dietary exposure is calculated for choline esterase inhibiting substances (organophosphates and carbamates).

The easiest way to perform estimation for the cumulative exposure is to use the Hazard Index (HI) approach. In this approach the Hazard Quotient (HQ) is estimated for each substance by dividing the exposure by the ADI of that substance. Then all the HQs are summed to give the HI. If HI is less than one the mixture is not expected to constitute a risk.[109]

Using the Toxic Equivalent Factor (TEF; also called Relative Potency Factor [RPF]) approach, exposures of a group of common mechanism chemicals with different potencies are normalized to yield a total equivalent exposure to one of the chemicals, the so-called 'index compound' (IC). TEFs are obtained as the ratio of the toxic potency at the chosen toxicological end-point of the IC to that of each of the other members in the group. This means that a substance with a toxic potency ten times the IC is assigned a TEF value of ten. The exposure to each chemical is then multiplied by the appropriate TEF, e.g., ten, to express all exposures in terms of the IC. Summation of these values provides a total combined exposure to all chemicals in terms of the IC. In EFSA work is ongoing to define common assessment groups for different effects to be used in the cumulative assessment.[109]

To assess the cumulative risk of the exposure in the United States the total Margin of Exposure (MOE)[110] is used. MOE for a single chemical is the ratio of the effect dose level (ED) at the chosen toxicological end-point to the level of dietary exposure.

$$MOE = \frac{ED}{Exposure}$$

The combined MOE is:

$$Combined\ MOE = \frac{1}{\dfrac{1}{MOE1} + \dfrac{1}{MOE2} + \dfrac{1}{MOE3}} \quad etc.$$

The greater the MOE, the lower the risk. In the assessments a target value of 100 is acceptable. MOEs less than 100 are undesirable.

11.3.2 Intake Calculations of Pesticide Residues

In this section intake calculations or dietary exposure from different part of the world using different approaches are presented.

11.3.2.1 Deterministic Approach

The results from the EU coordinated monitoring program are every year used to calculate both the chronic and acute intake.[14] The acute intake is estimated for the food products covered by the EU control program so the food products change from year to year. In 2015 the ARfD was exceeded for chlorpyrifos, imazalil, acrinathrin, ethephon, and lambda-cyhalothrin in one or more commodities. The chronic exposure was estimated for the pesticides found in the most commonly consumed food products included in the EU control program. For most of the pesticides the estimated intake was lower than 10% of the ADI. Only for dichlorvos was the ADI exceeded (143%) and only under the UB estimation, for example, that all results reported as below LOQ were set to LOQ. If all results reported as below LOQ were set to '0' the estimated intake amounted to 5% of the ADI. This is a good example of the fact that use of different figures for the content of pesticides can have a great impact on the estimation of the intake.

For some studies the pesticide intakes are limited to include specific pesticides or foods.[111] In Kazakhstan[112] a total of 80 grain samples were analyzed for 180 pesticides. Of these, ten were detected. The highest chronic intake was estimated for aldrin in wheat (789% of ADI) and pirimiphos-methyl in wheat and rye (49.8% of ADI). For the acute intake the highest intakes were calculated for aldrin and tebuconazole in wheat (316 and 98.7% of ARfD). In Ghana[113] an assessment was performed of the intake of 11 organochlorines in fruits and vegetables. These substances are forbidden and can also be seen as contaminants. Altogether, 320 samples of locally produced pawpaw and tomatoes as well as imported apples were analyzed. Food consumption was estimated on the background of the fruit consumption in Ghana and a bodyweight of 10 kg was used for children and 70 kg for adults. The residues varied from < 0.01 mg/kg to 0.11 mg/kg. The reference doses were exceeded for heptachlor, heptachlor epoxide, endrin aldehyde, and endrin ketone for one or more of the foods.

11.3.2.2 Total Diet and Duplicate Diet Studies

In India 1999–2002, a kind of duplicate diet study was performed for men aged 19–24 years.[114] Every month vegetarian and non-vegetarian total diet samples comprising breakfast, lunch, and dinner were collected. Lindane was the pesticide most widely found but the frequency decreased throughout the study, from about 90% in 1999 to about 25% in 2002. The Codex ADI for lindane (γ-HCH) of 0.008 mg/kg bw was exceeded in 1999 for the vegetarian diet and in 1999 and 2000 for the non-vegetarian diet. An explanation for the high contribution from lindane could be that about 21% of the consumption in the study came from milk and milk products, which another study showed could be highly contaminated with lindane.

In Lebanon the intake of Lebanese adults of 47 pesticides from food of plant origin and drinks was estimated. The study was conducted as a total diet study and was performed in two areas of the country.[115] A total of 1860 individual foods were collected, prepared, and cooked before analysis. The types and quantities of foods that made up the diets were based on food consumption data from the two areas. The intake was estimated under both LB and UB assumptions for the 16 pesticides where at least one positive was detected. A body weight of 72.8 kg and 73.6 kg was used for respectively the rural and semi-rural population. For the LB estimates all intakes amounted to less than 6% of the ADIs while the UB estimates the intakes amounted up to 101% of the ADI for dieldrin. Table 11.7 shows some of the results from the paper for the rural population. From here it appears that the intake of dieldrin exceeds the ADI for the UB estimate; it amounts to 0% for the LB estimate, showing the great impact the assumptions have on the intake. In this paper the intakes are compared with the intake estimates from total diet studies performed in France, Cameroon, New Zealand, and China published in the period 2006–2009.

TABLE 11.7

Results for the Five Pesticides That Contribute Most to the UB Intake (µg/day) and ADI (%) Lebanese Total Diet Study for the Rural Population

Pesticide	UB intake	% of ADI (UB)	LB intake	% of ADI (LB)
Dieldrin	7.33	100.69	0	0
Diazinone	7.33	50.34	0	0
Dimethoate	10.49	14.42	3.14	4.3
Endosulfan	28.24	6.47	0.58	0.1328
Ethion	7.33	5.03	0	0

11.3.2.3 Cumulative Exposure

The EPA has performed cumulative risk assessment for four groups of pesticides[116] namely organophosphates, N-methylcarbamates, triazines, chloroacetanilides, and pyrethrins/pyretroids. In these assessments not only is the dietary exposure calculated but also the exposure from water and residential uses. For triazines, exposure through food was not considered as relevant and for chloroacetanilides only two pesticides were included.

For both organophosphates and the carbamates, residue data were primarily obtained from the USDA PDP program collected from 1993 to 2003 or 1993 to 2004. The consumption data are from the USDA's Continuing Survey of Food Intakes by Individuals, 1994–1996/1998. In this survey ca 21,000 participants were interviewed over two discontinuous days. Processing factors are included in the estimates; undetectable residues were set to zero.

In the study the MOE approach as well as the probabilistic approach were used. For the organophosphates the IC was methamidophos, while for the carbamates oxamyl was chosen as IC (Table 11.8).

The foods that contributed most to children's (1–2 years) exposure to carbamates were strawberry and potato while the pesticides that contributed most were

TABLE 11.8

Exposure and MOE at the 99.9th Percentiles for Children 1–2 Years and Children 3–5 years Which Have the Lowest MOEs as Well as for Adults 20–49 years

	Organophosphates		Carbamates	
	Exposure (µg/kg bw)	MOE	Exposure (µg/kg bw)	MOE
Children 1–2 years	2.6	30	3.8	37
Children 3–5 years	2.3	34	3.7	42
Adults 20–49 years	1.1	75	1.3	110

TABLE 11.9

Cumulative Intake (μg/kg bw/day) of Cholinesterase Inhibitors at the 99.9% and 99.99% Percentiles*

		Brazil			
		Children (0–6 years)		Total Population	
	ARfD (μg/kg bw)	99.9% Percentile	99.99% Percentile	99.9% Percentile	99.99% Percentile
Methamidophos	10	8.02	30.7	3.36	13.5
Acephate	50	84.5	359	35.1	134

* The figures in the table are from Caldas et al. (2006)

methomyl and aldicarb. For organophosphates the food that contributed most to the intake for children 3–5 years was snap bean and the pesticides that contributed most methamidophos and phorate.

In a study from Brazil[117] the TEF as well as the probabilistic approach were used. In this study methamidophos and acephate were used as index compounds. The food consumption data used in this study were obtained from the Brazilian Household Budget Survey, 2002–2003. Data were collected from 45,348 households

TABLE 11.10

Cumulative Intakes of Pesticides Found in Denmark

	Exposure		
Correction for Undetected Residues:	**No Correction**	**½ LOR for Non-Detects; Correction Factor Limited to 25**	**½ LOR for Non-Detects**
	(μg/kg bw/day)		
Adults, no reduction for peeling	1.4	2.6	3.0
Children, no reduction for peeling	2.9	5.7	6.6
Adults	0.9	1.9	2.3
Children	2.0	4.5	5.3
	HI		
Adults, no reduction for peeling	7%	23%	49%
Children, no reduction for peeling	14%	56%	124%
Adults	4%	18%	42%
Children	10%	44%	108%

corresponding to 174,378 individuals. Each household recorded the amount of food entered in a diary over seven consecutive days and this was considered as eaten. For each individual, the weekly consumption was decomposed into daily consumption patterns over the seven days. Residue data were obtained from the Brazilian national program on pesticide residues. A total of 4001 samples of tomato, potato, carrot, lettuce, orange, apple, banana, papaya, and strawberry were analyzed for their contents of pesticides. Samples with non-detectable levels were assigned a value of ½ LOQ for the index compound. Processing factors were included. In Table 11.9, the intake at the 99.9% and 99.99% percentiles for both index compounds are shown. Tomato was, independent of IC, definitely the crop that contributed most to the total intake (more than 65%).

Cumulative risk assessment has been conducted for the Danish population for the chronic intake of all pesticides.[53] Residue data were from the Danish monitoring program 2004–2011 while the consumption data were from the Danish nationwide food consumption survey 2003–2008, so consumption and residue data were from the same period. Not all pesticides interact but the HI was calculated as a 'worst-case' situation regarding the cumulative intake. Average values were used for residue levels.

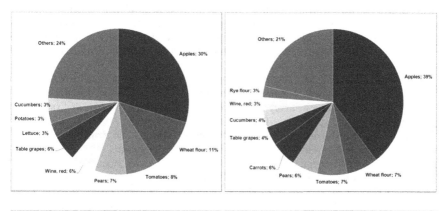

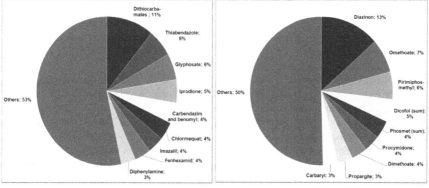

FIGURE 11.4 The nine commodities (two upper diagrams) or nine pesticides (two lower diagrams) that contributed most to the cumulative intake for the Danish population of pesticides. Left: μg/kg bw; Right: HI.

The intake was estimated for LB (non-detects = 0), UB (non-detects = LOR), as well as using an approach with ½ LOR but with correction. Limit of reporting (LOR) is used sometimes within pesticides analysis adaptations and reflects the LOQ that can be achieved at most times in the laboratory. The method of calculation as well as the inclusion of processing factors (peeling) has a great influence on the estimated intake (Table 11.10). The method using ½ LOR with correction was used for further estimations and it has a great influence on the way the intake, in µg/kg or as HI, is estimated considering pesticides or commodities that contributed most to the intake (Figure 11.4).

REFERENCES

1. Codex Alimentarius. Codex pesticides residues in food online database. http://www.fao.org/fao-who-codexalimentarius/codex-texts/dbs/pestres/en
2. European Commission. EU legislation on MRLs. https://ec.europa.eu/food/plant/pesticides/max_residue_levels/eu_rules_en
3. European Parliament, Council of European Union. Regulation (EU) No 609/2013 of the European Parliament and of the Council of 12 June 2013. https://eur-lex.europa.eu/legal-content/EN/ALL/?uri=celex%3A32013R0609
4. EPA. Pesticides. http://www.epa.gov/pesticides/regulating/tolerances.htm
5. Government of Canada. Maximum residue limits for pesticides. https://www.canada.ca/en/health-canada/services/consumer-product-safety/pesticides-pest-management/public/protecting-your-health-environment/pesticides-food/maximum-residue-limits-pesticides.html
6. Australia, Department of Agriculture and Water Resources. National residue survey databases. http://www.agriculture.gov.au/ag-farm-food/food/nrs/databases
7. FRSP. Food Residues Survey Programme. https://www.mpi.govt.nz/food-safety/food-monitoring-and-surveillance/food-residues-survey-program/
8. Food Safety and Standards Authority of India. Food safety and standards regulations. http://www.fssai.gov.in/home/fss-legislation/fss-regulations.html
9. The Japan Food Chemical Research Foundation. https://www.ffcr.or.jp/en/
10. https://www.gov.za/ (search for foodstuffs pesticides).
11. ANVISA. Regularização de produtos - agrotóxicos monografias autorizadas. http://portal.anvisa.gov.br/registros-e-autorizacoes/agrotoxicos/produtos/monografia-de-agrotoxicos/autorizadas
12. New Zealand: Food Safety Authority. Pesticide maximum residue level legislation around the world. https://www.mpi.govt.nz/growing-and-harvesting/plant-products/pesticide-maximum-residue-levels-mrls-for-plant-based-foods/pesticide-maximum-residue-level-legislation-around-the-world/
13. US FDA (US Food and Drug Administration). 2018. Pesticide Residue Monitoring Program Fiscal Year 2015 Pesticide Report, https://www.fda.gov/Food/FoodborneIllnessContaminants/Pesticides/ucm2006797.htm
14. European Food Safety Authority. 2017. The 2015 European Union report on pesticide residues in food. *Journal*, 15(4), 4791. http://www.EFSA.Europa.eu/EFSAjournal
15. European Food Safety Authority. 2018. Monitoring data on pesticide residues in food: results on organic versus conventionally produced food. EFSA Supporting publication 2018, EN-1397. http://www.efsa.europa.eu/publications
16. Monitoring of pesticide residues in Products of Plant Origin in the European Union. Norway, Iceland and Liechtenstein 2004. Report from the European Commission, SEC(2006)1416. http://ec.europa.eu/food/fvo/specialreports/pesticide_residues/report_2004_en.pdf

17. European Food Safety Authority. 2014. The 2012 European Union Report on pesticide residues in food. *EFSA J.*, 12(12), 3942, 156 pp. doi:10.2903/j.efsa.2014.3942

18. Mutengwe, M.T., Chidamba, L., Korsten, L. 2016. Pesticide residue monitoring on south african fresh produce exported over a 6-year period. *J. Food Prot.*, 79, 1759–1766.

19. Jardim, A.N.O., Caldas, E.D. 2012. Brazilian monitoring programs for pesticide residues in food - results from 2001 to 2010. *Food Control*, 25, 607–616.

20. Chang, J.M., Chen, T.H., Fang, T.J. 2005. Pesticide residue monitoring in marketed fresh vegetables and fruits in central Taiwan (1999–2004) and an introduction to the HACCP system. *J. Food Drug Anal.*, 13, 368.

21. Parveen, Z., Riazuddin Iqbal, S., Khuhro, M.I., Bhutto, M.A., Ahmed, M. 2011. Monitoring of multiple pesticide residues in some fruits in Karachi, Pakistan. *Pak. J. Bot.*, 43, 1915–1918.

22. Skrettebjerg, L.G., Lyrån, B., Holen, B., Jansson, A., Fohgelberg, P., Siivinen, K., Andersen, J.H., Jensen, B.H. 2015. Pesticide residues in food of plant origin from Southeast Asia -a Nordic project. *Food Control*, 51, 225–235.

23. Reynolds, S.L., Hill, A.R.C, Thomas, M.R., Hamey, P.Y. Occurrence and risks associated with chlormequat residues in a range of foodstuffs in the UK. *Food Addit. Contam.*, 21, 457, 2004.

24. EU. 2018. The EU pesticide database. http://ec.europa.eu/food/plant/pesticides/eu-pesti cides-database/public/?event=pesticide.residue.CurrentMRL&language=EN&pestR esidueId=51

25. BfR (Bündesinstitut für Risikobewertung). 2018. Database on processing factors. http://www.BfR.bund.de

26. Scholz, R., Herrmann, M., Michalski, B. 2017. Compilation of processing factors and evaluation of quality controlled data of food processing studies. *J. Consum. Prot. Food Saf.*, 12, 3–14.

27. FAO/WHO, Pesticides in Food. 2005. Report of the joint meeting of the FAO panel of Experts on Pesticide Residues in Food and the Environment and the WHO Core Assessment Group. FAO Plant Production and Protection Paper, 183. http://www.who.int/ipcs/publications/jmpr/en

28. Holland, P.T., Hamilton, D., Ohlin, B., Skidmore, M.W. 1994. Effects of storage and processing on pesticide residues on plant products. *Pure Appl. Chem.*, 66, 335.

29. Andersson, A. et al.1998. Beräknat intag av bekämpningsmedel från vissa frukter och grönsager. Report no 7. (Stockholm: National Food Administration).

30. Rasmussen, R.R., Poulsen, M.E., Hansen, H.C.B. 2003. Distribution of multiple residues in apple segments after home processing. *Food Addit. Contam.*, 20, 1044.

31. Chavarri, M.J., Herrera, A., Ariño, A. 2005. The decrease in pesticides in fruit and vegetables during commercial processing. *Int. J. Food Sci. Tech.*, 40, 205.

32. Nitz, S. et al. 1984. Fate of ethylenebis(dithiocarbamates) and their metabolites during the brew process. *J. Agric. Food Chem.*, 32, 600.

33. Ankumah, R.O., Marshall, W.D. 1984. Persistence and fate of ethylenethiourea in tomato sauce and paste. *J. Agric. Food Chem.*, 32, 1194.

34. Cabras, P., Conte, E. 2003. Pesticide residues in grapes and wine in Italy. *Food Addit. Contam.*, 18, 880.

35. Bar-L'Helgouach'h, C. 2004. *Quality Control of Cereals and Pulses.* Arvalis, Paris, ISBN:2.86492.618.0.

36. Babu, G.S., Farooq, M., Ray, R.S., Joshi, P.C., Viswanathan, P.N., Hans, R.K. 2003. DDT and HCH residues in basmati rice (Oryza sativa) cultivated in Dehradun (India). *Water Air Soil Pollut.*, 144, 149.

37. Bakore, N., John, P.J., Bhatnagar, P. 2004. Organochlorine pesticide residues in wheat and drinking water samples from Jaipur, Rajasthan, India. *Environ. Monit. Assess.*, 98, 381.

38. Toteja, G.S., Mukherjee, A., Diwakar, S., Singh, P., Saxena, B.N. 2003. Residues of DDT and HCH pesticides in rice samples from different geographical regions of India: a multicentre study. *Food Addit. Contam.*, 20, 933.

39. Bai, Y., Zhou, L., Li, J. 2006. Organochlorine pesticide (HCH and DDT) residues in dietary products from Shaanxi Province, People's Republic of China. *Bull. Environ. Contam. Toxicol.*, 76, 422.

40. Xu, J. Zhang, J, Dong, F., Liu, X, Zhub, G, Zheng, Y. 2015. A multiresidue analytical method for the detection of seven triazolopyrimidine sulfonamide herbicides in cereals, soybean and soil using the modified QuEChERS method and UHPLC-MS/MS. *Anal. Methods*, 23, 9791–9799.

41. Li, P., Yang, X., Miao, H., Zhao, Y., Liu, W., Wu, Y. 2013. Simultaneous determination of 19 triazine pesticides and degradation products in processed cereal samples from Chinese total diet study by isotope dilution–high performance liquid chromatography–linear ion trap mass spectrometry. *Anal. Chim. Acta*, 781, 63–71.

42. Cho, T.H., Kim, B.S., Jo, S.J., Kang, H.G., Choi, B.Y., Kim, M.Y. 2009. Pesticide residue monitoring in Korean agricultural products, 2003-05. *Food Addit. Contam. Part B Surveill.*, 2, 27–37. https://doi.org/10.1080/02652030902783350)

43. Betsy, A., Vemula, S.R., Sinha, S., Mendu, V.V.R., Polasa, K. 2014. Assessment of dietary intakes of nineteen pesticide residues among five socioeconomic sections of Hyderabad-a total diet study approach. *Environ. Monit. Assess.*, 186(1), 217–228. https://doi.org/10.1007/s10661-013-3367-0

44. Kumari, B.R., Rao, G.V.R., Sahrawat, K.L., Rajasekhar, P. 2012. Occurrence of insecticide residues in selected crops and natural resources. *Bull. Environ. Contam. Toxicol.*, 89(1), 187–192. https://doi.org/10.1007/s00128-012-0660-5

45. Zhou, P. et al. 2012. Dietary exposure to persistent organochlorine pesticides in 2007 Chinese total diet study. *Environ. Int.*, 42(1), 152–159. https://doi.org/10.1016/j.envint.2011.05.018

46. Haque, R., Inaoka, T., Fujimura, M., Ahmad, A.S., Ueno, D. 2017. Intake of DDT and its metabolites through food items among reproductive age women in Bangladesh. *Chemosphere*, 189, 744–751. https://doi.org/10.1016/j.chemosphere.2017.09.041

47. Chaiyarat, R., Sookjam, C., Eiam-Ampai, K., Damrongphol, P. 2015. Organochlorine pesticide levels in the food web in rice paddies of Bueng Boraphet wetland, Thailand. *Environ. Monit. Assess.*, 187(5). https://doi.org/10.1007/s10661-015-4469-7.

48. U.S. FDA. 2012. Pesticide Monitoring Program: Fiscal Year 2012 Pesticide Report. http://www.fda.gov/downloads/Food/FoodborneIllnessContaminants/Pesticides/UCM432758.pdf

49. U.S. FDA. 2013. Pesticide Monitoring Program: Fiscal Year 2013 Pesticide Report. http://www.fda.gov/downloads/Food/FoodborneIllnessContaminants/Pesticides/UCM432758.pdf

50. U.S. FDA. 2014. Pesticide Monitoring Program: Fiscal Year 2014 Pesticide Report. http://www.fda.gov/downloads/Food/FoodborneIllnessContaminants/Pesticides/UCM432758.pdf

51. U.S. FDA. 2015. Pesticide Monitoring Program: Fiscal Year 2015 Pesticide Report. http://www.fda.gov/downloads/Food/FoodborneIllnessContaminants/Pesticides/UCM432758.pdf

52. Juhler, R.K., Lauridsen, M.G., Christensen, M.R., Hilbert, G. 1999. Pesticide residues in selected food commodities: results from the Danish National Pesticide Monitoring Program 1995–1996. *J. AOAC Int.*, 82, 337.

53. Petersen, A. Jensen, B.H.; Andersen, J.H.; Poulsen, M.E.; Christensen, T.; Nielsen, E. 2013. *Pesticide Residues, Results from the Period 2004–2011.* DTU Food, Lyngby, Denmark.

54. Granby, K., Vahl, M. 2001. Investigation of the herbicide glyphosate and the plant growth regulators chlormequat and mepiquat in cereals produced in Denmark. *Food Addit. Contam.*, 18, 898.
55. European Food Safe Authority. 2013. The 2011 European Union report on pesticide residues in food. *EFSA J.*, 12(5). https://doi.org/10.2903/j.efsa.2014.3694
56. European Food Safe Authority. 2015. The 2013 European Union report on pesticide residues in food. *EFSA J.EFSA J.*, 13(3), 4038. https://doi.org/10.2903/j.efsa.2015.4038
57. European Food Safe Authority. 2016. The 2014 European Union report on pesticide residues in food. *EFSA J.*, 14(10), e04611. https://doi.org/10.2903/j.efsa.2016.4611
58. Uygun, U., Koksel, H., Atli, A. 2005. Residue levels of malathion and its metabolites and fenitrothion in post-harvest treated wheat during storage, milling and baking. *Food Chem.*, 92, 643.
59. Lalah, J.O., Wandiga, S.O. 2002. The effect of boiling on the removal of persistent malathion residues from stored grains, *J. Stored Prod. Res.*, 38 1.
60. Van Den Berg, H., Manuweera, G., Konradsen, F. 2017. Global trends in the production and use of DDT for control of malaria and other vector-borne diseases. *Malar J. BioMed. Central*, 16(1), 1–8.
61. Kalantzi, O.I. et al. 2001. The global distribution of PCBs and PCPs in butter. *Environ. Sci. Technol.*, 35, 1013.
62. Fromberg, A., Granby, K., Højgård, A., Fagt, S., Larsen, J.C. 2011. Estimation of dietary intake of PCB and Organochlorine pesticides for children and adults. *Food Chem.*, 125, 1179–1187.
63. Mallatou, M., Pappas, C.P., Kondyli, E., Albanis, TA. 1997. Pesticide residues in milk and cheeses from Greece. *Sci. Total Environ.*, 196, 111.
64. Pandit, G.G. 2002. Persistent OCP residues in milk and dairy products in India. *Food Addit. Contam.*, 19, 153.
65. Kumar, A. Dayal P., Singh G., Prasad F.M., Joseph P.E. 2005. Persistent organochlorine pesticide residues in milk and butter in Agra City, India: a case study, *Bull. Environ. Contam. Toxicol.*, 75, 175.
66. Battu, R.S., Singh, B., Kang, B.K. 2004. Contamination of liquid milk and butter with pesticide residues in the Ludhiana district of Punjab state, India. *Ecotoxicol. Environ. Saf.*, 59, 324.
67. Glynn, A.W., Wernroth, L., Atuma, S., Linder, C.E., Aune, M., Nilsson, I., Darnerud, P.O. 2000. PCB and chlorinated pesticide concentrations in swine and bovine adipose tissue in Sweden 1991–1997: Spatial and temporal trends. *Sci. Total Environ.*, 246, 195.
68. Frenich, A.G., Martinez Vidal, J.L., Cruz Sicilia, A.D., Gonzalez Rodriguez, M.J., Plaza, B. 2006. Multiresidue analysis of organochlorine and organophosphorus pesticides in muscle of chicken, pork and lamb by gas chromatography-triple quadrupole mass spectrometry. *Anal. Chim. Acta*, 558, 42.
69. Naccari, F., Giofrè, F., Licata, P., Martino, D., Calò, M., Parisi, N. 2004. OCPs and PCBs in wild boars from Calabria (Italy). *Environ. Monit. Assess.*, 96, 1991.
70. Fromberg, A. et al. 2005. *Pesticides – Food Monitoring, 1998–2003*. Part 1. 1st Edition. http://www.dfvf.dk/Default.aspx?ID=10875
71. Petersen, A., Fromberg, A., Andersen, J.H., Sloth, J.J., Granby, K., Duedahl-Olesen, L., Rasmussen, P.H., Cederberg, T.L. 2013. *Chemical Contaminants, Food Monitoring 2004–2011*. DTU Food, Lyngby, Denmark.
72. Weiss, J., Päpke, O., Bergman, Å. 2005. A worldwide survey of polychlorinated dibenzo-p-dioxins, dibenzofurans, and related contaminants in butter. *Ambio*, 34, 589.
73. Badia-Vila, M., Ociepa, M., Mateo, R., Guitart, R. 2000. Comparison of residue levels of persistent organochlorine compounds in butter from Spain and from other European countries. *J. Environ. Sci. Health*, B35, 201.

74. Yentür, G., Kalay, A., Öktem, A.B. 2001. A survey on OCP residues in butter and cracked wheat available in Turkish markets. *Nahrung/Food*, 45, 40.

75. Matsumoto, H. Kuwabara K., Murakami Y., Murata H. 2006. Survey of PCB and organochlorine pesticide residues in meats and processed meat products collected in Osaka, Japan. *J. Food Hygienic Soc. Jap.*, 47, 127.

76. Fromberg, A., Cleemann, M., Carlsen, M. 1999. Review of persistent organic pollutants in the environment of Greenland and Faroe Islands. *Chemosphere*, 38, 3075.

77. NIFES (Norwegian Institute of Marine Research). 2018. Seafood database on contaminants and nutrients. https://sjomatdata.nifes.no

78. Sun, F. Wong, S.S., Li, G.C., Chen, S.N. 2006. A preliminary assessment of consumer's exposure to pesticide residues in fisheries products. *Chemosphere*, 62, 674.

79. Chang, G.-R. 2018. Persistent organochlorine pesticides in aquatic environments and fishes in Taiwan and their risk assessment. *Environ. Sci. Pollut. Res.* 25, 7699–7708.

80. Chatterjee, N.S., Banerjee, K., Utture, S., Kamble, N., Rao, B.M., Panda, S.K., Mathew, S. 2016. Assessment of polyaromatic hydrocarbons and pesticide residues in domestic and imported pangasius (*Pangasianodon hypophthalmus*) fish in India. *J. Sci. Food Agric.* 96, 2373–2377.

81. Omwenga, I., Kanja, L., Nguta, J., Mbaria, J., Irungu, P. 2016. Organochlorine pesticide residues in farmed fish in Machakos and Kiambu counties, Kenya. *Cogent Environ. Sci.*, 2, 1153215.

82. Sinkkonen, S., Paasivirta, J. 2000. Polychlorinated organic compounds in the Arctic cod liver: trends and profiles. *Chemosphere*, 40, 619.

83. Wong, M.H., Leung, A.O.W., Chan, J.K.Y., Cho, M.P.K. 2005. A review on the usage of POP pesticides in China, with emphasis on DDT loadings in human milk. *Chemosphere*, 60, 740.

84. Minh, N.H., Someya, M, Minh, T.B., Kunisue, T. 2004. Persistent organochlorine residues in human breast milk from Hanoi and Hochiminh City, Vietnam; contamination, accumulation kinetics and risk assessment for infants. *Environ. Pollut.*, 129, 431.

85. Sundaryanto, A. Kunisue, T., Kajiwara, N., Iwata, H., Adibroto, T.A., Hartono, P., Tanabe, S. 2006. Specific accumulation of organochlorines in human breast milk from Indonesia: levels, distribution, accumulation kinetics and infant health risk. *Environ. Pollut.*, 139, 107.

86. Damgaard, I.N. et al. 2006. Persistent pesticides in human breast milk and cryptorchidism. *Environ. Health Perspec.*, 114, 1133.

87. Schade, G., Heinzow, B. 1998. Organochlorine pesticides and polychlorinated biphenyls in human milk of mothers living in northern Germany: current extent of contamination, time trend from 1986 to 1997 and factors that influence the levels of contamination. *Sci. Total Environ.*, 215, 31.

88. Wong, C.K.C. 2002. Organochlorine hydrocarbons in human breast milk collected in Hong Kong and Guangshou. *Arch. Environ. Contam. Toxicol.*, 43, 364.

89. Chao, H.W. et al. 2006. Levels of OCPs in human milk from central Taiwan. *Chemosphere*, 62, 1774.

90. Johnson-Restrepo, B., Addink, R., Wong, C., Arcaro, K., Kannan, K. 2007, Polybrominated diphenyl ethers and organochlorine pesticides in human breast milk from Massachusetts, USA. *J. Environ. Monit.*, 9(11), 1205–1212.

91. Lu, D., Wang, D., Ni, R., Lin, Y., Feng, C., Xu, Q. et al. 2015. Organochlorine pesticides and their metabolites in human breast milk from Shanghai, China. *Environ. Sci. Pollut. Res.*, 22(12), 9293–9306.

92. Müller, M.H.B. et al. 2017. Organochlorine pesticides (OCPs) and polychlorinated biphenyls (PCBs) in human breast milk and associated health risks to nursing infants in Northern Tanzania. *Environ. Res.* [Internet]. Academic Press; April 1 [cited 2018 June 22], 154, 425–434. https://www.sciencedirect.com/science/article/pii/

93. Chikuni, O., Nhachi, C.F., Polder, A., Bergan, S., Nafstud, I., Skaare, J.U. 2002. Effects of DDT on paracetamol half-life in highly exposed mothers in Zimbabwe. *Toxicol. Lett.*, 134, 147.

94. Fång, J., Nyberg, E., Winnberg, U., Bignert, A., Bergman, Å. 2015. Spatial and temporal trends of the Stockholm Conventional POPs in mothers' milk — a global review. *Environ. Sci. Pollut. Res.*, 22(12), 8989–9041.

95. Gelardi, R.C., Mountford, M.K. 1993. Infant formulas: evidence of the absence of pesticide residues. *Reg. Toxicol. Pharmacol.*, 17, 181–192.

96. Lackmann, G.M., Schaller, K.H., Angerer, J. 2004. Organochlorine compounds in breast-fed vs. bottle-fed infants: preliminary results at six weeks of age. *Sci. Total Environ.*, 329, 289.

97. Kalra, R.L. et al. 2001. Surveillance of DDT and HCH residues in infant formula samples and their implications on dietary exposure: a multicentre study. *Pestic. Res. J.*, 13 2, 147.

98. Poulsen, M.E. et al. 2005. *Pesticides – Food Monitoring, 1998–2003*. Part 2. 1st Edition. http://www.dfvf.dk/Default.aspx?ID=9410

99. U.S. Food and Drug Administration Center for Food Safety and Applied Nutrition Pesticide Program: Residue Monitoring 2000. May 2002. Food and drug administration pesticide program – Residue monitoring 2000. http://www.cfsan.fda.gov/~dms/pes00rep.html#surveys

100. Cressey, P.J., Vannoort, R.W. 2003. Pesticide content of infant formulae and weaning foods available in New Zealand. *Food Addit. Contam.*, 20, 57.

101. Australia New Zealand Food Authority. The 19th Australian total diet survey – A total diet survey of pesticide residue and contaminants. http://www.foodstandards.gov.au/_sr cfiles/19th%20ATDS.pdf

102. European Food Safety Authority. The EFSA comprehensive European food consumption database. https://www.efsa.europa.eu/en/food-consumption/comprehensive-database

103. World Health Organization. Food safety databases. http://www.who.int/foodsafety/databases/en/

104. Food and Agriculture Organizations of the United Nations. Plant Production and Protection. http://www.fao.org/ag/agp/agpp/pesticid/jmpr/Download/2002jmprreport2.doc

105. Pieters, M.N., Ossendorp, B.C., Bakker, MI, Slob, W. 2005. Probabilistic modelling of dietary intake of substances, the risk management question governs the method, RIVM, Report 3200110012005.

106. EPA. 1998. Guidance for submission of probabilistic human health exposure assessments to the office of pesticides program. http://www.epa.gov/fedrgstr/EPA-PEST/199 8/November/Day-05/o-p29665.htm

107. EFSA Panel on Plant Protection Products and their Residues (PPR). 2012. Guidance on the Use of Probabilistic Methodology for Modelling Dietary Exposure to Pesticide Residues. *EFSA J.*, 10(10), 2839, 95 pp. doi:10.2903/j.efsa.2012.2839.

108. Boon, P.E. et al. 2015. Cumulative dietary exposure to a selected group of pesticides of the triazole group in different European countries according to the EFSA guidance on probabilistic modelling. *Food Chem. Toxicol.*, 79, 2015.

109. EFSA Panel on Plant Protection Products and their Residues (PPR). 2014. Scientific opinion on the identification of pesticides to be included in cumulative assessment groups on the basis of their toxicological profile (2014 update). *EFSA J.*, 2013, 11(7), 3293, 131 pp. doi:10.2903/j.efsa.2013.3293

110. Wilkinson, C.F., Christoph, G.R., Julien, E., Kelley, J.M., Kronenberg, J., McCarthy, J., Reiss, R. 2000. Assessing the risks of exposures to multiple chemicals with a common mechanism of toxicity: how to cumulate? *Regul. Toxicol. Pharmacol.*, 31, 30.

111. Reffstrup, T.K., Larsen, J.C., Meyer. O. 2010. Risk Assessment of mixtures of pesticides. Current approaches and future strategies. *Regul. Toxicol. Pharmacol.*, 56, 174.

112. Lozowicka, B., Kaczynski, P., Paritova, C.A., Kuzembekova, G.B., Abzhalieva, A.B., Sarsembayeva, N.B., Alihan, K. 2014. Pesticide residues in grain from Kazakhstan and potential health risk associated with exposure to detected pesticides. *Food Chem. Toxicol.*, 64, 238.

113. Bempah, C.K. et al. 2011. A preliminary assessment of consumer's exposure to organochlorine pesticides in fruit and vegetables and the potential health risk in Accra Metropolis, Ghana. *Food Chem.*, 128, 1058.

114. Battu, R.S., Singh, B., Kang, B.K., Joia, B.S. 2005. Risk assessment through dietary intake of total diet contaminated with pesticide residues in Punjab, India, 1999–2002. *Ecotoxicol. Environ. Saf.*, 62, 132.

115. Nasreddine, L., Rehaime, M., Kassaify, Z., Rechmany, R., Jaber, F. 2016. Dietary exposure to pesticide residues from foods of plants origin and drinks in Lebanon. *Environ. Monit. Assess.*, 188, 485.

116. EPA. Cumulative Assessing of Risk from Pesticides. https://www.epa.gov/pesticide-science-and-assessing-pesticide-risks/cumulative-assessment-risk-pesticides

117. Caldas, E.D., Boon, P.E., Tressou, J. 2006. Probabilistic assessment of the cumulative acute exposure to organophosphorus and carbamates insecticides in the Brazilian diet. *Toxicology*, 222, 132.

12 Monitoring and Assessment of Pesticides and Transformation Products in the Environment

Ioannis Konstantinou, Dimitra Hela, Dimitra Lambropoulou, and Triantafyllos Albanis

CONTENTS

12.1 Introduction 366
12.2 Monitoring programs...366
    12.2.1 Purpose and Design of Pesticide Monitoring Programs 367
    12.2.2 Selection of Pesticides for Monitoring...369
    12.2.3 Types of Monitoring 370
        12.2.3.1 Air Monitoring..370
        12.2.3.2 Water Monitoring...377
        12.2.3.3 Soil and Sediment Monitoring....................................384
        12.2.3.4 Biological Monitoring 387
    12.2.4 Water Framework Directive and Monitoring Strategies.................389
12.3 Environmental Exposure and Risk Assessment 390
    12.3.1 Environmental Exposure 390
        12.3.1.1 Point and Non-Point Source Pesticide Pollution...............390
        12.3.1.2 Environmental Parameters Affecting Exposure...............391
        12.3.1.3 Pesticide Parameters Affecting Exposure 391
        12.3.1.4 Modeling of Environmental Exposure............................392
    12.3.2 Risk Assessment 393
        12.3.2.1 Preliminary Risk Assessment – Pesticide Risk
                 Indicators (PRIs) – Classification Systems 395
        12.3.2.2 Risk Quotient – Toxicity Exposure Ratio Method
                 (Deterministic – Tier 1) 395
    12.3.3 Probabilistic Risk Assessment (PRA) (Tier 2) 400
12.4 Limitations and Future Trends of Monitoring and Ecological Risk
    Assessment for Pesticides 405
References...406

12.1 INTRODUCTION

Worldwide pesticide usage has increased dramatically during the last decades coinciding with changes in farming practices and increasingly intensive agriculture. This widespread use of pesticides for agricultural and non-agricultural purposes has resulted in the presence of their residues in various environmental matrices. Numerous studies have highlighted the occurrence and transport of pesticides and their metabolites in rivers [1], channels [2], lakes [1,3], seas [4], air [5,6], soils [7,8], sediments [9], ground water [10,11], and even drinking water [12,13] proving the high risk of these chemicals to human health and environment.

In recent years, the growing awareness of the risks related to the intensive use of pesticides has led to a more critical attitude from society toward the use of agrochemicals. At the same time, many national environmental agencies have been involved in the development of regulations to eliminate or severely restrict the use and production of a number of pesticides (Directive 91/414/EEC, Regulation EC 33/2008, Regulation EU 540/2011) [14–16]. Despite these actions, pesticides continue to be present causing adverse effects on man and the environment. Monitoring of pesticides in different environmental compartments has proven a useful tool in order to quantify the amount of pesticides entering the environment and to monitor ambient levels for trends and potential problems, and different countries have undertaken, or are currently undertaking, campaigns with various degrees of intensity and success [17]. Although numerous local and national monitoring studies have been performed around the world providing nationwide patterns on pesticide occurrence and distribution, there are still several gaps. For example, only limited retrospective monitoring data are available in all compartments and there is a lack of monitoring data for many pesticides both in space and time [4,18]. In addition, there is little consistency in the majority of these studies in terms of site selection strategy, sampling methodologies, collection time and duration, selected analytes, analytical methods, and detection limits [17]. Therefore, dedicated efforts are needed for comprehensive monitoring schemes not only for pesticide screening but also for the establishment of cause–effect relationships between the concentration of pesticides and the damage, and to assess the environmental risk in all compartments.

12.2 MONITORING PROGRAMS

Environmental monitoring programs are essential to develop extensive descriptions of current concentrations, spatiotemporal trends, emissions and flows, to control the compliance with standards and quality objectives, and to provide early warning detection of pollution. Furthermore, environmental monitoring provides a viable basis for efficacious measures, strategies, and policies to deal with environmental problems at a local, regional, or global scale. Similar terms often used are "surveys" and "surveillance". A survey is a sampling program of limited duration for specific pesticides such as an intensive field study or exploratory campaign. Surveillance is a more continuous specific study with the aim of environmental quality reporting (compliance with standards and quality objectives) and/or operational activity reporting (e.g., early warning and detection of pollution).

12.2.1 Purpose and Design of Pesticide Monitoring Programs

In general pesticide monitoring is used to investigate and to gain knowledge that allows authorities to tentatively assess the quality of the environment, to recognize threats posed by these pollutants, and to assess whether earlier measures have been effective [17,19]. Whichever the objectives of a monitoring program may be, it is important that they are well-defined before sampling takes place in order to select suitable sampling and analysis methods and to plan the project adequately. Another important characteristic of a monitoring program is that data produced are often used to implement and to regulate existing directives concerning pesticides in the environment [4].

Because of the great number of parameters (pesticide physicochemical properties, and climatic and environmental factors) affecting the exposure of pesticides, monitoring of a single medium will not provide sufficient information about the occurrence of pesticides in the environment. A multimedia approach that involves tracking pesticides from sources through multiple environmental media such as air, water, sediment, soil, and biota provides data for understanding the fate and partitioning of pesticides and for the validation of environmental models [18].

A basic problem in the design of a pesticide monitoring program is that each of the above reasons for carrying out monitoring demands different answers to a number of questions. Thus, when a monitoring program consists of sampling, laboratory analysis, data handling, data analysis, reporting, and information exploitation, its design will necessarily have to include a wide range of scientific and management concepts, thus making it a large and difficult task [19]. Therefore, cost-effective monitoring programs should be based on clear and well thought out aims and objectives and should ensure, as far as possible, that the planned monitoring activities are practicable and that the objectives of the program will be met. There are a number of practical considerations to be dealt with, when designing a monitoring program, that are generic, regardless of the compartment being monitored (Figure 12.1).

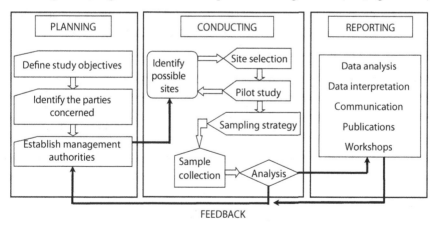

FIGURE 12.1 Phases in planning, conducting, and reporting of a monitoring program. (Modified from Calamari D., et al. Monitoring as an indicator of persistence and long-range transport. In Klecka G. et al., Eds. Evaluation of persistence and long range transport of organic chemicals in the environment, SETAC, 2000).

For pesticide monitoring programs, some general guidelines should be taken into consideration including the clear statement of the objectives, the complete description of the area, as well as the locations and frequency of sampling, and the number of the samples. The geographical limits of the area, the present and planned water or land uses, and the present and expected pesticide pollution sources should be identified. Background information of this type is of great help in planning a representative monitoring program covering all the sources of the spatial and temporal variability of the pesticide environmental concentration. Appropriate statistical analysis can be applied to determine probability distributions that may be used to select locations for further sampling programs and for risk assessment. The fieldwork associated with the collection and transportation of samples will also account for a substantial section of the plan of a monitoring program. The development of meaningful sampling protocols has to be planned carefully taking into account the actual procedures used in sample collection, handling, and transfer. The design of a sampling should target the representativeness of the samples, which is related to the number of samples and the selection of sampling stations intended within the objectives of the study. The sampling process of taking random grab samples and individually analyzing each sample is very common in environmental monitoring programs and it is the optimal plan when a measurement is needed for every sample. However, the process of combining separate samples and analyzing this pooled sample is sometimes beneficial. Such a composite sampling process is generally used under flow conditions and in situations where concentrations vary over time (surface water or air sampling), when samples taken from varying locations as well as when the representativeness of samples taken from a single site needs to be improved by reducing inter-sample variance effects. Composite sampling is also used to increase the amount of material available for analysis, the sensitivity, as well as to reduce the cost of analysis. However, certain limitations must be taken into account and it should be used only when the researcher fully understands all aspects of the plan of choice [17].

Apart from sampling, the selection and the performance of the analytical method used for the determination of pesticides is a very critical subject. Earlier chapters of this book discuss the various methods that can be successfully applied to monitor pesticides in various environmental compartments. Another point that should be considered in the planning stage concerns the quality assurance/control (QA/QC) procedures in order to produce reliable and reproducible data. These quality issues relate to the technical aspects of both sampling and analysis. The quality of the data generated from any monitoring program is defined by two key factors: the integrity of the sample and the limitations of the analytical methodology. The QA/QC procedures should be designed to establish intra-laboratory controls of sample collection and preparation, instrument operation and data analysis and should be subjected to "Good Analytical Practices". Laboratories should participate in a series of inter-calibration exercises and chemical analysis cross-validations in order to avoid false positives [18,20].

As already mentioned, the whole planning of a monitoring program is aimed at the generation of reliable data but it is acknowledged that simply generating good data is not enough to meet monitoring objectives. The data must proceed and be presented in a manner that aids understanding of the spatial and temporal patterns, taking into consideration the characteristics of the study areas, and must allow the human impact to be understood and the consequences of management action to be predicted. Thus, different statistical approaches are usually applied to designing, adjusting, and quantifying the informational value of monitoring data. However, because data are often collected at multiple locations and time points, correlation among some, if not all, observations is inevitable, making many of the statistical methods taught to be applied. Thus, in the last 15 years geographic information systems (GIS) and computer graphics have been used that have enhanced the ability to visualize patterns in data collected in time and space, and these have been gaining considerable attention [21]. In summary, statistical methods, including chemometric methods, coupled to GISs are preferably used to display the most significant patterns in pesticide pollution [17].

Finally, one of the major parameters of the monitoring plan should be the cost of the program. A cost estimate should be prepared for the entire program, including laboratory and field activities. The major cost elements of the monitoring program include the following: personnel cost; laboratory analysis cost; monitoring equipment costs; miscellaneous equipment costs; data analysis; and reporting costs.

As a conclusion based on the above arguments, monitoring activities must imply a long-term commitment and can be summarized as follows [17,18]: (a) establishment of monitoring stations for different environmental compartments to fill spatiotemporal data; (b) intensive monitoring over wider areas, and continuation of existing time trend series; (c) establishment of standardized sampling and analytical methods; (d) follow-up of improved quality assurance/quality control protocols; (e) adequate reporting of the results in the more meaningful manner; and (f) estimation of the monitoring program cost.

12.2.2 SELECTION OF PESTICIDES FOR MONITORING

The number and nature of pesticides monitored depends upon the objectives of the monitoring study. Some studies have concentrated on a limited number of target pesticides while others have performed a broad screening of different compounds. Research is usually focused on the most commonly used pesticides either in the agricultural area around the studied sites or in the country concerned. The selection of pesticides for monitoring is also based on pesticide properties (e.g., toxicity, persistence, and input), the cost, as well as on special directives and regulations [22].

The diversity of aims and objectives for the various monitoring programs has resulted in a variety of active ingredients and metabolites monitored in the studies performed.

For instance, until the beginning of the 1990s, halogenated, nonpolar pesticides were the focus of interest. As the environmental fate of hydrophobic pesticides became more generally understood, and new, more environmentally friendly, pesticide products were introduced in the market there has been an increase in monitoring studies that focused on current-use pesticides known to be present in the environment. Whereas environmental concentrations of halogenated, nonpolar pesticides have generally declined during the past 20 years, and whereas current concentrations in surface water are below the drinking water standards, concerns nevertheless remain, because these substances persist in the environment and accumulate in the food chain, and thus they continue to be included in the list for investigation. Current screening strategies have also included pesticides with endocrine disruption action due to their newly discovered ecotoxicological problems on human health and environment. Among the most studied chemical classes of pesticides are the s-triazines, acetamides, substituted ureas, and phenoxyacids from the group of herbicides and organophosphorus and carbamates from the group of insecticides. Currently modern fungicides have gain attention since their uses have been increased and new compounds have been introduced in the market.

Although that all new compounds or new uses of existing pesticides are carefully scrutinized, the list of pesticide of interest for monitoring programs is not getting shorter and there is a continuing need for development of new criteria that allow the prediction of which pesticides could be of concern for monitoring.

12.2.3   TYPES OF MONITORING

Pesticides can occur in all compartments of the environment or in other words in any or all of the solid, liquid, or gaseous phases. The environment is not a simple system and consequently pesticide monitoring should be carrying out in a specific phase (e.g., volatile pesticides in air) or may encompass two or more phases and/or media (e.g., water and sediment in the marine environment). Primary environmental matrices that are usually sampled for pesticide investigations include water, soil, sediment, biota, and air. However, each of these primary matrices includes many different kinds of samples. A brief description of each type of monitoring is given in the next paragraphs.

12.2.3.1   Air Monitoring

Historically, water contamination has garnered the lion's share of public attention regarding the ultimate fate of pesticides. In contrast, atmospheric monitoring is less expanded since the atmospheric residence time of a pesticide is very variable. However, in last 15 years, air quality has become a very important concern as more and more studies have shown the great impact of atmospheric pesticide pollution on environment and health and as non-agricultural pesticide usage in urban environments is increased. Pesticides can be potential air pollutants that can be carried by wind, and deposited through wet or dry deposition processes. They can revolatilize repeatedly and, depending on their persistence in the environment, can travel tens, hundreds, or thousands of kilometers [23]. For example, currently used organochlorine pesticides (OCPs) like endosulfans and lindane have been detected in Arctic samples [24] where, of course, they have never been used.

During spray application of pesticides, the most common agronomic practice for pesticide usage, a substantial fraction of the dosage applied to the target area can be transported short distances and deposited onto adjacent non-target fields, while another fraction can enter the atmosphere (spray drift) and be transported over varying distances. Spray drift depends on pesticide formulation characteristics, such as volatility, adjuvant content and viscosity, the application equipment (type of sprayer, spray nozzles) and technique, the climatological conditions (e.g., temperature, wind speed and direction, relative humidity), and operator skills and environmental awareness [25]. Pesticide losses into the air via spray drift and volatilization during the application can range from a few percent to 20 to 30%. Emission of pesticides after application involves volatilization from soil and plant surfaces and wind erosion of soil particles containing sorbed pesticides. The dominant factors affecting volatilization from soil are physicochemical properties of pesticidal compounds (vapor pressure, solubility, adsorption coefficient, persistence), soil properties (texture, organic matter content, density, pH, temperature, humidity), meteorological conditions (air temperature, wind, humidity), and agricultural practices (application rate, formulation type) [26]. The vapor pressure (V_p) is the most common predictor of pesticide losses from soil and crops through volatilization. Compounds with $V_p > 10^{-4}$ Pa (20°C) for soil applications, and $V_p > 10^{-5}$ Pa (20°C) for crop applications, have been considered as potential air pollutants. Volatilization from soil surfaces could be higher than 20% of the applied dose, and the major fraction of losses take place within a few days of application [27]. Cumulative losses from drift and volatilization could be as high as 50% of the initial applied dose [26].

After emission in the atmosphere, pesticides are partitioned between gas (G) and particle (P) phases depending on their physicochemical properties (e.g., vapor pressure, partition coefficients, water solubility), environmental factors (e.g., temperature, concentration, and content of suspended particles).

After emission and distribution of pesticides in the atmosphere, they can undergo different processes such as transport, degradation, or deposition. The dispersion process usually takes place much faster (i.e., minutes to hours depending on the wind speed) than transformation and deposition processes. Transformation processes involve mainly photolysis and oxidation reactions initiated by $HO^{\bullet}$ and $NO_3^{\bullet}$ radicals attack and ozone or other oxidant species [28]. Deposition processes are classified into wet and dry deposition. In general, compounds adsorbed to particulate phase are mostly found in wet deposition while compounds in gas phase are mostly divided between wet and dry deposition.

The design of monitoring networks for air pollution has been treated in several different ways. For example, monitoring sites may be located in areas of the severest public health effects, which involves consideration of pesticide concentration, exposure time, population density, and age distribution. Alternatively, the frequency of occurrence of specific meteorological conditions and the strength of sources may be used to maximize monitor coverage of a region with limited sources.

Air concentrations of pesticides may vary over the scales of hours, days, and seasons since they respond to air mass direction and depositional events. The sampling methods of pesticides in air may be divided into active (pump or vacuum-assisted sampling) or passive techniques (passive by diffusion gravity or other unassisted means). The sampling interval may be integrated over time or it may be continuous, sequential, or instantaneous (grab sampling). Measurements obtained from grab sampling give only an indication of concentration at the time of sampling. However, they can be useful for screening purposes and provide preliminary data needed for planning subsequent monitoring strategies. Probably, the collection of pesticides by using passive air samplers (PAS) is the most common sampling method for air samples. PAS continuously integrate the air burden of pesticides and give real-time or near real-time assessment of the concentration of pesticide in air [5,29].

12.2.3.1.1　Occurrence and Pesticide Levels in Air Monitoring Studies

Numerous investigations around the world consistently find pesticides in air, wet precipitation, and even fog. Monitoring programs have been established in many countries for the study of the spatial and temporal distribution of persistent OCPs such as DDTs, HCHs, and cyclodienes [18]. While many of the newer, current-use pesticides are less persistent than their predecessors, they also contaminate the air and can travel many miles from target areas. The monitoring of persistent OCPs has been a matter of numerous past studies while the monitoring data for currently used or banned pesticides have been increased during the last 10–15 years. An overview of pesticides concentrations in air samples around the world is given in Tables 12.1 and 12.2.

Regarding herbicides, in general, *s*-triazines (i.e., atrazine, simazine, terbuthylazine, etc.), acetanilides (i.e., alachlor, metolachlor, acetochlor, etc.), substituted carboxylic acids and esters (mecoprop, fluazifop, etc.), triazinones (metribuzin), and dinitroanilines (i.e., pendimethalin, trifluralin) are among the most frequently looked for and detected in air (Table 12.1) and precipitation [52]. As far as modern and lately banned insecticides, organophosphorus compounds (malathion, diazinon, dichlorvos, and chlorpyrifos), carbamates (e.g., carbofuran, pirimicarb, carbaryl, etc.), and pyrethroids (e.g., cypermetrhin, permethrin, delthamethrin, etc.) have been looked for most often. Finally, azoles (tebuconazole, cyproconazole, triadimefon, etc.), phthalimides (e.g., folpet, etc.), and amides-anilides (e.g., folpet, prochloraz, fenhexamid, etc.). The occurrence of other groups of pesticides in air and rain has generally been sporadically investigated [53]. Commonly detected concentration levels of pesticides in air range from a few (1–10) pg m^3 to a few (1–15) ng m^3 (Tables 12.1–12.2). In rain, concentrations have been measured from a few ng/L to several μg L^{-1}. Current-use fungicides and pyrethroids are new classes of pesticides that are frequently determined in rainwater [54]. The 95th percentile concentrations across Nebraska's rainwater basin wetlands were found to be 0.07 mg L^{-1} for bifenthrin, 0.28 mg L^{-1} for pyraclostrobin, 0.28 mg L^{-1} for azoxystrobin, and < 0.14 mg L^{-1} for other analytes such as propiconazole, trifloxystrobin, and metconazole [54]. However, concentrations in precipitation depended not only on the amount of pesticides present in the atmosphere, but also on the amount, intensity, and timing of rainfall [53]. Concentrations in fog are even higher. Deposition levels are in the order of several mg/ha/yr to a few g/ha/yr [6].

In general, air monitoring studies have been conducted on an *ad hoc* basis and are characterized by a relatively small number of sampling sites, covering limited geographical areas and time periods. In the US and Canada [6], however, some large nationwide studies have been conducted. In contrast, most European monitoring studies have been focused on regional air pollution. So far, over 100 pesticides have been detected in precipitation in Europe and 50 in air [53, Tables 12.1–12.2]. However, the lack of consistency in sampling and analytical methodologies holds for both US and European studies.

As an example of characteristic pesticide monitoring programs in air and rainwater, the Canadian Atmospheric Network for Current Use Pesticides (CANCUP, 2003) can be mentioned [55]. Other examples are the Air Monitoring Network (AMN),

TABLE 12.1

Representative Concentrations of Herbicides in Air Samples

Compound	Chemical Class	Regulatory EU Status	VP^a (mPa)	Average	Range	Site/Country	Phase	Reference
				Concentration (ng m^3)				
Acetochlor	Chloroacetamide	Out	0.022	4.6	ND–53.5	Rural/Iowa	G	[30]
Alachlor	Chloroacetamide	Out	2.9	1.32	0.30–8.87	Rural-urban/France	G + P	[31]
				1.1	ND–8.5	Iowa/USA	G	[30]
				0.468	0.125–0.9	Rural/Canada	G + P	[32]
				0.64	0.12–6.03	Rural-urban/ France	G + P	[31]
Atrazine	Triazine	Out	0.039	0.294	ND–1.4	Remote/France	G + P	[33]
				0.3	ND–0.688	Urban/France	G + P	[34]
				1.121	0.645–1.905	Rural/Canada	G + P	[35]
Butachlor	Chloroacetamide	Out	0.24	0.09	ND–0.15	Rural/USA	G	[30]
DEA/DIA	Metabolites	–	–	0.51	ND–1.3	Rural/USA	G	[30]
Desmedipham	Carbamate	In	4.0E-05	0.15	0.03–0.39	Urban/France	G + P	[36]
Diuron	Urea	In	1.15E-03	0.524	ND–1.6	Remote/France	G + P	[33]
				3.214	ND–12.8	Rural/France	G + P	[34]
Ethofumesate	Benzofuran	In	0.36	0.92	0.54–1.16	Urban-rural/France	G + P	[31]
Fluazifop-P-butyl	Phenoxy propionate	In	0.12	0.04	ND–0.07	Rural/Canada	G + P	[37]
Isoproturon	Urea	In	5.5E-03	0.861	ND–3.3	Rural/France	G + P	[33]
Metazachlor	Chloroacetamide	In	0.093	0.93	0.17–3.13	Urban-rural/France	G + P	[31]
Metribuzin	Triazinone	In	0.121	0.79	ND–0.96	Rural/Canada	G + P	[37]
				0.050		Urban/Canada	G	[38]
Metsulfuron	Triazinylsulfonylurea	In	4E-08		ND–11	Urban/France	G + P	[39]
Pendimethalin	Dinitroaniline	In	3.34	3.18	0.32–7.83	Urban/France	G + P	[36]

(Continued)

TABLE 12.1 (CONTINUED)
Representative Concentrations of Herbicides in Air Samples

Compound	Chemical Class	Regulatory EU Status	VP^a (mPa)	Concentration (ng m^3) Average	Range	Site/Country	Phase	Reference
				0.480	ND–3.309	Rural/Spain	G+P	[40]
				0.410	ND–1.5	Urban-rural/Italy	G+P	[41]
Propyzamide	Benzamide	In	0.058	0.118	ND–3.208	Rural/Spain	G+P	[40]
Terbuthylazine	Triazine	In	0.12	0.0419	0.007–0.946	Urban-rural/Spain	P	[42]
				3.66	ND–0.202	Rural/Spain	G+P	[40]
Trifluralin	Dinitroaniline	Out	9.5	1.93	0.4–7.98	Remote/France	G + P	[33]
					0.12–58.79	Urban-rural/France	G + P	[31]
					ND–0.03	Urban-rural/Italy	G+P	[40]

^{a} Vapor pressure, data extracted from Pesticides Properties Database (PPDB), https://sitem.herts.ac.uk/aeru/ppdb/en/ [43].

TABLE 12.2

Representative Concentration Levels of Insecticides and Fungicides in Air Samples

Compound	Chemical Class	Regulatory EU Status	VP^a (mPa)	Concentration (ng m^3) Average	Range	Site/Country	Phase	Reference
Azinphos-methyl	Organophosphorus	Out	5.00E-04	14.5	ND–16.5	Rural/Canada	G + P	[37]
Bifenthrin	Pyrethroid	In	0.0178	0.15	23–49	Urban/Hawaii	G+P	[44]
Carbofuran	Carbamate	Out	0.08	2.85	ND–8.1	Rural/France	G + P	[33]
Chlorpyrifos	Organophosphorus	In	1.43	0.391	0.038–1.068	Rural/Turkey	G + P	[45]
					0.0727–2.901	Urban/China	G + P	[46]
					ND–14.6	Urban/Chile	G+P	[47]
Chlorpyrifos oxon	Metabolite	–	–		0.196–0.189	Rural/Canada	G + P	[48]
α-Cypermethrin	Pyrethroid	In	6.78E-03	0.45	0.11–1.02	Urban/France	G + P	[36]
Deltamethrin	Pyrethroid	In	1.24E-05	27.41	5.80–79.00	Urban/France	G + P	[36]
Diazinon	Organophosphorus	Out	11.97	6.7	ND–59.1	Rural/USA	G	[30]
				1.128	ND–1.171	Rural/Canada	G + P	[35]
Diazinon oxon	Metabolite	–	–		ND–0.696	Rural/Canada	G + P	[35]
Dichlorvos	Organophosphorus	Out	2100	0.84	ND–2.3	Rural/USA	G	[30]
Dimethoate	Organophosphorus	In	0.247	0.045	20–70	Italy	G	[49]
Malathion	Organophosphorus	In	3.1		ND–0.280	Urban-rural/Italy	G + P	[41]
Omethoate	Metabolite	Out	19		10–30	Italy	G	[49]
Imidacloprid	Neonicotinoid	In	4E-07		0.012–0.014	Urban-rural/Spain	P	[42]
Permethrin	Pyrethroid	Out	0.007		ND–0.47	Urban-rural/France	G + P	[50]
Boscalid	Carboxamide	In	0.00072	0.53	0.35–0.81	Urban/France	G + P	[36]
Chlorothalonil	Chloronitrile	In	0.076	12.15	0.11–107.93	Urban-rural/France	G + P	[31]
Cyproconazole	Triazole	In	0.026	11.08	1.47–20.50	Urban/France	G + P	[36]
Cyprodinil	Anilinopyrimidine	In	0.51	0.55	0.12–3.29	Urban-rural/France	G + P	[31]

(Continued)

TABLE 12.2 (CONTINUED)
Representative Concentration Levels of Insecticides and Fungicides in Air Samples

Compound	Chemical Class	Regulatory EU Status	VP[a] (mPa)	Concentration (ng m^3) Average	Concentration (ng m^3) Range	Site/Country	Phase	Reference
Imazalil	Imidazole	In	0.158	0.105	0.037–0.215	Urban–rural/Spain	P	[42]
Myclobutanil	Triazole	In	0.198	5.78	1.1–8.7	Urban	G + P	[51]
Penconazole	Triazole	In	0.366	5.78	1.22–19.41	Urban/France	G + P	[36]
Tebuconazole	Triazole	In	1.3E-03	1.49	0.23–5.03	Urban/France	G + P	[36]
Tolylfluanid	Sulphamide	Out	0.2	6.92	0.10–86.42	Urban–rural/France	G + P	[31]
Trifloxystrobin	Strobilurin	In	3.40E-03	2.62	0.49–4.58	Urban/France	G + P	[36]

[a] Vapor pressure, data extracted from Pesticides Properties Database (PPDB), https://sitem.herts.ac.uk/aeru/ppdb/en/ [43].

which has been implemented since 2011 by the California Department of Pesticide Regulation in the United States [56]; Lig'Air-Association de surveillance de la qualité de l'air en région Centre-Val de Loire (France) [57]; and the Flemish Environmental Agency (FEA) for the monitoring of pesticides in rainwater (Flanders, Belgium, [58]).

12.2.3.2 Water Monitoring

The principal reason for monitoring water quality has been, traditionally, the need to verify whether the observed water quality is suitable for intended uses. However, monitoring has also evolved to determine trends in the quality of the aquatic environment and how the environment is affected by the release of pesticides and/or by waste treatment operations. Currently, spot (bottle or grab) sampling, also called active sampling, is the most commonly used method for aquatic monitoring of pesticides. With this approach, no special water sampling system is required and water samples are usually collected in precleaned amber glass containers. Although spot sampling is useful there are drawbacks to this approach in environments where contaminant concentrations vary over time, and episodic pollution events can be missed. Moreover, it requires a relatively large number of samples to be taken from any one location over the entire duration of sampling and therefore is time-consuming and can be very expensive. In order to provide a more representative picture and to overcome some of these difficulties, either automatic sequential sampling to provide composite samples over a period of time (e.g., 24 hours), or frequent sampling can be used. However, the former involves the use of equipment that requires a power supply, and needs to be deployed in a secure site, and the latter would be expensive because of transport and labor costs.

In the last two decades, an extensive range of alternative methods that yield information on environmental concentrations of pesticides have been developed. Of these, passive sampling methods, which involve the measurement of the concentration of an analyte as a weighted function of the time of sampling, avoid many of the problems outlined above, since they collect the target analyte *in situ* and without affecting the bulk solution. Passive sampling is less sensitive to accidental extreme variations of the pesticide concentration thus giving more adequate information for long-term monitoring of aqueous systems. Comprehensive reviews on the use of equilibrium passive sampling methods in aquatic monitoring as well on the currently available passive sampling devices has been recently published [59,60]. Despite the well-established advantages, passive sampling has some limitations such as the effect of environmental conditions (for example temperature, air humidity, and air and water movement) on analyte uptake. Despite such concerns, many users find passive sampling an attractive alternative to more established sampling procedures. To gain more general appeal, however, broader regulatory acceptance would probably be required.

Other technologies available for water sampling include continuous, on-line monitoring systems. In such installations water is continuously drawn from water input and automatically fed into an analytical instrument (i.e., LC–MS). These systems provide extensive, valuable information on levels of pesticides over time, however they require a secure site, they are expensive to install, and they have a significant maintenance cost [60].

Finally, another approach available and already in use for monitoring water quality are sensors. A wide range of sensors for use in pesticide monitoring of water has been developed in recent years, and some are commercially available. These are based on electrochemical or electroanalytical technologies and many are available as miniaturized screen-printed electrodes [61]. They can be used as field instruments for spot measurements, or can be incorporated into on-line monitoring systems. However, some of these methods do not provide high sensitivity, and in some cases specificity; they can be affected by the matrix and by environmental conditions. Thus, it is necessary to define closely the conditions of use [62].

12.2.3.2.1 Occurrence and Pesticide Levels in Water Samples

The majority of the pesticide monitoring effort goes into monitoring surface freshwaters (including rivers, lakes, and reservoirs) and monitoring programs for pesticides in marine waters and groundwaters have received less attention. Within Europe, the contamination of surface waters by pesticides follows comparable concentration levels and patterns as recorded in most countries (Table 12.3). Among the most commonly encountered herbicide compounds in European freshwaters are s-triazines (e.g., atrazine, simazine, terbuthylazine), chloroacetamides (e.g., metolachlor, alachlor), and glyphosate and its metabolite AMPA. s-Triazine herbicides are widely applied herbicides in Europe for pre- and post-emergence weed control among various crops as well as for non-agricultural purposes. In some studies acetamide herbicides alachlor and metolachlor (which are also used to control grasses and weeds in a broad range of crops) were also detected at levels comparable to those of the triazines. Concerning insecticide concentrations in European freshwaters, mainly organophosphates and organochlorine insecticides have been detected in the past while neonicotinoids and pyrethroids have been encountered in the last number of years. Diazinon, malathion, chlorpyrifos, and chlothianidin were the most frequently detected compounds [1, Table 12.3]. OCPs continue to be present in freshwaters, but at low levels due to their high hydrophobicity. Among them lindane was the most frequently detected compound. Other OCPs detected include α-endosulfan and aldrin. An extended review on the levels of pyrethroids in surface waters worldwide [70] revealed concentration levels in the range of ng L^{-1} to hundreds of µg L^{-1} with 29.72 and 470 µg L^{-1} the maximum concentrations in rivers and constructed wetland waters, respectively. Regarding fungicides, they constitute a significant class of pesticidal pollutants in recent years with azoles (e.g., tebuconazole, cyproconazole, etc.) and carboxamides (e.g., boscalid) being among the most frequently detected compounds.

It is underlined that water monitoring studies around the world have routinely focused on tracing parent compounds rather than their metabolites. Thus, little data are available on the occurrence of pesticide transformation products in freshwaters, including mainly transformation products of high-use herbicides, such as acetamide and triazine compounds. For example, desethylatrazine, a metabolite of atrazine, has been detected in rivers of both the United States and Europe [63,64,66,71].

Agricultural uses result in distinct seasonal patterns in the occurrence of a number of compounds, particularly herbicides and fungicides, in freshwaters. As regards rivers, critical factors for the time elapse between the period of pesticide application

TABLE 12.3

Representative Concentration Levels of Selected Past and Current-Use Pesticides in Surface Waters Worldwide during the Last Ten Years

Compound	Chemical Class	Koc^a (mL g^{-1})	DT50^a (field)	DT50 (water)	Max. Concentration (μg/L) (% Freq.)	Country/Type	Reference
Alachlor	Chloroacetamide	124	14	21	0.807 (50.5)	Greece/Lake Amvrakia	[63]
Atrazine	Triazine	100	60	47–193	0.328 (88.7)	Greece/Lake Amvrakia	[63]
					0.0122 (20)	Spain/ River Ebro	[64]
					0.0198 (5–51)	Spain/Turia, Júcar Rivers	[64]
Boscalid	Carboxamide	–	118	–	0.109 (75)	USA/Surface waters	[65]
Chlorpyrifos	Organophosphate	8151	50	5	0.101 (31.8)	Greece/River Acheloos	[66]
					0.0412 (72–100)	Spain/Turia, Júcar Rivers	[64]
Clothianidin	Neonicotinoid	123	121.3	40.3	0.399 (0–100)	Canada/Ontario surface	[67]
Cyproconazole	Azole	309	129	300	0.724 (31.8)	Greece/River Acheloos	[66]
DEA	Metabolite	16	36	–	0.559 (85.6)	Greece/Lake Amvrakia	[63]
					0.0588 (29)	Spain/ River Ebro	[64]
DIA	Metabolite	–	–	–	0.0135 (16)	Spain/ River Ebro	[64]
Diazinon	Organophosphate	643	18.4	4.3	0.0786 (70.3)	Greece/River Acheloos	[66]
					0.0370 (60–100)	Spain/Turia, Júcar Rivers	[64]
Diuron	Substituted urea	813	89	8.8	0.0245 (12)	Spain/ River Ebro	[64]
Glyphosate	Phosphonoglycine	1424	23.79	9.9	2.8	Italy/Surface waters	[68]
					27.8 (44)	USA/Midwestern streams	[69]
Imazalil	Azole	–	6.4	7.8	0.683 (45–93)	Spain/Turia, Júcar Rivers	[64]
Imidacloprid	Neonicotinoid	–	174	30	10.4	Canada/Ontario surface	[67]
					0.207 (13–82)	Spain/Turia, Júcar Rivers	[64]
s-Metolachlor	Chloroacetamide	226.1	21	9	0.0205 (24.5)	Greece/River Acheloos	[66]

(Continued)

TABLE 12.3 (CONTINUED)

Representative Concentration Levels of Selected Past and Current-Use Pesticides in Surface Waters Worldwide during the Last Ten Years

Compound	Chemical Class	Koc[a] (mL g⁻¹)	DT50[a] (field)	DT50 (water)	Max. Concentration (μg/L) (% Freq.)	Country/Type	Reference
Pendimethalin	Dinitroaniline	15744	90	4	0.0196 (16.5)	Greece/Lake Amvrakia	[63]
Prochloraz	Azole	500	16.7	2	0.486 (34–100)	Spain/Turia, Júcar Rivers	[64]
Pyrimethanil	Anilinopyrimidine	301	29.5	16.5	0.408 (74.2)	Greece/Lake Amvrakia	[63]
Simazine	Triazine	130	90	46	0.0561 (4.1)	Greece/Lake Amvrakia	[63]
Thiamethoxam	Neonicotinoid	56.2	39	30.6	1.34 (0–100)	Canada/Ontario surface	[67]
Terbuthylazine	Triazine	–	22.4	6	0.010 (50)	Spain/ River Ebro	[63]

[a] Koc=organic carbon partition coefficient, DT50=half-life, data extracted from Pesticides Properties Database (PPDB) [43].

in cultivation and their occurrence in rivers include the characteristics of the catchment (size, climatological regime, type of soil, or landscape) as well as the chemical and physical properties of the pesticides [72]. The size of the drainage basin affects the pesticide concentration profile and Larson and co-workers showed that in large rivers the integrating effects of the many tributaries result in elevated pesticide concentrations that spread out over the summer months. In rivers with relatively small drainage basins (50000–150000 km^2) pesticide concentrations increased abruptly and the periods of elevated concentrations were relatively short – about one month – as pesticides were transported in run-off from local spring rains in the relatively small area [73]. Although for the smaller drainage basins of the Mediterranean area short periods of increased pesticide concentrations would be expected, more spread out pesticide concentration profiles are observed. This is probably due to delayed leaching from soil as a result of dry weather conditions, which is reflected by the low mean annual discharges [1]. Generally, low concentrations were observed during the winter months because of dilution effects due to high-rainfall events and the increased degradation of pesticides after their application. Thus, pesticides were flushed to the surface water systems as pulses in response to late spring and early summer rainfall as reported also elsewhere [66].

The character of the landscape in combination with the type of cultivation in the catchment area may well affect the temporal variations in riverine concentrations of pesticides. For example, for the relatively large basin of the river Rhone, the concentration of triazines display a short peak from late April until late June with relatively constant concentrations during the rest of the year [74], due to the fact that herbicides are used in vineyards situated on mountain slopes which promotes rapid run-off. Finally, similar trends and temporal variations were observed also in lakes. The only difference is that residues were detected during a longer period as a result of the lower water flushing and renewal time compared to rivers.

Several pesticides and their metabolites have been also identified in groundwater (Table 12.4). However, fewer pesticide measurements are available around the world, located mainly in the areas of the US and Europe (Table 12.4). In previous published studies that summarized the groundwater monitoring data for pesticides, researchers reported that at least 40 pesticides have been detected. Other pesticides sporadically detected include isoproturon, fenuron propyzamide, propiconazole, oxadixyl, imazalil, metsulfuron-methyl, dimethenamid, triclopyr, dicamba, bromoxynil, procymidone, and oxadiazon metalaxyl [65,68,75,77]. The greater part of these pesticides were herbicides and fungicides. Recently reported concentrations reached up to 30 μg/L. Cohen et al. [82] have compiled the chemodynamic properties of the pesticides detected in groundwater and concluded that most of these chemicals had aqueous solubility in excess of 30 mg L^{-1} and degradation half-lives longer than 30 days.

In EU countries, as in the case of the US, commonly used pesticides such as triazines (atrazine and simazine) and the ureas (diuron and chlortoluron), which are used in relatively large quantities, are often detected in raw water sources. As a result of the restriction on the use of products containing these active ingredients and according to recent assessments a statistically significant downward trend in the contamination of groundwater with atrazine and its metabolites was recorded in a number of European countries [12]. However, in some cases, concentrations of other

TABLE 12.4

Representative Concentration Levels of Pesticides with Occurrences > 5% in Groundwater Worldwide

Compound	Chemical Class	Regulatory EU Status	Gus Index[a]	Concentration (μg L^{-1})		Country/Type	Reference
				% Frequency	Max. Conc.		
Alachlor	Chloroacetamide	Out	1.08		10.2	Italy/Various	[68]
Atrazine	Triazine	Out	3.20		>4	England–Wales/Various	[75]
				56	0.253	Europe/Various	[76]
				67	0.335	USA/Various	[65]
Bentazone	Benzothiazinone	In	2.89	32	0.11	Europe/Various	[76]
Boscalid	Carboxamide	In	2.66	58	2.12	USA/Various	[65]
Chlortoluron	Phenylurea	In	3.02		≈60	England–Wales	[75]
Chloridazondesphenyl	Metabolite	–	–	16.5	13	Europe/Various	[76]
DEA	Metabolite	–	4.50	55	0.487	Europe/Various	[76]
DET	Metabolite	–	3.80	49	0.266	Europe/Various	[76]
DIA	Metabolite	–			>4	England–Wales/Various	[75]
				8.1	0.210	France/Various	[77]
Diazinon	Organophosphorus	Out	1.14	9.1	1	Europe/Various	[76]
Diuron	Phenylurea	In	1.83	29	0.279	Europe/Various	[76]
Glyphosate	Phosphonoglycine	In	−0.25	41	2.5	Spain	[78]
				32	1.4	Sweden	[79]
				5.8	2.0	USA/Various	[80]
Isoproturon	Substituted urea	In	2.07	20.1	0.022	Europe/Various	[76]
Mecoprop	Aryloxyalkanoid	Out	2.29	13.4	0.785	Europe/Various	[76]
MCPA	Aryloxyalkanoid	In	2.94	7.9	0.036	Europe/Various	[76]
Metolachlor	Anilide	Out	2.10	20.7	0.209	Europe/Various	[76]
					12.5	Italy/Various	[68]

(Continued)

TABLE 12.4 (CONTINUED)

Representative Concentration Levels of Pesticides with Occurrences > 5% in Groundwater Worldwide

Compound	Chemical Class	Regulatory EU Status	Gus Index[a]	Concentration (µg L^{-1})		Country/Type	Reference
				% Frequency	Max. Conc.		
Simazine	Triazine	Out	2.00		1.54	Greece	[81]
				43	0.147	Europe/Various	[76]
				8	0.140	USA/Various	[65]
					221	Italy/Various	[68]
Terbuthylazine	Triazine	In	3.07	37	0.716	Europe/Various	[76]

[a] GUS index, data extracted from Pesticides Properties Database (PPDB), https://sitem.herts.ac.uk/aeru/ppdb/en/ [43].

pesticides show an upward trend [12]. As an example of a groundwater monitoring program, the Pesticides in European Groundwaters (PEGASE) is a detailed study of representative aquifers. Furthermore, the Pesticide National Synthesis Project, which is a part of the US Geological Survey's National Water Quality Assessment Program (NAWQA) with the aim of long-term assessment of the status and trends of water resources including pesticides as one of the highest priority issues, is also a nice example of a water monitoring program (http://ca.water.usgs.gov/pnsp/).

As mentioned previously, limited monitoring data are available for the occurrence of pesticides in marine waters. Mainly estuarine environments, ports, and marinas have been monitored for pesticides loadings. Nice examples of such monitoring programs are the Fluxes of Agrochemicals into the Marine Environment (FAME) project, supported by the European Union, which provides information for the Rhone (France), Ebro (Spain), Louros (Greece), and Western Scheldt (the Netherlands) river/estuary systems [83], and the MEDPOL program for monitoring priority fungicides in estuarine areas of the Mediterranean region [84]. In addition, the Assessment of Antifouling Agents in Coastal Environments (ACE) project of the European Commission (1999–2002) provides data concerning contamination and effects/risks of the most popular biocides currently used in antifouling paints to prevent fouling of submerged surfaces in the sea as alternatives to tributyltin compounds. A number of booster biocides have been detected in many European countries including Irgarol 1051, diuron, sea nine, and chlorothalonil. The occurrence, fate, and toxic effects of antifouling biocides have been also reviewed [85].

12.2.3.3   Soil and Sediment Monitoring

Soil and sediment compartments might also be regarded as reservoirs for many types of pesticides. Although high amounts of pesticide as well as a complex pattern of their metabolites are usually present in soils, this matrix is not generally monitored on a regular basis and there is a gap in knowledge on a national and global level regarding the pesticide residue levels. The majority of the investigation studies were carried out on researchers' initiative or licensing of new substances or under the frame of funded projects. As regards Europe, recent discussions have taken place to consider regulation of persistence of soil pesticide residues beyond the guidelines given in Directive 91/414/EEC [14]. In this regard, stronger emphasis should be given to soil monitoring programs such as Monitoring the state of European soils (MOSES) (http://projects-2004.jrc.cec.eu.int/) and Environmental Indicators for Sustainable Agriculture (ELISA) (http://www.ecnc.nl/CompletedProjects/ Elisa_119.html).

In contrast to soils, sediments are usually monitored for pesticide contamination. Sediments from river, lake, and sea waters provide a habitat for many benthic and epibenthic organisms and are a significant element of aquatic ecosystems. Many pesticide compounds, because of their hydrophobic nature, such as OCPs, are known to associate strongly with natural sediments and dissolved organic matter and high concentrations of pesticides are frequently found in bed sediments, both freshwater and coastal [86]. Monitoring studies using sediment core stratification also have the advantage of providing information on the chronologies of accumulation rates of persistent pesticides. This information is important in order to evaluate the rate of emission from probable sources, and to relate specific rates of pesticide accumulation and rates of ecosystem response. Sediment monitoring is also a task for the correct implementation of the Water Framework Directive (WFD) in order to assess any changes in the status of water bodies.

Soils and sediments are typically very inhomogeneous media; thus, a large number of samples may be required to characterize a relatively small area. Sampling sites can be distributed spatially at points of impact, reference sites, areas of future expected changes, or other areas of particular interest. Selection of specific locations is subject to accessibility, hydraulic conditions, and other criteria. The devices used for soil and sediment sampling are usually grab samplers and corers. Grab samplers are available for operation at surficial depths. Box corers or multicorers can be employed if more data on the chronologies of accumulation rates of the analytes is needed.

12.2.3.3.1 Occurrence and Pesticide Levels in Soils and Sediments

In view of the current concern about the assessment of soil quality, some pesticide monitoring studies have been conducted within Europe and the US [7,8,87,88]. According to the results a variety of pesticides, mainly herbicides (such as triazines and anilides) and insecticides (such as organophosphates), appeared consistently as contaminants of the tested soil samples. The high frequency of detection in the studied soils corresponds to their environmental properties, strongly sorbed to soil (log-K_{OC} of 3–4) and with low-to-moderate water solubility and low degradability.

Multiple pesticide residues have been frequently (e.g., 51% of soils with $\geq$ 5 pesticides) detected in arable soils from central Europe (e.g., Czech Republic) with noticeable concentration levels (e.g., 36% of soils with $\geq$ 3 pesticides exceeding the threshold of 0.01 mg kg^{-1}). Triazine herbicides (terbuthylazine, atrazine, and simazine in 89% of soils) were the dominant compounds, followed by azole fungicides (in 73% of soils) (i.e., epoxiconazole (48% of soils), tebuconazole (36%), flusilazole (23%), prochloraz (21%), propiconazole (13%), and cyproconazole (8%)) which showed also high concentration levels (53% soils with total azoles above 0.01 mg kg^{-1}). Chloroacetanilide herbicides and TPs were also detected less frequently (25% of soils), followed by fenpropidin (20%) and diflufenican (17%) [89].

In the Mediterranean area, pesticide residues in agricultural soils from intensive agricultural areas of Spain have been recently reported [90]. More than 25 pesticides have been detected in total, and the levels (mean concentration, range) for the pesticides detected in > 20% of the samples were as follows: bifenthrin (13, 2–38 ng g^{-1}), bupirimate (50, 2–228 ng g^{-1}), cyproconazole (10, 2–36 ng g^{-1}), cyprodinil (14, 2–52 ng g^{-1}), dimetomorph (67, 8–208 ng g^{-1}), procymidone (32, 3–337 ng g^{-1}), triadimefon (10, 2–35 ng g^{-1}), and Imidachloprid (39, 2–118 ng g^{-1}). Other residues included mainly compounds from the azole fungicide group such as difeconazole, diniconazole, penconazole, and tetraconazole. Low but not negligible levels of dinitroanilines were detected in plains from Central Greece, ranging from 0.01 to 0.21 mg g^{-1} d.w. for trifluralin and 0.01 to 0.048 mg g^{-1} d.w. for pendimethalin. Trifluralin was the herbicide most frequently detected (44.4%) [91]. Finally, residues of pyrethroid insecticides (resmethrin and cyfluthrin) were detected in Mediterranean paddy fields at concentrations $\leq$ 57.0 ng g^{-1} before plow and $\leq$ 62.3 ng g^{-1} during rice production [92]. An extended review on the levels of pyrethroids in soils worldwide [70] revealed maximum concentrations of 1184 ng g^{-1} for farmland soils in Pakistan (Punjab) [93].

Recently, glyphosate and AMPA were detected in more than 90% of sediment samples in seven sites in Indiana and Mississippi, US, with concentrations frequently exceeding 10 ng g^{-1}. The median and maximum glyphosate concentrations in these samples were 9.6 and 476 ng/g, respectively, whereas the median and maximum AMPA concentrations were 18 and 341 ng g^{-1}, respectively [80].

The monitoring studies performed on sediments show a large number of detected pesticides over the last 40 years. Most of the target analytes detected were OCPs and their transformation products despite the fact that most of these were banned or severely restricted by the mid-1970s in the US and EU. This reflects both the environmental persistence of these compounds and the limited target analytes list. DDT and metabolites, chlordane compounds, α-, β-, γ-HCH, and dieldrin were the most detected pesticides in bed sediments. Other OCPs that were sometimes detected included endosulfan compounds, endrin and metabolites, heptachlor and heptachlor epoxide, methoxychlor, and toxaphene [94].

Recent studies in sediment cores have shown that concentration levels of OCPs have a relative steady state for DDTs, with a slight decrease in the top layers, suggesting a slight decline in their concentrations due to restrictions in their usage [95]. Besides the OCPs, a few compounds in other pesticide classes were detected in some studies. Most of these pesticides contained chlorine or fluorine substituents and had medium hydrophobicity. Current-use pesticides detected in sediments included compounds with extensive usage and medium to high hydrophobicity. From the herbicide family s-triazines, acetanilides, substituted ureas, and dinitroanilines are the dominant chemical classes usually detected. For example, herbicide occurrence (frequency of detection and maximum concentration (MC), dry weight (d.w.)) has been reported as follows: trifluralin (40%, 7.6 ng g^{-1}), pendimethalin (22%, 123 ng g^{-1}), metolachlor (25%, 5.2 ng g^{-1}), US [96]; terbuthylazine (11–80%, 2.47 ng g^{-1}), chlortoluron (20–44%, 2.57 ng g^{-1}), diuron (22–50%, 3.28 ng g^{-1}), propanil (11%, 0.44 ng g^{-1}), Alqueva reservoir, Portugal [97]; terbuthylazine (MC 250 ng g^{-1}) Guadalquivir river basin [98]; and alachlor (0.65 ng g^{-1}), atrazine (0.18 ng g^{-1}), simazine (0.29 ng g^{-1}), trifluralin (n.d.–4.54 ng g^{-1}), different EU coastal lagoons [99].

Among insecticides, pyrethroids have received the greater attention for sediment pollution because they are the third most applied group of insecticides worldwide since they are extensively used in agricultural and non-agricultural applications. In addition, pyrethroids exhibit high hydrophobicity (logK$_{ow}$ 4.5–6.90) and pseudo-persistence due to continuous input, thus sediments are vulnerable to their accumulation (logK$_{oc}$=3.5–5.85) [9]. Two very informative reviews on previously reported sediment concentrations of pyrethroids and the associated toxicity to benthic invertebrates on a global scale have been recently published [9,70]. The trends revealed by the review [70] study pointed to North America, followed by Asia, Europe, Australia, and Africa as the most studied areas; pyrethroid occurrence in both agricultural and urban sediments; and bifenthrin and cypermethrin as the compounds mainly contributing to the toxicity in benthic organisms. Bifenthrin, cyfluthrin, lambda-cyhalothrin, cypermethrin, esfenvalerate, and permethrin were the most frequently detected fungicides in all sediment samples reviewed, with detection percentages reaching 78%, 37%, 49%, 48%,43% and 57%, respectively. The concentration levels (80% of the detected concentrations) of the above six pyrethroids in sediments varied in the range of 0.1 to 100 ng g^{-1} dry weight (d.w.) [9]. Maximum concentrations of pyrethroids in worldwide sediments reached levels as high as 375.7 µg g^{-1} [70]. It is also found that the occurrence of pyrethroids in sediments is influenced by the land use patterns, e.g., the concentrations of the above-mentioned pyrethroids in urban sediments were significantly greater than those in agricultural areas. Apart from pyrethroids, compounds from the OPs family are also frequently detected in sediments. For example, OPs occurrence (frequency of detection, range or maximum concentration (MC), dry weight (d.w.)) has been reported as follows: chlorpyrifos (30%, 444 ng g^{-1}) [96], chlorpyrifos (n.d.–280 ng g^{-1}), diazinon (n.d. –12.9 ng g^{-1}), [100] US; diazinon (10%, 0.06 ng g^{-1}), Alqueva reservoir, Portugal [132]; chlorpyrifos (20%, 0.7–15.9 ng g^{-1}), diazinon (17%, 0.2–175.5 ng g^{-1}) Guadalquivir River Basin [98]; methidathion (100%, 0.0061–0.158 ng g^{-1}), dichlorvos (100%, 0.450–4.65 ng g^{-1}), diazinon (100%, 0.0036–1.03 ng g^{-1}), parathion (100%, 0.0027–0.555 ng g^{-1}), chlorfenvinphos methyl (94%, n.d. –0.1723 ng g^{-1}), phosalone (89%, n.d. –0.109 ng g^{-1}), fenitothion (72%, n.d. –0.0283 ng g^{-1}), malathion (89%, n.d. –0.0256 ng g^{-1}), a eutrophic lake of China [101]; and finally, chlorpyrifos (n.d.–14.3 ng g^{-1}), chlorfenvinphos (n.d.–41.2 ng g^{-1}), dichlorvos (10 ng g^{-1}), different EU coastal lagoons [99].

As regards fungicides, compounds from strobilurins, carboxamide, and anilides are considered among the most frequently detected in sediments. For example, the occurrence of fungicides (frequency of detection, range or maximum concentration (MC), dry weight (d.w.)) in three geographic areas across the United States reported pyraclostrobin (75%, 198 ng g^{-1}), boscalid (53%, 44.5 ng g^{-1}), chlorothalonil (41%, 4.2 ng g^{-1}), azoxystrobin (20%, 2.5 ng g^{-1}), and pyrimethanil (15%, 0.4 ng g^{-1}) [96]. In another study conducted in United States' natural bed sediment samples, azoxystrobin presented a detection frequency between 12 and 43% and maximum concentration levels up to 3.80–12.60 (ng g^{-1}) [102].

Of pesticides from other chemical classes most were targeted at relatively few sites. Examples of specific uses include the booster biocides such as irgarol, diuron, and chlorothalonil which were detected in coastal marine sediments [85].

12.2.3.4 Biological Monitoring

A lot of biological organisms, from flora and fauna, as well as human beings, are monitored to determine the amounts of pesticides that are present in the environment and evaluate the associated hazard and risk. This type of monitoring is an essential part of pesticide pollution studies and is known as biological monitoring or biomonitoring. Another important facet of environmental biomonitoring is the emerging field of environmental specimen banking. A specimen bank acts as a bridge connecting real-time monitoring with future trends-monitoring activities.

In general, biomonitoring overcomes the problem of achieving a snapshot of the quality of the environment, and can provide a more representative picture of average conditions over a period of weeks to months. However, the use of biomonitors has limitations since some compounds are metabolized or eliminated at a rate close to the rate of uptake, and thus are not accumulated. Moreover, because of cost, the monitoring may be carried out only on a limited number of species and there is no guarantee that important species will be selected. Not all pesticides are amenable to biological monitoring. Pesticides that are rapidly absorbed and are neither sequestered nor metabolized to a significant extent are usually good candidates. Pesticides that have a high tendency to bioaccumulate, such as OCL pesticides, are the most commonly detected in biota samples.

Sample collection methods must be selected considering both the organisms to be collected and the conditions that will be encountered. Organisms that can be deployed for extended periods of time, during which they passively bioaccumulate pesticides in the surrounding environment, are usually selected. Plankton, bacteria, periphyton, benthos, fish, and fish-eating birds are the most commonly selected specimens for monitoring the aquatic compartment. Analysis of the tissues or lipids of the test organism(s) can give an indication of the equilibrium level of waterborne pesticide contamination. Adipose tissues, eggs, and liver have been recognized as accumulators of lipophilic pesticides and they are usually monitored in order to quantify the threat of pesticide contamination in species of wildlife. Apart from aquatic organisms and wildlife species, increasing attention is being focused on the monitoring and assessment of human exposure to pesticides throughout the world. Urine, blood, and exhaled air are the most used specimens for routine biological monitoring of human beings. Other biological media include adipose tissue, liver, saliva, hair, placenta, and bodily involuntary emissions such as nasal accretions, breast milk, and semen. However, many of these media have some serious problems (e.g., matrix effects, insufficient dose–effect relationships) and they do not necessarily provide as consistent results as those from blood, urine, or breathe.

12.2.3.4.1 Occurrence and Pesticide Levels in Biota

Several studies have been conducted around the world on the general topic of biological monitoring of pesticides. As in the case of sediments, most of the studies reveal the presence of OCPs and their transformation products. These compounds have been detected in different human specimens such as human milk, saliva, urine, adipose tissues, and liver [103,104]. DDT and its metabolites are still the most frequently determined compounds, especially in samples from developing countries. Other OCPs determined were cyclodienes such as dieldrin, aldrin, endrin, heptachlor and its epoxide, chlordane as well as isomers of hexachlorocyclohexane. Moreover, endosulfan I and II and the sulfate metabolite have been detected in fatty and non-fatty tissues and fluids from women of reproductive age and children in Southern Spain [103]. Apart from OCPs, currently used pesticides have also been detected in different human biological samples. Examples include bromophos in blood, fenvalerate, malathion, terbufos, and chlorpyrifos-methyl in urine, paraquat, 2,4-D, and pentachlorophenol in urine and blood, carbaryl, atrazine, and ethion in saliva, and DDT in blood and adipose tissue, etc. [104]. From the current used pesticides, organophosphorus pesticides (OPPs) are among the most frequently detected in different human biological fluids. Apart from the parent compounds, the measurement of dialkyl phosphate metabolites has been frequently used to study exposure to a wide range of OPPs. These metabolites have been detected in urine samples from exposed workers as well as from people who had no occupational exposure to OPPs. For example, four metabolites of OPPs (namely dimethylphosphate, dimethylthiophosphate, diethylphosphate, and diethylthiophosphate) have been determined in urine samples from adult Japanese females with concentrations up to 2770.5 µg g^{-1} creatinine [105]. In addition, metabolites of carbamates (carbaryl, carbofuran) and pyrethroids (e.g., cypermethrin, deltamethrin, permethrin) have been also detected in urine samples [104–105]. A summary of concentrations of pyrethroids and their metabolite residues in the human body worldwide is presented in Tang et al. 2018 [70] where levels up to 2.21 ng g^{-1} have been reported [106]. Finally, the occurrence of neonicotinoid (imidachloprid, clothianidin, dinotefuran, nitenpyram, thiamethoxam, thiacloprid) insecticides in urinary samples from adult Japanese females with concentrations up to 21.4 µg g^{-1} creatinine (dinotefuran) has been reported elsewhere [105].

Excepting human biological samples, the accumulation pattern of OCPs in aquatic organisms as well terrestrial wildlife has been reported. For example, concentration levels of DDT and its metabolites have been detected in different species of Arctic wildlife such as terrestrial animals, fish, seabirds, and marine mammals [107]. Extensive results have also reported for various bird species [108], fish, and amphibians [109] as well as mammals [110] when adipose tissues, liver, or eggs of these organisms have been analyzed. *p,p′*-DDE, a major metabolite of DDT, continues to be the dominating OCP burden in almost all the tested species, whereas cyclodienes and HCHs occurred at lower concentrations. Apart from OCPs, several currently used or banned pesticides (despite their lower bioaccumulation) such as trifluralin, chlorothalonil, parathion-methyl, phosalone, disulfoton, diazinon, dimethoate, chlorpyrifos, and fipronil have been also detected in biota samples [111,112]. For example, in a recent study analyzing eel muscle tissues, concentrations of fipronil and its metabolites fipronil desulfinyl and fipronil sulfone ranged from 0.04 to 0.32 ng g^{-1} ww, 0.02 to 0.13 ng g^{-1} ww, and 0.52 to 11.24 ng g^{-1} ww, respectively [112]. In the last ten years, pyrethroid insecticides have gained the focus and a growing number of studies dealing with their concentration levels in aquatic and land biota have been published. A comprehensive review on the presence of pyrethroids in the global environment summarizes well the up-to-date findings [70]. Concentrations as high as 9901.4 ng g^{-1} have been detected in aquatic products (fish, shrimp, crab) [113], dolphins (87% detection frequency, 300 ng g^{-1} lw ± 932 mean total concentration, range 2.7–5200 ng g^{-1} lw) [114], and 114.5 ng g^{-1} for land animal tissues [115]. It is underlined that a high variability in the concentrations of pesticides within the same species was observed and this was related to sampling location, age, and sex, and with condition and stage of the life cycle (starvation/feeding, lactation, illness/disease) of the analyzed organisms.

An example of a monitoring program that reported a range of diverse invertebrate, vertebrate, and human relevant tests is the Comparative Research on Endocrine Disrupters – Phylogenetic Approach and Common Principles focusing on Androgenic/Antiandrogenic Compounds (COMPRENDO) project [116].

12.2.4 WATER FRAMEWORK DIRECTIVE AND MONITORING STRATEGIES

The potential adverse consequences deriving from the use of pesticides have led to the development of special regulations. For instance, in the European Union, several directives and regulations have been issued with the aim of safeguarding human health and the environment from the undesirable effects of these chemicals (i.e., dangerous substances (76/464/EC) [117], groundwater (80/68/EEC) [118], and pesticide

(91/414/EEC) [14] directives). The Water Framework Directive (WFD, 2000/60/EC) [119] is widely recognized as one of the most ambitious and comprehensive pieces of European environmental legislation. Its aim is to improve, protect, and prevent further deterioration of water quality at the river-basin level across Europe. The term "water" within the WFD encompasses most types of water bodies. Furthermore, to monitor the progressive reduction in contaminants, trend studies, whether spatial or geographical, should be envisaged through the measurement of contaminants in sediment and biota. The Directive aimed to have achieved and ensured "good quality" status of all water bodies throughout Europe by the year 2015, with extended deadlines set for 2021 and 2027, and this is to be attained by implementing management plans at the river basin level. The WFD foresees that water quality should be monitored on a systematic and comparable basis. Thus, technical specifications should follow a common approach (e.g., the standardization of monitoring, sampling, and methods of analysis). Chemical monitoring is expected to intensify and will follow a list of 33 priority chemicals (inorganic and organic pollutants including pesticides) [119] that will be reviewed every four years [120]. Forty-five compounds including 20 pesticides were listed as priority compounds and Environmental Quality Standards (EQS) were determined [121]. The concentrations of the priority substances in water, sediment, or biota must be below the Environmental Quality Standards (EQSs): this is expressed as "compliance checking". EQSs for these substances including pesticides were set by Directive 2008/105/EC and subsequent amendments. The derivation of EQSs through a risk assessment procedure will be presented later in this chapter.

The implementation of the WFD is based on a three-level monitoring system which will form part of the management plans and was to be implemented from December 2006 [119,122]. This includes: a) surveillance monitoring aimed at assessing long-term changes in natural conditions; b) operational monitoring aimed at providing data on water bodies at risk or failing environmental objectives of the WFD; and c) investigative monitoring aimed at assessing the causes of such failure and the effects.

For the monitoring and assessment of the ecological and chemical status of water bodies, the WFD requires water samplings at least monthly (12 samplings a year) and compliance comparisons with the established EQS. The EU Directive on priority substances [121] specifies both maximum allowable and annual average concentrations for surface waters, which must be met in order to reach good ecological status.

Comprehensive reviews focused on the principal monitoring requirements of the WFD as well as on emerging techniques and methods for water quality monitoring have been published in order to identify and outline the tools or techniques that may be considered for water quality monitoring programs necessary for the implementation of the WFD [21,122].

12.3   ENVIRONMENTAL EXPOSURE AND RISK ASSESSMENT

12.3.1   Environmental Exposure

12.3.1.1   Point and Non-Point Source Pesticide Pollution

Environmental exposure of pesticides can occur by point and non-point sources. Point source can be any single identifiable source of pollution from which pesticides are discharged such as the effluent pipes, careless storage and disposal of pesticide containers, accidental spills, and overspray. Pesticide movement away from the targeted application site is defined as non-point source pollution and can occur through run-off, leaching, and drift. Non-point source pollution occurs over broad geographical scales and because of its diffuse nature typically yields relatively uniform environmental concentrations of pesticides in surface waters, sediments, and groundwater. Runoff is the surface movement of pesticide in water or bound to soil particle percolation, while leaching is the downward movement of a pesticide through the soil by water. Drift is the off-target movement of pesticide by wind or air currents and can be in the form of spray droplet drift, vapor drift, or particle (dust) drift.

12.3.1.2   Environmental Parameters Affecting Exposure

The environmental parameters that affect pesticide exposure could be classified as follows:

1. *Soil characteristics and field topography:* Texture composition and pH are the main soil properties that affect pesticide fate and transport, while topographic characteristics of the fields like watershed size, slope, drainage pattern, and permeability of soil layers affect greatly the potential to generate run-off water or leachates.
2. *Weather and climate:* Climatic factors such as the amount and timing of rainfall, the duration and intensity, temperature, and air movement influence the degree to which pesticides are mobilized by runoff, leaching, and drift. In addition, temperature and sunlight affects all abiotic and biotic transformation of pesticides reactions [123,124].

12.3.1.3   Pesticide Parameters Affecting Exposure

The pesticide factors affecting exposure could be organized in three main sets:

1. *Application factors*: These include the application site (crop or soil surface) and method, the type of use (agricultural, non-agricultural applications, indoor pest management, etc.), the formulation (e.g., granules or suspended powder or liquid), and the application amount and frequency; the application time does affect its possible routes of transport in the environment.
2. *Partitioning and mobility of pesticides in the environment*: The main physicochemical properties of pesticides that affect their mobility are the water solubility, vapor pressure, and soil–water partition coefficient (K_{oc}). K_{oc} defines the potential for the pesticide to bind to soil particles. Off-target movement by drift depends also on the spray droplet size and the viscosity of the liquid pesticide, while plant uptake from the soil is another important pathway in determining the ultimate fate of pesticide residues in the soil [123,124].
3. *Persistence in the environmental compartments*: Persistence is usually expressed in terms of half-life, that is, the time required for one-half of the pesticide to decompose to products other than the parent compound. The longer a pesticide persists within the environment, the greater the risk it

poses to it. Hydrolysis, direct and indirect photolysis, and biodegradation are the principal pesticide degradation processes and their rates depend on pesticide chemistry, as well as on environmental conditions [124].

12.3.1.4 Modeling of Environmental Exposure

Monitoring data and environmental modeling are interconnected to each other. Monitoring can provide the correct input data to models for calibration and validation or can be devoted to collecting data on the timing and magnitude of loadings. Mathematical models that simulate the fate of pesticides in the environment are used for developing Environmental Estimated Concentrations (EECs) or Predicted Environmental Concentrations (PECs). A complete presentation of environmental models describing the exposure of pesticide in the environment is presented elsewhere [125] and it is outside the scope of the present chapter. Thus, only some common environmental models that are used to estimate environmental exposure concentrations for aquatic systems in the context of current risk assessing techniques will be presented.

The Generic Estimated Environmental Concentration (GENEEC) model, developed by the EPA, determines generic EEC for aquatic environments under worst case conditions (i.e., application on a highly erosive slope with heavy rainfall occurred just after the pesticide application, the treatment of the entire area – essentially ten acres of surface area with uniform slope – with the pesticide, and the assumption that all runoff drains directly into a single pond). The model utilizes environmental fate parameters derived from laboratory studies under standard procedures as well as soil and weather parameters. The outputs of the model are the pesticide runoff and environmental concentration estimates [125]. This model can be used as first tier approach since it is based on a single event and a high-exposure scenario. In a higher tier approach (second and third), models that can account for multiple weather conditions and/or multiple sites are used. Such models are the Pesticide Root Zone Model (PRZM), edge-of field runoff/leaching the Exposure Analysis Modeling System (EXAMS), fate in surface water, and AgDrift (spray drift) that uses additional parameters, more descriptive of the site studied. PRZM simulates the leaching, runoff, and erosion from an agricultural field and EXAMS simulates the fate in a receiving water body. The water body simulated is a static pond, adjacent to the crop of interest. Typical conditions of the site including the soil characteristics, hydrology, crop management practices, and weather information are used. The output of this higher tier analysis is to define the EEC that can be reasonably expected under variable site and weather conditions. The model yields an output of annual maxima distributions of peak, 96 hour, 21 day, 60 day, 90 day, and yearly intervals. AgDrift includes generic data for screening level assessments including pesticide formulation, drop height, droplet size, nozzle type, and wind speed. The above approaches are used by pesticide registrants to address environmental exposure concerns and are frequently combined with geographical information systems (GIS) to produce regional maps.

The fugacity approach has also proven particularly suited for describing the behavior of pesticides in the environment. A tiered system of fugacity models has been introduced which distinguishes four levels of complexity, depending on whether the system is closed or in exchange with the surrounding environment. The four levels are: Level (I) close system equilibrium; Level (II) equilibrium steady state; Level (III) non-equilibrium steady state; and Level (IV) non-equilibrium non-steady state. Levels I and II are used in lower tier approaches while Level III is widely used in higher tiers to obtain exposure concentrations due to emission flux into a predefined standard environment. A detailed introduction to fugacity-based modeling can be found in Mackay [126].

For evaluating the impact of management practices on potential pesticide leaching the Groundwater Loading Effects of Agricultural Management Systems (GLEAMS) is a widely used, field scale model. GLEAMS assume that a field has homogeneous land use, soils, and precipitation. It consists of four major components: hydrology, erosion, pesticide transport, and nutrients. GLEAMS estimates leaching, surface runoff, and sediment losses from the field and can be used as a tool for comparative analysis of complex pesticide chemistry, soil properties, and climate. The model output data are daily, monthly, and annual pesticide mass and concentrations in runoff and sediment.

Finally, a fourth tier approach can be used based on watershed site assessments. These assessments are very complex since the landscape studied has a very high surface area, high diversity of soils and weather conditions, varied proximities of agricultural lands to receiving waters, and various water bodies. Thus, GIS are commonly used in order to distinguish high-risk versus low-risk areas on a watershed basis. Finally, modeling and monitoring often are combined within tier 4 to provide a more accurate distribution of pesticide exposure.

12.3.2 RISK ASSESSMENT

In order to evaluate the negative impact of pesticides on ecosystems, the environmental risk assessment is necessary. It is known that the environmental impact of a pesticide depends on the degree of exposure and on its toxicological properties. The risk assessment procedure involves three main steps: a formulation of the problem to be addressed, followed by an appraisal of toxicity and exposure, and concluding with the characterization of risk. A typical framework for ecological risk assessment is shown in Figure 12.2 [127]. The objective of the exposure assessment is to describe exposure in terms of source, intensity, and spatial and temporal distribution, evaluating secondary stressors (metabolites) in order to derive exposure profiles. Usually exposure assessment involves the measured environmental concentrations (MECs) derived from monitoring studies or the developing and application of models as discussed previously.

The toxicity assessment identifies concentrations that, when administered to surrogate organisms result in a measurable adverse biological response. Toxicological assessment is commonly based on laboratory studies with the aim of determination of the relationship between the magnitude of exposure and extent of observed effects, commonly referred as the dose–response relationship. Toxicity impacts are usually studied by indicator species selected to represent various trophic levels within an ecosystem. Representative groups of organisms are assessed for risk from pesticides, including, fish, aquatic invertebrates, algae and plants from the aquatic environment,

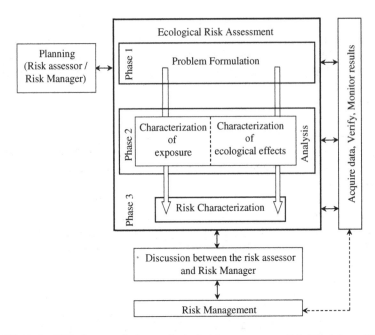

FIGURE 12.2 EPA framework for ecological risk assessment. (Modified from US EPA (U.S. Environmental Protection Agency), Framework for ecological risk assessment. Risk Assessment Forum, U.S. Environmental Protection Agency. Washington, DC, 1992).

and birds, mammals, bees and beneficial arthropods, earthworms, soil micro-organisms, and non-target plants from the terrestrial environment. All these organisms are assessed in Europe under 91/414/EEC [14], whereas the US EPA concentrates on birds and mammals, bees, non-target plants, and aquatic organisms. It is impossible and inadvisable to test every species (abundant, threatened, endangered) with every pesticide but the need for more toxicological data is acknowledged. Chosen organisms like *Daphnia* sp. for freshwater zooplankton or rainbow trout for fresh water fish categories typically should satisfy some basic criteria like ecological significance, abundance and wide distribution, susceptibility to pesticide exposure, and availability for laboratory testing.

Stressor-response analysis can be derived from point estimates of an effect (i.e., Lethal Concentration or Effect Concentration for 50% of the organism population LC50 or EC50) or from multiple-point estimates (Hazardous Concentration for 5% of the species, HC5) that can be displayed as cumulative distribution functions (Species Sensitivity Distributions, SSD). In addition, the establishment of cause-and-effect relationships from observational evidence or experimental data can be performed.

In a third phase, the risk characterization takes place, defining the relationship between exposure and toxicity. Two different approaches are usually applied for this purpose. The first is a deterministic approach that is based on simple exposure and toxicity ratios and the second is a probabilistic approach in which the risk is expressed as the degree of overlap between the exposure and effects. Apart from these methods, numerous Pesticide Risk Indicators (PRIs) based on classification systems have been developed for fast preliminary assessments and comparative purposes. All methods will be analyzed in detail below.

The last step in the assessment of risk is the weight-of-evidence analysis. Strengths, limitations, and uncertainties as well as magnitude, frequency, and spatial and temporal patterns of previously identified adverse effects and exposure concentrations are discussed in the weight-of-evidence analysis.

The assessment of the pesticide risks usually follows a tiered approach. Tiers are normally designed such that the lower tiers are more conservative while the higher tiers are more realistic with assumptions more closely approaching reality. Tier 1 is essentially a screen, thereby identifying low-risk uses, or those groups of organisms at low risk [128–131]. Higher tier approaches aim to refine the risk, i.e., a procedure (method, investigation, evaluation) performed to characterize in more depth the pesticide risks arising from the preliminary (tier 1) risk assessment. The risk refinement is triggered by increasingly more realistic and/or comprehensive sets of data, assumption, and models and/or mitigation options. Thus, if the assessment fails to "pass" tier 1 then a more detailed risk assessment is required.

12.3.2.1 Preliminary Risk Assessment – Pesticide Risk Indicators (PRIs) – Classification Systems

A preliminary estimation of the environmental impact of pesticides use can be performed through the development and use of PRIs, which are indices that combine the hazard and exposure characteristics for one or several environmental compartments that are assessed separately. PRIs make use of the physicochemical and biological properties of pesticides and have been used over the years by a large number of organizations for the purposes of selecting pesticide compounds for further regulatory actions. Details on all PRI approaches and systems are well described and compared in recently published articles and reports [132–135]. These indicators leave room for users and scientists to select the most appropriate approach according to the considered environmental effects and the environmental specific conditions at national or regional level. However, a harmonized scientific framework is highly recommended.

12.3.2.2 Risk Quotient – Toxicity Exposure Ratio Method (Deterministic – Tier 1)

At present the usual approaches to decide the acceptability of environmental risks are generally based on the concept of risk ratios expressed as the toxicity–exposure ratio (TER) adopted by the EU (Equation 12.1) or the risk quotient (RQ) adopted by the US EPA (Equation 12.2). This methodology usually involves comparing an estimate of toxicity, derived from a standard laboratory test with a worst-case estimate of exposure, EEC or PEC, from model applications or peak measured concentrations, for the US and EU respectively.

$$RQ = \frac{Exposure}{Toxicity} \qquad (12.1)$$

$$\text{TER} = \frac{\text{Toxicity}}{\text{Exposure}} \qquad (12.2)$$

Since the term risk implies an element of likelihood which is usually reported as probabilities it is more correct that the risk quotient should be better expressed as hazard quotient (HQ). However, both terms are used in several studies with the same meaning. Examples of toxicity measurements used in the calculation of RQs are: LC50 (fish and amphibians, birds); LD50 (birds and mammals); EC50 (aquatic plants and invertebrates); EC25 (terrestrial plants); and EC05 or non-observed effect concentration (NOEC) (endangered plants).

One standard procedure for the risk assessment in aquatic systems is the determination of RQ method for three taxonomic groups (i.e., algae, zooplankton, fish) at two effect levels (i.e., acute level, using LC50 or EC50 values, and chronic level, using NOEC or predicted non-effect concentration (PNEC) values) according to Directive 414/91/EEC [14].

For assessing the risk in sediments, if results from whole-sediment tests with benthic organisms are available, the PNEC_{sed} has to be derived from these tests. In the case that not enough reliable ecotoxicological data for sediment-dwelling organisms are known, the equilibrium partitioning method can be used [136] to derive PNEC_{sed} according to Equation 12.3:

$$\text{PNEC}_{sed} = \frac{\text{PNEC}_{wat} \times \text{K}_{susp\text{-}water}}{\text{RHO}_{susp}} \times 1000 \qquad (12.3)$$

where:
PNEC_{wat} is the PNEC calculated for the water compartment
$\text{K}_{susp\text{-}water}$ is the sediment/water partition coefficient and
RHO_{susp} is the bulk density of sediment

The same methodology can be applied for deriving PNEC values for soil using the corresponding Kp_{soil} (soil–water) partition coefficient.

For terrestrial systems the estimate of the distribution of exposure is separated into the chemical/physical and biological components. The first component of dose estimate is the environmental and chemical variables that influence the distribution of residue levels. The major variables that influence the biological component are species dependent including: (a) food, water, and soil ingestion rates; (b) dermal and inhalation rates; (c) dietary diversity; (d) habitat requirements and spatial movement; and (e) direct ingestion rates. These variables are combined into Equation 12.4 to estimate the distribution of total dose:

$$\text{Dose}_{total} = \text{Dose}_{oral} + \text{Dose}_{dermal} + \text{Dose}_{inhal} \qquad (12.4)$$

The oral dose can be further analyzed as follows:

$$\text{Dose}_{oral} = \text{Dose}_{food} + \text{Dose}_{water} + \text{Doses}_{oil} + \text{Dose}_{preening} + \text{Dose}_{granular} \qquad (12.5)$$

For each of these sources of oral exposure, the equations which can be used to estimate the dose are reported elsewhere [137]. Frequently for birds and mammals it is assumed that exposure is through eating treated food items and residue concentrations (w/w) in mg kg^{-1} are compared to dietary LC50, NOEC.

12.3.2.2.1 The Use of Assessment Factors for the Characterization of Uncertainty

For many substances the available toxicity data that can be used to predict ecosystem effects are very limited, thus, empirically derived assessment factors must be used depending on the confidence with which a PNEC can be derived from the existing data. The proposed assessment factors according to EC guidelines [136] are presented in Table 12.5 for water and sediment.

If the database on SSDs from long-term tests for different taxonomic groups is sufficient, statistical extrapolation methods may be used to derive a PNEC. In such methods the long-term toxicity data are log-transformed and fitted according to the distribution function and a prescribed percentile of that distribution is used as criterion. Kooijman [138] and Van Straalen and Denneman [139] assume a log-logistic function and Newton et al. [140] a Gompertz distribution. Newman et al. [141] proposed to bootstrap the data as a non-parametric alternative while Van der Hoeven [142] proposed a non-parametric method to estimate HC5 without any assumption about the distribution and without bootstrapping. The uncertainty in the calculation of HC5 can be refined with 50% of confidence level usually applied, while applying a 95% confidence level provides more strict values. According to the aforementioned, a PNEC value can be calculated as

$$PNEC = \frac{5\%SSD\left(50\%\,c.i.\right)}{AF} \tag{12.6}$$

TABLE 12.5
Assessment Factors to Derive a PNEC$_{aquatic}$ [from ref. 136].

Available Data	Assessment Factor
At least one short-term L(E)C50 from each of three trophic levels of the base-set (fish, Daphnia, and algae)	1000[a]
One long-term NOEC (either fish or Daphnia)	100[b]
Two long-term NOECs from species representing two trophic levels (fish and/or Daphnia and/or algae)	50[b]
Long-term NOECs from at least three species (fish, Daphnia, and algae) representing three trophic levels	10[b]
Species sensitivity distribution (SSD) method	5–1
Field data or model ecosystems	Reviewed on case by case basis

[a] A factor of 100 could be used for pesticides subject to intermittent release;

[b] the same assessment factors are used for derivation of PNEC in sediments using appropriate species.

AF is an appropriate assessment factor between 5 and 1 (as proposed in Table 12.5) reflecting the further uncertainties identified. Confidence can be associated with a PNEC derived by statistical extrapolation if the database contains at least ten NOECs (preferably more than 15) for different species covering at least eight taxonomic groups [136]. SSDs have been also used to derive environmental quality standard (EQS) concentration levels.

Uncertainty arises from an incomplete knowledge of the system being assessed and it is associated with the following aspects: measurement errors (accuracy), inherent variability, model error both conceptual and mathematical, assumption errors, and lack of data. As already mentioned, the characterization of risk at the first level of assessment is typically highly conservative, both from an exposure and an effects characterization perspective and thus it is characterized by high uncertainty. This means that even values of RQ that are below one are quite likely to be capable of causing an effect. Usually, a safety factor is applied to risk quotients to cover uncertainty. The factor can vary between 1 and 100, depending on the organisms being assessed and whether the toxicity endpoint is acute, based on short-term effects (LD/LC/EC50), or chronic, based on NOEC [83,131].

Therefore, as a final step in the risk characterization procedure, the results of the RQ are compared to acceptable levels designated by the particular jurisdiction. These regulatory triggers used to categorize the potential risk are defined as Levels of Concern (LOC). An example of LOCs of RQ values that can be used for terrestrial and aquatic risk assessments is shown in Table 12.6 [143]. In the EU, TERs for terrestrial acute effects must be $\geq$ 10 and aquatic short-term effects $\geq$ 100. Chronic and subchronic TERs $\geq$5 and 10 for terrestrial and aquatic species, respectively, are acceptable.

Descriptive uncertainty analysis is usually performed in the lower tiered risk assessments while sensitivity analysis and more complex model (i.e., Monte Carlo)

TABLE 12.6
EPA Established Risk Quotients and Levels of Concern for Different Environmental Applications [from ref. 143]

End Point and Scenario	Risk Quotient	Non-Endangered	Endangered
Mammalian acute (granular)	EEC/LD50/FT^2	0.5	0.1
Mammalian acute (spray)	EEC/LC50	0.5	0.1
Mammalian chronic (spray)	EEC/NOEC	1.0	1.0
Avian acute (granular)	EEC/LD50/FT^2	0.5	0.1
Avian dietary (spray)	EEC/LC50	0.5	0.1
Avian chronic (spray)	EEC/NOEC	1.0	1.0
Aquatic acute	EEC/LC50	0.5	0.05
	EEC/EC50		
Aquatic chronic	EEC/NOEC	1.0	1.0
Terrestrial plants	EEC/EC25	1.0	1.0
Aquatic plants	EEC/EC50	1.0	1.0

simulation are usually completed in higher tier assessments. Monte Carlo simulations can be performed by using a risk quotient approach by using randomly selected toxicity values from the generated SSDs and dividing these by the environmental concentrations randomly selected from their specified distributions to produce RQ or TER values. Such an approach when repeated thousands of times builds up a distribution of RQ or TER values and provides information on the risk assessment uncertainty, as more environmentally realistic assumptions are introduced [144].

In conclusion, if consideration of the "worst case" scenario results in TERs or RQs that are acceptable when compared with LOC, then no further risk assessment is needed. If the tier 1 assessment does not pass the risk criteria, then the assessment needs to be refined and iterated back to the initial exposure and toxicity characterization but using a higher tier procedure.

12.3.2.2.2 Risk Refining and Hazard of Pesticide Mixtures

Risk refinement must be a tiered process so that more realistic and/or comprehensive sets of data, assumptions, and models are used to re-examine the potential risk. There is a tendency to jump straight from tier 1 to chemical monitoring in the environment and generate "real world" data. However, this approach has its limitations since it provides only a snapshot in time and rarely gives sufficient information about concentrations over time, which is often necessary to determine exposure. For tier 1 the US EPA uses the GENEEC exposure model and for tier 2 modeling the PRZM/EXAMS modeling systems which are specific to a particular crop and region [131]. Currently used models that are used in risk assessment approaches were presented in Section 3.1.4. Refinement of toxic effects is usually obtained through the application of probabilistic approaches, presented below in this chapter.

Until now the relative risk of single pesticide compounds has been discussed. However, as already reported in the first sections of the chapter, multiresidues of pesticides are usually detected in the different environmental compartments. For the estimation of pesticide mixture effects, the quotient addition method is generally applied. The quotient addition approach assumes that toxicities are additive or approximately additive and that there are no synergistic, antagonistic, or other interactions. The additive response of a mixture of pesticides with the same toxicological mode of action can be assessed according to the so-called Loewe additivity model as described in Equation 12.7. The sum of the toxic quotients of all compounds detected gives an estimate of the total toxicity of the sample with respect to the compounds determined.

$$TU_{mix} = \sum_{i=1}^{n} TU_i \qquad (12.7)$$

Where $TU_i = C_i/EC_i$ are the toxic units of individual pesticides calculated as TERs or RQs.

This assumption may be most applicable when the modes of action of chemicals in a mixture are similar (as for carbamates and phosphate esters), but there is evidence that even with chemicals having dissimilar modes of action, additive or near-additive interactions are common [129]. This approach provides an estimate of the contribution of the compound of interest to the total toxicity of the water sample analyzed to a certain taxonomic group.

12.3.2.2.3 Limitations of the Method

The risk quotient is a useful tool because it provides the risk managers a screening method to facilitate the rapid identification of pesticides that are not likely to pose an ecological risk. However, the risk quotient cannot address issues related to magnitude, probability, and species diversity. A common error in the interpretation of RQs is the assumption that the RQ itself is proportional to the risk. Since the concept of risk incorporates an element of probability, the RQ is biased because it assumes that the conditions exist on every occasion and in every location, and that there is a 100% probability of co-occurrence of the stressor and the most sensitive organism.

Thus, major limitations of the quotient method for ecological risk assessment are that it fails to consider variability of exposures among individuals in a population, ranges of sensitivity among species, and the ecological function of species, assuming that is a keystone organism in the environment.

12.3.3 Probabilistic Risk Assessment (PRA) (Tier 2)

The use of probabilistic approaches allows the quantification of likelihood of effects, which by definition is the quantification of risk. In probabilistic approaches the risk is expressed as the degree of overlap between the exposure and the effects that is acceptable for a certain level of protection that would be attained [145]. PRA approaches use SSD combined with distributions of exposure concentrations to better describe the likelihood of exceedances of effect thresholds and thus the risk of adverse effects. The frequency of occurrence of levels of exposure (return frequencies) can be classified as follows: typical case (50th percentile); reasonable worst case (90th percentile); and extreme worst case (99th percentile). From the resulting SSDs, exposure levels that would protect 90, 95%, or indeed any percentage of the species, can be determined. Of course, there are a number of concerns such as what level, if any, of species affected might be acceptable, which species might be affected, how might they be affected, and are they economically, ecologically, or otherwise important.

Hart [146] in his summary of an EU-funded workshop on pesticide PRA, identified several strengths and weaknesses of PRA within the context of EC Directive 91/414/EEC [14]. Strengths of PRA include: (a) the ability to quantify the type, magnitude, and frequency of toxic effects and communicate more "meaningful" outputs to decision-makers and the public; (b) the ability to quantify variability, uncertainty, and model sensitivity; (c) the better use of available information by taking into account all available toxicity data in order to quantify variation between species and not just the more sensitive or representative organism for the ecosystem only; and (d) finally, probabilistic methods are also more prone to be coupled with new approaches such as geographical information systems and population modeling. Potential weaknesses include the greater complexity that could lead in misleading results, the requirement of more toxicity data and thus the increased animal testing, the lack of available expertise and guidance, and the lack of established criteria for decision-makers [146].

PRAs could be applied for all organisms as well as for human health and have been recommended for regulatory assessment of pesticides [128]. The general concepts have been reviewed and discussed [145-147]. The different PRA methods are similarly developed but they may be used for different purposes. Some uses include the setting of environmental quality objectives and criteria while others are used for assessing risks of known exposure. As a first example of a PRA method of risk assessment the inverse method of Van Straalen and Denneman [148] is presented. The method is based on the assumption that the frequency distribution of effect end-points for different species is log-logistic. The parameters describing the distribution could be estimated for the mean and the standard deviation of the ln-transformed data set of a number of toxicity end-points of a given pesticide reported in the literature. From this distribution a concentration is calculated that is hazardous for 5% of the species in an ecosystem (HC5, Equation 12.8) which is an acceptable level for protecting aquatic ecosystems [83].

$$HC5 = \exp\left(x_m - k_L s_m\right) \tag{12.8}$$

where:

m = the number of the test species

x_m = the mean of the ln-transformed toxicity end-points (LC50 or EC50 or NOEC)

s_m = the standard deviation of the ln-transformed effect levels

k_L = the extrapolation constant.

The 95% confidence level provides a strict or safe HC5 while the 50% confidence level provides the most probable or mean. In most studies the 50% confidence level was used.

The hazard or ecological risk is estimated by defining it as the probability, Φ, that a random species will be affected by the measured field concentrations (C).

$$\Phi = \left[1 + \exp\left\{\frac{x_m - \ln C}{k_L / \ln(95/5)\, s_m}\right\}\right]^{-1} \tag{12.9}$$

This method has been followed by several researchers in pesticide risk assessment in aquatic systems [3,83]. For estimating the combined risk ($\Sigma\Phi$) from pesticide mixtures the equation of the addition of probabilities can be used as follows:

$$\Phi\left[A_1 + A_2 + \ldots + A_n\right] = \sum_{i=1}^{n} \Phi\left[Ai\right] - \sum_{i_1 < i_2} \Phi\left[Ai_1 Ai_2\right] + \ldots$$

$$+ (-1)^{r+1} \sum_{i_1 < i_2 < \ldots < i_r} \Phi\left[Ai_1 Ai_2 \ldots Ai_r\right] + \ldots \tag{12.10}$$

$$+ (-1)^{(n+1)} \Phi\left[A_1 A_2 \ldots A_n\right]$$

The summation $\displaystyle\sum_{i_1 < i_2 < ... < i_r} \Phi\left[Ai_1 Ai_2 ... Ai_r\right]$ is taken over all of $\binom{n}{r}$ possible subsets of

the ecological risk r of the compounds $\{1, 2, ..., n\}$. The equation does not account for synergistic or antagonistic interactions.

A second generic method of PRA is presented below. The method has been used by a number of authors [149–150] and is currently being implemented by the US EPA [137,151]. Toxicity data for all species are combined to produce a distribution curve of effects concentration where appropriate data for all species are fitted to log-normal distributions, while other models or bootstrapping models can be also used. The exposure data (measured values from monitoring programs or estimated by modeling) are plotted on the same axes as the effects data. The extent of overlapping between the two curves indicates the probability of exceeding an exposure concentration associated with a particular probability of effects of the studied pesticide. For plotting cumulative percentage (or cumulative probability) of the total distribution, both the acute and chronic toxicity data and the environmental pesticide concentrations are separately sorted into ascending order and ranked. These data are then converted to a cumulative percentage of the total distribution using Equation 12.11 [149]:

$$\text{Cumulative Percent} = \left[\text{rank}/(n+1)\right] \times 100 \tag{12.11}$$

where n is the total number of environmental concentration or toxicity data used to perform the quantitative assessment. These percentiles were plotted against the log-transformed concentration, and a linear regression was performed to characterize each distribution (Figure 12.3A and B). Alternatively, straight-line transformations of probability functions are obtained by probit transformation according to the equation:

$$\int (x, \mu, \sigma) = \frac{1}{\sqrt{2\pi\sigma}} e^{-(x-\mu)/2\sigma} \tag{12.12}$$

where μ is the distribution mean and σ is the distribution standard deviation [145].

Approaches for handling data below the detection limits include the assigning of values as zero or one-half the detection limit or the detection limit. Alternatively, non-detected concentrations are assumed to be distributed along a lower extension of the distribution (Figure 12.3A). The use of distribution curves for exposure and toxicity data allows the application of a joint probability method (Figure 12.3C) to perform the environmental risk assessment. In this way, any level of effect is associated with an exposure concentration and inversely for any concentration level a probability of exceedance of this level can be determined [145]. In the example provided (Figure 12.3C), the concentration at which 5% of species toxicity values will be exceeded is approximately 50 µg/L. Approximately 90% of all water concentrations would be expected to be ≤ 50µg/L or in other words this concentration would be exceeded 10% of the time. The final step in the probabilistic approach is to generate a joint probability plot of the exceedance of data (exceedance profile). This can be performed by solving the functions describing the probability of exceeding both an

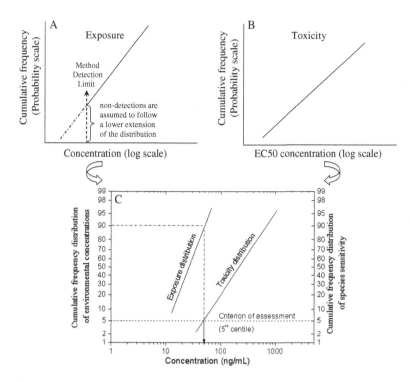

FIGURE 12.3 Graphical representation of combination of (A) exposure and (B) toxicity data expressed as linearized probability distributions (C) for the probabilistic risk estimation.

exposure and an effect concentration with appropriate fitted regression models or from Monte Carlo modeled data [128]. The graphical representation of a joint probability curve (JPC) which describes the probability of exceeding the concentration associated with a particular degree of effect is shown in Figure 12.4. In such a type of representation the closer the joint probability curve to the axes, the less probability of adverse effect [128].

There is a debate over which value from a range of species sensitivities is most appropriate to protect the various environmental compartments. The 5th percentile value is a generically applicable level of species protection used by US EPA [151], European [152], and Australian [153] quality criteria.

PRA methods gained acceptance and have been used for the ERA of pesticides in natural resources by several researchers (e.g., [154]) as well as regulatory agencies (e.g., [155]) over the past fifteen years. For PRA however, a harmonization of the methods used through calibration and validation should be established for their appropriate use in environmental risk assessment. For example, PRA can potentially underestimate risks if the substantial uncertainties arising from the extremely high variability of pesticide exposure data have not been taken into account [156]. In addition, methods for dealing with spatial and temporal variation and regional scenarios are recommended to be developed and validated with the help of GIS approaches.

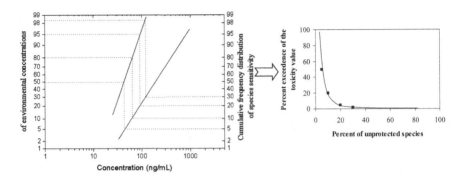

FIGURE 12.4   Graphical representation of the derivation of a joint probability curve (exceedance profile from exposure and toxicity probability functions). (Modified from Solomon K., Giesy J., Jones P. Probabilistic risk assessment of agrochemicals in the environment, *Crop Prot.*, 19, 649, 2000.)

As a generic conclusion, the methods that can be followed for pesticide risk assessment in the environmental compartments can include comparisons between point estimates and/or distributions of exposure and toxicity data depending on the data available and the questions being addressed in the assessment (Table 12.7). A summary of integrative methodologies, applied in contaminated estuarine systems, for the assessment of pesticides and predicted or measured adverse biological effects was reviewed recently [157]. An example of a tiered approach for risk characterization is showed in Table 12.8. Once risk has been characterized, it is necessary to follow basic guidelines for risk management and communication strategies [158].

TABLE 12.7
Methods for the Risk Assessment of Pesticides in the Environment Depending on the Data Availability and the Questions being Addressed in the Assessment

Method	Exposure	Effects	Output
Point estimate quotients	Point estimation	Point estimation	A ratio of exposure/toxicity
Distribution-point estimation comparison	Distribution	Point estimation	Probability of exposure exceeding the effect levels
Exposure and effect distribution comparison	Cumulative frequency distribution	Distribution	Probability of certain effects occurring when a fixed exposure level is exceeded
Distribution based quotients	Distribution (Monte Carlo simulation)	Distribution	Probability distribution of quotients (probability that exposure exceeds toxicity)
Integrated exposure and effects distribution	Distribution (Monte Carlo simulation)	Distribution (Monte Carlo simulation)	Probability and magnitude of effect occurring

TABLE 12.8
Tiered Risk Assessment Scheme [modified from ref. 158]

Tier Level	Exposure Assessment	Risk Assessment
Tier 1 (Deterministic)	Screening level EEC based on a high-exposure scenario	Is the EEC < point estimates of toxicity for the most sensitive species (L(E)C50 or NOEC)?
	If "yes" → No further assessment necessary; if "no" → Tier 2 or mitigate	
Tier 2 (Probabilistic)	Reasonable high-exposure EECs based on improved model simulations	Is the upper 10th percentile of the distribution of EEC < the lower 10th percentile of the distribution of toxicity estimates (L(E)C50 or NOEC)?
	If "yes" → No further assessment necessary; if "no" → Tier 3 or mitigate	
Tier 3	More specific scenarios for defining geographical and climate driven EECs	As for Tier 2
	If "yes" → No further assessment necessary; if "no" → Tier 4 or mitigate	
Tier 4	Site-specific EECs (pulsed exposures) or landscape modeling confirmed by environmental monitoring	As for Tier 2 or use more realistic toxicity tests
	If "yes" → No further assessment necessary; if "no" → mitigate	

12.4 LIMITATIONS AND FUTURE TRENDS OF MONITORING AND ECOLOGICAL RISK ASSESSMENT FOR PESTICIDES

Although many monitoring studies have been conducted in the past, several gaps need to be completed. Monitoring has been preoccupied with measuring environmental levels rather than describing exposure and fate and/or determining the possible adverse effects and/or evaluating the efficiency of mitigation methods. Metabolites have not been included in many monitoring programs and also novel pesticides should be studied since the patterns of pesticide use are constantly changing as the popularity of existing compounds rises and falls and as new compounds are introduced into farming. In addition, available monitoring data rarely are comparable due to variability of analytical methods followed and the objectives targeted. It should also be noted that monitoring strategies which follow reduced or sporadic sampling largely underestimate pesticide concentrations and fluxes and fail to assess realistic pesticide exposures. An event-based sampling design is more appropriate due to the very transient nature of pesticide exposures. Traditionally applied fixed-interval sampling schemes fail to accurately depict the low-frequency/high-risk exposure patterns of pesticides. Well-structured monitoring programs based on multimedia monitoring approaches and conducted according to QA/QC in sampling and analysis should be developed covering regional, national, and/or global patterns.

Concerning the current approaches of ecological risk assessment, one criticism is that some groups of organisms are not represented. The most commonly mentioned are wild mammals, reptiles, and amphibians. Ecological risk assessment for pesticides concentrates on direct impacts on exposed species. If direct effects are anticipated, the potential for indirect effects does need to be considered and should be addressed by community level studies. In addition, research is needed to characterize the response of organisms to pulsed exposures and site-specific conditions. Furthermore, ecological risk assessment for pesticides often concentrates on single compounds, whereas in the environment organisms might be exposed to more than one pesticide or other chemical mixtures including natural toxins and a variety of other stressors. In fact, complex pesticide mixtures in aquatic environments generate a combined effect that exceeds that of each individual compound. Additive effects are usually considered for chemicals having the same mode of action. Greatest concern should be expressed for synergistic action, that is, the effect of the chemicals together is greater than that predicted from the parts, while the converse of this (antagonistic effect) is not likely to be a priority concern in a risk assessment.

It must be highlighted that, for a standard risk assessment procedure applicable for screening and comparison purposes, a database of exposure and toxicity for major pesticides and metabolites using a standard methodology is required. In many cases, risk assessments are hampered by the lack of data, especially of toxic effects, thus more ecotoxicological studies, which will result in less conservative hazard limits with less uncertainty, are needed.

In conclusion, monitoring of the environment for pesticide residues and ecological risk assessment must continue based on harmonized methodologies and systematic studies. Monitoring data are critical elements in quantitative evaluations of environmental and human hazards and risk. We must be vigilant for early warning signs of damage of ecological systems. The ecological risk assessment approach could thus contribute to debate and give invaluable help in defining environmental guidelines for pesticides. To achieve the goal of environmental sustainability, the continuous and deeper scientific knowledge obtained from the monitoring and risk assessment procedures constitutes a powerful tool.

REFERENCES

1. Konstantinou I.K., Hela D.G., Albanis T.A. The status of pesticide pollution in surface waters (rivers and lakes) of Greece. Part I. Review on occurrence and levels, *Environ. Pollut.*, 141, 555, 2006.
2. Müller J.F. et al. Pesticides in sediments from Queensland irrigation channels and drains, *Mar. Pollut. Bull.*, 41, 294, 2000.
3. Hela D.G. et al. Determination of pesticide residues and ecological risk assessment in lake Pamvotis, Greece, *Environ. Toxicol. Chem.*, 24, 1548, 2005.
4. Albaigés J. *Persistent Organic Pollutants in the Mediterranean Sea*, Handbook of Environmental Chemistry, Vol. 5, Part K, DOI:10.1007/b107145, Springer-Verlag, Berlin, 2005.
5. Harner T. et al. Global pilot study for persistent organic pollutants (POPs) using PUF disk passive air samplers, *Environ. Pollut.*, 144, 445, 2006.
6. Tuduri L. et al. A review of currently used pesticides in Canadian air and precipitation: Part 1: Lindane and endosulfans, *Atmos. Environ.*, 40, 1563, 2006.

7. Sánchez-Brunete C., Albero B., Tadeo J.L. Multiresidue determination of pesticides in soil by gas chromatography–mass spectrometry detection, *J. Agric. Food Chem.*, 52, 1445, 2004.

8. Goncalves C. et al. Chemometric interpretation of pesticide occurrence in soil samples from an intensive horticulture area in north Portugal, *Anal. Chim. Acta*, 560, 164, 2006.

9. Li H. et al. Occurrence, seasonal variation and inhalation exposure of atmospheric organophosphate and pyrethroid pesticides in an urban community in South China, *Chemosphere*, 95, 363, 2014.

10. Guzzella L., Pozzoni F., Giuliano G. Herbicide contamination of surficial groundwater in Northern Italy, *Environ. Pollut.*, 142, 344, 2006.

11. Sahoo G.B. et al. Application of artificial neural networks to assess pesticide contamination in shallow groundwater, *Sci. Total Environ.*, 367, 234, 2006.

12. Bartram J. *Water and Health in Europe: A Joint Report from the European Environment Agency and the WHO Regional Office for Europe*, World Health Organization (WHO), Regional publications European series, No. 93. 2002.

13. Maloschik E. et al. Monitoring water-polluting pesticides in Hungary, *Microchem. J.*, 85, 88, 2007.

14. European Economic Community. Council Directive concerning the placing of plant protection products on the market. 91/414/EEC, OJ L 230. Brussels, Belgium, 1991.

15. Commission Regulation (EC), No 33/2008. Official Journal of the European Union L 15/5-L15/12, 18.1.2008.

16. Commission Implementing Regulation (EU). No 540/2011 of 25 May 2011, implementing Regulation (EC) No 1107/2009 of the European Parliament and of the Council as regards the list of approved active substances, Official Journal of the European Union L 153/1-L153/186, 11.6.2011.

17. Wiersma B.G. *Environmental Monitoring*, CRC Press, Boca Raton, FL, 2004.

18. Calamari D. et al. Monitoring as an indicator of persistence and long-range transport. In Klecka G. et al. (Eds.), *Evaluation of Persistence and Long Range Transport of Organic Chemicals in the Environment*, SETAC, 2000.

19. Minsker B. *Long-Term Groundwater Monitoring Design: The State of the Art*, American Society of Civil Engineers, Reston, VA, 2003.

20. Quevauviller Ph. *Quality Assurance for Water Analysis*, John Wiley & Sons, Chichester, UK, 2002.

21. Usländer T. Trends of environmental information systems in the context of the European Water Framework Directive, *Environ. Model. Soft.*, 20, 1532, 2005.

22. Coquery M. et al. Priority substances of the European Water Framework Directive: analytical challenges in monitoring water quality, *Trends Anal. Chem.*, 24, 117, 2005.

23. Shen L. et al. Atmospheric distribution and long range transport behavior of organochlorine pesticides in North America, *Environ. Sci. Tech.*, 39, 409, 2005.

24. Garbarino J.R. et al. Contaminants in Arctic snow collected over northwest Alaskan sea ice. *Water Air Soil Pollut.*, 139, 183, 2002.

25. Gil Y., Sinfort, C., Emission of pesticides to the air during sprayer application: a bibliographic review, *Atmos. Environ.*, 39, 5183, 2005.

26. Bedos C. et al. Rate of pesticide volatilization from soil: an experimental approach with a wind tunnel system applied to trifluralin, *Atmos. Environ.*, 36, 5917, 2002.

27. Ferrari F. et al. Predicting and measuring environmental concentration of pesticides in air after soil application, *J. Environ. Qual.*, 32, 1623, 2003.

28. Socorro J. et al. The persistence of pesticides in atmospheric particulate phase: an emerging air quality issue, *Sci. Rep.*, 6, Article No. 33456, 2016.

29. Klánova J. et al. Passive air sampler as a tool for long-term air pollution monitoring: Part 1. Performance assessment for seasonal and spatial variations, *Environ. Pollut.*, 144, 393, 2006.

30. Peck A.M., Hornbuckle K.C. Gas-phase concentrations of current-use pesticides in Iowa, *Environ. Sci. Technol.*, 39, 2952, 2005.

31. Coscolla C. et al. Occurrence of currently used pesticides in ambient air of Centre Region (France), *Atmos. Environ.*, 44, 3915, 2010.

32. Hayward J. et al. Levels and seasonal variability of pesticides in the rural atmosphere of Southern Ontario, *J. Agric, Food Chem.*, 58, 1077, 2010.

33. Sanusi A. et al. Comparison of atmospheric pesticide concentrations measured at three sampling sites: local, regional and long-range transport, *Sci. Total Environ.*, 263, 263, 2000.

34. Scheyer A. et al. Variability of atmospheric pesticide concentrations between urban and rural areas during intensive pesticide application, *Atmos. Environ.*, 41, 3604, 2007.

35. Aulagnier F. et al. Pesticides measured in air and precipitation in the Yamaska Basin (Québec): occurrence and concentrations in 2004, *Sci. Total Environ.*, 394, 338, 2008.

36. Schummer C. et al. Temporal variations of concentrations of currently used pesticides in the atmosphere of Strasbourg, France, *Environ. Pollut.*, 158, 576, 2010.

37. White L.M. et al. Ambient air concentrations of pesticides used in potato cultivation in Prince Edward Island, Canada, *Pest Manage. Sci.*, 62, 126, 2006.

38. Gouin T. et al. Atmospheric concentrations of current-use pesticides across south-central Ontario using monthly-resolved passive air samplers, *Atmos. Environ.*, 34, 8096, 2008.

39. Baraud L. et al. A multi-residue method for characterization and determination of atmospheric pesticides measured at two French urban and rural sampling sites, *Anal. Bioanal. Chem.*, 377, 1148, 2003.

40. Carratala A. et al. Occurrence and seasonal distribution of polycyclic aromatic hydrocarbons and legacy and current-use pesticides in air from a Mediterranean coastal lagoon (Mar Menor, SE Spain), *Chemosphere*, 167, 382, 2017.

41. Estellano V.H. et al. Assessing levels and seasonal variations of current-use pesticides (CUPs) in the Tuscan atmosphere, Italy, using polyurethane foam disks (PUF) passive air samplers, *Environ. Pollut.*, 205, 52, 2015.

42. Coscolla C. et al. LC–MS characterization of contemporary pesticides in PM10 of Valencia Region, Spain, *Atmos. Environ.*, 77, 394, 2013.

43. Pesticides Properties Database (PPDB). https://sitem.herts.ac.uk/aeru/ppdb/en/, 2018.

44. Wang J. A case study of air quality - pesticides and odorous phytochemicals on Kauai, Hawaii, USA, *Chemosphere*, 189, 143, 2017.

45. Sofuoglu A. et al. Temperature dependence of gas-phase polycyclic aromatic hydrocarbon and organochlorine pesticide concentrations in Chicago air, *Atmos. Environ.*, 35, 6503, 2001.

46. Li H. et al. Occurrence, seasonal variation and inhalation exposure of atmospheric organophosphate and pyrethroid pesticides in an urban community in South China, *Chemosphere*, 95, 363, 2014.

47. Pozo K. et al. Occurrence of chlorpyrifos in the atmosphere of the Araucanía Region in Chile using polyurethane foam-based passive air samplers, *Atmos. Pollut. Res.*, 7, 706, 2016.

48. Raina R., Sun L. Trace level determination of selected organophosphorus pesticides and their degradation products in environmental air samples by liquid chromatography-positive ion electrospray tandem mass spectrometry, *J. Environ. Sci. Health B*, 43, 323, 2008.

49. Russo M.V. et al. Sampling of organophosphorus pesticides at trace levels in the atmosphere using XAD-2 adsorbent and analysis by gas chromatography coupled with nitrogen-phosphorus and ion-trap mass spectrometry detectors, *Anal. Bioanal. Chem.*, 404, 1517, 2012.

50. Air PACA. *Observatoire des Residus de Pesticides en PACA Marseille*, Certified Associations of Air Quality Monitoring - Association ATMO, France, 2012.

51. Raeppel C. et al. Coupling ASE, sylilation and SPME-GC/MS for the analysis of current-used pesticides in atmosphere, *Talanta*, 121, 24, 2014.

52. Huston R. et al. Characterisation of atmospheric deposition as a source of contaminants in urban rainwater tanks, *Water Res.*, 43, 1630, 2009.

53. Dubus I.G., Hollis J.M., Brown C.D. Pesticides in rainfall in Europe, *Environ. Pollut.*, 110, 331, 2000.

54. Mimbs IV W.H. et al. Occurrence of current-use fungicides and bifenthrin in rainwater basin wetlands, *Chemosphere*, 159, 275, 2016.

55. Tuduri L. et al. *CANCUP: Canadian Atmospheric Network for Currently Used Pesticides*, Managing Our Waters, Cornwall, Canada, 2004.

56. California Department of Pesticide Regulation. USA, 2018. http://www.cdpr.ca.gov/do cs/emon/airinit/air_network.htm

57. Lig'Air-Association de surveillance de la qualité de l'air en région Centre-Val de Loire. https://www.ligair.fr/publication-et-outils-pedagogiques/nouveaux-polluants/pesticides

58. Quaghebeur D. et al. Pesticides in rainwater in Flanders, Belgium: results from the monitoring program 1997–2001, *J. Environ. Monit.*, 6, 182, 2004.

59. Namiesnik J. et al. Passive sampling and/or extraction techniques in environmental analysis: a review, *Anal. Bioanal. Chem.*, 381, 279, 2005.

60. Vrana B. et al. Passive sampling techniques for monitoring pollutants in water, *Trends Anal. Chem.*, 24, 845, 2005.

61. Hanrahan G., Patil D., Wang J. Electrochemical sensors for environmental monitoring: design, development and applications, *J. Environ. Monit.*, 6, 657, 2004.

62. Tschmelak J., Proll G., Gauglitz G. Optical biosensor for pharmaceuticals, antibiotics, hormones, endocrine disrupting chemicals and pesticides in water: assay optimization process for estrone as example, *Talanta*, 65, 313, 2005.

63. Thomatou A.-A. et al. Determination and risk assessment of pesticide residues in lake Amvrakia (W. Greece) after agricultural land use changes in the lake's drainage basin, *Intern. J. Environ. Anal. Chem.*, 93, 780, 2013.

64. Ccanccapa A. et al. Pesticides in the Ebro River basin: occurrence and risk assessment, *Environ. Pollut.*, 211, 414, 2016.

65. Reilly T.J. et al. Occurrence of boscalid and other selected fungicides in surface water and groundwater in three targeted use areas in the United States, *Chemosphere*, 89, 228, 2012.

66. Stamatis N. et al. Spatiotemporal variation and risk assessment of pesticides in water of the lower catchment basin of Acheloos River, Western Greece, *Sci. World J.*, 2013, Article ID 231610, 16 pages.

67. Struger J. et al. Factors influencing the occurrence and distribution of neonicotinoid insecticides in surface waters of southern Ontario, Canada, *Chemosphere*, 169, 516, 2017.

68. Meffe R., de Bustamante I. Emerging organic contaminants in surface water and groundwater: a first overview of the situation in Italy, *Sci. Total Environ.*, 481, 280, 2014.

69. Mahler B.J. et al. Similarities and differences in occurrence and temporal fluctuations in glyphosate and atrazine in small Midwestern streams (USA) during the 2013 growing season, *Sci. Total Environ.*, 579, 149, 2017.

70. Tang W. et al. Pyrethroid pesticide residues in the global environment: an overview, *Chemosphere*, 191, 990, 2018.

71. Claver A. et al. Study of the presence of pesticides in surface waters in the Ebro River basin (Spain), *Chemosphere*, 64, 1437, 2006.

72. Capel P.D., Larson S.J., Winterstein T.A. The behaviour of 39 pesticides in surface waters as a function of scale, *Hydrol. Proces.*, 15, 1251, 2001.

73. Larson S.J. et al. Relations between pesticide use and riverine flux in the Mississippi river basin, *Chemosphere*, 31, 3305, 1995.

74. Steen R.J.C.A. Fluxes of Pesticides into the Marine Environment: Analysis, Fate and Effects. PhD Thesis, Vrije Universiteit, Amsterdam, The Netherlands, 2002.

75. Manamsa K. et al. A national-scale assessment of micro-organic contaminants in groundwater of England and Wales, *Sci. Total Environ.*, 568, 712, 2016.

76. Loos R. et al. Pan-European survey on the occurrence of selected polar organic persistent pollutants in ground water, *Water Res.*, 44, 4115, 2010.

77. Lopez B. et al. Screening of French groundwater for regulated and emerging contaminants, *Sci. Total Environ.*, 518, 562, 2015.

78. Sanchis J. et al. Determination of glyphosate in groundwater samples using an ultrasensitive immunoassay and confirmation by on-line solid-phase extraction followed by liquid chromatography coupled to tandem mass spectrometry, *Anal. Bioanal. Chem.*, 402, 2335, 2012.

79. Börjesson E., Torstensson L. New methods for determination of glyphosate and (aminomethyl)-phosphonic acid in water and soil, *J. Chromatogr. A*, 886, 207, 2000.

80. Battaglin W.A. et al. Glyphosate and its degradation product AMPA occur frequently and widely in U.S. soils, surface water, groundwater, and precipitation, *J. Am. Water Res. Assoc.*, 50, 275, 2014.

81. Vryzas Z. et al. Occurrence of pesticides in transboundary aquifers of North-eastern Greece, *Sci. Total Environ.*, 441, 41, 2012.

82. Cohen S.Z. et al. A ground water monitoring study for pesticides and nitrates associated with golf courses on Cape Cod, *Ground Water Monit. Rev.*, 10, 160, 1990.

83. Steen R.J.C.A. et al. Ecological risk assessment of agrochemicals in European estuaries, *Environ. Toxicol. Chem.*, 18, 1574, 1999.

84. FAO/IAEA/UNEP. Report of the FAO/IAEA/UNEP Consultation Meeting on the Fungicides Pilot Survey (Ioannina, Greece, 1993). UNEP, Athens, 9 pp. (mimeo), 1993.

85. Konstantinou I.K. (Eds.) *Antifouling Paint Biocides*, Handbook of Environmental Chemistry, Vol. 5, Part O, DOI:10.1007/11555148, Springer-Verlag, Berlin, Heidelberg, 2006.

86. Long J.L.A. et al. Micro-organic compounds associated with sediments in the Humber rivers. *Sci. Total Environ.*, 210–211, 229, 1998.

87. Vega A.B. et al. Monitoring of pesticides in agricultural water and soil samples from Andalusia by liquid chromatography coupled to mass spectrometry, *Anal. Chim. Acta*, 538, 117, 2005.

88. Harner T. et al. Residues of organochlorine pesticides in Alabama soils, *Environ. Pollut.*, 106, 323, 1999.

89. Hvězdová M. et al. Currently and recently used pesticides in central European arable soils, *Sci. Total Environ.*, 613–614, 361, 2018.

90. Padilla-Sánchez J.A. et al. Residues and organic contaminants in agricultural soils in intensive agricultural areas of Spain: a three years survey. *Clean – Soil Air Water*, 43, 746, 2015.

91. Karasali H. et al. Occurrence and distribution of trifluralin, ethalfluralin, and pendimethalin in soils used for long-term intensive cotton cultivation in central Greece, *J. Environ. Sci. Health B*, 52, 719, 2017.

92. Aznar R. et al. Spatio-temporal distribution of pyrethroids in soil in Mediterranean paddy fields, *J. Soils Sediments*, 17, 1503, 2017.

93. Rafique N. et al. Monitoring and distribution patterns of pesticide residues in soil from cotton/wheat fields of Pakistan, *Environ. Monit. Assess.*, 188, 695, 2016.

94. Wurl O., Obbard J.P. Organochlorine pesticides, polychlorinated biphenyls and poly-brominated diphenyl ethers in Singapore's coastal marine sediments, *Chemosphere*, 58, 925, 2005.

95. Covaci A. et al. Polybrominated diphenyl ethers, polychlorinated biphenyls and organo-chlorine pesticides in sediment cores from the Western Scheldt river (Belgium): ana-lytical aspects and depth profiles, *Environ. Int.*, 31, 367, 2005.

96. Smalling K.L. et al. Occurrence and persistence of fungicides in bed sediments and suspended solids from three targeted use areas in the United States, *Sci. Total Environ.*, 447, 179, 2013.

97. Palma P. et al. Occurrence and potential risk of currently used pesticides in sediments of the Alqueva reservoir (Guadiana Basin), *Environ. Sci. Pollut. Res.*, 22, 7665, 2015.

98. Masiá A. et al. Screening of currently used pesticides in water, sediments and biota of the Guadalquivir River Basin (Spain), *J. Hazard. Mater.*, 263P, 95, 2013.

99. Pinto M.I. et al. Priority pesticides in sediments of European coastal lagoons: a review, *Mar. Pollut. Bull.*, 112, 6, 2016.

100. Delgado-Moreno L. et al. Occurrence and toxicity of three classes of insecticides in water and sediment in two southern California coastal watersheds, *J. Agric. Food Chem.*, 59, 9448, 2011.

101. Wu Y. et al. Residues of organophosphorus insecticides in sediment around a highly eutrophic lake, Eastern China, *J. Soils Sediments*, 15, 436, 2015.

102. Rodrigues E.T. et al. Occurrence, fate and effects of azoxystrobin in aquatic ecosys-tems: a review, *Environ. Int.*, 53, 18, 2013.

103. Cerrillo I. et al. Endosulfan and its metabolites in fertile women, placenta, cord blood, and human milk, *Environ. Res.*, 98, 233, 2005.

104. Franklin C.A., Worgan J.P. *Occupational and Residential Exposure Assessment for Pesticides*, Wiley & Sons, New York, USA, 28–45, 2005.

105. Ueyama J. et al. Temporal levels of urinary neonicotinoid and dialkylphosphate con-centrations in Japanese women between 1994 and 2011, *Environ. Sci. Technol.*, 49, 14522, 2015.

106. Fiedler N. et al. Neurobehavioral effects of exposure to organophosphates and pyre-throid pesticides among Thai children, *Neurotoxicology*, 48, 90, 2015.

107. Fisk A.T. et al. An assessment of the toxicological significance of anthropogenic con-taminants in Canadian arctic wildlife, *Sci. Total Environ.*, 351–352, 57, 2005.

108. Sakellarides Th. et al. Accumulation profiles of persistent organochlorines in liver and fat tissues of various waterbird species from Greece, *Chemosphere*, 63, 1392, 2006.

109. Klemens J.A. et al. A cross-taxa survey of organochlorine pesticide contamination in a Costa Rican wildland, *Environ. Pollut.*, 122, 245, 2003.

110. Naso B. et al. Organochlorine pesticides and polychlorinated biphenyls in European roe deer Capreolus capreolus resident in a protected area in Northern Italy, *Sci. Total Environ.*, 328, 83, 2004.

111. Sapozhnikova Y. et al. Evaluation of pesticides and metals in fish of the Dniester River, Moldova, *Chemosphere*, 60, 196, 2005.

112. Michel N. et al. Fipronil and two of its transformation products in water and European eel from the river Elbe, *Sci. Total Environ.*, 568, 171, 2016.

113. Jiang C.Z. et al. Investigation on the pesticide residues of aquatic products in Ningbo, *Chin. J. Hyg. Res.*, 36, 692, 2007.

114. Aznar-Alemany O. et al. Insecticide pyrethroids in liver of striped dolphin from the Mediterranean Sea, *Environ. Pollut.*, 225, 346, 2017.

115. Gao Z.X. et al. Determination of tetramethrin and cyhalothrin residues in mutton tis-sues by RP-HPLC with solid phase extraction, *Anal. Abst.*, 31, 116, 2012.

116. COMPRENDO, final publishable report. Executive Summary on the Project Results, Energy, Environment and Sustainable Development, March 2006.

117. Directive N. 76/464/EEC of the Council of the European Community of 4 May 1976. Pollution caused by certain dangerous substances discharged into the aquatic environment of the community, *Off. J. Eur. Comm.*, 18.05, 1976, 7.

118. EEC. Council Directive 80/68/EEC of 17 December 1979 on the protection of groundwater against pollution caused by certain dangerous substances, *Off. J. Eur. Comm. L*, 020, 1980, (26/01/1980).

119. European Parliament. Directive 2000/60/EC of the European Parliament and of the Council of 23 October 2000 establishing a framework for Community action in the field of water policy, *Off. J. Eur. Comm.*, L327, (2000), 1.

120. EU. Commission implementing decision (EU) 2015/495 of 20 March 2015 establishing a watch list of substances for union-wide monitoring in the field of water policy pursuant to Directive 2008/105/EC of the European Parliament and of the Council, *Off. J. Eur. Union*, 78, 40–42, 2015.

121. EU. Directive 2013/39/EU of the European Parliament and of the Council of 12 August 2013 amending Directive 2000/60/EC and 2008/105/EC as regards priority substances in the field of water policy, *Off. J. Eur. Union*, 226, 1–17, 2013.

122. Allan I.J. et al. Strategic monitoring for the European Water Framework Directive, *Trends Anal. Chem.*, 25, 704, 2006.

123. Wheeler W. *Pesticides in Agriculture and the Environment*, CRC Press, New York, USA, 2002.

124. Schnoor J.L. *Fate of Pesticides and Chemicals in the Environment*, John Wiley & Sons, New York, USA, 1992.

125. Williams P.R.D. An overview of exposure assessment models used by the US environmental protection agency. In Hanrahan G. (Ed.), *Modelling of Pollutants in Complex Environmental Systems*, Vol. II, ILM Publications, 61–131, 2010.

126. Mackay D. *Multimedia Environmental Models. The Fugacity Approach*, Lewis Publ, Chelsea, MI, 257, 1991.

127. USEPA (U.S. Environmental Protection Agency). Framework for ecological risk assessment. Risk Assessment Forum, U.S. Environmental Protection Agency, Washington, DC, 1992.

128. ECOFRAM (Ecological Committee on FIFRA Risk Assessment Methods). Aquatic and Terrestrial Final Draft Reports, USEPA, 1999. www.epa.gov/oppefed1/ecorisk/index.htm

129. USEPA, U.S. Environmental Protection Agency. Guidelines for ecological risk assessment. Fed Reg 63:26846–26924, Washington, DC, 1998.

130. Ritter L. et al. Sources, pathways and relative risks of contaminants in surface water and groundwater: a perspective prepared for the Walkerton inquiry, *J. Toxicol. Environ. Health*, 65, 1, 2002.

131. Hamer M. Ecological risk assessment for agricultural pesticides, *J. Environ. Monit.*, 2, 104N, 2000.

132. Reus J. et al. Comparison and evaluation of eight pesticide environmental risk indicators developed in Europe and recommendations for future use, *Agric. Ecosyst. Environ.*, 90, 177, 2002.

133. Finizio A., Villa S. Environmental risk assessment for pesticides. A tool for decision making, *Environ. Impact Assess. Rev.*, 22, 235, 2002.

134. Finizio A., Calliera M., Vighi M. Rating systems for pesticide risk classification on different ecosystems, *Ecotoxol. Environ. Saf.*, 49, 262, 2001.

135. Van Bol V. et al. Pesticide indicators, *Pestic. Outlook*, 14, 159, 2003.

136. European Commission. Technical Guidance Document on risk assessment in support of Council Directive 93/67/EEC for new notified substances and commission regulation 1488/94 on risk assessment for existing substances and Directive 98/8/EC of the European Parliament and the Council concerning the placing of biocidal products of the market. EU, JRC, Brussels, Belgium, 2002.

137. USEPA. Technical Progress Report: Implementation Plan for Probabilistic Ecological Assessments- Terrestrial Systems, USEPA, 2000. www.epa.gov/scipoly/sap/.

138. Kooijman S.A.L.M. A safety factor for LC50 values allowing for differences in sensitivity among species, *Wat. Res.*, 21, 269, 1987.

139. Van Straalen N.M., Denneman C.A.J. Ecotoxicological evaluation of soil quality criteria, *Ecotoxol. Environ. Saf.*, 18, 269, 1989.

140. Newton M.C. et al. Applying species-sensitivity distributions in ecological risk assessment: assumptions of distribution type and sufficient numbers of species, *Environ. Toxicol. Chem.*, 19, 508, 2000.

141. Newman M.C. et al. Applying species sensitivity distributions in ecological risk assessment: assumptions of distribution type and sufficient number of species, *Environ. Toxicol. Chem.*, 19, 508, 2000.

142. Van der Hoeven N. Estimating the 5-percentile of the species sensitivity distributions without any assumptions about the distribution, *Ecotoxicology*, 10, 25, 2001.

143. US EPA. Technical Overview of Ecological Risk Assessment: Risk Characterization. https://www.epa.gov/pesticide-science-and-assessing-pesticide-risks/technical-overview-ecological-risk-assessment-risk

144. Crane M. et al. Evaluation of probabilistic risk assessment of pesticides in the UK: chlorpyrifos use on top fruit, *Pest Manag. Sci.*, 59, 512, 2003.

145. Solomon K., Giesy J., Jones P. Probabilistic risk assessment of agrochemicals in the environment, *Crop Prot.*, 19, 649, 2000.

146. Hart A. Probabilistic Risk Assessment For Pesticides in Europe: Implementation & Research Needs. Report of European workshop on probabilistic risk assessment for the environmental impacts of plant protection products, Netherlands 2001.

147. Postuma L., Traas T., Suter G.W. *Species Sensitivity Distributions in Risk Assessment*, SETAC Press, Pensacola, FL, 2001.

148. Van Straalen N.M., Schobben J., Traas T.P. The use of ecotoxicological risk assessment in deriving maximum acceptable half-lives of pesticides, *Pestic Sci.*, 34, 227, 1992.

149. Giesy J.P. et al. Chlorpyrifos: ecological risk assessment in North American aquatic environments, *Rev. Environ. Contam. Toxicol.*, 160, 1, 1999.

150. Hall L.W.J. et al. A probabilistic ecological risk assessment of tributyltin in surface waters of the Chesapeake Bay, *Human Ecol. Risk Assess.*, 6, 141, 2000.

151. USEPA. Technical Progress Report of the Implementation Plan for Probabilistic Ecological Assessments - Aquatic Systems, USEPA, 2000. www.epa.gov/scipoly

152. Crommentuijn T. et al. Maximum permissible and negligible concentrations for some organic substances and pesticides, *J. Environ. Manag.*, 58, 297, 2000.

153. ANZECC (Australian and New Zealand Environment and Conservation Council). Australian and New Zealand guidelines for fresh and marine water quality. Volume 2, Aquatic Ecosystems – Rationale and Background Information. Canberra, 2000.

154. Hall L.W. Analysis of diazinon monitoring data from the Sacramento and Feather River watersheds: 1991–2001, *Environ. Monit. Assess.*, 86, 233, 2003.

155. Starner K. et al. Pesticides in Surface Water from Agricultural Regions of California 2006–2007. Report 238. Sacramento: California Environmental Protection Agency, California Department of Pesticide Regulation, 2011.

156. Stehle S. et al. Probabilistic risk assessment of insecticide concentrations in agricultural surface waters: a critical appraisal, *Environ. Monit. Assess.*, 185, 6295, 2013.

157. Cuevas N. et al. Risk assessment of pesticides in estuaries: a review addressing the persistence of an old problem in complex environments, *Ecotoxicology*. https://doi.org/10.1007/s10646-018-1910-z

158. SETAC. Pesticide Risk and Mitigation. Final Report of the Aquatic Risk Assessment and Mitigation Dialog Group, SETAC Foundation for Environmental Education, Pensacola, FL, 220, 1994.

Index

A

AASQA, *see* Air quality survey networks
Acaricides, 209
Accelerated Solvent Extraction (ASE), 307
Acceptable daily intake (ADI), 4, 330–332, 352–355
Acetolactate synthase (ALS), 7, 36
Acetone, 179–180
Acetonitrile, 177–179
Acetylcholinesterase (AChE), 37
Active ingredient, 2
Acute reference dose (ARfD), 330
Acute toxicity, 38
Admissible daily intake (ADI), 4
Affinity sorbents, 58–61
Air monitoring, 370–377
Air Monitoring Network (AMN), 372
Air quality survey networks (AASQA), 305
Algicides, 209
ALS, *see* Acetolactate synthase
Amide herbicides, 5–6
Aminomethylphosphonic acid (AMPA), 11, 378, 386
AMN, *see* Air Monitoring Network
AMPA, *see* Aminomethylphosphonic acid
Animal origin food, pesticides in, 209–211
    application to real samples
        carbamates, 231–232
        multiresidue analysis of pesticides, 230
        neonicotinoids, 233
        organochlorine pesticides, 231
        organophosphorus pesticides, 231
        pyrethroids, 232
    gas chromatography coupled to mass spectrometry, 212–214, 224–226
    internal quality control, 228–229
    IQC measures, 229–230
    liquid chromatography coupled to mass spectrometry, 215–220, 226–228
    new trends and issues, 233–235
    sample preparation, 211
        cleanup and fractionation, 223–224
        extraction techniques, 221–223
        pretreatment, 211, 221
Antifouling Agents in Coastal Environments (ACE) project, 384
APCI, *see* Atmospheric pressure chemical ionization

APPI, *see* Atmospheric pressure photoionization
ARfD, *see* Acute reference dose
ASE, *see* Accelerated Solvent Extraction
ATD Conditions, 317
Atmosphere and pesticides, 301–303
    in ambient air, 304–308
    analysis by gas chromatography, 321–322
    analysis by HPLC, 323–324
    derivatization, 322–323
    monitoring, 304
    in rainwater samples, 308–314
    soil–air transfer evaluation, 314–321
Atmospheric pressure chemical ionization (APCI), 80, 89, 192
Atmospheric pressure photoionization (APPI), 91
Azoles fungicides, 30–31

B

Back pressure regulator (BPR), 91–92
Benzimidazoles fungicides, 32, 38
Benzoic acid herbicides, 6–7
Benzoylurea insecticides, 19, 21
Biological monitoring, 387–389
Biosensors, 117
    descriptions, 117–120
    electrochemical, 119
    fiber optic, 119
    methods for pesticides
        conductive polymers, 124–125
        piezoelectric measurements, 123–124
        potentiometric, light addressable potentiometric sensor (LAPS), and amperometric detection, 122–123
        surface plasmon resonance, 124
    microarrays, 120–122
    optical, 119
Bipyridylium, 13
BPR, *see* Back pressure regulator
Brazilian national program on pesticide residues, 357
Bromoxynil, 10

C

CAC, *see* Codex Alimentarius Commission
Calibration curve, 230

415

California Department of Food and Agriculture
    (CDFA) method, 95, 177–178
Canadian Atmospheric Network for Current Use
    Pesticides (CANCUP), 372
Capillary electrophoresis (CE), 293
Carbamate insecticides, 7, 19–22
Carbamates (CBs), 209
Carbon nanotubes, 66–67
Carcinogenesis, 39
CCPR, see Codex Committee on Pesticide
    Residues
Celite, 183
Certified Reference Materials (CRMs), 229
Chemical ionization (CI), 78–79
Chlormequat, 338, 340
Chloroacetamides, 5
Chloroacetanilide herbicides, 385
Chlorpyriphos in apples, 336–337
Chromatography–mass spectrometry
    techniques, 84
    application in pesticide residue, 93–97
    comprehensive two-dimensional gas
        chromatography–time of flight mass
        spectrometry (GC×GC–TOFMS),
        92–93
    future trends, 101
    gas chromatography-mass spectrometry
        (GC-MS), 74, 84–86
        multi-residue pesticides in vegetables,
            97–98
        qualitative analysis, 86–87
        quantitative analysis, 87–89
    liquid chromatography-mass spectrometry
        (LC-MS)
        multi-residue pesticides in vegetables,
            97–98
    liquid chromatography–mass spectrometry
        (LC–MS), 74, 89–91
    overview, 74
    supercritical fluid chromatography–mass
        spectrometry (SFC–MS), 91–92
    surface water samples using liquid
        chromatography–high resolution mass
        spectrometry (LC–HRMS), 99–100
CI, see Chemical ionization
CID, see Collision Induced Dissociation
Codex Alimentarius Commission (CAC), 330
Codex Committee on Pesticide Residues
    (CCPR), 330
Collision Induced Dissociation (CID), 322
Column adsorption chromatography, 182, 184
Comprehensive two-dimensional gas
        chromatography–time of flight mass
        spectrometry (GC×GC–TOFMS),
        92–93
Contaminants, 209
Conventional gas chromatography, 321

Cost-effective monitoring programs, 367
CRM, see Certified Reference Materials
Crop Protection Products, 303
Cumulative exposure, 352–353

D

Danish Food Monitoring Program, 349
Danish nationwide food consumption survey
    2003–2008, 357
DDT, see Dichlorodiphenyltrichloroethane
Denmark, pesticides intake, 356–357
Deposition, 302
Dermal toxicity test, 38
Diatoms, 183
Dichlorodiphenyltrichloroethane (DDT), 2, 23,
    67, 182, 338–339, 342–350
Diffusive sampler, 306
Dispersive liquid-liquid microextraction
    (DLLME), 54, 222, 275–279
Dispersive solid-phase extraction (dSPE), 184
Dithiocarbamate fungicides, 32–33
Dithiocarbamates, 333, 349
DLLME, see Dispersive liquid-liquid
    microextraction
Dutch mini-Luke (NL) method, 179

E

ECD, see Electron capture detectors
Ecological risk assessment, 394
EEC, see Environmental Estimated
    Concentrations
EFSA, see European Food Safety Authority
EI, see Electron ionization
Electrochemical biosensors, 119
Electroconductive polymer sensors, 120
Electrolytic conductivity detection (ECD), 178
Electron capture detectors (ECD), 310, 321
Electronic pressure control (EPC)
    devices, 188
Electron ionization (EI), 77–78
Electrospray ionization (ESI), 80
ELISA, see Environmental Indicators
        for Sustainable Agriculture;
        Enzyme-linked immunosorbent assay
Enhanced solvent extraction (ESE)
    methods, 110
Environmental Estimated Concentrations
    (EECs), 392
Environmental Indicators for Sustainable
    Agriculture (ELISA), 384
Environmental monitoring programs, 366
    cost-effective monitoring programs, 367
    and GIS, 369
    monitoring types
        air monitoring, 370–377

biological monitoring, 387–389
soil and sediment, 384–387
water monitoring, 377–384
pesticides selection for monitoring, 369–370
purpose and design, 367–369
Environmental Quality Standards (EQS), 390
Enzyme-linked immunosorbent assay (ELISA), 107–108, 110, 115–117
direct competitive, 109
indirect competitive, 108
methods for pesticides, 110–114
EQS, *see* Environmental Quality Standards
Ergosterol-inhibitor fungicides, 38
ESI, *see* Electrospray ionization
Ethyl Acetate, 180–181
Ethylenthiourea (ETU), 337
European Food Safety Authority (EFSA), 352
European Reference Laboratory (EURL) method, 181
2015 European Union Report on Pesticide Residues in Food, 333
Exactive-Orbitrap detector, 224
Exposure Analysis Modeling System (EXAMS), 392
Extraction techniques, 274

F

FAB, *see* Fast atomic bombardment
FAME project, *see* Fluxes of Agrochemicals into the Marine Environment project
Fast atomic bombardment (FAB), 79–80
FEA, Flemish Environmental Agency (FEA)
FI, *see* Field ionization
Fiber optic biosensors, 119
Field ionization (FI), 79
Fipronil, 28
Flame photometric detection (FPD), 178
Flemish Environmental Agency (FEA), 309, 377
Fluxes of Agrochemicals into the Marine Environment (FAME) project, 384
FOCUS-Air group, 302
Food and environmental sample handling, 42
future trends, 68
liquid–liquid extraction, 50–51
liquid-phase microextraction, 51–54
nanoparticles-based extractions, 65–66
carbon nanotubes, 66–67
graphene, 67
magnetic nanoparticles, 67–68
metal–organic frameworks, 66
QuEChERS, 64–65
sample pretreatment, 42
drying, 43
homogenization, 43–44
solid–liquid extraction, 44–46

microwave-assisted extraction, 48
pressurized solvent extraction, 48–49
shaking, 45
Soxhlet extraction, 47
solid-phase extraction, 54–55
affinity sorbents, 58–61
ion-exchange sorbents, 58
non-polar sorbents, 56–58
polar sorbents, 55–56
solid-phase microextraction, 61
desorption, 63
extraction, 62–63
solid–solid extraction: matrix solid-phase dispersion, 63–64
stir-bar sorptive extraction, 64
supercritical fluid extraction, 49–50
Food Quality Protection Act of 1996, 106
Food safety and pesticides levels, 329–330
consumer exposure and risk assessment
dietary intake estimation, 350–353
intake calculations of pesticide residues, 353–358
monitoring residue levels
animal origin food, 342–346
cereals, 337–342
fruit and vegetables, 332–337
infant and baby food, 346–350
legislation, 330–331
programs, 331–332
Fourier transform ion cyclotron resonance analyzer (FTICR), 83–84
Fourier transform mass spectrometer (FTMS), 83–84
FQPA, *see* Food Quality Protection Act of 1996
FTICR, *see* Fourier transform ion cyclotron resonance analyzer
FTMS, *see* Fourier transform mass spectrometer
Fungicides, 209
azoles, 30–31
benzimidazoles, 32
cell division inhibitors, 38
dithiocarbamates, 32–33
ergosterol synthesis inhibitors, 38
market, 4
miscellaneous, 34–36
morpholines, 33–34
sulfhydryl reagents, 37–38

G

GAP, *see* Good agricultural practice
Gas chromatography-mass spectrometry (GC-MS), 74, 84–86
multi-residue pesticides in vegetables, 97–98
qualitative analysis, 86–87
quantitative analysis, 87–89

Gel permeation chromatography (GPC), 180,
        223–224
Generic Estimated Environmental Concentration
        (GENEEC) model, 392
German Institute for Risk assessment, 336
G.K. Walter Eigenbrodt Environmental
        Measurements Systems, 308
GLEAMS, *see* Groundwater Loading Effects
        of Agricultural Management
        Systems
Glufosinate, 11
Glutamine synthetase (GS), 36
Glyphosate, 11, 260
Good agricultural practice (GAP), 330
GPC, *see* Gel permeation chromatography
Graphene, 67
"Green Chemistry" principles, 278
Green solvents, 278
Groundwater Loading Effects of Agricultural
        Management Systems (GLEAMS), 393

H

Hard ionization technology, 77
Hazard Index approach, 352–353
Hazard quotient (HQ) approach, 353, 396
Heated pneumatic nebulizer interface, 90
Herbicides, 5, 209
        amides, 5–6
        amino acid synthesis inhibitors, 36
        benzoic acids, 6–7
        carbamates, 7
        cell division inhibitors, 36
        imidazolinones, 7–10
        market, 4
        nitriles, 10
        nitroanilines, 10–11
        organophosphorus, 11
        phenoxy acids, 11–13
        photosynthesis inhibitors, 36–37
        pyridazines and pyridazinones, 15–16
        pyridines and quaternary ammonium
                compounds, 13–15
        triazines, 16–17
        ureas, 18–19
HF-LPME, *see* Hollow fiber membrane liquid
        phase microextraction
High performance liquid chromatography
        (HPLC), 321, 323–324
High resolution mass spectrometry (HRMS),
        224, 226
Hollow fiber membrane liquid phase
        microextraction (HF-LPME),
        280–281
HQ, *see* Hazard quotient (HQ) approach
HRMS, *see* High resolution mass spectrometry

I

Imidazolinone herbicides, 7–10
Immunoassays, 107
        data analysis, 115–117
        Elisa methods for pesticides, 110–114
        method development, 109–110
        overview, 107–109
Immunosorbent (IS), 58
Index compound, 353
Inhalation toxicity test, 39
Insecticides, 19, 209
        benzoylureas, 19
        carbamates, 19–21
        chitin synthesis inhibitors, 37
        cholinesterase inhibitors, 37
        market, 4
        miscellaneous, 28–30
        neonicotinoids, 21–23
        organochlorines, 23–25
        organophosphorus, 25–27
        pyrethroids, 28
        signal interference in the nervous system, 37
Integrative sampling techniques, 306
Internal quality control (IQC), 228–229
        measures, 229–230
        in pesticide residue analysis, 229
Ion-exchange sorbents, 58
Ion trap mass analyzer, 81–82
Ioxynil, 10
IQC, *see* Internal quality control (IQC)

J

Joint FAO/WHO Meeting on Pesticide Residues
        (JMPR), 336

K

Kuderna-Danish (K-D) evaporator, 177–180

L

LAPS, *see* Light Addressable Potentiometric Sensor
Laser desorption (LD), 81
Lebanese, 354
Levels of pesticides in food, 329–330
        consumer exposure and risk assessment
                dietary intake estimation, 350–353
                intake calculations of pesticide residues,
                        353–358
        monitoring residue levels
                animal origin food, 342–346
                cereals, 337–342
                fruit and vegetables, 332–337
                infant and baby food, 346–350

legislation, 330–331
  programs, 331–332
Light Addressable Potentiometric Sensor
  (LAPS), 122
Limit of detection (LOD), 333, 351
Limit of quantification (LOQ), 230, 276, 351, 354
Limit of reporting (LOR), 358
Lindane, 354
Lipophilic pesticides, 336
Liquid chromatography-mass spectrometry
  (LC-MS), 74
  multi-residue pesticides in vegetables, 97–98
Liquid chromatography–mass spectrometry
  (LC–MS), 89–91
Liquid-liquid extraction (LLE), 50–51, 274–275
Liquid-liquid microextraction (LLME)
  methods, 273
  DLLME, 275–279
  HF-LPME, 280–281
  SDME, 279–280
Liquid-phase microextraction (LPME), 52–53
LLE, see Liquid-liquid extraction
LLME, see Liquid-liquid microextraction
  (LLME) methods
LOD, see Limit of detection
LOQ, see Limit of quantification
LOR, see Limit of reporting
Low-resolution mass spectrometry (LRMS),
  224–225
LPME, see Liquid-phase microextraction
LRMS, see Low-resolution mass
  spectrometry

M

MAE, see Microwave-assisted extraction
Magnetic nanoparticles, 67–68
Magnetic sector mass analyzer, 82–83
Magnetic solid-phase extraction (mSPE)
  methods, 287–289
MALDI, see Matrix assisted laser desorption
  ionization
Margin of Exposure (MOE), 353
Mass spectrometry, 74
  detector, 84
  ion source
    atmospheric pressure chemical ionization
      (APCI), 80
    chemical ionization (CI), 78–79
    electron ionization (EI), 77–78
    electrospray ionization (ESI), 80
    fast atomic bombardment (FAB),
      79–80
    field ionization (FI), 79
    laser desorption (LD), 81
  mass analyzer

Fourier transform ion cyclotron resonance
    analyzer (FTICR), 83–84
  ion trap, 81–82
  magnetic sector, 82–83
  single quadruple, 81
  time-of-flight (TOF), 83
  triple quadrupole, 82
performance parameters to characterize
  mass measurement range, 75
  resolution, 76
  sensitivity, 76
  vacuum system, 77
  working principle, 74–75
Matrix assisted laser desorption ionization
  (MALDI), 81
Matrix solid-phase dispersion (MSPD),
  63–64, 221
Maximum residue levels (MRLs) of pesticides,
  211, 329–331, 333–335
MECK, see Micellar electrokinetic
  chromatography
MEPS, see Microextraction in packed sorbent
Metal–organic frameworks, 66
Methanol, 181
Micellar electrokinetic chromatography
  (MECK), 293
Microextraction in packed sorbent (MEPS), 222
Microwave-assisted extraction (MAE), 48
MIP, see Molecularly imprinted polymer
Miscellaneous fungicides, 34–35
Miscellaneous insecticides, 28, 30
MISPE, see Molecularly imprinted solid-phase
  extraction
Modes of action, 210–211
  fungicides
    cell division inhibitors, 38
    ergosterol synthesis inhibitors, 38
    sulfhydryl reagents, 37–38
  herbicides
    amino acid synthesis inhibitors, 36
    cell division inhibitors, 36
    photosynthesis inhibitors, 36–37
  insecticides
    chitin synthesis inhibitors, 37
    cholinesterase inhibitors, 37
    signal interference in the nervous
      system, 37
MOE, see Margin of Exposure
Molecularly imprinted polymer (MIP), 58–59,
  125, 285, 289
Molecularly imprinted solid-phase extraction
  (MISPE), 59, 289–290
Monitoring the state of European soils
  (MOSES), 384
Monte Carlo modeled data, 403
Morpholines fungicides, 33–34

MOSES, *see* Monitoring the state of European soils
MRL, *see* Maximum residue levels (MRLs) of pesticides
MSPD, *see* Matrix solid-phase dispersion
mSPE, *see* Magnetic solid-phase extraction (mSPE) methods
Mucous membrane and eye toxicity test, 39
Müller, Paul, 2
Multiresidue, 262

N

National Water Quality Assessment Program (NAWQA), 384
Neonicotinoids insecticides, 21, 23–24, 209–210, 261
Nitrile herbicides, 10
Nitroaniline herbicides, 10–12
Non-polar sorbents, 56–58
Normalization method, 88

O

OCP, *see* Organochlorine pesticides
OPP, *see* Organophosphorus pesticides
Optical biosensors, 119
Oral toxicity test, 39
Orbitrap mass analyzers, 190
Organochlorine pesticides (OCPs), 344, 346, 370, 372, 378, 386, 388–389
Organochlorines insecticides, 23, 25, 37
Organophosphates (OPs), 209
Organophosphorus herbicides, 11–12
Organophosphorus insecticides, 25–27
Organophosphorus pesticides (OPPs), 388

P

Passive air samplers (PAS), 371
Passive sensors, 306
PEC, *see* Predicted environmental concentration
PEGASE, *see* Pesticides in European Groundwaters
Pentafluorobenzyl bromide (PFBBr), 322–323
Perkin Elmer Perkin-Elmer Corp., 317
Pest, 2
Pesticide, 2–4
    fungicides (*see* Fungicides)
    herbicides (*see* Herbicides)
    insecticides (*see* Insecticides)
    modes of action (*see* Modes of action)
    toxicity and risk assessment, 38–40
    world market and sales, 3–4
Pesticide Manual, The, 4
Pesticide National Synthesis Project, 384
Pesticide residues, 84, 332

application of chromatography-mass spectrometry techniques, 93–97
cereals, 337–342
comprehensive two-dimensional gas chromatography–time of flight mass spectrometry (GC×GC–TOFMS), 92–93
food of animal origin, 342–346
fruit and vegetables, 332–337
future trends, 101
gas chromatography-mass spectrometry (GC-MS), 74, 84–86
    multi-residue pesticides in vegetables, 97–98
    qualitative analysis, 86–87
    quantitative analysis, 87–89
infant and baby food, 346–350
in Korean agricultural products, 338
liquid chromatography-mass spectrometry (LC-MS), 74
    multi-residue pesticides in vegetables, 97–98
liquid chromatography–mass spectrometry (LC-MS), 89–91
monitoring, 334–335
supercritical fluid chromatography–mass spectrometry (SFC–MS), 91–92
surface water samples using liquid chromatography–high resolution mass spectrometry (LC-HRMS), 99–100
Pesticide risk indicators (PRIs), 394–395
Pesticide Root Zone Model (PRZM), 392
Pesticides and products transformation in environment, 366
    environmental exposure, 390–393
    purpose and design of monitoring programs, 367–369
    risk assessment, 393–395
        limitations and future trends, 405–406
        pesticide risk indicators (PRIs), 395
        probabilistic risk assessment (PRA), 400–405
        risk quotient – toxicity exposure ratio method, 395–400
    selection of pesticides for monitoring, 369–370
    types of monitoring
        air monitoring, 370–377
        biological monitoring, 387–389
        soil and sediment monitoring, 384–387
        water monitoring, 377–384
    water framework directive and monitoring strategies, 379–380
Pesticides in atmosphere, 301–303
    in ambient air, 304–308
    analysis by gas chromatography, 321–322
    analysis by HPLC, 323–324

derivatization, 322–323
  monitoring, 304
  in rainwater samples, 308–314
  soil–air transfer evaluation, 314–321
Pesticides in European Groundwaters
      (PEGASE), 384
Pesticides in soil, 246
  application to samples
    glyphosate, 260
    multiresidue, 262
    neonicotinoids, 261
    organophosphorus, 261
    pyrethroids, 261
    pyrimethanil and kresoxim-methyl
        fungicides, 262
    sulfonylureas, 260
    triazines, 260–261
  determination of residues, 256–260
  future trends, 262–263
  sample preparation, 247
    cleanup, 252–254
    derivatization, 254–256
    extraction, 247–251
Pesticides intake in Denmark, 356
Pesticides in water, 273–274
  actual concentration, 303
  liquid-liquid extraction, 274–275
  liquid-liquid microextraction
    DLLME, 275–279
    HF-LPME, 280–281
    SDME, 279–280
  other techniques, 290–291
  solid-phase extraction
    MISPE, 289–290
    mSPE, 287–289
    SBSE, 286–287
    SPME, 283–286
    standard SPE, 281–283
  techniques for pesticide determination,
      292–293
Pesticides levels in food, 329–330
  consumer exposure and risk assessment
    dietary intake estimation, 350–353
    intake calculations of pesticide residues,
        353–358
  monitoring residue levels
    animal origin food, 342–346
    cereals, 337–342
    fruit and vegetables, 332–337
    infant and baby food, 346–350
    legislation, 330–331
    programs, 331–332
Pesticides recoveries from Tenax, 318
PFBBr, see Pentafluorobenzyl bromide (PFBBr)
Phenoxy acid herbicides, 11, 13–14
Phenyl urea herbicides, 18
Phenylureas, 18

PLE, see Pressurized liquid extraction
PNEC, see Predicted non-effect concentration
      (PNEC) values
Polar sorbents, 55–56, 281
Polychlorinated biphenyls (PCBs), 282
Polydimethylsiloxane (PDMS) fibers, 284
Polystyrene divinylbenzene copolymer (PLRP-s)
      sorbent, 283
Polytron homogenizer, 188
Polyurethane foam (PUF), 304, 306
PRA, see Probabilistic Risk Assessment
Predicted environmental concentration (PEC),
      40, 392
Predicted non-effect concentration (PNEC)
      values, 40, 396–397
Pressurized liquid extraction (PLE), 221
Pressurized solvent extraction (PSE), 48–49
PRI, see Pesticide Risk Indicators
Probabilistic modeling, 351–352
Probabilistic risk assessment (PRA), 400–405
Proficiency testing (PT), 228
PRZM, see Pesticide Root Zone Model
PSE, see Pressurized solvent extraction
PUF, see Polyurethane foam
Pyrethroids, 209, 261
Pyrethroids insecticides, 28, 37
Pyrethroids in sediments, 387
Pyridazines and pyridazinones herbicides, 15–16
Pyridine herbicides, 13, 15
Pyrimethanil and kresoxim-methyl
      fungicides, 262

Q

QuEChERS, see Quick, Easy, Cheap, Effective,
      Rugged, and Safe
Quenchbody, 120
Quick, Easy, Cheap, Effective, Rugged, and Safe
      (QuEChERS), 64–65, 67, 110, 178,
      185–187, 189–194, 222–223, 231–234
Quick Polar Pesticide (QuPPe) method, 181,
      186–187

R

Raw agricultural commodities (RAC), 330,
      335, 350
Relative Potency Factor (RPF), 353
Reliability and quality of residue data, factors
      affecting, 138–139
    packing of samples, 143–144
    plant and animal origin sampling, 141–142
    sample preparation and processing, 144–146
    sample quality, 139–140
    soil sampling, 142–143
    stability of residues
      analytical standards, 149–152

sample processing, 148–149
storage of samples, 146–148
surface or groundwater sampling, 142
transport, shipping, and receiving of
samples, 144
Repeated dose test, 39
Residue measurement results, random errors
calculation of combined uncertainty, 156–159
characterization of uncertainty results,
155–156
determination of uncertainties
analysis, 165–166
sample processing, 163–165
sampling, 159–161
subsampling, 161–163
systematic errors, 166–167
Risk quotient – toxicity exposure ratio method,
395–400
Rodenticides, 209
RPF, *see* Relative Potency Factor

S

Sample drying/freezing, 43
SBSE, *see* Stir bar sorptive extraction
SDME, *see* Single-drop microextraction
Selected reaction monitoring mode (SRM), 188
SFE, *see* Supercritical fluid extraction
Silent Spring (Carson), 2
SIM, *see* Single ion monitoring mode
Single-drop microextraction (SDME), 279–280
Single ion monitoring mode (SIM), 321
Single quadruple mass analyzer, 81
Soft ionization method, 78
Soil and pesticides, 246
application to samples
glyphosate, 260
multiresidue, 262
neonicotinoids, 261
organophosphorus, 261
pyrethroids, 261
pyrimethanil and kresoxim-methyl
fungicides, 262
sulfonylureas, 260
triazines, 260–261
determination of residues, 256–260
future trends, 262–263
sample preparation, 247
cleanup, 252–254
derivatization, 254–256
extraction, 247–251
Soil and sediment monitoring, 384–387
Solid-liquid extraction (SLE) technique, 22,
44–45, 54–55, 60, 110, 221
affinity sorbents, 58–61
ion-exchange sorbents, 58
microwave-assisted extraction, 48

non-polar sorbents, 56–58
polar sorbents, 55–56
pressurized solvent extraction, 48–49
shaking, 45
Soxhlet extraction, 47
SPE–LC coupling, 60–61
techniques, 46
Solid-phase extraction (SPE)
MISPE, 289–290
mSPE, 287–289
SBSE, 286–287
SPME, 283–286
standard SPE, 281–283
Solid-phase microextraction (SPME), 61, 283–286
desorption, 63
extraction, 62–63
Solid–solid extraction: matrix solid-phase
dispersion, 63–64
Soxhlet extraction, 47, 307–308
SPE, *see* Solid-phase extraction
Spirotetramat, 28
SPME, *see* Solid-phase microextraction
Stir bar sorptive extraction (SBSE), 64, 286–287
Stressor-response analysis, 394
Sulfhydryl reagents, 37–38
Sulfonylurea herbicides, 18, 20
Sulfonylureas, 260
Supercritical fluid chromatography–mass
spectrometry (SFC–MS), 91–92
Supercritical fluid extraction (SFE), 49–50
Surface water samples using liquid
chromatography–high resolution mass
spectrometry (LC–HRMS), 99–100
Swedish Ethyl Acetate (SweEt) method, 181, 194

T

Tebuconazole stereoisomers, 96
TEF, *see* Toxic Equivalent Factor
Tenax, 318–319
pesticides recoveries from, 318
retention efficiency, 319–320
TER, *see* Toxicity-exposure ratio
Thermal desorption principle, 315
Time-of-flight (TOF) mass analyzer, 83, 190
Time of flight mass spectrometry (TOFMS), 92
TOF mass analyzer, *see* Time-of-flight (TOF)
mass analyzer
TOFMS, *see* Time of flight mass spectrometry
TOPO, *see* Trioctylphosphine oxide
Toxic Equivalent Factor (TEF), 353
Toxicity-exposure ratio (TER), 395–400
Toxicity in humans, 273
Toxicological risk assessment, 39
Transformation pathways for pesticides, 302
Triazine herbicides, 385
Triazines, 260–261

Triazines herbicides, 16–17
Trioctylphosphine oxide (TOPO), 53
Triple quadrupole mass analyzer, 82

U

Ultra-high performance liquid chromatography
    (UPLC) technology, 90, 189
Ultrasonic bath, 308
Ultrasound-assisted emulsification-
    microextraction (USAEME), 277
UPLC, *see* Ultra-high performance liquid
    chromatography (UPLC) technology
Urea herbicides, 18–19
USAEME, *see* Ultrasound-assisted
    emulsification-microextraction
US Environmental Protection Agency (EPA), 106
US Food and Drug Administration (USFDA),
    332, 338
US Geological Survey's National Water Quality
    Assessment Program (NAWQA), 384

V

VALLME, *see* Vortex-assisted liquid-liquid
    microextraction
Vegetal origin food, pesticides in
    challenges and considerations, 193–195
    extraction of pesticides, solvent extract,
        176–181
        acetone, 179–180
        acetonitrile, 177–179
        ethyl acetate, 180–181
        methanol, 181
    gas chromatography–mass spectrometry,
        187–189
    liquid and gas chromatography–high
        resolution mass spectrometry,
        190–193
    liquid chromatography–mass spectrometry,
        189–190
    QuEChERS, 184–187
    sample extract cleanup, 182–184
Vortex-assisted liquid-liquid microextraction
    (VALLME), 290

W

Water and pesticides, 273–274
    liquid-liquid extraction, 274–275
    liquid-liquid microextraction
        DLLME, 275–279
        HF-LPME, 280–281
        SDME, 279–280
    other techniques, 290–291
    solid-phase extraction
        MISPE, 289–290
        mSPE, 287–289
        SBSE, 286–287
        SPME, 283–286
        standard SPE, 281–283
    techniques for pesticide determination,
        292–293
Water Framework Directive (WFD),
    385, 390
Water monitoring, 377–384
Wet-only rainwater collector, 309
WFD, *see* Water Framework Directive

X

XAD-2 passive sampler, 307

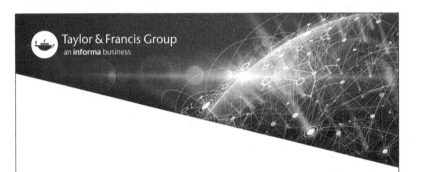